T0207096

NEUROMETHODS

Series Editor
Wolfgang Walz
University of Saskatchewan
Saskatoon, SK, Canada

For further volumes:
http://www.springer.com/series/7657

Brain Morphometry

Edited by

Gianfranco Spalletta

IRCCS Santa Lucia Foundation, Laboratory of Neuropsychiatry, Rome, Italy
Division of Neuropsychiatry, Menninger Department of Psychiatry and Behavioral Sciences
Baylor College of Medicine, Houston, TX, USA

Fabrizio Piras

IRCCS Santa Lucia Foundation, Laboratory of Neuropsychiatry, Rome, Italy

Tommaso Gili

IMT School for Advanced Studies–Lucca, Lucca, Italy
IRCCS Santa Lucia Foundation, Laboratory of Neuropsychiatry, Rome, Italy

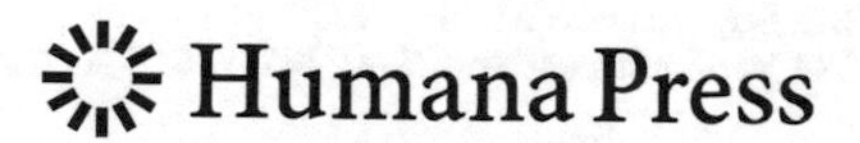

Editors
Gianfranco Spalletta
IRCCS Santa Lucia Foundation
Laboratory of Neuropsychiatry
Rome, Italy

Division of Neuropsychiatry
Menninger Department of Psychiatry
and Behavioral Sciences
Baylor College of Medicine
Houston, TX, USA

Fabrizio Piras
IRCCS Santa Lucia Foundation
Laboratory of Neuropsychiatry
Rome, Italy

Tommaso Gili
IMT School for Advanced Studies–Lucca
Lucca, Italy

IRCCS Santa Lucia Foundation
Laboratory of Neuropsychiatry
Rome, Italy

ISSN 0893-2336 ISSN 1940-6045 (electronic)
Neuromethods
ISBN 978-1-4939-9253-9 ISBN 978-1-4939-7647-8 (eBook)
https://doi.org/10.1007/978-1-4939-7647-8

Printed on acid-free paper

This Humana Press imprint is published by Springer Nature
The registered company is Springer Science+Business Media, LLC
The registered company address is: 233 Spring Street, New York, NY 10013, U.S.A.

Preface to the Series

Experimental life sciences have two basic foundations: concepts and tools. The *Neuromethods* series focuses on the tools and techniques unique to the investigation of the nervous system and excitable cells. It will not, however, shortchange the concept side of things as care has been taken to integrate these tools within the context of the concepts and questions under investigation. In this way, the series is unique in that it not only collects protocols but also includes theoretical background information and critiques which led to the methods and their development. Thus, it gives the reader a better understanding of the origin of the techniques and their potential future development. The *Neuromethods* publishing program strikes a balance between recent and exciting developments like those concerning new animal models of disease, imaging, in vivo methods, and more established techniques, including immunocytochemistry and electrophysiological technologies. New trainees in neurosciences still need a sound footing in these older methods in order to apply a critical approach to their results.

Under the guidance of its founders, Alan Boulton and Glen Baker, the *Neuromethods* series has been a success since its first volume published through Humana Press in 1985. The series continues to flourish through many changes over the years. It is now published under the umbrella of Springer Protocols. While methods involving brain research have changed a lot since the series started, the publishing environment and technology have changed even more radically. *Neuromethods* has the distinct layout and style of the Springer Protocols program, designed specifically for readability and ease of reference in a laboratory setting.

The careful application of methods is potentially the most important step in the process of scientific inquiry. In the past, new methodologies led the way in developing new disciplines in the biological and medical sciences. For example, physiology emerged out of anatomy in the nineteenth century by harnessing new methods based on the newly discovered phenomenon of electricity. Nowadays, the relationships between disciplines and methods are more complex. Methods are now widely shared between disciplines and research areas. New developments in electronic publishing make it possible for scientists who encounter new methods to quickly find sources of information electronically. The design of individual volumes and chapters in this series takes this new access technology into account. Springer Protocols makes it possible to download single protocols separately. In addition, Springer makes its print-on-demand technology available globally. A print copy can therefore be acquired quickly and for a competitive price anywhere in the world.

Saskatoon, Canada ***Wolfgang Walz***

Preface

The structure and function of the human brain are the result of numerous biochemical and biophysical processes interacting across multiple scales in space and time. Variation in total gray and white matter volumes of the human brain is initially genetically determined. Most of the genes known to control these processes during brain development, maturation, and aging are highly conserved and participate to a complex process that leads the brain, during development, to a dramatic transformation from a simple tubular structure to a highly convoluted shape. However, this does not necessarily mean that genes influence different focal brain structures in the same manner and that a common genetic origin associated with general cognition and behavior is shared by all structures. In addition, life events and experiences during childhood impact directly on neurodevelopment. In line with this, a variety of studies have linked early and late events with brain morphology and disrupted neurodevelopment.

Pronounced differences, both at the morphological and at the cognitive/behavioral levels, abound among individuals. Macroscopic variations in brain anatomy are sufficiently maintained to grant comparative investigations. Indeed, morphological analyses that compare brains at different healthy or pathological stages can reveal important information about the progression of normal or abnormal development. Characterizing focal brain morphology and its association with development, functioning, and age-related neurodegenerative processes in healthy humans as well as local morphological alterations as found in psychiatric disorders and neurological diseases is crucial for the development of modern neuroscience.

Brain morphometry as a discipline is mainly concerned with the development of tools and strategies for the measurement of brain structural properties according to the kind of imaging data used, whether ontogenetic, pathological, or phylogenetic issues are targeted, and the spatial scales of interest. Further, shape feature comparisons have long been constrained to simple and mainly volume- or slice-based measures but benefited enormously from the digital revolution, as now all sorts of shapes in any number of dimensions can be handled numerically.

Magnetic resonance imaging (MRI), the state of the art of structural neuroimaging, provides a spatial representation of the brain and its components and allows for the calculation of several parameters of interest, related to the morphological features of brain regions. The fast evolution in terms of spatial resolution and signal-to-noise ratio in MRI scanners and the improvements on new imaging techniques and data processing algorithms have helped in developing studies able to detect and to quantify initially gross but even subtle structural abnormalities that appear when comparing different populations.

The most popular MRI-based approaches typically used to investigate morphological properties of gray matter structures both in healthy people and in patients are voxel-based morphometry (VBM), deformation-based morphometry (DBM), pattern-based morphometry (PBM), and surface-based morphometry (SBM). Analogously, the axonal architecture of white matter is generally obtained by diffusion tensor imaging (DTI) and diffusion spectrum imaging (DSI) fiber-tracking techniques. All of these methods are able to discriminate, noninvasively, the different tissues of the brain, thanks either to their molecular composition (i.e., local magnetic properties) or to their intercellular organization (i.e., local hydrodynamic regimes). In fact, if on one hand the first four approaches are performed commonly using the image contrast between water and fat (T1-weighted images), on the other hand, the second two aim to identify the molecular diffusion processes of water occurring along dominant axonal tracts.

Brain morphometric studies clarified that the largest changes within an individual generally occur during early development, more subtle ones follow during adulthood, and, again, dramatic changes occur in the last part of the human life: the aging. Currently, however, most applications of MR-based brain morphometry have a clinical target. They help to diagnose and monitor neuropsychiatric disorders. In fact, advances in neuroimaging progressively moved the scientific community toward a new understanding of neurological diseases as well as of psychiatric disorders based on their underlying neurobiology, facilitating the diagnostic classification, improving our ability to predict treatment outcome, and enhancing our understanding of the genetic and environmental causes of these disorders.

The rationale behind the realization of this book has been to offer to a broad audience, from expert neuroscientists to neuroimaging beginners, an overview of the state of the art of gray matter morphometry, as it can be derived from magnetic resonance images. Firstly, the book takes on the main topics about the technical procedures that underlie a morphometric study, from the registration and segmentation steps to the statistical analysis of structural parameters, passing through the most advanced methods for the measurement of cortical shaping and its multilayered structure. Subsequently, we intended to go through clinical and nonclinical applications. The nonclinical part of the book includes the characterization of normal development, aging, and the interplay between morphometry and genetics. Advanced approaches for the cytoarchitectonic description of brain structures and the impact that multicenter studies can have on morphometry results thanks to the inclusion of massively large cohorts are presented together with possible integrations with functional MRI data. The clinical part highlights the importance of brain morphometry in the improvement of the neurobiological characterization of major psychiatry disorders (schizophrenia, bipolar disorder, unipolar depression, obsessive-compulsive disorder, personality disorders, and suicide) and neuropsychiatric/neurological diseases (Alzheimer disease, non-Alzheimer dementias, Parkinson disease, multiple sclerosis, traumatic brain injury, and epilepsy).

The strength of the present book lies not only in the collection of the up-to-date methodological approaches thought to improve the characterization of the human brain structural properties and of a consistent coverage of brain dysfunctional and non-dysfunctional neuroanatomical variations but also in the multi-register harmonization obtained across its chapters. In fact, the final product is suitable for a specific update at the student, clinician, and researcher levels.

As it happens for every scientific approach to the study of a system, morphometry is a discipline in continuous evolution. Certainly, the access to high-performance informatics has notably improved the quality of results, as well as the technological amelioration of MRI signals equalized data reliability and stimulated the production of sophisticated algorithms for analysis. However, there is a long road ahead, with the connection between the microscopic and the mesoscopic levels of our knowledge about the structure of the brain still distant. Some basis has been placed, with the introduction of ultra-high-field MRI, which not only allows for the detection of structures at the micron scale but will open also the chance to use novel sources of contrast. Contemporarily, a better understanding of brain physiology due to the development of techniques able to detect, noninvasively, spatially precise local tissue metabolism will benefit the characterization of shapes, thickness, and gyrification of cortex also in terms of their role in the whole system. Finally, the recent application of complex network science to clinical and nonclinical aspects of neuroscience will surely help the characterization of structural differences across age, medication, or dysfunction at the population level, thanks to its ability in unearthing elusive patterns and trends in big data.

We give our thanks to all the authors of the chapters in this book. They are leading investigators in their respective fields and kindly offered their insights into different aspects of brain morphometry. It goes without saying that, without their help, this book would not exist.

Rome, Italy

Gianfranco Spalletta
Tommaso Gili
Fabrizio Piras

Contents

Contributors

CARLO A. ALTAMURA • *Department of Neurosciences and Mental Health, Institute of Psychiatry, Fondazione IRCCS Ca' Granda, Ospedale Maggiore Policlinico, University of Milan, Milan, Italy*

ELISA AMBROSI • *NESMOS Department (Neurosciences, Mental Health, and Sensory Organs), Sapienza University of Rome, Rome, Italy; School of Medicine and Psychology, Sant' Andrea Hospital, Rome, Italy*

JOSE ARAUJO • *University of Toronto, Toronto, Canada*

BRIAN B. AVANTS • *Department of Radiology, University of Pennsylvania, Philadelphia, PA, USA*

ALI BANI-FATEMI • *University of Toronto, Toronto, Canada*

GAETANO BARBAGALLO • *Institute of Neurology, University Magna Graecia of Catanzaro, Catanzaro, Italy*

NEDA BERNASCONI • *Neuroimaging of Epilepsy Laboratory, McConnell Brain Imaging Center, Montreal Neurological Institute and Hospital, McGill University, Montreal, QC, Canada*

ANDREA BERNASCONI • *Neuroimaging of Epilepsy Laboratory, McConnell Brain Imaging Center, Montreal Neurological Institute and Hospital, McGill University, Montreal, QC, Canada*

PREMIKA S.W. BOEDHOE • *Department of Psychiatry, VU University Medical Center, Amsterdam, The Netherlands; Department of Anatomy and Neurosciences, VU University Medical Center, Amsterdam, The Netherlands; Amsterdam Neuroscience, Amsterdam, The Netherlands*

ALI K. BOURISLY • *Biomedical Engineering Unit, Department of Physiology, Faculty of Medicine, Kuwait University, Kuwait City, Kuwait*

PAOLO BRAMBILLA • *Department of Neurosciences and Mental Health, Institute of Psychiatry, Fondazione IRCCS Ca' Granda, Ospedale Maggiore Policlinico, University of Milan, Milan, Italy; Department of Psychiatry and Behavioural Neurosciences, University of Texas at Houston, Houston, TX, USA*

VINCE D. CALHOUN • *The Mind Research Network, Albuquerque, NM, USA; Department of Electrical and Computer Engineering, University of New Mexico, Albuquerque, NM, USA*

CHIARA CHIAPPONI • *Neuropsychiatry Laboratory, IRCCS Santa Lucia Foundation, Rome, Italy*

NICHOLAS C. CULLEN • *Department of Psychiatry, Perelman School of Medicine, University of Pennsylvania, Philadelphia, PA, USA; Department of Electrical & Systems Engineering, University of Pennsylvania, Philadelphia, PA, USA*

DEBORA CUTULI • *IRCCS Santa Lucia Foundation, Rome, Italy; Department of Psychology, University Sapienza of Rome, Rome, Italy*

ROBERT DAHNKE • *Departments of Neurology and Psychiatry, Jena University Hospital, Jena, Germany*

EMANUELA DANESE • *NESMOS Department (Neurosciences, Mental Health, and Sensory Organs), Sapienza University of Rome, Rome, Italy; School of Medicine and Psychology, Sant' Andrea Hospital, Rome, Italy*

GIUSEPPE DELVECCHIO • *Scientific Institute IRCCS "E. Medea", Bosisio Parini, Italy*

NICOLA DUSI • *Psychiatry Unit, Department of Mental Health, Azienda Socio-Sanitaria Territoriale Nord Milano, Milan, Italy*
CHRISTIAN GASER • *Departments of Neurology and Psychiatry, Jena University Hospital, Jena, Germany*
TOMMASO GILI • *IMT School for Advanced Studies—Lucca, Lucca, Italy; IRCCS Santa Lucia Foundation, Laboratory of Neuropsychiatry, Rome, Italy*
SAVANNAH N. GOSNELL • *Baylor College of Medicine and Michael E DeBakey VA Medical Center, Houston, TX, USA*
COTA NAVIN GUPTA • *The Mind Research Network, Albuquerque, NM, USA; Department of Biosciences and Bioengineering, Indian Institute of Technology, Guwahati, India*
ODILE A. VAN DEN HEUVEL • *Department of Psychiatry, VU University Medical Center, Amsterdam, The Netherlands; Department of Anatomy and Neurosciences, VU University Medical Center, Amsterdam, The Netherlands; Amsterdam Neuroscience, Amsterdam, The Netherlands*
MARIANGELA IORIO • *IRCCS Santa Lucia Foundation, Laboratory of Neuropsychiatry, Rome, Italy*
LUTZ JANCKE • *Institute of Psychology, Department of Neuropsychology, University of Zurich, Zurich, Switzerland*
DELFINA JANIRI • *NESMOS Department (Neurosciences, Mental Health, and Sensory Organs), Sapienza University of Rome, Rome, Italy; School of Medicine and Psychology, Sant' Andrea Hospital, Rome, Italy*
RICARDO E. JORGE • *Michael E. DeBakey VA Medical Center, Houston, TX, USA; Menninger Department of Psychiatry and Behavioral Sciences, Baylor College of Medicine, Houston, TX, USA*
ANAND A. JOSHI • *Signal and Image Processing Institute, University of Southern California, Los Angeles, CA, USA; Brain and Creativity Institute, University of Southern California, Los Angeles, CA, USA*
FLORIAN KURTH • *School of Psychology, University of Auckland, Auckland, New Zealand*
DANIELA LARICCHIUTA • *IRCCS Santa Lucia Foundation, Rome, Italy; Department of Psychology, University Sapienza of Rome, Rome, Italy*
JASON P. LERCH • *Program in Neurosciences and Mental Health, The Hospital for Sick Children, Toronto, ON, Canada; Department of Medical Biophysics, The University of Toronto, Toronto, ON, Canada*
ILONA LIPP • *Institute of Psychological Medicine and Clinical Neurosciences, Cardiff University School of Medicine, Cardiff, UK; School of Psychology, Cardiff University Brain Research Imaging Centre (CUBRIC), Cardiff, UK*
VINCENZO DE LUCA • *University of Toronto, Toronto, Canada*
EILEEN LUDERS • *School of Psychology, University of Auckland, Auckland, New Zealand*
MATTEO DE MARCO • *Department of Neuroscience, Medical School, University of Sheffield, Sheffield, UK*
HIROSHI MATSUDA • *Integrative Brain Imaging Center, National Center of Neurology and Psychiatry, Tokyo, Japan*
DAVID L. MOLFESE • *Baylor College of Medicine and Michael E DeBakey VA Medical Center, Houston, TX, USA*
NILS MUHLERT • *School of Psychology, Cardiff University Brain Research Imaging Centre (CUBRIC), Cardiff, UK; School of Psychological Sciences, University of Manchester, Manchester, UK*

Federico Nemmi • *Toulouse NeuroImaging Centre (ToNIC), Inserm, UPS, Université de Toulouse, Toulouse, France*

Ylva Østby • *Department of Psychology, University of Oslo, Oslo, Norway; National Centre for Epilepsy, Oslo University Hospital, Oslo, Norway*

Isabella Panaccione • *NESMOS Department (Neurosciences, Mental Health, and Sensory Organs), Sapienza University of Rome, Rome, Italy; School of Medicine and Psychology, Sant' Andrea Hospital, Rome, Italy; Centro Lucio Bini, Rome, Italy*

Margherita Di Paola • *Department of Mental Health, King Faisal Specialist Hospital & Research Center—KFSH&RC, Riyadh, Saudi Arabia; Morphology and Morphometry for Neuroimaging Lab, Department of Clinical and Behavioral Neurology, IRCCS Santa Lucia Foundation, Rome, Italy*

Patrice Péran • *Toulouse NeuroImaging Centre (ToNIC), Inserm, UPS, Université de Toulouse, Toulouse, France*

Cyril R. Pernet • *Centre for Clinical Brain Sciences, The University of Edinburgh, Edinburgh, UK*

Laura Petrosini • *IRCCS Santa Lucia Foundation, Rome, Italy; Department of Psychology, University Sapienza of Rome, Rome, Italy*

Eleonora Picerni • *IRCCS Santa Lucia Foundation, Rome, Italy; Department of Psychology, University Sapienza of Rome, Rome, Italy*

Fabrizio Piras • *IRCCS Santa Lucia Foundation, Laboratory of Neuropsychiatry, Rome, Italy; Centro Fermi—Museo Storico della Fisica e Centro Studi e Ricerche "Enrico Fermi", Rome, Italy*

Federica Piras • *IRCCS Santa Lucia Foundation, Laboratory of Neuropsychiatry, Rome, Italy*

Pietro De Rossi • *Neuropsychiatry Laboratory, IRCCS Santa Lucia Foundation, Rome, Italy; NESMOS Department, Faculty of Medicine and Psychology, "Sapienza" University of Rome, Rome, Italy; Department of Neurology and Psychiatry, "Sapienza" University of Rome, Rome, Italy*

Chiara Rovera • *Department of Neurosciences and Mental Health, Institute of Psychiatry, Fondazione IRCCS Ca' Granda, Ospedale Maggiore Policlinico, University of Milan, Milan, Italy*

Ramiro Salas • *Baylor College of Medicine and Michael E DeBakey VA Medical Center, Houston, TX, USA*

Gabriele Sani • *NESMOS Department (Neurosciences, Mental Health, and Sensory Organs), Sapienza University of Rome, Rome, Italy; School of Medicine and Psychology, Sant' Andrea Hospital, Rome, Italy; Centro Lucio Bini, Rome, Italy*

Tanya Santos • *University of Toronto, Toronto, Canada*

Dewi S. Schrader • *Neuroimaging of Epilepsy Laboratory, McConnell Brain Imaging Center, Montreal Neurological Institute and Hospital, McGill University, Montreal, QC, Canada*

Alessio Simonetti • *NESMOS Department (Neurosciences, Mental Health, and Sensory Organs), Sapienza University of Rome, Rome, Italy; School of Medicine and Psychology, Sant' Andrea Hospital, Rome, Italy; Centro Lucio Bini, Rome, Italy*

Gianfranco Spalletta • *IRCCS Santa Lucia Foundation, Laboratory of Neuropsychiatry, Rome, Italy; Division of Neuropsychiatry, Menninger Department of Psychiatry and Behavioral Sciences, Baylor College of Medicine, Houston, TX, USA*

Christian K. Tamnes • *Department of Psychology, University of Oslo, Oslo, Norway*

Samia Tasmim • *University of Toronto, Toronto, Canada*

BRIAN A. TAYLOR • *Michael E. DeBakey VA Medical Center, Houston, TX, USA; Department of Radiology, Baylor College of Medicine, Houston, TX, USA; Department of Physical Medicine and Rehabilitation, Baylor College of Medicine, Houston, TX, USA*

PAUL M. THOMPSON • *Imaging Genetics Center, Institute for Neuroimaging & Informatics, Keck USC School of Medicine, Marina del Rey, CA, USA*

VALENTINA TOMASSINI • *Institute of Psychological Medicine and Clinical Neurosciences, Cardiff University School of Medicine, Cardiff, UK; School of Psychology, Cardiff University Brain Research Imaging Centre (CUBRIC), Cardiff, UK*

JESSICA A. TURNER • *The Mind Research Network, Albuquerque, NM, USA; Department of Psychology and Neuroscience Institute, Georgia State University, Atlanta, GA, USA*

DANIELA VECCHIO • *IRCCS Santa Lucia Foundation, Laboratory of Neuropsychiatry, Rome, Italy; Department of Psychology, Sapienza University of Rome, Rome, Italy*

ANNALENA VENNERI • *Department of Neuroscience, Medical School, University of Sheffield, Sheffield, UK; Royal Hallamshire Hospital, Sheffield, UK; IRCCS Fondazione Ospedale San Camillo, Venice, Italy*

KONRAD WAGSTYL • *Brain Mapping Unit, Department of Psychiatry, University of Cambridge, Cambridge, UK*

ELISABETH A. WILDE • *Michael E. DeBakey VA Medical Center, Houston, TX, USA; Department of Neurology, Baylor College of Medicine, Houston, TX, USA; Department of Radiology, Baylor College of Medicine, Houston, TX, USA; Department of Physical Medicine and Rehabilitation, Baylor College of Medicine, Houston, TX, USA*

Part I

Brain Morphometry: Methods

Check for updates

Chapter 1

Registration

Anand A. Joshi

Abstract

The goal of image registration is to find a 1-1 point-wise correspondence between two images, a subject image and a target image. Knowing the pointwise correspondence between two brain images allows comparison of structural and functional imaging data such as regions of interest, functional data (e.g., fMRI, EEG, MEG, DTI), and geometric shapes. The image registration process also allows creation of probabilistic anatomical atlases (Mazziotta et al., Neuroimage 2(2):89–101, 1995; Thompson, J Comput Assist Tomogr 21(4):567–581, 1997; Thompson et al., Detecting disease-specific patterns of brain structure using cortical pattern matching and a population-based probabilistic brain atlas. In: IPMI2001. Lecture notes in computer science, pp 488–501, 2001), automatic segmentation by label transfer, modality fusion, morphological analysis (Hua, Neuroimage 43(3):458–469, 2008), and many other applications. Image registration techniques strive to find a one-to-one correspondence between subject and target images to perform this task. This correspondence is defined by a smooth deformation field. This deformation field captures the geometric variations in the two images. In this chapter, we will review various techniques for image registration that are specifically designed for human brain.

Key words Image registration, Brain, MRI

1 Introduction

Registration of brain images is an essential step in multi-subject and multi-modality brain image analysis. Studies of anatomical changes in the brain over time or differences in brain anatomy between populations require that the data first be transformed to a common coordinate system in which anatomical structures in the brain scans are aligned. Similarly, longitudinal studies within subject or group analyses of functional data also require that the brain images are anatomically aligned. The morphological differences in the brains can be analyzed using techniques such as MRI-volumetry [1] that uses segmentations of regions of interests (ROIs), either manually or automatically, voxel based morphometry (VBM) [2] that studies voxel-wise intensity statistics, deformation based morphometry (DBM) [3], which analyzes spatial position differences, and tensor based morphometry (TBM) [4] that analyzes the deformation

Gianfranco Spalletta et al. (eds.), *Brain Morphometry*, Neuromethods, vol. 136,
https://doi.org/10.1007/978-1-4939-7647-8_1, © Springer Science+Business Media, LLC 2018

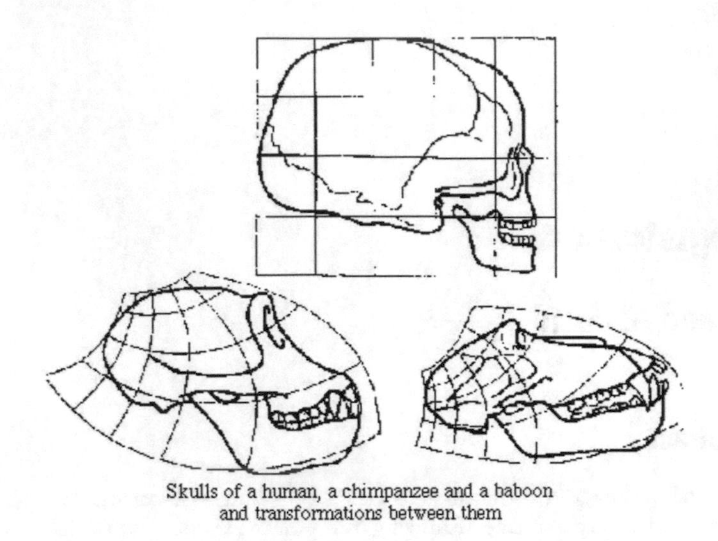

Fig. 1 Image registration as change of coordinates

tensors at every voxel computed from Jacobian of the deformation field. These techniques rely on the 1-1 pointwise correspondence established by image registration as well as the resulting deformation fields that aligns the images.

The goal of image registration is to find a one-to-one correspondence between biological homologous points between two brain imaging scans. This correspondence is represented by a mathematical transformation called deformation or warping field. The idea of image registration originates from the continuum mechanics and perhaps was first applied for biological structures in the works of D'Arcy Thompson in his book *On Growth and Form* [5]. Figure 1 shows his application of deformation grids for warping coordinate grids to deform skulls of primates and humans.

The human brain can be analyzed either as a surface represented by the cerebral cortex or the volume that contains cortical as well as sub-cortical structures. Since the cerebral cortex of the brain is highly folded, it is often convenient to model it as a 2D surface and use surface registration techniques. On the other hand, volumetric registration techniques are often applied where we are interested in cortical as well as sub-cortical structures.

In this chapter, we will focus on anatomical registration of brain images based on T1 MRI. We will first describe the surface based registration techniques that focus on alignment of the cerebral cortex. Next we will describe volumetric registration techniques that try to align brain volumes and in the end we will describe surface-constrained volumetric registration techniques (combined registration techniques) that try to align cortical surface as well as the sub-cortical structures. Finally, we will review existing intensity based linear and nonlinear image registration approaches commonly used.

2 Surface Registration

Human cerebral cortex is often modeled as a highly convoluted sheet of gray matter. A triangular mesh representation of cortical surfaces is generated using software such as BrainSuite (http://brainsuite.org), FreeSurfer (https://surfer.nmr.mgh.harvard.edu/), BrainVisa (http://brainvisa.info/), etc. Since it is often not possible to align this highly convoluted cortex in the 3D space, a necessary first step for cortical registration is cortical surface parameterization that maps the cortical surface to a sphere as in FreeSurfer or BrainVisa or squares as in BrainSuite (Fig. 2). Inter- and intra-subject comparison involving anatomical changes over time or differences between populations requires the spatial alignment of the cortical surfaces, such that they have a common coordinate system that is anatomically meaningful. Sulcal curves are fissures in the cortical surface and are commonly used as surrogates for the cytoarchitectural boundaries in the brain. Therefore, there is also great interest in direct analysis of the geometry of these curves for studies of disease propagation, symmetry, development, and group differences (e.g. [6, 7]). Labels of cortical regions of interest (ROIs) or sulcal curves that are required for these studies can be produced using manual [8] or automatic delineation [9, 10]. The manual delineation is often performed using interactive software tools [8] which, however, can be a tedious and subjective task that also requires substantial knowledge of neuroanatomy and is therefore confounded by intra- and inter-rater variability. This variability is reduced to some extent using rigorous definitions of a sulcal tracing protocol and extensive training as described in [8, 11].

An alternative approach to this problem is to use automatic surface registration to align surface curvature or sulcal depth [12].

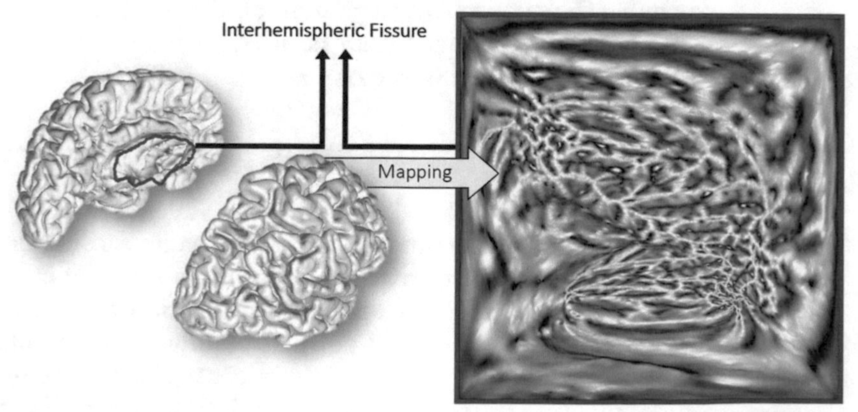

Fig. 2 Surface parameterization of the cortex is done by constraining the interhemispheric fissure to the unit square and rest of the cortex is mapped inside the unit square by using *p* harmonic maps

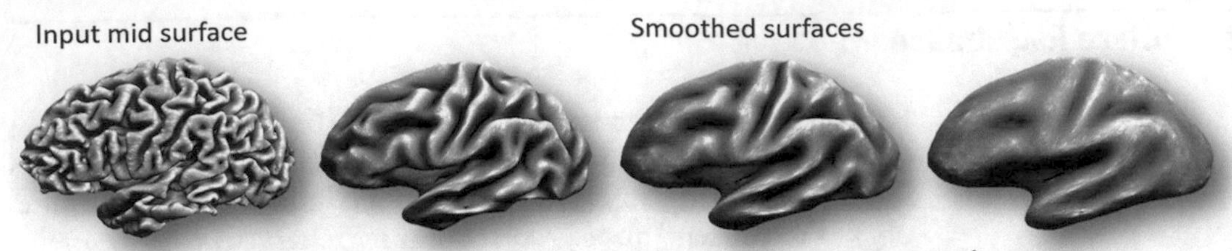

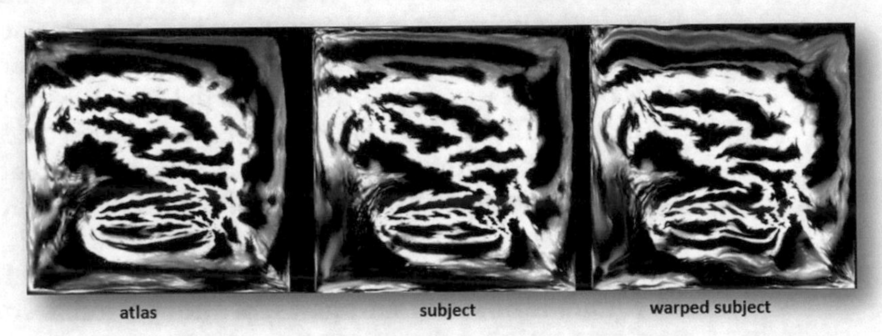

Fig. 3 Curvature alignment is performed by first generating a multiresolution representation of the mean curvature (top row) and then performing the alignment of the curvature in the flat 2D space

The mean curvature is used as it represents the sulci fundi with negative values and gyral crowns with positive values; therefore, its alignment leads to accurate registration of the cortex. The curvature maps generated are then transferred to the unit square using the point correspondence established by the p-harmonic maps. The alignment of the curvature maps is then performed by minimizing the cost function which is a weighted sum of a curvature matching penalty and a 3D coordinate matching penalty, regularized by an elastic energy. This step, as shown in Fig. 3, establishes a 1-1 correspondence between the subject and target cortical surfaces such that the sulcal and gyral patterns on the two brains are aligned. This correspondence can then be used to transfer data or labels from one brain image to the other.

3 Volume Registration

The most popular approach for brain image registration is the volumetric registration based on intensity information in anatomical T1 image. A piecewise affine transformation termed Talairach normalization [13, 14] was the first commonly used volumetric alignment technique. This method is constrained to be piecewise affine and uses a restricted set of anatomical landmarks. Therefore it results in a relatively poor alignment across subjects. Automated intensity-based registration methods overcome this constraint and also allow non-rigid deformations [15, 16].

Intensity-based volumetric image registration can be formulated in terms of an optimization problem:

$$\hat{h} = \arg\min_{h} \left(C_{\text{sim}}(T, S, h) + \lambda C_{\text{reg}}(h) \right)$$

where the transformation h minimizes the weighted sum of a similarity function C_{sim}, which defines a metric between corresponding features in the pair of images being matched, and a regularization function C_{reg}, which resolves the ambiguities among the set of transformations that minimize the similarity function by selecting a smooth transformation. The regularization parameter λ determines the degree of the smoothing imposed by the regularizer. Most of the image registration techniques can be placed in this framework. With this formulation, we now consider each of the three major components that define the most registration methods: feature selection and similarity metrics, transformation parameterization, choice of regularizing function. Table 1 lists many of the similarity measures, parameterizations, and regularization operators that have been used to produce many of the commonly used image registration algorithms.

Small deformation models such as polynomial warps and linear elastic deformations do not guarantee preservation of topology for larger deformations [17, 18]. The viscous fluid approach [18] and more recent approaches using large-deformation diffeomorphic metric mapping [19–21] were developed to address the problem of ensuring diffeomorphic maps and can register the objects whose alignment requires large deformations while conserving their topology.

Table 1
Registration methods

Corresponding feature (dimensionality)	Transformation parameterization (dimensionality)	Regularization/constraints (deformation type)
Landmark (0-D)	Rigid (low)	None
Contour (1-D)	Affine (low)	Thin-plate spline (small)
Surface (2-D)	Talairach (low)	Differential operators (small)
Sub-volume (3-D)	Polynomial (low-medium)	Prior distributions (small)
Intensity difference (N-D)	B-spline(medium-high)	Elasticity (small or large)
Cross correlation (N-D)	Thin-plate spline (medium)	Viscous fluid (large)
Intensity demons (N-D)	Discrete cosine (medium-high)	Inverse consistency (small or large)
Intensity vectors (N-D)	Fourier series (medium-high)	
Intensity variance (N-D)	Wavelet (medium-high)	
Mutual information (N-D)	Discrete lattice (high)	

One important consideration, apart from topology preservation, is inverse consistency of image registration [17]. A deformable image registration is called inverse consistent, if the correspondence between two images, obtained by reciprocal registration, is invariant to the order of the choice of source and target. More precisely, let S and T be the source and target images, and h and g be the forward and backward transformations obtained by a given registration method. Therefore $S \xrightarrow{h} T$ and $T \xrightarrow{g} S$; then an inverse consistent registration satisfies $h \circ g = I_d$ and $g \circ h = I_d$, where I_d is the identity map. The property of inverse consistency is applied explicitly by minimizing the difference between $h \circ g$ as well as $g \circ h$ to I_d [17] or by modifying the cost function such that the resulting forward map, generated by minimization of that cost function is inverse consistent [22]. Additional constraints such as transitivity can be imposed to get more desirable results.

In addition, the choice of a regularizer is an important consideration and can significantly affect the quality of the registration. A regularizer is used to constrain the transformation and ensure that the deformation is smooth and invertible. One common way to select a regularizer is to assign a continuum mechanical law to the deforming image medium. For instance, in elastic registration [23], the image is regarded as embedded in an elastic medium. A force is applied based on the chosen similarity function that pulls the template into agreement with the study, while linear elastic forces attempt to restore it to its original shape. Elastic registration is not guaranteed to produce 1-1 mapping and often is unsuitable for large deformations. In fluid registration, for instance, the image is treated as a viscous fluid that follows the Navier–Stokes equation, with a velocity field that is the derivative of the deformation field. The fluid flow allows large deformations while ensuring 1-1 mapping. These approaches often are computationally very expensive and can take hours for registering brain scans. Therefore, a demons algorithm [24, 25] was proposed that models image registration as a diffusion process. The demons algorithm and its variants (e.g., [26]) have become increasingly more popular than the fluid registration due to their speed.

4 Combined Approaches

Since cytoarchitecture and function of the cortex is closely related to the folding pattern of the cortex, it is important when comparing brain anatomy and function in two or more subjects that their cortical surfaces are aligned. For this reason, there has been an increasing interest in development of volumetric brain registration algorithms that also align the cortical surface accurately. Similarly, in inter-subject longitudinal studies or group analyses of functional data such as fMRI and DTI it is important that the cortical surfaces

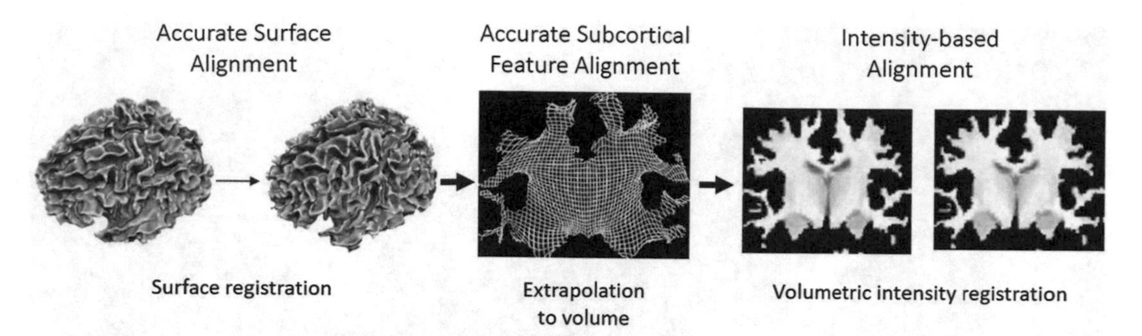

Fig. 4 Surface-constrained volumetric registration

of the subjects are aligned when brain registration is performed. Several methods have been developed that perform the surface constrained volumetric registration [27–31]. Here we describe our approach to brain image registration based on harmonic maps that combines the surface and volumetric registration approaches producing a volumetric alignment in which there is also an accurate one-to-one correspondence between points on the two cortical surfaces [29, 32]. This approach is implemented in SVReg software available with BrainSuite (http://brainsuite.org).

We perform volumetric alignment of brains by first extending the surface registration to the entire volume [29] (Fig. 4). This begins with two surfaces that have been aligned using the surface registration process. For each brain, the unit-square representations of the brain surfaces are mapped onto the unit sphere. The interior brain volume is then mapped to the unit ball. This is achieved by extrapolating from the surface to the interior of the sphere while minimizing the harmonic energy of the map from brain to sphere. The harmonic mapping gives one-to-one correspondence between the two brains based on their respective maps to the unit sphere. This provides an initial registration based solely on the initial alignment of cortical geometry. Such initialization ensures that cortical features are aligned. However, since the interior mapping is based solely on geometry of the cortex, subcortical features tend to be misaligned. A refinement of this mapping is computed by minimizing the elastic bending energy driven by an intensity matching forces. The output of this process is a one-to-one point correspondence between the two brain volumes. The cortical constraint ensures that one cortical surface maps precisely onto the other (Fig. 5).

5 Conclusion

Image registration is a very active and fast moving field of research that produced a number of software tools available on regular basis in the public domain. Most registration approaches are included in some popular software for brain imaging investigation as:

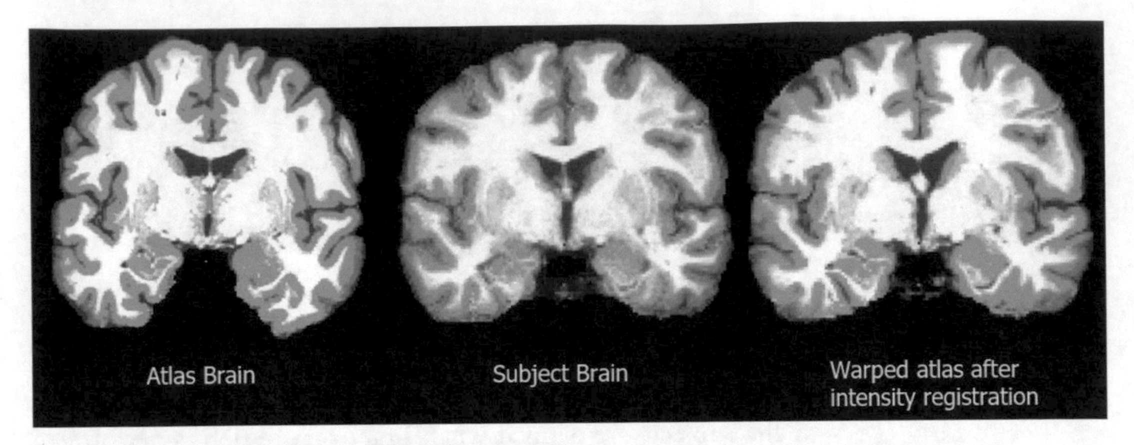

Fig. 5 SVReg registration showing good alignment at the cortical as well as sub-cortical structures

Automated Image Registration (AIR), ANTs, FreeSurfer, BrainSuite, FSL, LDDMM, ITK, CARET, and SPM. A comprehensive comparison of 14 of these software is performed in [33]. In this chapter, we reviewed basic techniques of brain image registration, which is an essential step for brain image analysis, trying to encompass the widest cohort of methods available.

References

1. Deblaere K, Boon PA, Vandemaele P, Tieleman A, Vonck K, Vingerhoets G, Backes W, Defreyne L, Achten E (2004) MRI language dominance assessment in epilepsy patients at 1.0 T: region of interest analysis and comparison with intracarotid amytal testing. Neuroradiology 46(6):413–420
2. Ashburner J, Friston KJ (2000) Voxel-based morphometry - the methods. Neuroimage 11 (6):805–821
3. Chung MK, Worsley KJ, Paus T, Cherif C, Collins DL, Giedd JN, Rapoport JL, Evans AC (2001) A unified statistical approach to deformation-based morphometry. NeuroImage 14(3):595–606
4. Hua X, Leow AD, Parikshak N, Lee S, Chiang M-C, Toga AW, Jack CR, Weiner MW, Thompson PM, Alzheimer's Disease Neuroimaging Initiative et al (2008) Tensor-based morphometry as a neuroimaging biomarker for Alzheimer's disease: an MRI study of 676 AD, MCI, and normal subjects. Neuroimage 43(3):458–469
5. Thompson DW et al (1942) On growth and form. Cambridge University Press, Cambridge
6. Narr K, Thompson P, Sharma T, Moussai J, Zoumalan C, Rayman J, Toga A (2001) Three-dimensional mapping of gyral shape and cortical surface asymmetries in Schizophrenia: gender effects. Am J Psychiatry 158 (2):244–255
7. Rettmann ME, Kraut MA, Prince JL, Resnick SM (2006) Cross-sectional and longitudinal analyses of anatomical sulcal changes associated with aging. Cereb Cortex 16(11):1584–1594
8. Shattuck DW, Joshi AA, Pantazis D, Kan E, Dutton RA, Sowell ER, Thompson PM, Toga AW, Leahy RM (2009) Semi-automated method for delineation of landmarks on models of the cerebral cortex. J Neurosci Methods 178(2):385–392
9. Tao X, Prince JL, Davatzikos C (2002) Using a statistical shape model to extract sulcal curves on the outer cortex of the human brain. IEEE Trans Med Imaging 21(5):513–524
10. Vaillant M, Davatzikos C (1997) Finding parametric representations of the cortical sulci using an active contour model. Med Image Anal 1(4):295–315
11. Pantazis D, Joshi A, Jiang J, Shattuck DW, Bernstein LE, Damasio H, Leahy RM (2010) Comparison of landmark-based and automatic methods for cortical surface registration. Neuroimage 49(3):2479–2493

12. Fischl B, Sereno MI, Tootell RBH, Dale AM (1998) High-resolution inter-subject averaging and a coordinate system for the cortical surface. Hum Brain Mapp 8:272–284
13. Collins DL, Neelin P, Peters TM, Evans AC (1994) Automatic 3D intersubject registration of MR volumetric data in standardized Talairach space. J Comput Assist Tomogr 18 (2):192–205
14. Desco M, Pascau J, Reig S, Gispert JD, Santos A, Benito C, Molina V, Garcia-Barreno P (2001) Multimodality image quantification using the Talairach grid. In: Medical imaging 2001. International Society for Optics and Photonics, Bellingham, WA, pp 1385–1392
15. Oliveira FPM, Tavares JMRS (2014) Medical image registration: a review. Comput Methods Biomech Biomed Eng 17(2):73–93
16. Sotiras A, Davatzikos C, Paragios N (2013) Deformable medical image registration: a survey. IEEE Trans Med Imaging 32(7):1153–1190
17. Christensen GE (1999) Consistent linear-elastic transformations for image matching. In: Information processing in medical imaging. Lecture notes in computer science, vol 1613. Springer, Berlin, pp 224–237
18. D'Agostino E, Maes F, Vandermeulen D, Suetens P (2003) A viscous fluid model for multimodal non-rigid image registration using mutual information. Med Image Anal 7 (4):565–575
19. Beg MF, Miller MI, Trouvé A, Younes L (2005) Computing large deformation metric mappings via geodesic flows of diffeomorphisms. Int J Comput Vis 61(2):139–157
20. Miller MI, Beg MF, Ceritoglu C, Stark C (2005) Increasing the power of functional maps of the medial temporal lobe by using large deformation diffeomorphic metric mapping. Proc Natl Acad Sci USA 102 (27):9685–9690
21. Zitova B, Flusser J (2003) Image registration methods: a survey. Image Vis Comput 21 (11):977–1000
22. Skrinjar O, Tagare H (2004) Symmetric, transitive, geometric deformation and intensity variation invariant nonrigid image registration. In: IEEE international symposium on biomedical imaging: nano to macro, 2004. IEEE, New York, pp 920–923
23. Ashburner J (2007) A fast diffeomorphic image registration algorithm. Neuroimage 38 (1):95–113
24. Thirion J-P (1998) Image matching as a diffusion process: an analogy with Maxwell's demons. Med Image Anal 2(3):243–260
25. Vercauteren T, Pennec X, Perchant A, Ayache N (2009) Diffeomorphic demons: efficient non-parametric image registration. NeuroImage 45(1):S61–S72
26. Vercauteren T, Pennec X, Perchant A, Ayache N (2008) Symmetric log-domain diffeomorphic registration: a demons-based approach. In: International conference on medical image computing and computer-assisted intervention. Springer, Berlin, pp 754–761
27. Auzias G, Glaunes J, Colliot O, Perrot M, Mangin J-F, Trouvé A, Baillet S (2009) Disco: a coherent diffeomorphic framework for brain registration under exhaustive sulcal constraints. In: International conference on medical image computing and computer-assisted intervention. Springer, Berlin, pp 730–738
28. Du J, Younes L, Qiu A (2011) Whole brain diffeomorphic metric mapping via integration of sulcal and gyral curves, cortical surfaces, and images. NeuroImage 56(1):162–173
29. Joshi AA, Shattuck DW, Thompson PM, Leahy RM (2007) Surface-constrained volumetric brain registration using harmonic mappings. IEEE Trans Med Imaging 26(12):1657–1669
30. Lederman C, Joshi A, Dinov I, Darrell J Horn V, Vese L, Toga A (2016) A unified variational volume registration method based on automatically learned brain structures. J Math Imaging Vision 55(2):179–198
31. Postelnicu G, Zollei L, Fischl B (2009) Combined volumetric and surface registration. IEEE Trans Med Imaging 28(4):508–522
32. Joshi AA, Shattuck DW, Thompson PM, Leahy RM (2005) A framework for registration, statistical characterization and classification of cortically constrained functional imaging data. In: Information processing in medical imaging. Lecture notes in computer science, vol 3565. Springer, Berlin, pp 186–196
33. Klein A, Andersson J, Ardekani BA, Ashburner J, Avants B, Chiang M-C, Christensen GE, Collins DL, Gee J, Hellier P et al (2009) Evaluation of 14 nonlinear deformation algorithms applied to human brain MRI registration. Neuroimage 46(3):786–802
34. Mazziotta JC, Toga AW, Evans A, Fox P, Lancaster J (1995) A probabilistic atlas of the human brain: theory and rationale for its

development: the international consortium for brain mapping (ICBM). Neuroimage 2 (2):89–101

35. Thompson PM, MacDonald D, Mega MS, Holmes CJ, Evans AC, Toga AW (1997) Detection and mapping of abnormal brain structure with a probabilistic atlas of cortical surfaces. J Comput Assist Tomogr 21(4):567–581
36. Thompson PM, Mega MS, Toga AW (2000) Disease-specific probabilistic brain atlases. In: Proceedings of IEEE international conference on computer vision and pattern recognition, pp 227–234
37. Thompson PM, Mega MS, Vidal C, Rapoport J, Toga AW (2001) Detecting disease-specific patterns of brain structure using cortical pattern matching and a population-based probabilistic brain atlas. In: IPMI2001. Lecture notes in computer science, pp 488–501

Check for updates

Chapter 2

Convolutional Neural Networks for Rapid and Simultaneous Brain Extraction and Tissue Segmentation

Nicholas C. Cullen and Brian B. Avants

Abstract

Convolutional neural networks are poised to become a standard technology in neuroimage analysis. This general purpose framework learns both low-level and high-level features directly from images, making them ideal for image segmentation. To highlight the potential of these tools, we present a novel convolutional-deconvolutional network architecture designed for efficient three-dimensional, supervised brain segmentation. We detail the problem definition, network design, evaluation and interpretation underlying this effort. We also provide evidence that such networks can achieve accuracy in a matter of seconds that rivals what traditional methods may take over an hour to compute.

Key words Deep learning, Segmentation, Convolutional, Brain, Neuroimaging, Deconvolutional

1 Introduction

Constantin von Economo's cytoarchitectonic map, a collaborative 13-year effort, is among the most detailed and influential works in quantitative segmentation [1]. The data sets of interest to the majority of practitioners today are collected in diverse populations during life and stored in images that preserve the three-dimensional structure of the brain. Contemporary work in biomedical segmentation seeks to automatically annotate such data in a structural or functional context. In the human brain, segmentation may involve very high-resolution labeling of neurons, classification of primary tissue classes from magnetic resonance imaging (MRI) [2], a detailed parcellation of cortical regions [3] or even fine-grained localization of hippocampal subfields [4].

Supervised segmentation approaches seek to directly model expert knowledge. Such methods may automatically reproduce the performance of highly-trained neuroanatomists, diagnosticians

Dr. Avants recently became a Biogen employee.

Gianfranco Spalletta et al. (eds.), *Brain Morphometry*, Neuromethods, vol. 136,
https://doi.org/10.1007/978-1-4939-7647-8_2, © Springer Science+Business Media, LLC 2018

or other computational methods at both lower cost and with greater speed. A supervised algorithm that performs near expert levels of accuracy allows that expert knowledge to be shared via the combination of software and data. A recent segmentation method, LINDA [5], was trained on an expert neurologist's annotations of post-stroke lesions in T1-weighted neuroimages and shown to provide comparable performance on unseen data sets from other clinical sites. The model itself is freely available for download and yet does not require the original annotations to be shared. Thus, supervised biomedical segmentation algorithms may be used to store and disseminate rarefied expertise thereby helping to standardize challenging biomedical quantification problems and establish new widely accessible pathways to knowledge. The combination of large data sets and powerful computational methods present the opportunity to perform, for the first time in human history, large studies of brain variability and longitudinal change.

The last two decades of algorithms designed for prior-driven (supervised) brain segmentation fall roughly into *probabilistic*, *multi-atlas* or *machine learning* categories. The boundaries between these categories are not stark and transition between them also correlates with increasing computing power. Probabilistic methods, such as those provided by the popular SPM package [6], tend to rely on a single atlas that summarizes population data with spatial probability maps for different anatomical classes. This era was computationally restricted and waned, more or less, in the second decade of the current millennium. Around 2010, multi-atlas labeling (MAL)—which often relies heavily on deformable registration—emerged as the premier technology for performing brain parcellation, in particular when spatial location is highly informative about the class of a given part of the brain [7]. Rather than averaging expert labels in a common template space before applying to a new brain, MAL propagates the full cohort of labels into each individual image space and performs aggregation within that space. The best of these methods also incorporate local patch-based similarity between the atlases and the target brain while accounting for redundancy between the atlases [4]. More recently, machine learning methods have become available that may capture nonlinear and highly multivariate information that may be leveraged to improve segmentation [5].

Despite these advances, there remains the possibility for further improvement in segmentation, in particular due to the size of training and unlabeled data sets grow which will greatly increase the degree of variability to which algorithms must adapt. The primary limitation of algorithms based on single probabilistic atlases is that they relied heavily on, for instance, Gaussian mixture models to customize the priors to individual data sets. The limitation for MAL may be similar in that they may be limited by design: most MAL methods use assumptions of locality and linearity. There are also significant improvements to be had with regard to speed. Newer methods, such as deep learning, have the ability to store—within their own architecture—a much greater degree of adaptability and non-linearity than is available from existing labeled data sets alone. Furthermore, deep learning software is built to exploit graphical processing units (GPUs), which may accelerate computations by an order of magnitude or more.

1.1 What Is Deep Learning?

Deep learning algorithms link and optimize computational layers in order to build predictive multi-scale data representations. The design of these general purpose computational machines is inspired by the layered and interconnected cortical columns of the mammalian brain. An excellent review of the field is available here [8] which describes deep learning as composing multiple simple but non-linear modules that, in aggregate, allow very complex functions to be learned. Some of the more famous examples of deep learning architectures include AlexNet (8 layers), VGG Net (19 layers), GoogLeNet (22 layers) and ResNet (152 layers), all of which pushed the performance envelope in the international ImageNet challenge. ImageNet winners currently compete with or exceed human performance [9] on an image-based 1000 class object identification problem, an achievement that is a testament to both decades of prior work as well as the value of public competition, communication and evaluation [10].

Current limitations of deep learning, in particular within the biomedical domain, include the perceived need for relatively large training data sets and the lack of interpretability. However, substantial progress is being made on both fronts, i.e. the visualization of deep learning [11] as well as practical methods for data augmentation and one-shot learning [12]. Furthermore, deep learning performance is accelerating with recent substantial investment from industry. In particular, Google broke new ground in machine translation [13], which now approaches human performance in several language pairs. While there are many challenges to broad adoption of this machinery, it is likely that the investments into deep learning infrastructure made by Google, Facebook, Baidu and Microsoft (among others) will only further improve the value of the underlying software [14]. It is therefore imperative that more scientific investigators—brain mappers, in particular—become not only familiar but also facile with deep learning. Toward that end, we discuss perhaps the most transparent application for deep learning to a problem that is commonly faced in neuroimaging: tissue segmentation of T1-weighted MRI as in [3].

1.2 Convolutional Networks and Brain Segmentation

Convolutional neural networks have image-specific layers, which consist of patch-like "local modules" that act as a set of spatially varying and interconnected feature representations. Shallow layers typically learn basic features such as edges, while deeper layers begin to aggregate information into more abstract representations. In the case of object detection, layers may model the class appearance itself [15]. This ability to represent nonlinear hierarchical *spatially constrained* information is thought to confer performance advantages over more traditional approaches (even non-convolutional networks.) The winning team for a connectomics challenge employed convolutional networks and "had no prior experience with EM (electron microscopic) images" [16], thus showing the power of convolutional networks in combination with well-curated training data in a supervised learning framework.

Convolutional networks are prevalent in brain segmentation research [17–23] and push performance standards in open challenges such as ISLES and BRATS [23]. Specifically, deconvolutional architectures [24–26] reach or exceed state-of-the-art in image segmentation problems in both computer vision and medical imaging. Importantly, such networks may not need enormous numbers of labeled individual subjects within training data sets in order to achieve clinically valuable results [27]. This is due to the fact that most neuroimages are, at several scales, highly redundant. That is, even a single image provides a relatively large sampling of variability in the appearance of local anatomy. Convolutional networks transform this structure into a spatially informed set of weight functions or patterns. By exploiting supervision, these expressive multi-scale patterns are *automatically customized* according to the prediction problem at hand. Thus, a single architecture can be repurposed for many different prediction problems [28].

In the remainder of this chapter, we will detail the steps involved in developing a convolutional-deconvolutional network for brain extraction and segmentation. These steps illustrate the efforts in this particular application's (recent) history but we also reflect on the more general lessons learned in engineering a practical machine learning system.

2 Materials

2.1 Imaging Data

We demonstrate our application on the T1-weighted neuroimages collected in the Dallas Lifespan Brain Study (DLBS) available (http://fcon_1000.projects.nitrc.org/indi/retro/dlbs.html) and described in [29]. The cohort subjects have average age 55.2 ± 20 years (min 20.6, max 89.0) and include 172 females and 103 males with mean educational attainment of 16.3 ± 2.30 years. The DLBS MRI data set includes 275 subjects with 1 mm^3 T1-weighted MPRAGE SENSE MRI collected on a 3T Philips Medical System machine as described here. The scanning session used a whole body coil to transmit the RF excitation and an 8-channel receive head coil with parallel imaging. We processed each T1 image through the ANTs cortical thickness pipeline leveraging a pre-existing template [3].

In order to define ground truth data for the DLBS, we took the six tissue segmentation produced by the pipeline which has been validated with respect to FreeSurfer [3] and other segmentation methods [30]. This segmentation procedure performs brain extraction followed by tissue segmentation within the brain mask. The brain masking algorithm involves bias correction, two registration steps, segmentation and morphological operations. The Atropos segmentation step uses N4 bias field correction [31] and on the order of 25 iterations of Atropos (depending on convergence speed), which optimizes a Markov Random Field regularized Gaussian mixture model. In total, these steps may take one to 2 h of CPU time, depending on the data, compilation of source code and degree of multi-threading. Using the final segmentation as ground truth training for our deep learning algorithm enables us to evaluate whether the convolutional-deconvolutional network can learn to reproduce the cumulative output of these complex and time consuming steps in much less computation time. As we see below, the proposed network indeed can simultaneously perform brain extraction and segmentation in a single shot in very little GPU time.

2.2 Software

We employ R version 3.3.1 ("Bug in Your Hair") as well as Python version 3.5.0. We also use ANTsR version 0.3.3 [32] for core image analysis, data organization and early development efforts. Deep learning software is both well documented and highly optimized for both large data sets (via incremental learning) and modern parallel computation (via graphical processing units—GPUs) in comparison to other machine learning methodologies. Our preferred framework is *TensorFlow*, which is entering its version 1.0 release, which will guarantee backward compatibility. TensorFlow features many of the latest advances in deep learning research as well as the unique (in current software) ability to distribute problems across multiple CPUs or GPUs on a given machine. In contrast, Caffe (another leading platform) is currently limited to using a single GPU. Thus, we recommend TensorFlow as the underlying platform for future deep learning implementations. TensorFlow is described in [14] but check the latest documentation for up to date features.

2.3 Hardware

We used NVIDIA K40 GPUs for performing this research which retail between $3 and 4 thousand. Although training deep neural networks with multiple GPUs at once is a highly active area of research at the moment, we trained each model on only one GPU at a time. Training our models on even a cluster of CPUs would be unfeasible. While the reliance on GPUs is certainly one drawback of deep learning, it may also be interpreted as a good thing for the end-user. Once our model is trained, it will almost never need to be altered again. This is in stark contrast to existing models for tissue segmentation, which must learn from scratch to segment each new image they see. This means that with our model, the computational burden falls on us—the researchers—who have access to the expensive equipment required for training deep neural networks. This significantly lessens the computational burden on the end-user, who simply has to run our model on their data. Moreover, this means that our model will operate in a deterministic manner by producing the same result each run—something that will increase reproducibility of results.

3 Methods

We detail, below, a new network for performing very rapid, simultaneous brain extraction and tissue segmentation for T1-weighted MRI. This network is trained to reproduce—with a single step—the output of both the brain extraction and tissue segmentation steps in the ANTs cortical thickness pipeline. As shown below, the performance of this network approaches that of the original algorithm but is achieved in 100×–500× speed in this dataset. In brief, what would take 3 h completes in under 30 s, by exploiting convolutional-deconvolutional architectures on the GPU. Our development process for this new technology roughly followed three steps: problem identification, network design and evaluation, and interpretation stages. These steps are quite general to most machine learning, and can thus serve as a guideline for future work in related problems.

3.1 Problem Identification

Traditionally, segmenting the entire brain from the head is performed before tissue segmentation. We believe these two steps can be fused and thus sought a system which could perform both tasks simultaneously and which could be applied to near raw T1 data. It is almost unheard-of for segmentation methods to be applied on raw brain images. Typically, processing steps such as bias correction, normalization, and registration to a template or standard space are performed prior to running the segmentation algorithm. We wanted a model that would be robust to brain images, which may not have undergone all of these steps, thereby minimizing the time to go from image collection at the scanner to having a tissue-segmented brain at the researcher's disposal.

The problem of segmenting brain MRI into distinct tissue classes using deep learning models is inherently a pixel- or voxel-wise classification problem. In this work, we construct a probabilistic voting ensemble of three 2D convolution-deconvolution architectures, each trained on full image slices belonging exclusively to either the axial, sagittal, or coronal plane. Our work is one of the few to actually train on entire image slices, whereas previous work in this area has mostly involved training on image patches. The slice-based approach is undoubtedly the preferred method, as the patch-based approach is computationally unfeasible for segmenting entire images due to the large number of patches, which must be sampled.

We also set out to develop a framework that would cross operating systems, programming languages, and computing hardware. Since the computational steps involved in the model are mostly matrix multiplication and image convolution—which are supported in every programming language—we can easily extract the model weights and plug them back in across languages using very lightweight, custom functions. This portability also means that end users do not have to spend computational resources on *training* (that is our burden) but only on *prediction*—which is extremely fast.

In summary, our goal was to establish and test a new approach to rapid brain extraction and segmentation that is practical and portable. Details of the methods follow.

3.2 Data Curation Phase

3.2.1 Choosing a Dataset

The DLBS is a well-curated dataset with no apparent failures or highly aberrant subjects. We provide additional notes on data curation in **Note 1**. These data span a broad age range and thus test the ability of the proposed network to perform well in a diverse population (*see* Fig. 1). Furthermore, the dataset is accompanied by demographics that include age, gender, educational attainment and scores on the mini mental state exam (MMSE). These additional variables allow a different look at the validity of the ground truth versus network-produced segmentation values, as discussed in **Note 2**.

3.2.2 Split Data into Training, Validation, Testing Sets

Common with best practices in machine learning, we split the data into training, validation, and test sets with each subject belonging to one distinct set. After the convolutional-deconvolutional segmentation model is trained on the training set, we evaluate the model fit on the validation set. If the score on the validation set is not sufficient, we revise the model architecture and parameters, then retrain the model on the training set. We repeat this process until the validation score is strongest among all model variations. Finally, we evaluate the model on a completely left-out testing set. The score on the testing set represents the true generalizability of our model, because the testing set is not involved in any part of the training/validation phases and thus has no undue influence on model outcome.

The DLBS dataset has 270 subjects, so we randomly split each subjects into the following configuration:

- 190 train subjects
- 30 validation subjects
- 50 test subjects

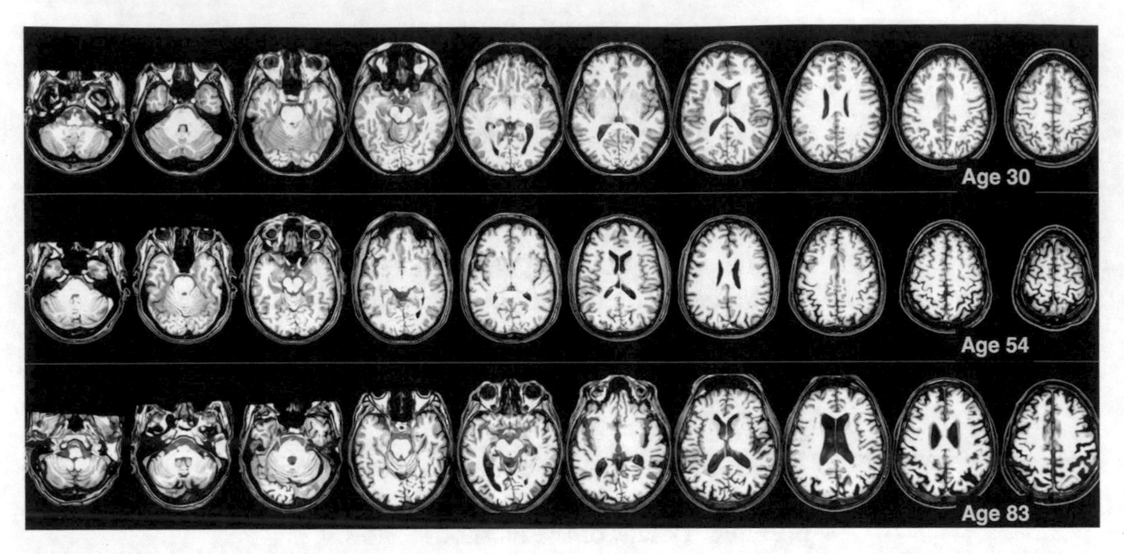

Fig. 1 Three images randomly selected from the DLBS cohort illustrate variability in the training dataset

Because the DLBS dataset has all healthy subjects and a diverse population, we are confident in randomly splitting the subjects into one of the three above categories. However, when working with a dataset involving healthy controls and diseased subjects, it is best practice to ensure the classes (healthy vs. control) are represented in equal proportions across the train/validation/test splits.

3.3 Batch Processing Pipeline

With the raw imaging data in hand, we must decide on how to process the images and how to feed images into our model. As far as processing the raw images, we trained our deep segmentation model on images with *zero* pre-processing aside from globally scaling each image to have a 0–1 scale. This scaling was done for each image, completely independent of any other image. As mentioned before, however, we found marginal improvements from using N4 bias correction and histogram equalization, so it is recommended to perform these steps for best fit.

Since each 2D convolutional-deconvolutional model in our three-model ensemble takes in 2D image slices, we had to formulate a method for sampling 2D slices from the 3D brain images. To do this, we wrote code to uniformly randomly sample a slice from the chosen axis for each input brain image (the axial-slice model only samples axial slices, the coronal-slice model only samples coronal slices, and so on). The only consideration during sampling was what to do about empty slices—slices with no brain in them. We decided not to allow any empty slices, as this would significantly bias our model training with no benefit to prediction. The class labels that we seek to predict include: background, cerebrospinal fluid, cortical gray matter, white matter, deep gray matter, brainstem and cerebellum.

3.4 Out-of-Memory Sampling

In a normal setting, we would randomly sample 2D slices from the images until the model converged in our standards. However, a major issue with medical imaging data is that it is very high-dimensional and so the entire dataset does not usually fit in memory. With this in mind, we implement an "out-of-memory" sampling algorithm, which does not explicitly load the entire dataset into memory at once. This is a common practice in convolutional neural network training, where standard datasets such as ImageNet have 1.2 million images and cannot fit in memory.

3.5 Building the Model: Construction Phase

Once the dataset was chosen and curated, we began building our model. We chose the convolutional-deconvolutional model because it has performed well on tasks related to semantic segmentation of natural images. Moreover, this is a "fully convolutional" model, meaning that it has relatively few parameters as compared to traditional convolutional models with dense, fully-connected layers. The relatively few parameters greatly increase the speed of model prediction, without compromising model complexity or generalizability due to the global sliding of the convolutional kernel across the image.

3.5.1 Model Architecture

The architecture of a convolutional-deconvolutional model generally starts with the input image being passed through some number of convolutional layers, which includes a number of square kernels, or filters, performing the convolution operation on the image. These so-called kernel maps are then passed through a non-linearity—a rectified linear unit, in our case—and a set of kernel activation is produced. Then, these kernel maps are either passed through another set of convolutional kernels, or are optionally down-sampled to smaller kernel maps using an average or max pooling operation (usually with a 2×2 neighborhood). This describes the "convolutional" portion of the architecture, and generally leads to a set of down sampled feature maps, which have been passed through numerous non-linearities and normalization functions.

Since the convolutional portion of the architecture leads to numerous (usually between 20 and 40) downsampled set of feature maps after starting from the normal input image, it is then necessary to build the feature maps back up in spatial resolution and combine these maps together to get back to the target image shape. That task is carried out by the "deconvolutional" portion of the architecture, where the downsampled kernel maps undergo a transposed convolution operation. These transposed kernel maps are then passed through more non-linear functions, and are optionally normalized or up-sampled until a single map, which has the same size as the target segmentation image, is produced. The symmetrical nature of the convolutional-deconvolutional architecture should be evident at this point, as shown in Fig. 2.

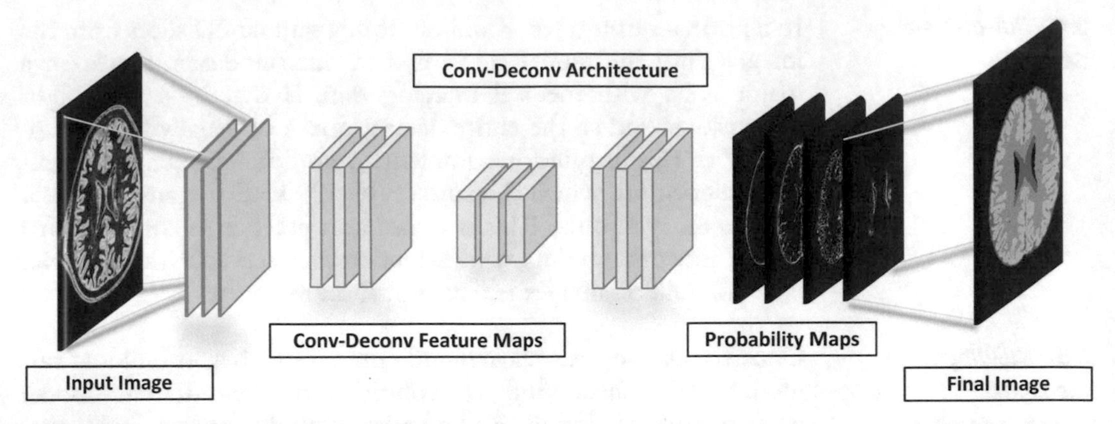

Fig. 2 The architecture of the convolutional-deconvolutional network. This illustration shows only the axial component of the network. At left we see the input T1-weighted MRI slice. At far right is the predicted probability maps as well as the final segmentation. The intermediate convolutional layers (left half of network) learns features whereas the right half of the network decodes these features into the shape of the prediction target (here, a seven class set of spatial probability maps)

Of course, there are many details left out in the above explanation of the convolutional-deconvolutional architecture. It suffices to say that for specifically deciding on an instantiation of the general model architecture, we had two main considerations:

- First, we decide how *deep* to make the model. The model depth is measured by how many successive convolutional layers exist in the architecture. With a very deep model, we are able to learn significantly more abstract and non-linear features, but we pay for that added complexity with increased computational cost and a higher risk for the model to simply "memorize" the training data rather than learn generalizable features of brain images—a problem generally known as over-fitting. Thus, there is a trade-off for model depth, which we found to be optimal at around three to four convolutional layers. The convolutional layers are transposed and symmetrically added as deconvolutional layers, so three convolutional layers corresponds to three additional deconvolutional layers and therefore six layers in total.
- Next, we decide how large to make the convolutional kernels. A kernel, or filter, abstractly corresponds to a local receptive field of a visual neuron. A larger kernel means that more of the image is "seen", but a large receptive field can lead to computational burden and, again, over-fitting. At the same time, a kernel, which is too small will lack the ability to identify even local features of the brain such as curves in gray matter or location of small brain structures. For instance, if the convolutional kernel is too small, it will not be able to reliably identify the boundary between skull, CSF, and gray matter. Through

experimentation on the training set and evaluation on the validation set, we found that relatively larger kernels worked better. That is, square kernels of size 9×9, 11×11, or 13×13 pixels led to better performing models than those that used kernels of size 3×3 or 5×5 pixels.

Besides the two main considerations of model depth and kernel size, there are various other hyper-parameters such as kernel stride, average or max pooling, activation functions, L1 and L2 regularization for layer weights and activations, and normalization methods such as batch normalization or dropout.

3.6 Objective (Loss) Function

The loss function of a deep convolutional network, or any machine learning model for that matter, tells us how well we are doing at the particular task in which we are interested. In our case, we are broadly interested in classifying pixels from 2D slices, and eventually 3D voxels, into its correct tissue class. It is therefore easy to naively believe that training a model to simply maximize classification accuracy would lead to strong performance. This turns out not to be the case, as the gradients of the loss function become highly volatile and difficult to calculate, leading to quick convergence to a highly sub-optimal solution.

Instead, we adopt the commonly used cross-entropy loss function, which instead measures an information theoretic notion of how far away our predicted class distribution lies from the true class value. Since the predicted distribution is valued between 0 and 1, and indeed sums to 1 across all possible classes for a given pixel/voxel, the class distribution essentially gives us a confidence value for our predictions. Our goal is to push this distribution to be perfectly sure of its predictions for each pixel/voxel, meaning it will have a 1 in the correct class index and a 0 elsewhere. The correspondence to the one-hot representation of the ground-truth image should now be clear.

The cross-entropy function can be represented as follows:

$$H(p, q) = -\Sigma_x p(x) \cdot \log(q(x)),$$

where $p(x)$ is the predicted probability distribution over all possible tissue classes for a single voxel and $q(x)$ is the true distribution represented as a one-hot vector with a one in the correct tissue class index and a zero everywhere else. Thus, by minimizing this function the optimization is driving the model to predict the correct class labels while allowing flexibility through use of a probabilistic distribution.

In practice, the more confident our class distribution is in its prediction, the better the model performs. Moreover, we can leverage low-confidence class predictions by using an ensemble of different models and hoping that an unsure prediction for one model in a given pixel/voxel will be "saved" by a high-confidence prediction in that same location by another model in the ensemble.

3.7 Optimizer & Initializations

Training a deep convolutional neural network involves navigating a high dimensional, non-linear, and extremely volatile parameter space filled with numerous local minima. Landing in a local minima is quite easy with a neural network, and when it happens the model performance will almost immediately stop improving. Moreover, a volatile part of the parameter space may cause performance to deteriorate in an unbounded manner. Besides previously mentioned ways for avoiding this scenario, such as appropriate model architecture and good data/augmentation, there are two more important considerations in our model.

First, the choice of optimizer is perhaps the most important factor in determining how well a model will train—all else equal. Since training proceeds by gradient descent with stochastic minibatches, it is easy for the model to go wild once it sees a completely new mini-batch. With that in mind, we found that the Adam optimizer is the most well-suited for our architecture. Indeed, the Adam optimizer is probably the most popular optimizer in the current deep learning literature, as it was found to out-perform regular stochastic gradient descent in nearly all cases. *See* **Note 3** for further comments on the Adam optimizer.

Additionally, the avoidance of volatile parts of the parameter space can be achieved through clever initialization schemes for the model parameters. The choice of initialization scheme is an enormously under-appreciated factor in the training of deep convolutional networks. For instance, if the model weights are initialized too large, then the final activation map values will explode and the loss value will shoot to infinity. If the model weights are too small, then no gradient will pass back-propagate through the network, leaving the network to simply stand in place and never change. This will also happen if all model weights are initialized to be very similar values. With that in mind, we chose a well-trusted initialization scheme called the *Xavier Initialization*. We believe that our choice of optimizer and parameter initialization scheme were vastly important factors in the success of our model to the problem of simultaneous brain extraction and tissue segmentation.

4 Training and Validating the Model: Execution Phase

With the data processed, the sampling scheme laid out, and the model architecture in place, we finally started training our models. The general flow of the execution proceeds by fitting the conv-deconv model on the training data, and once that fitting procedure is complete, we evaluate the model fit on a separate held-out validation dataset. If one encounters early failure in this process, it may indicate a bug in coding as discussed in **Note 4**.

The training procedure consisted of 150 epochs, and for each epoch we sampled 5000 image slices from the subjects in the training set. The image slices were sampled evenly from the subjects, so each subject contributed equally to the model fitting. In order to improve convergence—and also due to computational constraints—we fit our model using "mini-batches" of data, using a batch size of 30 image slices. Larger batch sizes allow the model to see more variance in the dataset, but at the cost of larger compute time and a higher risk of falling into sub-par local minima from which the model may never recover. Smaller batch sizes—usually anything less than 30—run faster overall and allow the model to escape from bad local minima, with the risk of finding a worse fit in the long run. We found that anything between 20 and 30 image slices was a good batch size, as it was fast but still led to a good model fit.

Training time took around 5–15 min per epoch, leading to total training time of anywhere from 1 to 2 days. This may seem like a large amount of time, but this is a true benefit of the deep learning paradigm. Once the model is trained, it rarely needs to be updated or trained again. More importantly, predicting on new images is incredibly fast because there is nothing to "learn" from new images. Compare this to traditional approaches, were the segmentation model must start from scratch for each image and the benefit of deep learning approaches becomes clear.

Training was performed using NVIDIA K20 and K40 GPUs, which represent the state-of-the-art in hardware for training deep neural networks. It would be basically unfeasible to train these models on normal CPU computers, or even CPU clusters, as training a single model instantiation would take on the order of weeks.

4.1 Determining When to Stop Training

The goal during training is to maximize the generalizability of the model, without actually memorizing the training set. After each epoch of model training, we evaluate the model on the held-out, unseen validation dataset. This allows us to determine in real-time when our model is finished training. We know that a model is fully trained when the validation loss no longer decreases or even begins to increase. Because of the nature of gradient descent, the complexity of the convolutional neural network framework, and machine learning in general, the training loss will continually decrease until the model has either completely memorized the training set or the complexity of its representation is fully utilized. Thus, seeing that the loss on the held-out validation set is no longer decreasing is a good sign that the generalizability of the model is no longer increasing and we should stop training. Of course, another method for avoiding over-fitting is to simply decrease model complexity by decreasing the number of layers or decreasing the number of kernels.

4.2 Deciding Whether to Refine the Model

Once the model is finished training, we again test the model on the validation dataset. If we are "happy" with how the model has fit the held-out validation set—here, by happy, we mean that the segmentation accuracy on unseen images is nearly perfect—then we finish the execution phase and move onto model testing. However, it is unlikely that the first machine learning model is the optimal model. In that case, we decide to make systematic changes to our model architecture or optimization procedure, and begin the execution (training and validation) procedure all over again. The refinement process is described in the next section.

5 Refining the Model: Refinement Phase

During the refinement process, we generally take clues from how the execution phase went to determine our next steps. There are many useful ideas to guide recent adopters of deep learning in the competitive machine learning market, as covered in **Note 5**.

5.1 Architecture and Optimization Refinements

If the training loss was very low but the validation loss was very high, we know that our model over-fitted the data by simply memorizing the training set instead of learning generalizable features of brain images. In this case, we decide to decrease model complexity by decreasing the number of layers or decreasing the number of convolutional kernels in each layer.

On the other hand, if the training loss never seemed to decrease very much, and neither did the validation loss, then we may be certain that our model complexity is not sufficient to learn the non-linear features common to brain images and the segmentation task at hand. When this happened, we increased the number of convolutional kernels in each layer and/or increased the number of convolutional/deconvolutional layers.

Another common problem is that the training loss will jump around from very low to very high numbers. As discussed before, this is typically a phenomenon, which occurs when the learning rate is too high. In that case, we simply decrease the learning rate and train the model until this no longer occurs.

5.2 Data Pre-processing Refinements

It may be the case that none of these model improvements increase the model fit and generalizability. In this case, it is best to look at the steps taken to pre-process the images and ask if there are any improvements to be made here. For instance, we discussed earlier how, although our model still performs well without any pre-processing, processing steps such as N4 bias correction and image normalization do improve model fit.

5.3 Data Augmentation

It is often the case with deep neural networks, as applied to the medical imaging domain, that over-fitting will almost always occur. This is typically because the datasets are not sufficiently large. As discussed above, convolutional networks typically perform best on computer vision tasks with more than a million available training images. This is a stark contrast to the typical brain imaging datasets, which have only a few hundred images at best. In order to alleviate this issue, we can perform clever data augmentation techniques to get the most out of our data. In the computer vision community, this typically involves applying some affine transform to the image or adding random noise to each mini-batch.

Our 2D slice sampling technique can be considered a type of data augmentation, since it allows us to extract more than just one individual training image for each subject. However, more sophisticated data augmentation techniques, which preserve the task goals but allow us to get the most out of brain images, are greatly needed in the medical imaging community. In general, data augmentation should be the first consideration for refining the model when no architectural or optimization refinement helps.

5.4 Unsupervised Pre-training and Transfer Learning

If no changes to model architecture and optimization procedure lead to better results, and no changes to the data pre-processing or augmentation procedure help, there is only one option—find more data. However, it is not so simple to find a great deal of labelled data, which is completely relevant to our task. In our case, there are only so many segmented MRI images publicly available.

In this case, unsupervised pre-training is a method, which may greatly improve model fit. For unsupervised pre-training, you successively train each layer to learn back a representation of the image input to that layer. We perform this in an iterative fashion, so the first layer learns a lower dimensional representation of the input image, the second layer learns a lower dimensional representation of the first layer features, and so on. The benefit of this procedure is that it does not require any labels (i.e. no segmentation), meaning it is completely unsupervised. The justification for this procedure is that by allowing the layers to train as such, they will still learn generalizable, low-level features for brain images. Low-level features, of course, are fairly common across all computer vision tasks—typically Gabor filters which act as edge detectors—and it is reasonable to believe that low-level features of brain images are common across tasks as well.

Another method, which has gained great traction in the computer vision community, is transfer learning. With transfer learning, we simply take the early convolutional layers from a model, which was trained adequately on a completely different task, and plug those layers into our own model. During the training of our own model, we hold those early layers to be fixed. That is, we do not allow those transferred layers to be altered at all. Again, the idea here is that low-level features should be approximately the same across all tasks, and by utilizing these low-level features we may be able to train a model on much smaller datasets than we otherwise could.

6 Testing Phase

After the refinement phase ended with our identification of a satisfactory set of models as evaluated on the validation dataset, we then moved on to the testing phase. In this phase, we used our ensemble of 2D slice models to fully segment the brain of each subject in the test set. The model has never seen the images in the test set, nor have the images in the test set played *any* part in our decisions to alter model structure in the refinement phase. In this sense, the model performance on the test set represents a true unbiased evaluation of generalizability to new data.

The testing phase is important because it allows us to determine whether our model is able to perform well on unseen data. The implications for this are important, since our goal is to build a segmentation model, which is not only uncannily fast, but is also able to perform accurately and robustly on brain images it has never seen.

6.1 3D Prediction from an Ensemble 2D Models

As mentioned above, we trained three separate models to segment brain tissue on 2D slices of the axial, coronal, and sagittal axes, respectively. For a single image of voxel dimensions $160 \times 276 \times 276$ (i.e. 1 mm^3 T1), we ran each slice model over each appropriate slice in the image. Since the raw output of the model is really a set of seven probability maps, we are left with an output of shape (160, 276, 276, 7, 3) for our ensemble model. That is, we now have three sets of probability maps. Combining these probability maps is done using a simple average for each voxel over the three models. This shrinks the output back to size (160, 276, 276, 7). Finally, we perform the argmax operation for each voxel to select the class whose probability value is highest over the set of seven probability maps. With that, we are now down to a single image whose shape is the same as the ground-truth segmentation model.

7 Interpretation Phase

With a set of test predictions and evaluation metrics, we finally moved on to the interpretation phase. For evaluation, we calculated the *Dice Coefficient* overlap for each tissue in the test image. The dice coefficient is a measure of classification accuracy for comparing segmentation predictions that takes into account both false positives and true negatives. This measure also weights class accuracy

Table 1
Dice overlap results in 50 testing images

	Background	Brainstem	Cerebellum	CSF	DGM	CGM	WM
Mean	0.9969	0.8656	0.9008	0.8972	0.8558	0.9314	0.9497
SD	0.0018	0.0783	0.0770	0.0215	0.0336	0.0342	0.0246

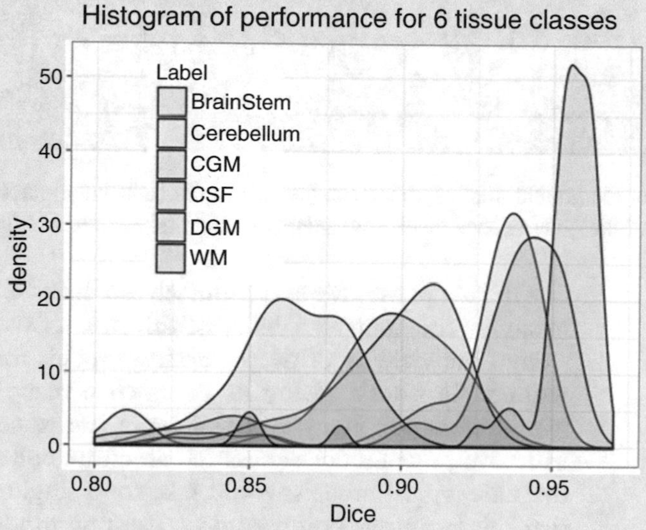

Histogram of performance for 6 tissue classes

based on the total number of voxels in each class. Results are summarized in Table 1 and the associated figure.

The results are quite promising, given the context:

- We performed minimal pre-preprocessing: no registration or bias correction;
- The segmentation method simultaneously classifies tissue while separating the cerebrum from extra-cerebral image regions;
- We employed a minimal number of deep learning "tricks"—regularization, data augmentation, etc.—on our model which are universally seen in the state-of-the-art models in computer vision tasks.

Most importantly, the ensemble of models consumes a total of between 15 and 20 s for a single subject. Since the model is already trained, it will *always* take between 15 and 20 s for new images. This is a main benefit of prediction, since if we can prove our model generalizes across datasets, then users will have an accurate and fast segmentation model that will apply to any arbitrary dataset. A characteristic result is in Fig. 3. *See* **Note 2** for other alternatives to evaluation.

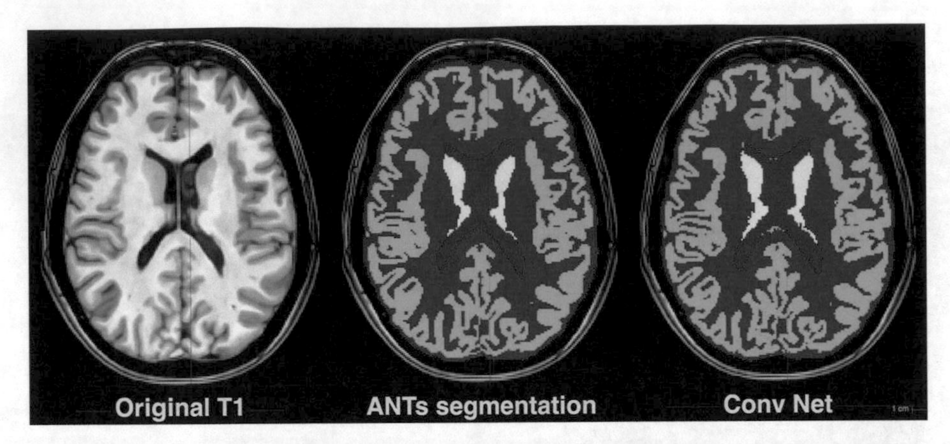

Fig. 3 A comparison of the ground truth segmentation (center) to the convolution-deconvolutional network's output (right). Note that the conv-deconv result both segments and brain extracts in one step

In this phase, we asked ourselves what the evaluation results mean for the general viability of this model. We certainly demonstrated the viability of deep learning models for a complex tissue segmentation task, giving us confidence that this type of model would work for many types of brain image segmentation tasks. Moreover, our model segmented an entire collection of images in the time typical models would take for a single image. We believe that this fact alone points to deep learning models as the future of brain image segmentation.

Our use of a large testing set (relative to the size of the training set) shows that this model can generalize to new images beyond the DLBS and is not simply memorizing the training set. However, we did not perform any rigorous validation of our model on a completely new dataset. This is certainly future work and is necessary to obtain the confidence of the neuroscience community.

There are a few outlier subjects for which performance is as low as 0.79 for cortical gray matter. We therefore visually inspect these individual images and found that the initial images represent fairly unique cases, e.g. high atrophy and unrepresentative orientation. Even so, the conv-deconv results look plausible which may suggest some lack of quality in the original ground truth. This is a standard event that may be overcome by exposing the network to additional data (or better ground truth).

8 Distribution Phase

The final phase of our work, which we believe to be often overlooked in the development of novel neuroimaging tools, is the *distribution phase*. In this phase, we packaged our code base into an easy-to-use, flexible, and light-weight package. In fact, our code base can be easily modified to solve any segmentation problem of a similar nature—from skull-stripping, to parcellation, to specific ROI segmentation. This increases the probability that our model and computational framework will see actual usage in the neuroscience community—something that is quite important to us. However, this work is currently in progress.

Our reliance on code distribution ensures that we will be able to solicit feedback on the API, consistently improve model accuracy by incorporating new datasets, and drive research and development in novel convolutional architectures for neuroimaging segmentation. Our views are consistent with that of the deep learning community, which generally holds that making code and interfaces maximally available (with publishing considerations in mind) to the general community. Interested readers should contact the authors for information on this new and evolving deep learning framework.

9 Notes

1. One of the key issues leading to models that do not generalize is a failure to identify training data that accurately represents variability in data at large. To avoid this, one should check any relevant data parameters for outliers. When dealing with images, it is imperative to *visually inspect* the imaging data and potentially reject non-representative images.
2. We observe that, within the DLBS cohort, there is a reliable relationship between tissue volumes and the MMSE score. There is also a significant relationship between tissue volumes and age as well as age and MMSE. A very useful evaluation strategy might use tissue volumes as a surrogate measurement for "neurological age" or "brain age" [33]. Ideally, the convolutional network output will also reproduce this finding with the *predicted* segmentations. That is, we can run the same regression model but with the network's segmentations (in testing data) replacing the original ground truth segmentation volumes. While we do not report these results, the proposed architecture performs well, as expected.
3. We also note that training can be heavily affected by the learning rate of the optimizer. The learning rate of the optimizer determines how far of a step to take in the direction of the calculated loss gradient. Too high of a learning rate will cause the model to jump around in the parameter space without ever converging to a solution. A low learning rate, however, will cause the model to stand still. A common method for dealing with this problem is to start with a decently high learning rate and incrementally decrease it. Typically, the learning rate will have some exponential decay factor which decrease it slowly after each epoch, along with a step function

which halves the learning rate after a number of epochs. Thankfully, the Adam optimizer is an *adaptive* learning rate, so it takes care of the considerations on its own. This is a major benefit over Stochastic Gradient Descent.

4. Until network building and design becomes more automated (inevitably), there remains the possibility of inducing bugs that are not easy to detect but may manifest in poor performance. Common *naive* software bugs in network implementation include:
   - Bad optimizer, bad data or bad loss function;
   - Bug in coding (permuting the data);
   - Bug in data translation, e.g. from R to python which use different array indexing.
5. Deep learning is popular within the competitive machine learning communities, such as Kaggle. Exposure to Kaggle competition and ensuing discussions as the competition proceeds (or after they finish) provides a view of machine learning from the practical angle not often covered in academic papers. A few common themes that arise in the competitions include:
   - Nearly all competition winner use multiple models and voting or consensus;
   - It is fairly common for the eventual winner to come from outside the top 10 in the preliminary (training data) leaderboard; this indicates that many teams over-fit to the training data which is possible because teams can make many submissions and observe how their scores change with model variations;
   - Data augmentation is a second strategy that is common;
   - Finally, winning teams often carefully inspect the input and ground truth data and discover "problems" that may or may not be resolvable; in some cases, challenge organizers fix issues but this is not always feasible as the 'problems' identified by machine learning researchers may indeed represent real variability (or error rate) in human performance.

Acknowledgments

This work was supported by K01 ES025432-01.

References

1. von Economo CF, Koskinas GN (1925) Die cytoarchitektonik der hirnrinde des erwachsenen menschen. Springer, Wien
2. Mendrik AM, Vincken KL, Kuijf HJ et al (2015) MRBrainS challenge: online evaluation framework for brain image segmentation in 3T

MRI scans. Comput Intell Neurosci 2015:813696
3. Tustison NJ, Cook PA, Klein A et al (2014) Large-scale evaluation of ANTs and FreeSurfer cortical thickness measurements. NeuroImage 99:166–179
4. Wang H, Suh JW, Das SR et al (2013) Multi-atlas segmentation with joint label fusion. IEEE Trans Pattern Anal Mach Intell 35:611–623
5. Pustina D, Coslett HB, Turkeltaub PE et al (2016) Automated segmentation of chronic stroke lesions using LINDA: lesion identification with neighborhood data analysis. Hum Brain Mapp 37:1405–1421
6. Ashburner J (2012) SPM: a history. NeuroImage 62:791–800
7. Commowick O, Warfield SK (2010) Incorporating priors on expert performance parameters for segmentation validation and label fusion: a maximum a posteriori STAPLE. Med Image Comput Comput Assist Interv 13:25–32
8. LeCun Y, Bengio Y, Hinton G (2015) Deep learning. Nature 521:436–444
9. He K, Zhang X, Ren S et al (2015) Delving deep into rectifiers: surpassing human-level performance on imagenet classification. arXiv:1502.01852
10. Schoenick C, Clark P, Tafjord O et al (2016) Moving beyond the turing test with the Allen AI science challenge. arXiv:1604.04315
11. Mordvintsev A, Olah C, Tyka M (2015) Inceptionism: going deeper into neural networks. Google Research Blog. Accessed 20 June 2014
12. Santoro A, Bartunov S, Botvinick M et al (2016) One-shot learning with memory-augmented neural networks. arXiv:1605.06065
13. Wu Y, Schuster M, Chen Z et al (2016) Google's neural machine translation system: bridging the gap between human and machine translation. arXiv:1609.08144
14. Goldsborough P (2016) A tour of tensorflow. arXiv:1610.01178
15. Le QV, Ranzato M, Monga R et al (2011) Building high-level features using large scale unsupervised learning. International Conference in Machine Learning, Edinburgh, Scotland
16. Arganda-Carreras I, Turaga SC, Berger DR et al (2015) Crowdsourcing the creation of image segmentation algorithms for connectomics. Front Neuroanat 9:142
17. Moeskops P, Viergever MA, Mendrik AM et al (2016) Automatic segmentation of MR brain images with a convolutional neural network. IEEE Trans Med Imaging 35:1252–1261
18. Zhang W, Li R, Deng H et al (2015) Deep convolutional neural networks for multi-modality isointense infant brain image segmentation. NeuroImage 108:214–224
19. Choi H, Jin KH (2016) Fast and robust segmentation of the striatum using deep convolutional neural networks. J Neurosci Methods 274:146–153
20. Korfiatis P, Kline TL, Erickson BJ (2016) Automated segmentation of hyperintense regions in FLAIR MRI using deep learning. Tomography 2:334–340
21. Havaei M, Davy A, Warde-Farley D et al (2017) Brain tumor segmentation with deep neural networks. Med Image Anal 35:18–31
22. Xing F, Xie Y, Yang L (2016) An automatic learning-based framework for robust nucleus segmentation. IEEE Trans Med Imaging 35:550–566
23. Kamnitsas K, Ledig C, Newcombe VFJ et al (2017) Efficient multi-scale 3D CNN with fully connected CRF for accurate brain lesion segmentation. Med Image Anal 36:61–78
24. Zeiler MD, Krishnan D, Taylor GW et al (2010) Deconvolutional networks. In: Computer vision and pattern recognition (cvpr), 2010 ieee conference on, pp 2528–2535 IEEE
25. Zeiler MD, Fergus R (2014) Visualizing and understanding convolutional networks. In: European conference on computer vision, pp 818–833 Springer
26. Noh H, Hong S, Han B (2015) Learning deconvolution network for semantic segmentation. In: Proceedings of the IEEE international conference on computer vision, pp. 1520–1528
27. Brosch T, Tang LYW, Yoo Y et al (2016) Deep 3D convolutional encoder networks with shortcuts for multiscale feature integration applied to multiple sclerosis lesion segmentation. IEEE Trans Med Imaging 35:1229–1239
28. Janowczyk A, Madabhushi A (2016) Deep learning for digital pathology image analysis: a comprehensive tutorial with selected use cases. J Pathol Inform 7:29
29. Lu H, Xu F, Rodrigue KM et al (2011) Alterations in cerebral metabolic rate and blood supply across the adult lifespan. Cereb Cortex 21:1426–1434
30. Avants BB, Tustison NJ, Wu J et al (2011) An open source multivariate framework for n-tissue segmentation with evaluation on public data. Neuroinformatics 9:381–400

31. Tustison NJ, Avants BB, Cook PA et al (2010) N4ITK: improved n3 bias correction. IEEE Trans Med Imaging 29:1310–1320
32. Tustison NJ, Shrinidhi KL, Wintermark M et al (2015) Optimal symmetric multimodal templates and concatenated random forests for supervised brain tumor segmentation (simplified) with ANTsR. Neuroinformatics 13:209–225
33. Franke K, Ziegler G, Klöppel S et al (2010) Estimating the age of healthy subjects from t1-weighted MRI scans using kernel methods: exploring the influence of various parameters. NeuroImage 50:883–892

Check for updates

Chapter 3

Cortical Thickness

Konrad Wagstyl and Jason P. Lerch

Abstract

Multiple studies have measured cortical thickness from MRI since the advent of high contrast and 1 mm^3 resolution anatomical acquisitions and the development of advanced image processing algorithms. In this chapter we provide an overview of the methods for extracting cortical thickness, focusing on the two dominant packages in the field, FreeSurfer and CIVET. In addition, we review the confounds and artifacts that can bedevil cortical thickness studies. Lastly, we describe the potential of inferring microstructural changes based on the mesoscopic measurements acquired from standard structural acquisitions.

Key words Cortical thickness, Cerebral cortex, MRI, Image processing, Statistics, Microstructure

1 Introduction

1.1 Anatomy of the Cortex

The cerebral cortex is a multilayered structure encasing the outside of the cerebrum. It can be coarsely divided into the six-layered neocortex and the smaller, and evolutionarily older, three-layered allocortex. In primitive animals the neocortex is primarily occupied with receiving sensory input via the thalamus and coordinating motor activity [1]. As the cortex expanded in evolution, more and more cortical territory became occupied by the association cortices, which primarily receive their input from other cortical regions. Indeed, expansion of the neocortex is a hallmark of primate evolution; taking a simple progression index shows that prosimians have 14.5, simians have 45.5, and humans have 156 times the neocortex as would be expected from simply scaling up an insect brain [1, 2].

The anatomy of the cerebral cortex can be thought of along three dimensions: its surface area, folding pattern, and thickness. Each of these is governed by separate developmental processes. The cortex is organized, from inside to outside, into distinct layers, identified by their composition and density of neurons. While the neocortex maintains a pattern of six layers and the allocortex of three layers, the exact makeup of each layer varies across the cortex, leading to the ability to parcellate the cortex based on its laminar

Gianfranco Spalletta et al. (eds.), *Brain Morphometry*, Neuromethods, vol. 136,
https://doi.org/10.1007/978-1-4939-7647-8_3,

composition. The most well-known such parcellations are from Brodmann [3] and von Economo [4].

The vertical organization across the cortical layers is less well understood. A school of thought posits a columnar organization, wherein neurons are arranged in columns from the inside to the outside of the cortex, and each such column acts as a computational unit [5]. That such columns exist is certain; classic examples include whisker barrels in rodents and visual-ocular dominance columns. More contentious, however, is whether they exist throughout the cortex and even what constitutes a single column [1]. Still, for the purposes of mapping the cerebral cortex from MRI, columns present a useful concept, in so far as expansion of surface area can be thought of as increasing the number of columns in the cortex, whereas increasing cortical thickness can be thought of as alterations to the neuronal microcircuitry within a column.

Surface area and thickness are equally heritable; yet this heritability comes from different genetic determinants for each [6]. Nevertheless, the human cortex appears to be especially susceptible to environmental influence, with only 30% of variance in cortical measures ascribable to genetic variation, as compared to 71% in baboons [7, 8]. Moreover the degree of folding is less heritable than cortical volume, and, while it increases with brain size across, this is not so within a species [7]. Thus there is a wealth of potentially important neurobiological information to be found in the shape of the brain. This chapter outlines the development and validation computational techniques used to measure the morphology of the cortex, before describing the applications and neurobiological interpretation of these measures.

2 Materials and Methods for Measuring Cortical Thickness

Measuring the thickness of the cerebral cortex from MRIs is made inherently challenging by the limited resolution of most commonly used MRIs (typically 1 mm^3) and highly folded nature of the cortex. There are thus opposing gyri that appear to be touching with the sulcal CSF having all but disappeared. Separating these touching gyri is thus one of the key image processing challenges in accurately measuring cortical thickness. In addition, a coordinate system has to be established to allow for comparisons of thickness measures across subjects.

There are three classes of techniques for measuring cortical thickness: (1) manual, (2) surface based, and (3) voxel based. Manual measurements rely on the digital equivalents of calipers. Still the ground truth in many ways, it suffers somewhat from a reproducibility challenge in positioning the digital slice orthogonal to the cortex and, more importantly, is too labor intensive to be practical for anything beyond a few preselected regions of interest.

The advent of automated tools to measure cortical thickness has made purely manual measurements rare, yet interesting studies prior to those methods are numerous (c.f. [9]).

Surface-based measures of cortical thickness have become the dominant standard in the brain imaging community, led by the FreeSurfer [10, 11] and CIVET [12, 13] packages. They follow a common processing pipeline: (1) preprocessing including nonuniformity correction and registration to stereotaxic space, (2) masking the brain and tissue classification of cortical gray matter, (3) fitting two polyhedral meshes to the inside (white/gray) and outside (pial, gray/csf) boundaries of the cortex, (4) measuring cortical thickness as the distance between the two surfaces, (5) surface-based smoothing, and (6) surface-based alignment of the thickness maps (*see* Fig. 1 for an overview). Each of these steps will be discussed in more detail below.

Preprocessing. Initialization of the surface-based processing algorithms is greatly eased by being able to rely on templates in standard space. Both FreeSurfer and CIVET rely on linear registration to MNI space [14]. Concurrently, intensity inhomogeneity artifacts are removed [15, 16], and intensities optionally normalized.

Masking. This is where the cerebrum is separated from dura, skull, and background, with the cerebellum and brain stem removed as well. In FreeSurfer a deformable model is then expanded outward from the center of the brain toward a normalized intensity threshold [16]. In CIVET, the Brain Extraction Toolkit [17], similarly uses a deformable model with a set of locally adaptive model forces. Recent advances in multi-atlas and patch-based segmentation techniques [18–20] provide further refinement for brain extraction.

Tissue classification. Both FreeSurfer and CIVET begin by extracting the white matter surface. To identify which voxels represent white matter, FreeSurfer begins with a preliminary classification based on image intensities. Ambiguous voxels, containing a mix of gray and white matter, are then reassessed by computing a plane that best separates gray and white, and ambiguous voxels on the other side of the plane are re-labeled [16]. In CIVET, preliminary tissue classification is performed based on a trained neural-net classifier [21] and then modified through a partial volume estimation procedure [22].

Surface fitting. The key computational step for surface-based cortical thickness measurements is to extract a topologically correct mesh for the inside (white/gray boundary) and outside (pial, gray/csf boundary) of the cortex. Both FreeSurfer and CIVET begin by extracting the white matter surface, yet differ in their computational approaches. FreeSurfer first identifies a connected component of each hemisphere's white matter, then tessellates that surface, and finally refines that initial tessellation using a deformable

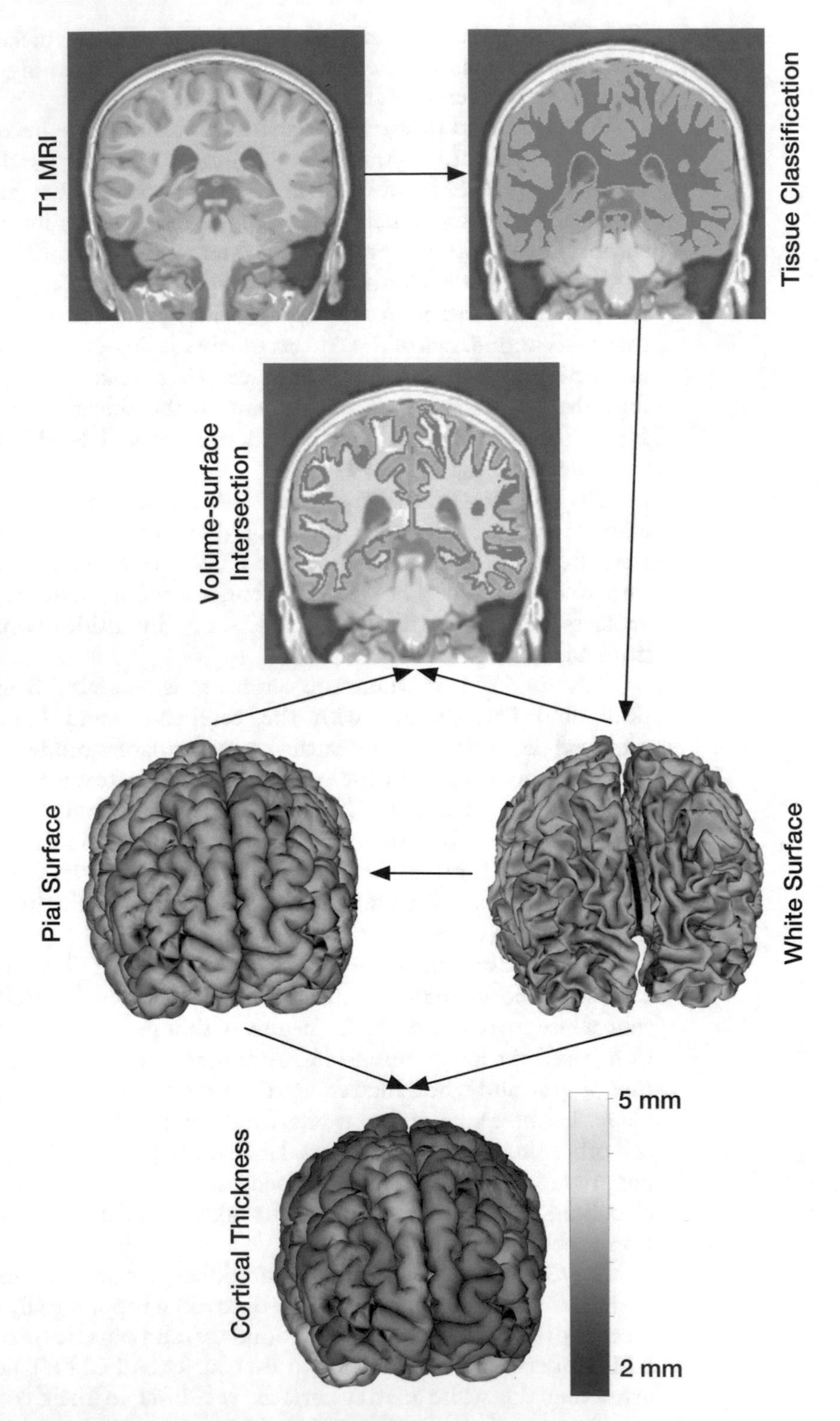

Fig. 1 An overview of the image processing pipeline for extracting cortical thickness. The T1 MRI is classified into white matter, gray matter, and spinal fluid, from which the inner and outer cortical surfaces are extracted. The distance between those surfaces is then measured as cortical thickness. Refer to the text for greater details on the methods

model algorithm [16]. CIVET, conversely, uses a deformable model to "shrink-wrap" the white matter surface [12, 13]. The FreeSurfer approach results in occasional topological errors that must be corrected, whereas the fully deformable model algorithm from CIVET must find ways of driving a surface through tight junctions (e.g., into the insula), risking geometric errors. Both algorithms then take the white matter surface as a starting point to deform toward the pial boundary [12, 13, 16].

Thickness measurements. Cortical thickness can be measured as the distance between the inner and outer surfaces. FreeSurfer's distance measure is to take the closest vertex on the opposite surface, then find that vertex' closest point, and average the two distances [10]. In CIVET, the default distance measurement is to follow the surface expansion and take the distance of the two linked vertices [23], though other distance measures, including nearest point, nearest point along the surface normal, and those based on Laplace's equation, are available [12, 13, 23, 24].

Surface-based smoothing. To reduce noise and increase comparability across subjects due to subtle misalignment, both FreeSurfer and CIVET employ surface-based smoothing. In each case an iterative nearest neighbor averaging procedure or a diffusion equation solved along the surface provides averages neighboring value within a geodesic distance of 7–30 mm [10, 23, 25]. The smoothing kernels employed are typically larger than for volumetric studies, though the use of geodesic distance implies very different coverage than the equivalent Euclidian distance [23].

Surface-based alignment. Improved alignment of cortical thickness maps increases the accuracy of localizing cortical thickness measures across subjects. Both FreeSurfer and CIVET perform these registrations on the sphere; in the case of FreeSurfer, by optimizing the overlap in an average convexity measure [26], whereas CIVET optimizes area-minimizing flows [27, 28]. These types of nonlinear surface-based registrations can increase localization accuracy across subjects by up to a factor of 3 [26].

Surface-based techniques for extracting the cortices and measuring cortical thickness are computationally expensive; pure volumetric techniques can be much faster, though at the cost of not having the same degree of model constraints to regularize thickness measurements. The most used of such algorithms implement variants of boundary partial differential equations [24, 29–31]. Here the brain is first segmented into cortex, outside of cortex (pial surface, cortical CSF, and background) and inside of cortex (white matter and deep gray matter), and then paths running between the inside and outside boundaries identified and their lengths measured. For comparing across subjects, thickness measures are summarized by ROI and/or projected back to a surface for use of a surface-based coordinate system [32].

Once thickness measures have been computed, statistics are usually computed in a massively univariate fashion, i.e., a separate statistical test at every vertex of the common surface or for every cortical ROI. Multiple comparisons are then controlled for using the false discovery rate [33] or one of the family-wise error rate (FWER) correction techniques, such as random field theory adapted to cortical meshes [34] or permutation tests [35]. More rarely, multivariate techniques such as partial least squares can be employed on thickness data as well (e.g., [36].).

In addition, cortical thickness across the surface within a population can be modeled as a network. Areas that are functionally and/or anatomically connected covary in cortical thickness, for example, the thicker ones Broca's area, the thicker Wernicke's area is likely to be [37]. Such covariances can then be modeled as a network, establishing a connection above a certain correlation and then testing for network properties such as small-worldness, hubs, etc. [38–40].

3 Validation of Cortical Thickness Measurements

The ability to measure the 1–5 mm thick cortex based on (usually) 1 mm^3 voxels has been met with skepticism, including by the authors of FreeSurfer and CIVET. Multiple validation studies have thus been carried out over the years. The early papers compared automatically derived measures against manual assessments on the same scans [41–43], generally finding good agreement. Comparisons to ex vivo samples or literature data based on ex vivo thickness measures were also conducted, also identifying strong concordance [10, 41, 44]. A consistent pattern that does emerge, however, is one of larger thickness estimates from CIVET than FreeSurfer, with CIVET measures, unlike FreeSurfer's, also generally exceeding those from the pathology-based literature [10, 41]. On that note, increasing resolution along with concomitant increases in field strength (to 7T) reduces estimated thickness by one sixth to one third [45].

Additional validation included assessments of scan-rescan reliability. Using repeated scans from the same individual, the average absolute thickness difference was found to be less than 0.15 mm, much lower than image resolution [10, 23, 46]. The most variable region for measurement of cortical thickness surrounds the central sulcus, likely attributable to its high cortical myelination, which makes identification of the gray/white boundary challenging [23, 46, 47]. Variability increases across scanner field strengths, but only very slightly across different installations of the same scanner platform, and not at all across console upgrades on the same installation [46]. Brain-behavior relations also were found to be stable across scanner platforms [48]. Increasing acceleration during acquisition (i.e., SENSE factors) resulted in underestimation of cortical thickness, though up to a SENSE factor of 3 the impact was minimal [49].

An artifact that bedevils analyses of structural MRIs, and cortical thickness in particular, is motion. The need for rejecting motion-corrupted scans has long been accepted, but the impact of even subtle motion on thickness measurements has only recently become clear. In an elegant experiment involving controlled motion in the scanner, Reuter et al. showed up to 1.5% decreases in cortical thickness per 1 mm/min motion (with average motion ranging from 3 to 12 mm/min) in a nonuniform manner across the cortex [50]. These trends persisted even if the normally accepted criteria of rejecting motion-affected scans was adopted, though highly stringent rejection criteria did ameliorate the effects on thickness estimates noticeably [50]. The effects of motion were found not due to failures in the processing tools but rather could be ascribed to subtle but consistent changes in the MR images themselves [50]. Encouragingly, modified acquisition sequences with navigators and prospective motion correction showed promise in significantly reducing this problem [51].

There has been some limited work comparing the different thickness algorithms. Lee et al. [52] used a cross-validation framework, wherein surfaces were extracted using each algorithm, then a new MR image simulated based on those surfaces, followed by applying each algorithm anew to simulated image. FreeSurfer and CIVET had the best results, with a slight edge to CIVET. Redolfi et al. [53] compared the ability of FreeSurfer and CIVET to identify disease state in a large Alzheimer's cohort; performance again was very similar. Ultimately, the comparison of algorithms is never complete, as each algorithm is regularly modified and improved, and different metrics of performance need to be evaluated. It appears, however, that currently the performance of CIVET and FreeSurfer is quite comparable.

4 Microstructure and Cortical Thickness

The real attraction of cortical thickness is that it is both measurable through automated neuroimaging tools and sensitive to in vivo changes in cortical structure. It can therefore be useful as a diagnostic marker [54] and to characterize developmental [55] or pathological [42, 56] changes. However, in order to identify the neurobiological processes which underlie these changes, we need to understand how cortical microstructure determines cortical thickness. Cortical microstructure encompasses both the structural components of neocortical tissue and how these components are arranged to form the layered structure of the cortex: the what and where of any thickness change.

4.1 What: Components of the Cerebral Cortex

The components of the cerebral cortex can be usefully subdivided into neuronal cell bodies, glia, dendrites, axons, dendritic spines, and extracellular space. Fortunately these components systematically covary such that cortical thickness changes can be a useful surrogate of a pattern of microstructural differences.

Neuronal cell bodies, together with the blood vessels and glia, make up only 16% of the cortical volume [57]. Importantly, neurogenesis is almost entirely prenatal, and the number of neurons in the neocortex remains largely constant postnatally [58, 59]. Furthermore, regional differences in neuronal density show a broadly inverse relationship with cortical thickness [3, 60]. Therefore increased cortical thickness is due not to an increase in the number of neurons but instead to volumetric increases in the other cortical microstructures, particularly in the neuropil.

Cortical neuropil—the axons, dendrites, and synapses—makes up around 84% of the cortical volume and are the means of communication or information processing within the cortex. The percentage of neuropil and the density of synapses per unit cortical volume are generally constant across cortical regions and species [57]. Therefore to increase the internal connectivity between neurons in the same region, the total volume of cortical neuropil must increase. More extensive dendritic arborization will allow a neuron to connect to more neurons. Moreover, the extent of dendritic arborization has been found to correlate with the diversity of distant neurons connected to [61–65]. Summarizing, increasing cortical thickness postnatally acts as a marker of increasing neuropil volume, decreasing neuronal density but increasing neuronal arborization [61–64]) and ultimately increasing intracortical connectivity.

Glia cells play a particularly active role in supporting intracortical connectivity, and their development mirrors the development of dendritic arborization and synapses [66]. Moreover, similar to neuropil, glial density is thought to be relatively conserved between species and cortical regions [67, 68], such that the number of glia cells per surface area probably increases with thickness and intracortical connectivity.

In summary, consideration of the constituent components of the cortex generates insight into what might contribute to a change in cortical thickness; however the effect of a thickness change is dependent on the functional role of the region in question.

4.2 Where: Regional and Laminar Differences in Thickness

Cortical thickness changes do not affect the full depth of the cortex uniformly. The cortex has an inhomogeneous laminar structure [3], and these layers exhibit regionally variability [4], differential development [69, 70], different susceptibility to pathology [71–73], differing electrophysiology [74], and differing functional roles [75]. Unfortunately, the limits of current MRI resolution prevent direct measurement of cortical layers, so it is not possible to directly evaluate in vivo the effect of changing laminar structure.

Nevertheless useful insights arise from the comparison of patterns of cortical thickness and postmortem measurement of regional laminar structure. For example, regional cortical thickness is systematically related to laminar structure [4, 76]. In particular, primary sensory areas tend to be thinner with a clearly defined laminar structure described as eulaminate, while association areas tend to be thicker with less clearly defined layers.

Similarly, cortical thickness and laminar structure are both affected by the curvature of the cortex [4, 77] (*see* Fig. 2 for an illustration). Even within the same cytoarchitectonic region, gyri differ from sulci in their laminar structure, axonal innervation, vascularization, neuronal density, and developmental trajectories [4, 69, 77, 78]. These differences amount to a consistent difference in overall MRI cortical thickness—on average gyri are measured as 0.5 mm thicker than sulci [10]. Therefore investigation of gyral and sulcal thickness trajectories [79, 80] may reveal differences that indicate a specific change in one of these components.

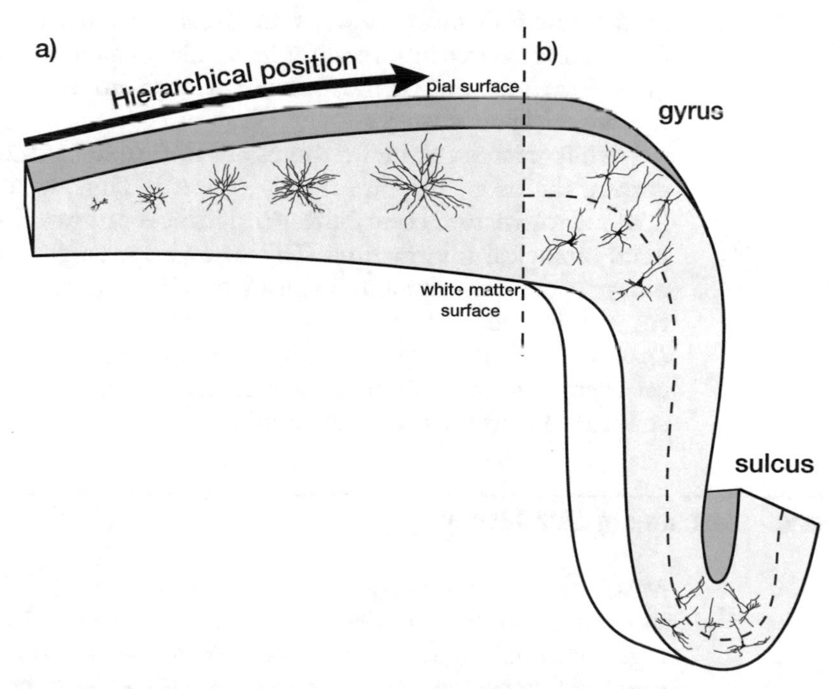

Fig. 2 Cytoarchitecturally driven differences in cortical thickness. (**a**) Increasing dendritic arborization of neurons contributes to increasing cortical thickness across sensory hierarchies. (**b**) Cortical thickness varies closely with morphological position. Corresponding differences in the underlying cytoarchitecture include the thickness of cortical layers—lower layers are thicker in gyri, upper layers are thicker in sulci—and in the morphology of neurons. Dendritic arborizations are larger and more radially oriented in gyral crowns but smaller and tangentially oriented in sulci. Cortical thickness can be used as an in vivo surrogate marker of these cytoarchitectural trends

5 Interpreting Thickness Differences

Drawing on the relationship between cortical microstructure and thickness can help to interpret both increases and decreases in cortical thickness. For example, cortical thickness varies across the human cerebral cortex [4, 10, 69, 77, 78], and regional cortical thickness is closely associated with functional specialization [76]. The functional significance of a thickness change is therefore regionally dependent. The primary motor cortex is generally very thick. Moreover, acrobatic training and exercise in rats is associated with increases in the thickness of motor regions that are associated with an improved ability to carry out motor tasks [81]. Increases in thickness are here attributed to increased motor neuronal connectivity required for the coordinated activation and inhibition of muscle groups. Thus, increasing thickness of the motor cortex may improve motor function. This is not the case for the primary visual cortex, which is generally very thin in humans; and individuals with thinner visual cortices have improved spatial discrimination [82]. Here, increased neuronal integration afforded by a thicker cortex is less desirable as high spatial discrimination requires neurons with narrow receptive fields. Hence microstructural understanding enables more mechanistic interpretation of the functional implications of regionally specific thickness changes.

While microstructural changes are not explicitly measured by cortical thickness, they are tightly coupled. Thus, understanding how microstructure contributes to thickness can reveal important neurobiological information. This knowledge might be used to interpret developmental and disease-specific patterns of observed cortical thickness change as resulting from specific microstructural change, enable development of more specific markers of pathological change, or contribute to our understanding of how cortical structure affects function and cognition.

6 Interrelated Measures of Morphology

Measures of morphology are interdependent. Their relationships must therefore be considered to accurately characterize morphological change and to correctly interpret their cause. At a basic level, group differences in gray matter volume (GMV) can be caused by independent changes in surface area and cortical thickness [83] but also the degree of folding and any changes to the pattern of folding [84]. Therefore a change in GMV is not specific to any single morphological change.

Moreover even among surface-based measures, there are different biological determinants and relationships depending on age and pathology. For example, in early life, gyrification is driven by the expansion of the cortex [77, 85, 86]; thus differences in the degree of folding are closely linked to the increases in cortical surface area [66, 87]. However during aging or neurodegeneration, changes in gyrification are more closely related to atrophy [88]. This can be due to cortical atrophy, particularly reflected in cortical thinning [89], or atrophy of the white matter indicated in changes to the shape of the underlying white matter [88, 90]. Therefore folding abnormalities must be interpreted not only in the context of age and likely pathology but also considering any differences in cortical thickness and surface area.

Similarly, analyses of cortical thickness can be affected by differences in other morphological parameters. For example, gyri and sulci have systematic thickness differences, gyri are thicker than sulci, and deep sulci are especially thin. Surface-based registration and smoothing can mitigate these differences [23, 91]; however a difference in folding could both obfuscate and create an apparent local change in cortical thickness. Furthermore, while mean cortical thickness is largely independent of variations in total cortical surface area between human individuals [83], surface area is related to the pattern and degree of folding [87]. Thus differences in overall surface area could lead to apparent focal changes in cortical thickness.

7 Conclusion

There is a rich and growing literature on measuring cortical thickness from MRI, which has expanded our understanding of brain development and aging and intrinsic organization of the cortex and shed insights into a multitude of neurological and psychiatric disorders. Going forward, key will be dealing with confounds—motion being the most prominent. This is especially important in cases where the extent of the artifact might segregate with the population being studied, i.e., patients more likely to move in the scanner than healthy controls [92]. In addition, greater understanding of the microstructural bases of cortical thickness, and cortical thickness change, will greatly improve our ability to interpret our results. Combining standard acquisitions with those adding biophysical models of water diffusion in gray matter [93–97] or using quantitative measures of myeloarchitectonics [98, 99] will be key. This can be further enhanced by the use of ultrahigh-field MRI. Ultimately, while acknowledging artifacts and confounds, we believe there is a bright future for cortical thickness mapping from MRI in the neurosciences.

Acknowledgments

Many thanks to Professor Paul C Fletcher for helpful feedback on this chapter. Many thanks to Christopher Hammil for his help in assembling the methods figure. JPL would like to acknowledge funding support from the CIHR, Brain Canada, Ontario Brain Institute, and Simons Foundation. KW is supported by the University of Cambridge MB/PhD Programme and the Wellcome Trust.

References

1. Nieuwenhuys R, Voogd J, van Huijzen C (2008) The human central nervous system. Springer, Berlin
2. Stephan H, Andy OJ (1964) Quantitative comparisons of brain structures from insectivores to primates. Am Zool 4:59–74
3. Brodmann K (1909) Vergleichende Lokalisationslehre der Grosshirnrinde in ihren Prinzipien dargestellt auf Grund des Zellenbaues. Barth, Leipzig
4. von Economo CF, Parker S (1929) The cytoarchitectonics of the human cerebral cortex. J Anat 63(Pt 3):389. Humphrey Milford
5. Mountcastle VB (1998) Perceptual neuroscience: the cerebral cortex. Harvard University Press, Cambridge, MA
6. Panizzon MS, Fennema-Notestine C, Eyler LT et al (2009) Distinct genetic influences on cortical surface area and cortical thickness. Cereb Cortex 19:2728–2735
7. Zilles K, Palomero-Gallagher N, Amunts K (2013) Development of cortical folding during evolution and ontogeny. Trends Neurosci 36:275–284
8. Rogers J, Kochunov P, Zilles K et al (2010) On the genetic architecture of cortical folding and brain volume in primates. NeuroImage 53:1103–1108
9. Meyer JR, Roychowdhury S, Russell EJ et al (1996) Location of the central sulcus via cortical thickness of the precentral and postcentral gyri on MR. AJNR Am J Neuroradiol 17:1699–1706
10. Fischl B, Dale AM (2000) Measuring the thickness of the human cerebral cortex from magnetic resonance images. Proc Natl Acad Sci 97:11050–11055
11. Fischl B (2012) FreeSurfer. NeuroImage 62:774–781
12. MacDonald D, Kabani N, Avis D, Evans AC (2000) Automated 3-D extraction of inner and outer surfaces of cerebral cortex from MRI. NeuroImage 12:340–356
13. Kim JS, Singh V, Lee JK et al (2005) Automated 3-D extraction and evaluation of the inner and outer cortical surfaces using a Laplacian map and partial volume effect classification. NeuroImage 27:210–221
14. Collins DL, Neelin P, Peters TM, Evans AC (1994) Automatic 3D intersubject registration of MR volumetric data in standardized Talairach space. J Comput Assist Tomogr 18:192–205
15. Sled JG, Zijdenbos AP, Evans AC (1998) A nonparametric method for automatic correction of intensity nonuniformity in MRI data. IEEE Trans Med Imaging 17:87–97
16. Dale AM, Fischl B, Sereno MI (1999) Cortical surface-based analysis. I. Segmentation and surface reconstruction. NeuroImage 9:179–194
17. Smith SM (2002) Fast robust automated brain extraction. Hum Brain Mapp 17:143–155
18. Eskildsen SF, Coupé P, Fonov V et al (2012) BEaST: brain extraction based on nonlocal segmentation technique. NeuroImage 59:2362–2373
19. Chakravarty MM, Steadman P, van Eede MC et al (2013) Performing label-fusion-based segmentation using multiple automatically generated templates. Hum Brain Mapp 34:2635–2654
20. Sabuncu MR, Yeo BTT, Van Leemput K et al (2010) A generative model for image segmentation based on label fusion. IEEE Trans Med Imaging 29:1714–1729
21. Zijdenbos AP, Forghani R, Evans AC (2002) Automatic "pipeline" analysis of 3-D MRI data for clinical trials: application to multiple sclerosis. IEEE Trans Med Imaging 21:1280–1291
22. Tohka J, Zijdenbos A, Evans A (2004) Fast and robust parameter estimation for statistical partial volume models in brain MRI. NeuroImage 23:84–97
23. Lerch JP, Evans AC (2005) Cortical thickness analysis examined through power analysis and a

population simulation. NeuroImage 24:163–173

24. Jones SE, Buchbinder BR, Aharon I (2000) Three-dimensional mapping of cortical thickness using Laplace's equation. Hum Brain Mapp 11:12–32
25. Chung MK, Robbins SM, Dalton KM et al (2005) Cortical thickness analysis in autism with heat kernel smoothing. NeuroImage 25:1256–1265
26. Fischl B, Sereno MI, Dale AM (1999) Cortical surface-based analysis. II: inflation, flattening, and a surface-based coordinate system. NeuroImage 9:195–207
27. Lyttelton O, Boucher M, Robbins S, Evans A (2007) An unbiased iterative group registration template for cortical surface analysis. NeuroImage 34:1535–1544
28. Boucher M, Whitesides S, Evans A (2009) Depth potential function for folding pattern representation, registration and analysis. Med Image Anal 13:203–214
29. Waehnert MD, Dinse J, Weiss M et al (2014) Anatomically motivated modeling of cortical laminae. NeuroImage 93(Pt 2):210–220
30. Dahnke R, Yotter RA, Gaser C (2013) Cortical thickness and central surface estimation. NeuroImage 65:336–348
31. Acosta O, Bourgeat P, Zuluaga MA et al (2009) Automated voxel-based 3D cortical thickness measurement in a combined Lagrangian-Eulerian PDE approach using partial volume maps. Med Image Anal 13:730–743
32. Lerch JP, Carroll JB, Dorr A et al (2008) Cortical thickness measured from MRI in the YAC128 mouse model of Huntington's disease. NeuroImage 41:243–251
33. Genovese CR, Lazar NA, Nichols T (2002) Thresholding of statistical maps in functional neuroimaging using the false discovery rate. NeuroImage 15:870–878
34. Worsley KJ, Taylor JE, Tomaiuolo F, Lerch J (2004) Unified univariate and multivariate random field theory. NeuroImage 23(Suppl 1): S189–S195
35. Nichols TE, Holmes AP (2002) Nonparametric permutation tests for functional neuroimaging: a primer with examples. Hum Brain Mapp 15:1–25
36. Yang J-J, Yoon U, Yun HJ et al (2013) Prediction for human intelligence using morphometric characteristics of cortical surface: partial least square analysis. Neuroscience 246:351–361
37. Lerch JP, Worsley K, Shaw WP et al (2006) Mapping anatomical correlations across cerebral cortex (MACACC) using cortical thickness from MRI. NeuroImage 31:993–1003
38. He Y, Chen ZJ, Evans AC (2007) Small-world anatomical networks in the human brain revealed by cortical thickness from MRI. Cereb Cortex 17:2407–2419
39. Evans AC (2013) Networks of anatomical covariance. NeuroImage 80:489–504
40. Alexander-Bloch A, Giedd JN, Bullmore E (2013) Imaging structural co-variance between human brain regions. Nat Rev Neurosci 14:322–336
41. Kabani N, Le Goualher G, MacDonald D, Evans AC (2001) Measurement of cortical thickness using an automated 3-D algorithm: a validation study. NeuroImage 13:375–380
42. Kuperberg GR, Broome MR, McGuire PK et al (2003) Regionally localized thinning of the cerebral cortex in schizophrenia. Arch Gen Psychiatry 60:878–888
43. Salat DH, Buckner RL, Snyder AZ et al (2004) Thinning of the cerebral cortex in aging. Cereb Cortex 14:721–730
44. Rosas HD, Liu AK, Hersch S et al (2002) Regional and progressive thinning of the cortical ribbon in Huntington's disease. Neurology 58:695–701
45. Lüsebrink F, Wollrab A, Speck O (2013) Cortical thickness determination of the human brain using high resolution 3T and 7T MRI data. NeuroImage 70:122–131
46. Han X, Jovicich J, Salat D et al (2006) Reliability of MRI-derived measurements of human cerebral cortical thickness: the effects of field strength, scanner upgrade and manufacturer. NeuroImage 32:180–194
47. Scholtens LH, de Reus MA, van den Heuvel MP (2015) Linking contemporary high resolution magnetic resonance imaging to the von Economo legacy: a study on the comparison of MRI cortical thickness and histological measurements of cortical structure. Hum Brain Mapp 36:3038–3046
48. Dickerson BC, Fenstermacher E, Salat DH et al (2008) Detection of cortical thickness correlates of cognitive performance: reliability across MRI scan sessions, scanners, and field strengths. NeuroImage 39:10–18
49. Park H-J, Youn T, Jeong S-O et al (2008) SENSE factors for reliable cortical thickness measurement. NeuroImage 40:187–196
50. Reuter M, Tisdall MD, Qureshi A et al (2015) Head motion during MRI acquisition reduces gray matter volume and thickness estimates. NeuroImage 107:107–115

51. Tisdall MD, Reuter M, Qureshi A et al (2016) Prospective motion correction with volumetric navigators (vNavs) reduces the bias and variance in brain morphometry induced by subject motion. NeuroImage 127:11–22
52. Lee JK, Lee J-M, Kim JS et al (2006) A novel quantitative cross-validation of different cortical surface reconstruction algorithms using MRI phantom. NeuroImage 31:572–584
53. Redolfi A, Manset D, Barkhof F et al (2015) Head-to-head comparison of two popular cortical thickness extraction algorithms: a cross-sectional and longitudinal study. PLoS One 10:e0117692
54. Hong S-J, Kim H, Schrader D et al (2014) Automated detection of cortical dysplasia type II in MRI-negative epilepsy. Neurology 83:48–55
55. Shaw P, Kabani NJ, Lerch JP et al (2008) Neurodevelopmental trajectories of the human cerebral cortex. J Neurosci 28:3586–3594
56. Lerch JP, Pruessner JC, Zijdenbos A et al (2005) Focal decline of cortical thickness in Alzheimer's disease identified by computational neuroanatomy. Cereb Cortex 15:995–1001
57. Braitenberg V, Schüz A (1991) Anatomy of the cortex: statistics and geometry. Springer-Verlag Publishing, Berlin; New York
58. Bhardwaj RD, Curtis MA, Spalding KL et al (2006) Neocortical neurogenesis in humans is restricted to development. Proc Natl Acad Sci U S A 103:12564–12568
59. Rakic P (1985) Limits of neurogenesis in primates. Science 227:1054–1056
60. la Fougère C, Grant S, Kostikov A et al (2011) Where in-vivo imaging meets cytoarchitectonics: the relationship between cortical thickness and neuronal density measured with high-resolution [18F]flumazenil-PET. NeuroImage 56:951–960
61. Scholtens LH, Schmidt R, de Reus MA, van den Heuvel MP (2014) Linking macroscale graph analytical organization to microscale neuroarchitectonics in the macaque connectome. J Neurosci 34:12192–12205
62. Collins CE, Airey DC, Young NA et al (2010) Neuron densities vary across and within cortical areas in primates. Proc Natl Acad Sci 107:15927–15932
63. Elston GN, Rosa MG (1998) Morphological variation of layer III pyramidal neurones in the occipitotemporal pathway of the macaque monkey visual cortex. Cereb Cortex 8:278–294
64. Elston GN, Rosa MG (1997) The occipitoparietal pathway of the macaque monkey: comparison of pyramidal cell morphology in layer III of functionally related cortical visual areas. Cereb Cortex 7:432–452
65. Jacobs B, Scheibel AB (1993) A quantitative dendritic analysis of Wernicke's area in humans. I. Lifespan changes. J Comp Neurol 327:83–96
66. Blinkov SM (1968) The human brain in figures and tables: a quantitative handbook. Basic Books, New York
67. Stolzenburg JU, Reichenbach A, Neumann M (1989) Size and density of glial and neuronal cells within the cerebral neocortex of various insectivorian species. Glia 2:78–84
68. Herculano-Houzel S (2014) The glia/neuron ratio: how it varies uniformly across brain structures and species and what that means for brain physiology and evolution. Glia 62:1377–1391
69. Conel JL (1947) The cerebral cortex of the 3-month infant. Anat Rec A Discov Mol Cell Evol Biol 97:382
70. Petanjek Z, Judaš M, Šimic G et al (2011) Extraordinary neoteny of synaptic spines in the human prefrontal cortex. Proc Natl Acad Sci U S A 108:13281–13286
71. Lewis DA, Campbell MJ, Terry RD, Morrison JH (1987) Laminar and regional distributions of neurofibrillary tangles and neuritic plaques in Alzheimer's disease: a quantitative study of visual and auditory cortices. J Neurosci 7:1799–1808
72. Harrison PJ (1999) The neuropathology of schizophrenia. Brain 122:593–624
73. Benes FM, Davidson J, Bird ED (1986) Quantitative cytoarchitectural studies of the cerebral cortex of schizophrenics. Arch Gen Psychiatry 43:31–35
74. Douglas RJ, Martin KA (1991) A functional microcircuit for cat visual cortex. J Physiol 440:735–769
75. Bastos AM, Usrey WM, Adams RA et al (2012) Canonical microcircuits for predictive coding. Neuron 76:695–711
76. Wagstyl K, Ronan L, Goodyer IM, Fletcher PC (2015) Cortical thickness gradients in structural hierarchies. NeuroImage 111:241–250
77. Welker W (1990) Why does cerebral cortex fissure and fold? In: Jones EG, Peters A (eds) Cerebral cortex. Springer US, Boston, MA, pp 3–136
78. Hilgetag CC, Barbas H (2006) Role of mechanical factors in the morphology of the primate cerebral cortex. PLoS Comput Biol 2: e22
79. Vandekar SN, Shinohara RT, Raznahan A et al (2015) Topologically dissociable patterns of

development of the human cerebral cortex. J Neurosci 35:599–609

80. Wagstyl K, Ronan L, Whitaker KJ et al (2016) Multiple markers of cortical morphology reveal evidence of supragranular thinning in schizophrenia. Transl Psychiatry 6:e780
81. Anderson BJ, Eckburg PB, Relucio KI (2002) Alterations in the thickness of motor cortical subregions after motor-skill learning and exercise. Learn Mem 9:1–9
82. Song C, Schwarzkopf DS, Kanai R, Rees G (2015) Neural population tuning links visual cortical anatomy to human visual perception. Neuron 85:641–656
83. Winkler AM, Kochunov P, Blangero J et al (2010) Cortical thickness or grey matter volume? The importance of selecting the phenotype for imaging genetics studies. NeuroImage 53:1135–1146
84. Palaniyappan L, Liddle PF (2012) Differential effects of surface area, gyrification and cortical thickness on voxel based morphometric deficits in schizophrenia. NeuroImage 60:693–699
85. Ronan L, Fletcher PC (2015) From genes to folds: a review of cortical gyrification theory. Brain Struct Funct 220:2475–2483
86. Tallinen T, Chung JY, Rousseau F et al (2016) On the growth and form of cortical convolutions. Nat Phys 12:588–593. https://doi.org/10.1038/nphys3632
87. Toro R, Perron M, Pike B et al (2008) Brain size and folding of the human cerebral cortex. Cereb Cortex 18:2352–2357
88. Im K, Lee J-M, Won Seo S et al (2008) Sulcal morphology changes and their relationship with cortical thickness and gyral white matter volume in mild cognitive impairment and Alzheimer's disease. NeuroImage 43:103–113
89. Hogstrom LJ, Westlye LT, Walhovd KB, Fjell AM (2013) The structure of the cerebral cortex across adult life: age-related patterns of surface area, thickness, and gyrification. Cereb Cortex 23:2521–2530
90. Magnotta VA, Andreasen NC, Schultz SK et al (1999) Quantitative in vivo measurement of gyrification in the human brain: changes associated with aging. Cereb Cortex 9:151–160
91. Fischl B, Sereno MI, Tootell RB, Dale AM (1999) High-resolution intersubject averaging and a coordinate system for the cortical surface. Hum Brain Mapp 8:272–284
92. Weinberger D, Radulescu E (2016) The inconvenient truth about MRI in psychiatric research. Psychiatrics News 51:1. https://doi.org/10.1176/appi.pn.2016.2a6
93. Tariq M, Schneider T, Alexander DC et al (2016) Bingham-NODDI: mapping anisotropic orientation dispersion of neurites using diffusion MRI. NeuroImage 133:207–223
94. Zhang H, Schneider T, Wheeler-Kingshott CA, Alexander DC (2012) NODDI: practical in vivo neurite orientation dispersion and density imaging of the human brain. NeuroImage 61:1000–1016
95. Jespersen SN, Leigland LA, Cornea A, Kroenke CD (2012) Determination of axonal and dendritic orientation distributions within the developing cerebral cortex by diffusion tensor imaging. IEEE Trans Med Imaging 31:16–32
96. Jespersen SN, Bjarkam CR, Nyengaard JR et al (2010) Neurite density from magnetic resonance diffusion measurements at ultrahigh field: comparison with light microscopy and electron microscopy. NeuroImage 49:205–216
97. Jespersen SN, Kroenke CD, Østergaard L et al (2007) Modeling dendrite density from magnetic resonance diffusion measurements. NeuroImage 34:1473–1486
98. Tardif CL, Gauthier CJ, Steele CJ et al (2016) Advanced MRI techniques to improve our understanding of experience-induced neuroplasticity. NeuroImage 131:55–72
99. Dinse J, Waehnert M, Tardif CL et al (2013) A histology-based model of quantitative T1 contrast for in-vivo cortical parcellation of high-resolution 7 tesla brain MR images. Med Image Comput Comput Assist Interv 16:51–58

Check for updates

Chapter 4

Surface and Shape Analysis

Robert Dahnke and Christian Gaser

Abstract

During evolution, the brain becomes more and more complex. With increasing volume, the surface area expands to a disproportionately greater extent through the development of a species-specific but individual folding pattern. As shaping of the brain is virtually complete in early development, this permits the adult brain to be the subject of shape analysis to investigate its development. Other surface properties such as thickness alter significantly over the entire lifetime and in diseases, and reflect the current state of the brain. This chapter offers an introduction to individual development theories and models, surface reconstruction techniques, and shape measures to describe surfaces properties.

Key words Surface, Shape, Measures, Folding, Gyrification, MRI, Brain, Thickness, Curvature, Development, Aging, Evolution, Morphometry, Structure

1 The Mammalian Brain

The beginning of systematic studies of the human brain in the nineteenth century raised questions about the link between anatomical structure and its function and how obvious folding affects its abilities [1–9]. During evolution and development, the enlargement of the brain coincides with increased and more individual folding that comprises a non-linear enlargement of surface area that correlates with increased intellectual capabilities [6–9]. The individual shape of the brain, especially for larger species, requires nonlinear registration techniques to compare different brain structures [10–12]. Besides highly individual pattern folding, population- and disease-specific pattern have been found that are the product of early development [8, 9, 13–15].

The brain is arranged in two major classes of tissue, gray matter (GM) and white matter (WM), which are surrounded by cerebrospinal fluid (CSF) and packed within the skull (Fig. 1a). The GM can be seen as the processing region with a large number of neurons that are connected by myelinated dendrites that form WM fiber tracts and allow high-speed connection between different regions.

Gianfranco Spalletta et al. (eds.), *Brain Morphometry*, Neuromethods, vol. 136,
https://doi.org/10.1007/978-1-4939-7647-8_4, © Springer Science+Business Media, LLC 2018

In contrast, CSF serves as a physical buffer that allows geometrical changes in brain development and aging. The surface area of the cortex, a strong folded ribbon of GM that surrounds the WM, is particularly increased during both individual and evolutional development [5–9, 14] (Fig. 1b, c). The cortex can be described as an

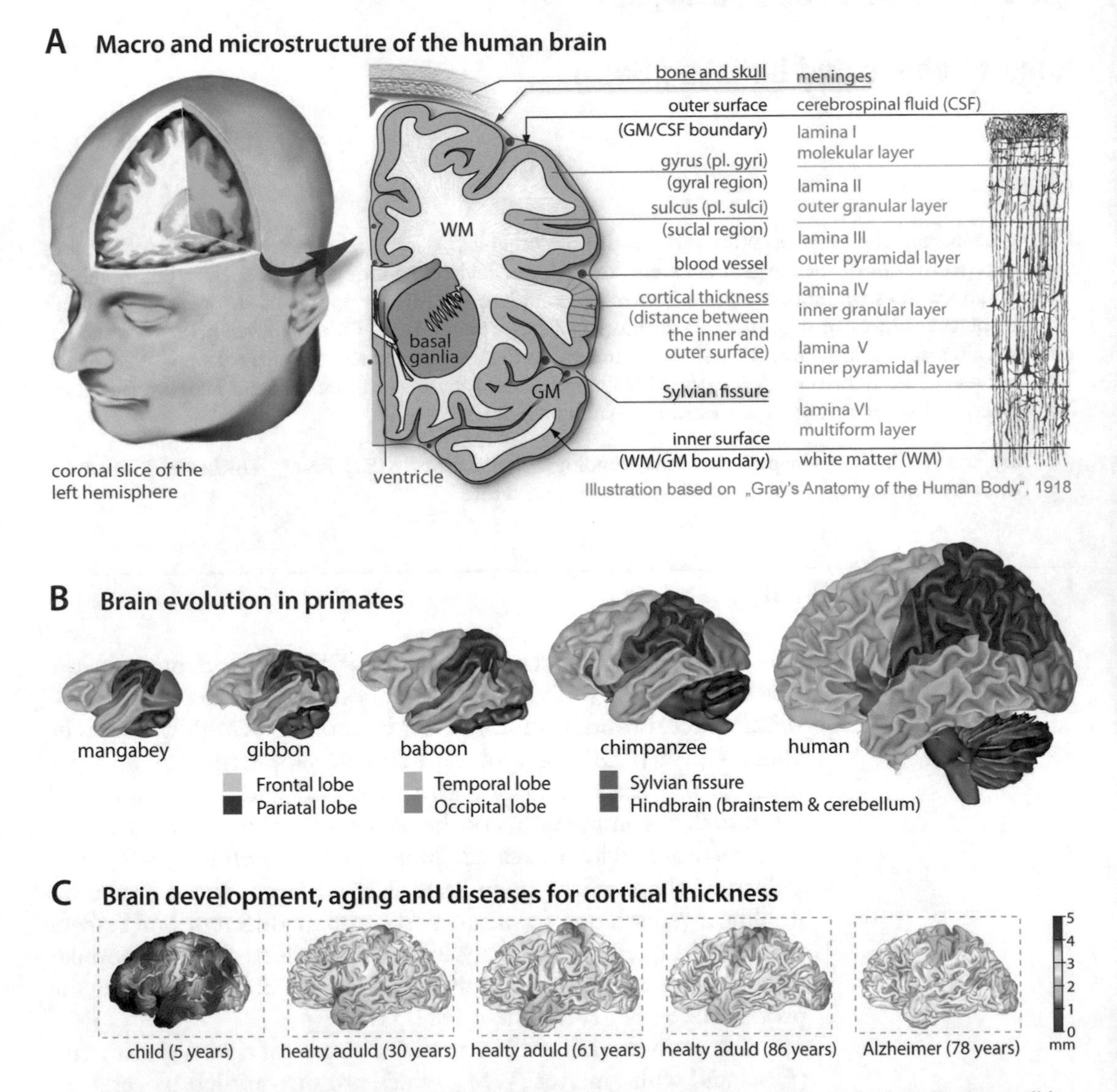

Fig. 1 The human cerebrum (**a**) is a highly folded structure that can be macroscopically described as a ribbon of gray matter (GM) that surrounds a core of white matter (WM). This GM ribbon (neocortex) is around 2–4 mm thick and organized into six regions- and function-specific layers that contain different types of neurons and can be simply described as a processing region, whereas the WM is a high-speed connection between different brain regions. With increasing size, the brain evolves in a species-specific folding pattern (**b**) with increased individual influences (**c**) that occur early during an individual's development and stay relatively constant over an individual's lifetime, whereas other parameters such as thickness change significantly during development and aging (**c**)

organized surface whose folding allows a large surface to fit compactly within the cranium [7, 13, 16–18]. The gyrification process that creates outward (gyri) and inward (sulci) folding during embryogenesis is still under discussion [9, 14, 15, 18, 19]. The closer connectivity within the gyri and the obvious similarities in the folding pattern of smaller species and major structures led to the expectation that the gyri process related things [8, 13, 18]. The cortex of the cerebrum (neocortex) is organized into six layers with regional variation in thickness and different functional processing. Its structure further depends on the local folding and compensate for the number of layer specific neurons, where imaginary cortical units contain the same amount of neurons per layer [1, 5, 13, 20] (GM blocks a, b, c in Fig. 2b). I.e., a cortical unit on top of a gyrus has a larger outer and smaller inner surface area with thicker inner and thinner outer layer (region c in Fig. 2c), whereas a cortical unit on the bottom of a suclus has a smaller outer and larger inner surface area with thicker outer and thinner inner layer (region b in Fig. 2c). It can therefore be expected that local folding only has a limited influence on function and can be seen as a simple product of energy-minimizing processes related to brain growth [14, 15, 19, 21].

Magnetic resonance imaging (MRI) and automatic preprocessing techniques allow in vivo analysis of the macroscopic brain structure in the field of computational morphometry of even large cohorts [10, 22]. Early regional manual measures were extended to automatic whole brain techniques such as voxel-based (VBM) [10], region-based (RBM) [23–25], deformation-based (DBM) [26], and surface-based morphometry (SBM) [12, 22, 23, 27] that allow the detection of even subtle changes in the brain structure. In the last decade, the volume of the GM in particular as well as the cortical thickness has become an important biomarker for development [28], aging [29, 30], plasticity [31], and a number of different diseases [32]. At this point, SBM allows essential improvements compared to VBM or DBM by (a) additional measures that describe the shape of the brain [18, 33, 34], (b) dissection of GM volume into thickness and area [35], (c) improved registration and partitioning (region alignment) [36], (d) correct anatomical smoothing [32, 37, 38], (e) mathematical shape modeling [14, 15, 18, 21, 39], and (f) combining different MRI modalities such as functional imaging (fMRI) that focuses on task-specific activation of cortical areas [38], diffusion imaging (dMRI) to analyze WM fiber tracts [40], and structural weightings such as T1, T2, PD, and quantitative imaging (qMRI) [41] to analyze tissue-specific properties such as myelination [42], WM hyperintensities or lesions in multiple-sclerosis [43]. Although VBM is very sensitive to subtle GM changes in brain plasticity, it lacks the function to describe complex folding pattern and its development, whereas

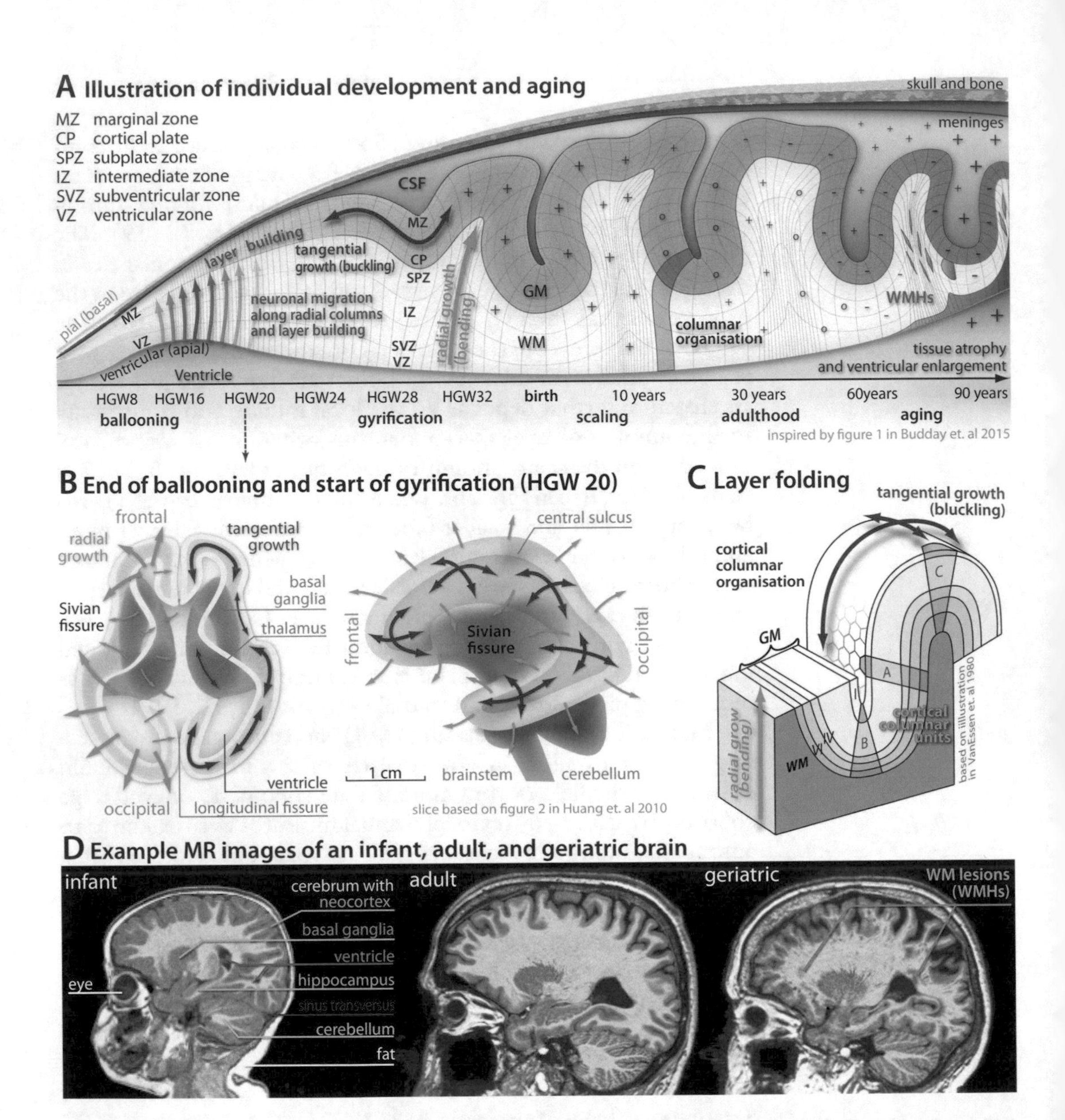

Fig. 2 An illustration of human brain development and aging (**a**). It is initiated with the ballooning phase that strongly increases the area of the ventricular zone by both radial and tangential growth, where neuroepithelial cells are generated by cell division and migrate to the marginal zone forming a columnar migration and cortical layer pattern [1, 5, 45]. The ongoing migration and initiation of the cortical connection increase the tangential growth by about HGW 20 (**b**) and gyrification shapes major structures such as the central sulcus. Because the Sylvian fissure lies hidden behind the subcortical structures, such as the basal ganglia and the thalamus, it profits less from neuronal migration and is finally overgrown by the surrounding brain regions (**b**). In humans, gyrification has nearly finished around birth and radial and tangential growth is balanced again, leading to a scaling of brain size with tissue growth and surface area enlargement (**a**). Over an individual's lifetime, the WM keeps growing up to the age of around 40 years, whereas the cortex shrinks slightly every year. In aging, the WM also shrinks and shows tissue degeneration that appears in MRIs as WM hyperintensities (WMHs) with GM-like intensities. Overall, the tissue atrophy is accompanied by an enlargement of the ventricle, that helps to keep the shape of the brain relatively constant. The local folding (bending and buckling) compresses and stretches the cortical layers shown in (C) by keeping the volumes of each layer of the imaginary cortical columnar units (**a**), (**b**), and (**c**) relatively similar and facilitates the increasing individual local folding pattern in higher species [1, 5, 45]. For comparison colorized real MR slices are shown in subfigure (**d**)

DBM partially covers folding differences as well as volume changes that impede analysis. RBM on the other hand allows the combination of different techniques but depends on the atlas maps.

Prior to the technical description of surface reconstruction, modification, and measures, a small introduction to brain development, its underlying biomechanical processes, and modeling will be described here.

2 Brain Development, Plasticity, and Aging

It is expected that brain folding follows the same biomechanical rules in all mammals, but the process itself is still undergoing significant research [6, 8, 9, 14, 19, 44]. The development of the cerebrum undergoes three major periods: (a) the ballooning stage, (b) the gyrification phase, and (c) a subsequent scaling in childhood and adolescence. Further changes in the healthy adult brain are recognized as plasticity (short-time) and aging (long-time). The early ballooning phase is relatively similar between species including an enlargement by radial and tangential tissue growth (Fig. 2), whereas gyrification is species-specific and shows higher tangential than radial growth that causes folding with more individual patterns in larger brains [8, 9, 14].

2.1 Phase I: Ballooning

The ballooning phase from human gestation week (HGW) 0–15 is described by an intensive radial enlargement of the ventricle that compensates the simultaneous tangential growth of the intermediate zone and increases the brain surface without significant folding, where only the longitudinal and Sylvian fissures become prominent by bending.[1] In HGW 5 to 20, neurons are generated in the ventricular zone and migrate to the skull, where they create the structure of the cortical layer. At this time, the cortex shows a radial dMRI pattern, indicating low connectivity within the cortex [9, 14, 44], with the first large fiber tracts becoming visible in the WM [40].

2.2 Phase II: Gyrification

After ballooning and layer building, the neurons in the cortex start forming connections and the radial dMRI pattern gets lost [40]. Without intensive ventricular enlargement, the tangential growth becomes prominent and causes buckling.[2] Gyrification starts with major structures such as the central sulcus [14]. External forces due to limitations of the skull and meninges were found to have minor effects [2, 9, 14, 15, 19], and it is presumed that gyrification depends on internal forces of WM connectivity (the

[1] Bending = forces *below* the developing cortex (in the WM)

[2] Buckling = forces *within* the developing cortex (in the GM)

axial tension theory [13]) or tangential growth of the GM (the buckling theory) [3, 7]. Recent experimental and computational growth models [15, 18, 19, 39] have shown promising results to explain the natural folding as an energy-minimizing process of surface expansion that relies on the stiffness of the inner core, the growing-rate, and local thickness, where thinner regions and faster growing rates increase folding and stiffer cores trigger more complex structures [15, 19, 39]. As far as the cortex, it has a lower limit of thickness of about 0.4 mm [6], gyrification generally only occurring for brains larger than 3 cm (about 10 cm^3).

2.3 Phase III: Further Scaling

The folding is nearly completed around birth in humans [46] and both tangential and radial growth is balanced again [47], whereas gyrification starts after birth in other species such as ferrets [19].

2.4 Adulthood and Aging

Over an individual's lifetime, the cortex shrinks slowly every year, whereas the WM continues to grow up to the age of around 40 years. The WM can show further degeneration as evidenced by MRI as WM hyperintensity with GM-like intensities in aging, as well as in diseases such as multiple sclerosis. Beside the global trend of tissue atrophy, brain plasticity allows an increase in local tissue volume. For elderly and people with neurodegenerative diseases such as Alzheimer's disease, accelerated tissue atrophy was reported [30]. Overall, tissue atrophy accompanies an enlargement of the ventricle and sulcal CSF that keeps the brain in a general shape within the skull.

2.5 Interim Conclusion

Finally, we can conclude that the gyrification of the cortex in most mammals occurs most significantly during the second and third trimester of pregnancy most likely by local tangential growth of GM tissue after initial lamination at the end of the first trimester. As far as the fact that the folding pattern stays relatively constant over an individual's lifetime, it is expected to be possible to understand developmental processes and diseases even in the adult brain.

3 Folding Theories and Models

Folding processes can be found in most biological structures that require area enlargement, and it was shown that brain folding is also driven by biomechanical concepts that can be described by mathematical models [3, 6, 15, 19]. It is assumed that the surface structure is driven by the organization of processing [13, 18, 48], that it is similar in mammals [6, 8, 9, 19], and that folding abnormality such as lissencephaly or polygyria can help to understand the gyrification process [3, 13, 14, 19]. A summary of mammal brain evolutional and abstract brain structure modeling is presented by Hofman [6], whereas a good introduction of up-to-date folding models can be found in previous reports [8, 9, 19]. There are two major types of gyrification theories: (a) the axonal tension theorem and (b) the active growth models.

The axonal tension model [13] is based on the idea that neurological processing is more strongly correlated to gyri than sulci and that both sides of a gyrus are strongly connected by fibers that trigger the folding process to minimize connectivity costs. Although this theory looks elegant and has garnered support [17], it has four major drawbacks: (a) the predicted radial connections have not been observed macroscopically [19], rather in diffusion images [49], where most fibers run in radial direction, rather between the opposing sides of gyri, (b) the predicted tension has not been observed in macroscopic cuts [21], (c) perforation of the WM after neuronal migration and before the onset of gyrification did not lead to less folding [2], and finally (d) mathematical folding models without the simulation of axonal fiber tensions [15, 18] have proved to be successful.

In active growth models, cortical folding is just a side product of cortical enlargement and external and internal constraints [3, 7, 9, 15, 18, 19, 39]. In recent years, different computational folding models were introduced with varying combinations of radial and tangential growth [14, 15, 18, 21, 39, 49], thickness [39], stiffness [19, 39], growing speed [19], and external constraints such as the skull or meninges [9, 49].

The work of Tallinen [15] was especially noteworthy and he investigated the development of specific folding patterns depending on WM stiffness, GM thickness, and the growing speed that allowed the creation of a naturally 3D folding pattern. It is further supported by the continuous work of the groups of Budday [14, 39], Bayly [19], Toro [18], and Nie [49]. The idea of folding prediction based on real MRIs that allows validation by longitudinal studies in neonates is also remarkable [49].

4 Surface Creation

The development of the brain as an organized surface has clearly outlined the potential of surface-based analysis, leading to the development of several software packages for automatic surface reconstruction and analysis of MRIs. Surface meshes are graph structures that describe a shape by a set of vertices V and faces F that connect the vertices. V is a $n_v \times 3$ vector of the xyz-coordinates of each point, whereas F describes the triangles by a $n_f \times 3$ vector of vertex-indices (Fig. 3):

$$S = [V, F] \tag{1}$$

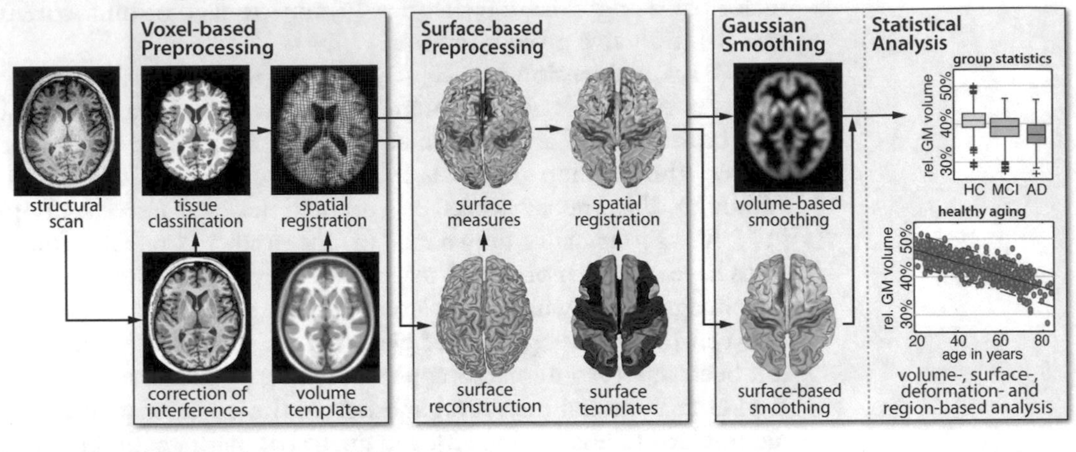

B Individual surface mesh

C Volume vs. surface-based smoothing

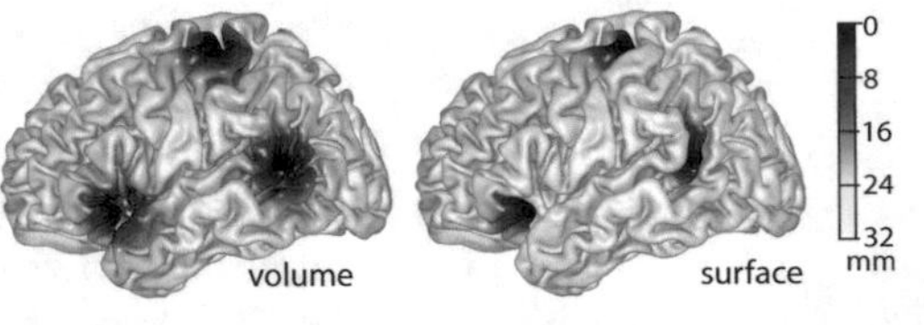

Fig. 3 The preprocessing of structural MRIs often contains a voxel-based part that classifies the tissues and registers each brain to a template (**a**). The processed images support the reconstruction of surfaces that facilitates further surface-based measures. Similar to the voxel-based processing, a registration to a template mesh is required. For the final analysis, the VBM, DBM, and SBM data are smoothed to reduce individual variance and guarantee Gaussian distribution for statistical testing or average region-wise RBM analysis. (**b**) Surface meshes consist of vertices that are connected by faces and include multiple surface measures. (**c**) Smoothing on the surfaces is closer to the anatomical structure of the cortex and can improve analysis, especially in regions with deep folds [32, 37]

Individual meshes can be generated on a regular volume grid by marching cubes or isosurface algorithms that generally require further pre- and post-processing. Surface measures are stored as vertex or face-wise vectors C that can be visualized as surface textures and analyzed similarly to VBM. Validation of surface reconstruction and measures is typically part of the method proposal and often includes simulated [50, 51], scan-rescan [51], expert-classification [36], or large-scale datasets [52]. The quality of the generated meshes and measures depends on the method used [27], the reconstructed structure and region [11, 20], as well as the quality of the input data [47, 53]. In general, structural data that is suitable for VBM analysis also allows an adequate SBM analysis. The generation and analysis of surface measures will be part of Sects. 4.5 and 4.6, as the focus in this chapter is on mesh generation, modification, and mapping. Surfaces are usually generated using volumetric scans and require three major processing steps:

(a) voxel-based preprocessing, (b) the generation and optimization of individual meshes, and (c) the registration to common templates (Fig. 3a).

4.1 Voxel-Based Preprocessing

The voxel-based preprocessing is required to estimate mappings between individual and common brain templates (registration, *see* Chap. 1), to classify different tissues (segmentation, *see* Chap. 2) and prepare data for surface reconstruction.

The classification of WM, GM, and CSF is driven by image intensity and a priori knowledge [10, 22, 54] and generally comprehends the extraction of the brain [10, 54], the handling of image interferences such as noise [10, 55] and inhomogeneity [10], and in some cases also the registration [54]. Popular software packages such as BrainSuite[3], FSL[4], MIPAV[5], SPM[6] [10, 54], and VBM8/CAT[7] applied common Gaussian-mixture, maximum-likelihood, maximum a posteriori probability, and expectation maximization models [10, 40, 54, 56]. To increase accuracy and stability, recent approaches use brain-specific properties such as topological constrains [57], multimodal input images [10, 54], longitudinal modeling [58], species or aging-specific templates and parameters [12, 58], or other concepts entirely [59]. The segmentation can further be used for intensity normalization of MRIs [43].

Spatial registration estimates a mapping between the individual brain and common templates [60]. They are typically realized as iterative processes and start with affine transformations and low frequency deformations that are systematically increased to reduce the anatomical variance of the subjects [46]. Atlas maps that partition brains into different regions are often manually obtained in the native (subject) space and mapped to an average template space [24] or are directly generated in the template space [25]. Besides manual-defined atlas maps, automatic parceling methods e.g., fMRI and dMRI connectivity maps have also been suggested [61, 62].

4.2 Mesh Generation

Shape analysis requires surfaces with identical topology with the same faces and a similar number of vertices that can be achieved in two manners. The direct approach (top-down) uses an existing template mesh and deforms it to the individual anatomy [63–65]. This type of surface deformation works well for simple unfolded objects such as the skull [66], but runs into problems in the case of strongly folded structures [27]. Therefore, bottom-up methods dominate surface reconstruction with the creation of

[3] https://www.nitrc.org/projects/brainsuite
[4] http://fsl.fmrib.ox.ac.uk/fsl/fslwiki/
[5] https://www.nitrc.org/projects/mipav
[6] http://www.fil.ion.ucl.ac.uk/spm
[7] http://dbm.neuro.uni-jena.de/cat

individual objects and registrations to an average mesh, typically a sphere [11, 12, 21, 22, 27, 51, 56, 67, 68].

Due to its wide set of cognitive function, the reconstruction of the neocortex of both cerebral hemispheres is most relevant and different reconstruction pipelines have been purposed, such as BrainSuite [69], BrainVoyager[8] [68], Caret[9] [12], CAT[10] ASP/CLASP[11] [27, 63], FreeSurfer[12] [11], and MIPAV[13] [67]. Most methods reconstruct the GM-WM (inner/WM) surface that allows a better initial representation of the folded brain than the GM-CSF (outer/Pial) boundary that is often blurred in sulcal regions [22, 27, 56, 63, 67, 68, 70]. They fixed and optimized the mesh topology and deformed it to the CSF-GM boundary to estimate cortical thickness [27, 37, 63, 71]. Some methods prefer the central surface to represent the cortex [12, 51, 67]. The central surface runs in the middle of the cortex and is the average of the inner and outer surface and is therefore less noisy compared to either the inner or outer surface.

Another approach is applied by BrainVisa[14] that uses the WM surface to create independent surfaces of the major sulci to estimate and compare their morphology [48, 69]. Besides the cortex, reconstruction of other brain structures such as ventricles [72], hippocampi [73], basal ganglia [73], or fiber tracts [74] have been proposed.

4.3 Mesh Modification

The modification of surface meshes is required to optimize the initial meshes, prepare the surface registration, and create modified meshes for specific shape measures. Surface meshes can be modified in different ways, with the most important including: (a) smoothing and inflation, (b) deformation, (c) remeshing, (d) decomposition, and (e) averaging (*see* Fig. 4).

4.3.1 Smoothing and Inflation

Smoothing of mesh geometry reduces noise and artifacts by averaging the coordinates of neighbored vertices. At the same time, it removes anatomical details and unfolds the surface with growing number of iterations [12, 37, 75].

4.3.2 Deformation

The movement of mesh vertices (deformation) allows small refinements by anatomical details, e.g., to handle longitudinal changes [49, 64], midscale deformation such as the transformation of the

[8] http://www.brainvoyager.com/
[9] http://brainvis.wustl.edu
[10] http://dbm.neuro.uni-jena.de/cat
[11] Not public available
[12] https://surfer.nmr.mgh.harvard.edu/fswiki
[13] http://mipav.cit.nih.gov/
[14] http://brainvisa.info/

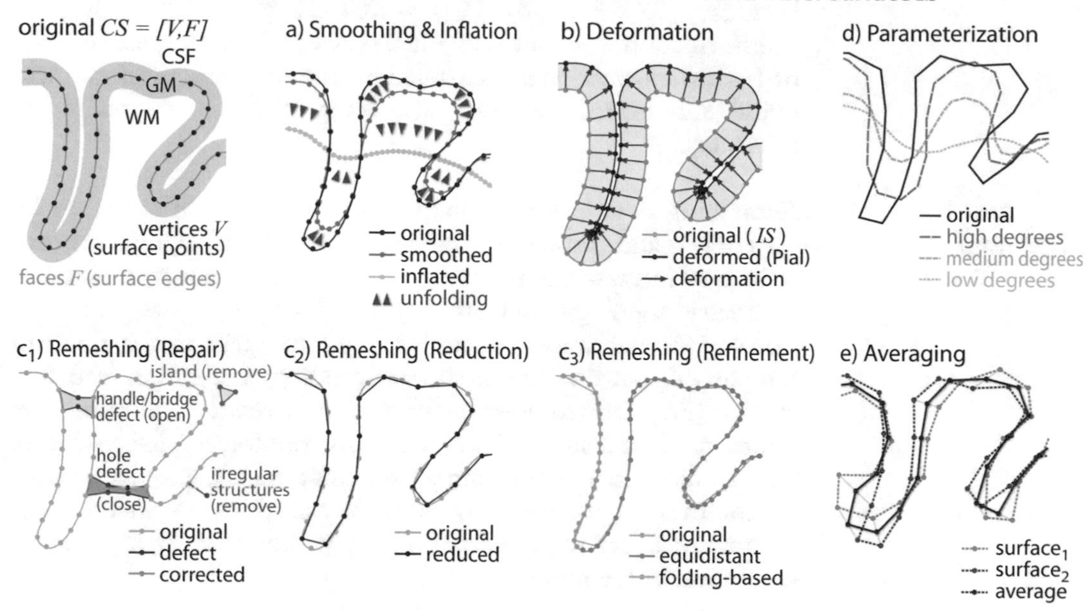

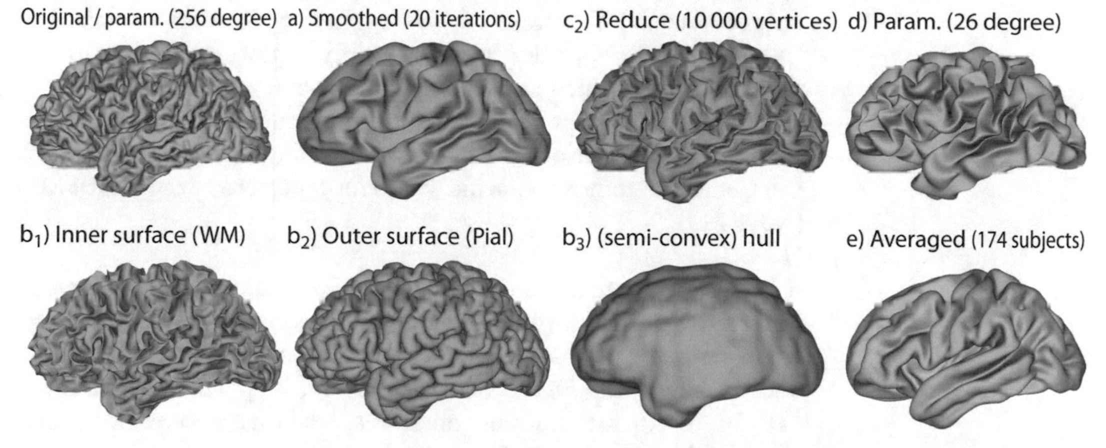

Fig. 4 The surface creation and many shape measures require modification of the surface, e.g., to create smoother unfolded versions, repair the topology, or reduce the resolution for faster processing. The most typical operations are illustrated here for the central surface *CS* in 2D (**a**) and 3D (**b**): smoothing averages the coordinates of each vertex with its neighbors and remove artifacts, anatomical details, or the folding pattern (**a**). Deformation moves the vertices based on internal (e.g., mesh connectivity) and external forces (e.g., tissue intensities) (**b**). Remeshing (reduction/refinement/repair) changes the complexity and topology of the mesh (**c**). Parameterization comprises the analysis and synthesis of signals by sums of simpler trigonometric functions (**d**). Averaging mix normalized meshes with different vertex positions but identical structures to create a common mesh (**e**)

brain surface position (e.g., from the GM-WM to the GM-CSF boundary [22, 27, 56, 67, 70]), as well as large changes such as the transformation from one individual surface to another one [63, 65, 66]. The deformation is controlled by internal (e.g., mesh connectivity) and external forces (e.g., vector fields based on image intensity).

4.3.3 Remeshing and Repairing

Remeshing describes the modification of the mesh structure by resolution and topology changes. Remeshing algorithms can reduce or increase the number of vertices and faces by preserving geometry, topology, and other properties to optimize computational and anatomical constrains, e.g., to guarantee a uniform sampling distance of the mesh after topology correction or deformation [76]. Due to noise, artifacts, blood vessels, and resolution limits, the initial surface often contains topological defects (holes and handles), islands (unconnected components), singular vertices or complex edges, gaps, overlaps, intersections, or inconsistent orientations that require repairing by geometrical or topological correction of the mesh [68, 77].

4.3.4 Parameterization

The Fourier analysis and synthesis describes the representation, approximation, and reconstruction of signals by sums of simpler (trigonometric) functions. It allows the application of spherical harmonics (a fast Fourier transformation on the sphere) for objects that can be simplified as a folded sphere such as the cortical hemispheres [33, 78]. The fraction of specific frequency can be used for shape characterization [78], specific folding measures (*see* Sect. 4.5), and to remove specific frequency patterns (e.g., artifacts) [33, 77].

4.3.5 Averaging

After surface registration (see next section), the relations between the vertices of multiple meshes allow the creation of an average mesh with the topology of one of the meshes and a mix of the coordinates of the linked vertices [63, 67, 75]. The average mesh can be used for folding measures, data representation, and visualization.

4.4 Spatial Normalization and Spherical Registration of Meshes

To compare individual meshes, a stable mapping to a common template (e.g., a sphere) is required [16, 36, 75]. The surface registration is the minimization of surface properties and shape features[15] for small (intra-individual) [28], medium (inter-individual) [16, 36, 75], or large (inter-species) folding patterns [16]. Although voxel-based registration works with high accuracy, surface-based registration profits by the improved characterization

[15] Mesh properties such as the area and inner angles of each face

of the cortex by surface measures and matching techniques with advanced alignment of individual structures.

5 Surface Measures

There are various ways to describe structural properties of one or more multiple shapes: (a) projection of volumetric data, (b) (cortical) thickness, (c) surface relations, (d) curvature, (e) depth, (f) (span) width, (g) parameterization, and (h) landmarks (*see* Fig. 5).

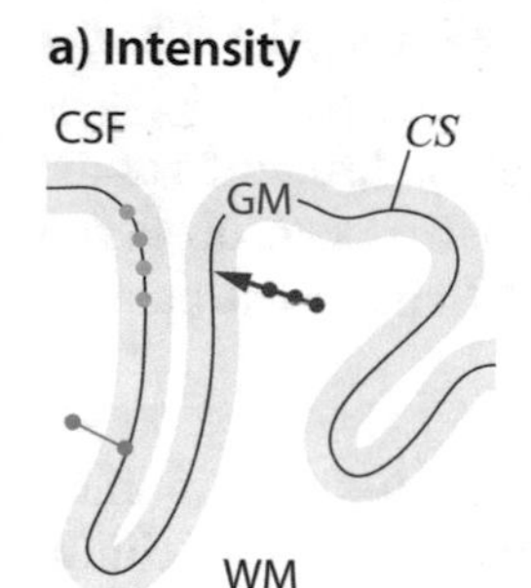

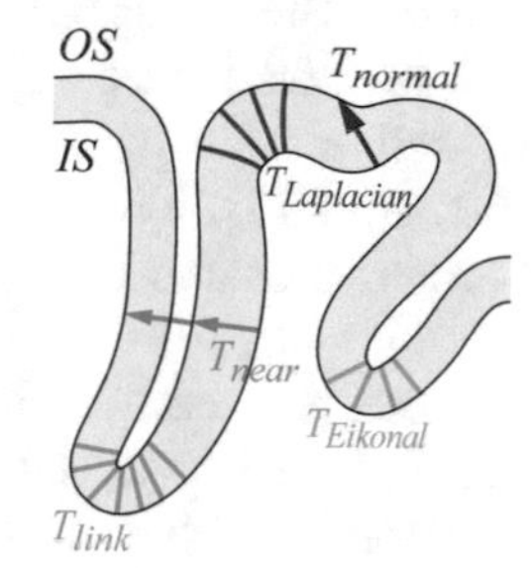

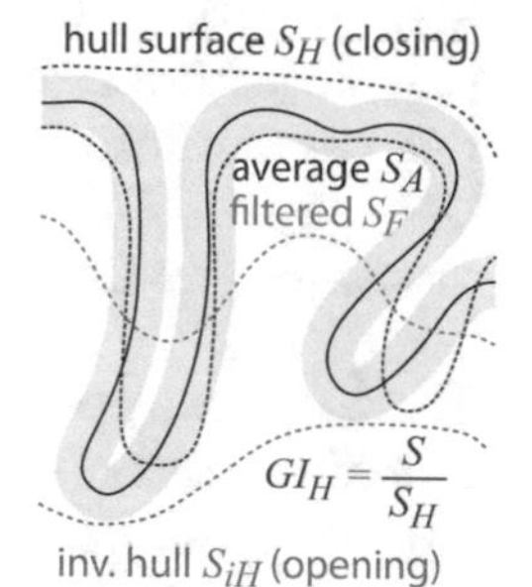

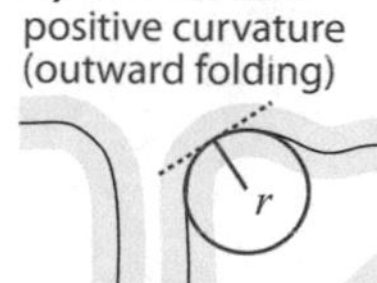

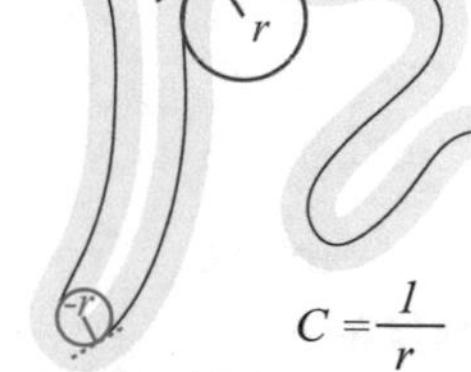

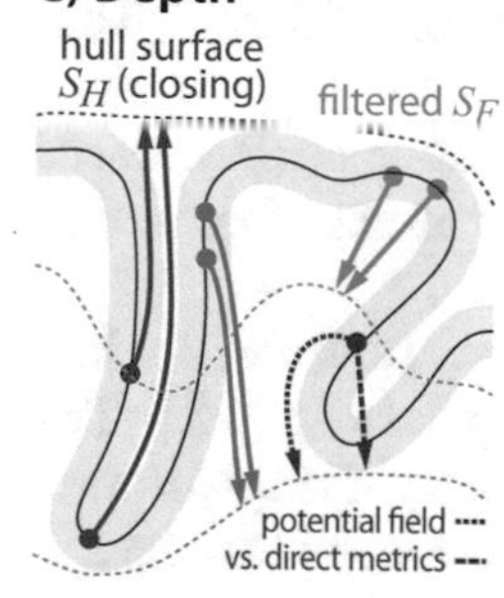

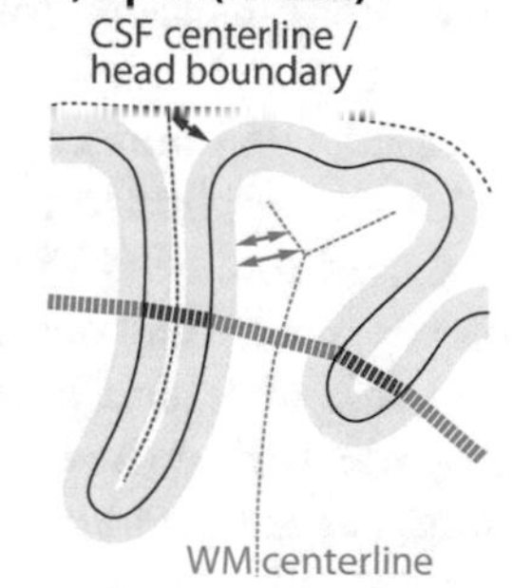

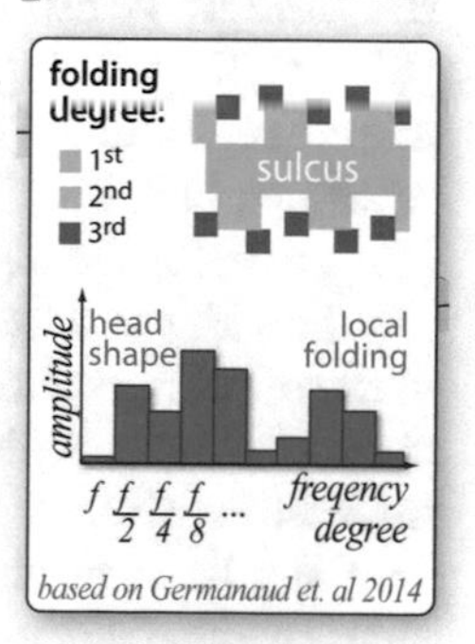

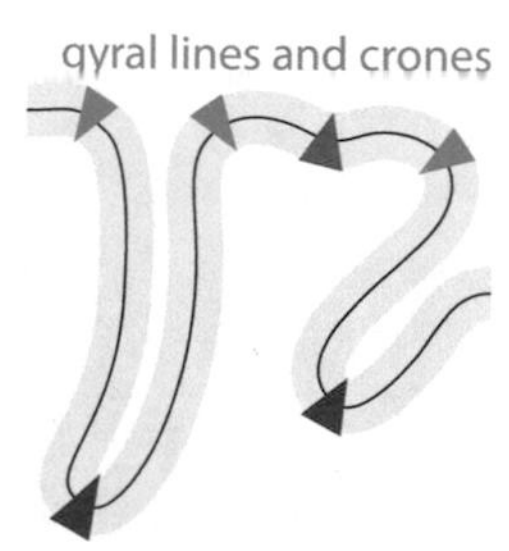

Fig. 5 Conceptual surface measures: (**a**) intensity, (**b**) thickness, (**c**) surface relation, (**d**) curvature, (**e**) depth, (**f**) span(width), (**g**) parameterization, and (**h**) landmarks. Shown is the 2D illustration of the central (CS), inner (IS), outer surface (OS), and unfolded versions such as the hull surface S_H, its counterpart S_{iH}, and the filtered unfolded surface S_F

5.1 Value extraction

The extraction of intensity can be used to process volumetric data from different MRI-modalities such as T1, T2, PD, dMRI, qMRI, or fMRI at different layer-specific positions, e.g., to characterize local myelinization [42], fiber orientation (DTI tensor field vs. surface normal) [79], fiber density [80], or tract geometry [81]. For further information and discussion, *see* Chap. 8 (cytoarchitectonic tissues and MRI-based signal intensities).

5.2 Thickness

One of the best known and most frequently used shape measures is the cortical thickness (sometimes also named cortical depth) that describes the width of the GM ribbon as the voxel- or surface-based distance between the inner and outer boundary. There are multiple metrics to estimate the thickness, most important are the (average) nearest neighbor T_{near} (FreeSurfer) [37, 63, 71], the surface normal T_{normal} [63], the coupled surface T_{link} [27, 37, 63], the Eikonal $T_{Eikonal}$ [51, 67], and the Laplacian metric $T_{Laplacian}$ [51, 82]. Although these metrics lead to slightly different results that should not be confused, similar patterns have been observed [29, 32, 35, 51, 52, 63, 80, 83]. For further information and discussion, *see* Chap. 3 (cortical thickness).

5.3 Surface Relations

The complexity of a shape can be measured in relation to simplified unfolded version(s) with removed local details by (a) smoothing, (b) morphologic operations such as closing or opening, (c) averaging, (d) down-sampling, or (e) other low-frequency representations such as spherical harmonics [84]. The most famous surface relation-based complexity measures are the gyrification index (GI) and the fractal dimension (FD).

The GI was first defined as the relation between the length of the folded contour and its envelope contour within a slice [85]. With growing computational possibilities, the GI was automated regional surface-based [86] and continuous surface-based measures [18, 87]. The GI was applied in the context of evolution [17], development, aging, and diseases [18].

The FD is a complexity ratio that describes how details in a pattern change with the scale at which it is measured [88]. The classic example is given by measuring the coastline of England that increases with finer scaling, recording more and more local details. In a similar way, the cortical folding of the brain can be partially characterized by describing the local enlargements by increased folding [89]. The FD of the brain can be defined by reducing volume [89] or mesh resolution [84]. FD has been applied to normal development and aging [89], as well as in the context of diseases [84].

The principle advantage of these measures is the intrinsic handling of the object size that allows simple comparisons for different individual and evolutional development stages [85, 88]. Interestingly, GI and FD end up with a similar complexity of about 2.5 for the human brain [84, 87].

5.4 Curvature

The local curvature of a surface can be illustrated in 2D as a circle that fits the local contour. In 3D, the so-called principal curvatures[16] are estimated for each vertex and allow the definition of a wide set of folding measures, with the four most prominent: (a) the (absolute) mean curvature[17] [90, 91], (b) the Gaussian curvature[18] [86, 90], (c) the shape index[19] [90], and (d) the curvedness[20] [90]. In most cases, the cortical curvature is described as the average of the curvature of the inner and outer surface that is equivalent to the curvature of the central surface [91]. Because the principle curvatures depend on brain size [78, 91], more complex measures try to incorporate normalization factors [86, 90]. Nevertheless, most curvature measures correlate strongly, and restriction to the best fitting and simplest measures is recommended. Curvature measures were successfully used to describe changes in normal development, aging, and various diseases [86, 90, 91].

5.5 Depth

The brain surface can be seen as a 3D signal [84] and its folding can be described by its frequency and amplitude. The amplitude can be characterized as the distance to a simplified surface, typically the hull surface of each hemisphere [82]. Similar to thickness, multiple distance metrics are available: the nearest neighbor [16], the Eikonal [67], the Laplacian [82], and the geodesic distance metric [92]. The nearest neighbor metric can cross sulci and gyri and therefore have lower values (especially in the Sylvian fissure), whereas the geodesic distance have the highest values [92]. Sulcal depth changes have been found in normal development, aging, and in various diseases [92].

5.6 (Span)Width

Besides the sulcal depth as a folding amplitude, the frequency of folds is also an interesting parameter that allows various measures including width, span, diameter, or thickness that describe the full or half distance between two sides of a gyrus or sulcus [6, 83, 93]. The width of the WM of a gyrus describes the local amount of myelinated fibers and how strong a region is connected to other regions [83], whereas the width of the CSF within a sulcus facilitates the investigation of local atrophy of WM and GM [93].

5.7 Parameterization

A more abstract way of describing the folding is given by the spectral analysis of shape features [46, 78]. Even complex signals

[16] The minimum and maximum fitting circle $k_{\min}$ and $k_{\max}$

[17] $C_{\text{mean}} = H = (k_{\min} + k_{\max})/2$

[18] $C_{\text{gaussian}} = K = k_{\min} \times k_{\max}$

[19] $C_{\text{shapeindex}} = 2/\pi \times \arctan(k_{\max} + k_{\min})/(k_{\max} - k_{\min})$

[20] $C_{\text{curvedness}} = \sqrt{(k_{\min} + k_{\max}^2)/2}$

can be described by simpler signals, e.g., the decomposition into a set of cosine or sine waves of different wavelength. This can be done by analyzing stepwise unfolded versions of the surface by Laplace-Beltrami [94], Spherical Harmonic [33, 34, 84], or Wavelet decomposition [34]. The spectral analysis of shape features allows a focus on specific spatial frequency bands that give the most important information to describe differences in the folding pattern [46, 78], where especially the second and third folding degree is relevant and not the basic shape of the brain or head [78]. It is important to mention that low folding reconstruction (Fourier synthesis, *see* Fig. 4b) creates an abstract pattern that supports no straightforward interpretation e.g., as a development pattern [95]. Parameterization has been applied in the context of development, aging, and various diseases [46, 78].

5.8 Landmarks

Besides global and continuous measurement, the subdivision of the cortex into gyral and sulcal regions [62, 96], or the extraction of surface landmarks such as sulcal bottom lines and pits, or gyral crones and peaks [97–99] were developed to support region-based analysis [69, 96], to extract further anatomical features [97, 98], or to improve registration accuracy [36]. The classification of special regions and structures can be further improved by other modalities such as dMRI [61] and fMRI [62]. In particular, BrainVisa focuses on the analysis of sulcal surfaces and allows the estimation of sulcus-specific measures of length, width, and folding [23, 93, 99].

5.9 Interim Conclusion

There are many approaches that describe different properties of the surface shape, reflecting new opportunities, as well as challenges for morphologic brain analysis, due to overlapping and similar measures, variable dependency of brain size (scaling invariance), and highly abstract measures that do not allow straightforward interpretation. A clear theory about the anatomical background of shape changes and the behavior of the applied measures is therefore essential.

6 Surface Analysis (SBM)

Surface analysis, especially the cortical thickness and folding measures, have become an important aspect of structural brain imaging. Similar to VBM, SBM can be evaluated globally, by regions, or continuously over the whole surface. Beyond that, it allows new and more subtle measures, anatomical correct registration and smoothing, and direct interaction with mathematical folding models.

In the previous chapter, several different types of surface measures were introduced. In particular, shape measures allow questions to be answered that VBM does not support. SBM allows the simplified decomposition of the GM volume V_{GM} into surface area A_{GM} and thickness T_{GM}:

$$V_{GM} = A_{GM} \cdot T_{GM}, \quad (2)$$

where the local folding can be neglected due to the expected compensation by the alteration of the cortical layers [1, 13, 20]. The decomposition of volume is especially important in brain development, with increasing surface area (tangential growth), but decreasing cortical thickness due to WM formation that impedes GM volume analysis.

The cortex is an organized surface [13, 18, 48] making surface registration preferable compared to volume-based methods. Besides, the registration and, in particular, the smoothing benefit from the surface-based organization of the brain, where the surface distance between the top of opposing gyri is in most cases more than twice the direct distance [32, 37, 38] and a typical 8 mm volume-based smoothing blurred opposing sulci and gyri [37, 71, 75] (*see* Fig. 3). Smoothing has the general effect of rendering the data to be more normally distributed and thereby increases the validity of the subsequent statistical tests and reduces outliers by noise, artifacts, or preprocessing errors [32, 37, 38].

Recent computational folding models demonstrated that gyrification depends on surface properties and that such models are capable of forecasting individual folding pattern development [15, 18, 19, 39, 49]. Hence, they also predict which circumstances lead to current folding patterns and can be used to understand developmental diseases such as autism spectrum disorder, or schizophrenia [18, 84]. Surface measures are therefore an important source of validating and improving cortical folding models. On the other hand, folding models can help to refine surface generation by further constraints or improve brain phantoms such as the brain web phantom [50] by supporting anatomical changing (longitudinal) phantoms for method evaluation.

The major drawbacks of SBM are: (a) the high complexity, which makes it vulnerable to noise, artifacts, and errors, (b) the considerable computational demands, and (c) the sophisticated interpretation of some folding measures. Surface preprocessing is more complex and therefore in general more error-prone and it is expected to be less sensitive (due to its constraints), as well as less robust (because of its complexity), especially for subtle changes in brain plasticity. On the other hand, constraints can improve the robustness and the increased complexity comes along with more characteristic measures, anatomical advanced registration and smoothing that may compensate this handicap [35, 84]. Because of the high amount of available measures, the challenge is to focus on measures that describe the expected changes or the use of big data techniques. A general limit of some gyrification measures is given by the arbitrary definition of unfolded structures, different metrics, and normalization factors. Many folding measures use unfolded structures that include the Sylvian but not the inter-hemispheric fissure, which might therefore bias the results. Furthermore, some folding measures are limited in describing the correct localization of changes that depend on deep WM tracts or the ventricle. Different metrics, e.g., for thickness or curvature, lead to slightly varying results, that limit the comparisons of different studies. It is also relevant to know if the used measures are intrinsic scaling invariant such as most relation measures that compare a folded and unfolded surface of the same subject, in contrast to most absolute measures, such as thickness, curvature, folding depth, and width, that depend on brain size and require covariates such as the total intra-cranial volume (TIV) for scaling normalization in the analysis [83].

Similar to VBM, SBM relies on the quality of the original data, with recent studies showing a clear influence of image quality on structural measures, with lower quality leading to GM underestimation [100] making quality assurance an important side aspect of the analysis [47, 53].

7 Conclusion

Shape properties are one of the key factors to understand the causes and effects of individual and evolutional folding development [14, 15, 18, 19, 21, 22]. Because folding is mostly affected by early development, shape measures have a high potential to investigate developmental dysfunctions even in the adult brain. Surfaces come along with a wide set of new or improved measures and an anatomical convenient registration and smoothing model. The description of surface characteristics by surface measures is essential for enhanced mathematical folding models that can simultaneously improve surface reconstruction, measures, and their validation [15, 18, 19, 84]. Surface analysis offers a number of new measures with various definitions and properties that require careful evaluation, especially of abstract shape measures [46, 78, 84].

References

1. Bok ST (1929) Der Einfluss der in den Furchen und Windungen auftretenden Krümmungen der Grosshirnrinde auf die Rindenarchitektur. Zeitschrift für die gesamte Neurologie und Psychiatrie 121:682–750. https://doi.org/10.1007/BF02864437
2. Barron DH (1950) An experimental analysis of some factors involved in the development of the fissure pattern of the cerebral cortex. J Exp Zool 113:553–581. https://doi.org/10.1002/jez.1401130304

3. Richman DP, Stewart RM, Hutchinson JW (1975) Mechanical mode of brain convolutional development. Science 189:18–21
4. Mietchen D, Gaser C (2009) Computational morphometry for detecting changes in brain structure due to development, aging, learning, disease and evolution. Front Neuroinform 3:25
5. Rakic P (2009) Evolution of the neocortex: a perspective from developmental biology. Nat Rev Neurosci 10:724–735. https://doi.org/10.1038/nrn2719
6. Hofman MA (1989) On the evolution and geometry of the brain in mammals. Prog Neurobiol 32:137–158
7. Welker W (1990) Why does cerebral cortex fissure and fold? vol 8B. Springer US, Boston, MA, pp 3–136
8. Lewitus E, Kelava I, Huttner WB (2013) Conical expansion of the outer subventricular zone and the role of neocortical folding in evolution and development. Front Hum Neurosci 7:424. https://doi.org/10.3389/fnhum.2013.00424
9. Striedter GF, Srinivasan S, Monuki ES (2015) Cortical folding: when, where, how, and why? Annu Rev Neurosci 38:291–307. https://doi.org/10.1146/annurev-neuro-071714-034128
10. Ashburner J, Friston KJ (2000) Voxel-based morphometry--the methods. NeuroImage 11:805–821. https://doi.org/10.1006/nimg.2000.0582
11. Fischl BR (2012) FreeSurfer. NeuroImage 62:774–781. https://doi.org/10.1016/j.neuroimage.2012.01.021
12. Van Essen DC et al (2001) An integrated software suite for surface-based analyses of cerebral cortex. J Am Med Inform Assoc 8:443–459
13. Van Essen DC (1997) A tension-based theory of morphogenesis and compact wiring in the central nervous system. Nature 385:313–318. https://doi.org/10.1038/385313a0
14. Budday S, Steinmann P, Kuhl E (2015) Physical biology of human brain development. Front Cell Neurosci 9:257. https://doi.org/10.3389/fncel.2015.00257
15. Tallinen T et al (2016) On the growth and form of cortical convolutions. Nat Phys 12:588–593. https://doi.org/10.1038/nphys3632
16. Van Essen DC (2004) Surface-based approaches to spatial localization and registration in primate cerebral cortex. NeuroImage 23(Suppl 1):S97–107. https://doi.org/10.1016/j.neuroimage.2004.07.024
17. Hilgetag CC, Barbas H (2006) Role of mechanical factors in the morphology of the primate cerebral cortex. PLoS Comput Biol 2:e22. https://doi.org/10.1371/journal.pcbi.0020022
18. Toro R (2012) On the possible shapes of the brain. Evol Biol 39(4):600–612
19. Bayly PV, Taber LA, Kroenke CD (2014) Mechanical forces in cerebral cortical folding: a review of measurements and models. J Mech Behav Biomed Mater 29:568–581. https://doi.org/10.1016/j.jmbbm.2013.02.018
20. Amunts K, Zilles K (2015) Architectonic mapping of the human brain beyond Brodmann. Neuron 88:1086–1107. https://doi.org/10.1016/j.neuron.2015.12.001
21. Xu G et al (2010) Axons pull on the brain, but tension does not drive cortical folding. J Biomech Eng 132:071013. https://doi.org/10.1115/1.4001683
22. Dale AM, Fischl BR, Sereno MI (1999) Cortical surface-based analysis. I. Segmentation and surface reconstruction. NeuroImage 9:179–194. https://doi.org/10.1006/nimg.1998.0395
23. Rivière D et al (2002) Automatic recognition of cortical sulci of the human brain using a congregation of neural networks. Med Image Anal 6:77–92
24. Hammers A et al (2003) Three-dimensional maximum probability atlas of the human brain, with particular reference to the temporal lobe. Hum Brain Mapp 19:224–247. https://doi.org/10.1002/hbm.10123
25. Shattuck DW et al (2008) Construction of a 3D probabilistic atlas of human cortical structures. NeuroImage 39:1064–1080. https://doi.org/10.1016/j.neuroimage.2007.09.031
26. Gaser C, Volz HP, Kiebel S, Riehemann S, Sauer H (1999) Detecting structural changes in whole brain based on nonlinear deformations-application to schizophrenia research. NeuroImage 10:107–113. https://doi.org/10.1006/nimg.1999.0458
27. Kim JS et al (2005) Automated 3-D extraction and evaluation of the inner and outer cortical surfaces using a Laplacian map and partial volume effect classification. NeuroImage 27:210–221. https://doi.org/10.1016/j.neuroimage.2005.03.036

28. Li G et al (2014) Measuring the dynamic longitudinal cortex development in infants by reconstruction of temporally consistent cortical surfaces. NeuroImage 90:266–279. https://doi.org/10.1016/j.neuroimage.2013.12.038
29. Fjell AM et al (2009) High consistency of regional cortical thinning in aging across multiple samples. Cereb Cortex 19:2001–2012. https://doi.org/10.1093/cercor/bhn232
30. Ziegler G, Ridgway GR, Dahnke R, Gaser C, Alzheimer's Disease Neuroimaging Initiative (2014) Individualized Gaussian process-based prediction and detection of local and global gray matter abnormalities in elderly subjects. NeuroImage 97:333–348. https://doi.org/10.1016/j.neuroimage.2014.04.018
31. Maguire EA et al (2000) Navigation-related structural change in the hippocampi of taxi drivers. Proc Natl Acad Sci U S A 97:4398–4403. https://doi.org/10.1073/pnas.070039597
32. Spjuth MS, Gravesen FH, Eskildsen SF, Østergaard LR (2007) Early detection of AD using cortical thickness measurements. Medical Imaging 6512:65120L–65129L. https://doi.org/10.1117/12.709806
33. Shen L, Chung MK (2006) Large-scale modeling of parametric surfaces using spherical harmonics. International Symposium on 3D Data Processing, Visualization, and Transmission. 294–301
34. Yu P et al (2007) Cortical surface shape analysis based on spherical wavelets. IEEE Trans Med Imaging 26:582–597. https://doi.org/10.1109/TMI.2007.892499
35. Winkler AM et al (2009) Cortical thickness or grey matter volume? The importance of selecting the phenotype for imaging genetics studies. NeuroImage 53(3):1135–1146. https://doi.org/10.1016/j.neuroimage.2009.12.028
36. Tardif CL et al (2015) Multi-contrast multi-scale surface registration for improved alignment of cortical areas. NeuroImage 111:107–122. https://doi.org/10.1016/j.neuroimage.2015.02.005
37. Lerch JP, Evans AC (2005) Cortical thickness analysis examined through power analysis and a population simulation. NeuroImage 24 (1):163–173
38. Anticevic A et al (2008) Comparing surface-based and volume-based analyses of functional neuroimaging data in patients with schizophrenia. NeuroImage 41:835–848. https://doi.org/10.1016/j.neuroimage.2008.02.052
39. Budday S, Raybaud C, Kuhl E (2014) A mechanical model predicts morphological abnormalities in the developing human brain. Sci Rep 4:5644. https://doi.org/10.1038/srep05644
40. Huang H (2010) Structure of the fetal brain: what we are learning from diffusion tensor imaging. Neuroscientist 16:634–649. https://doi.org/10.1177/1073858409356711
41. Weiskopf N et al (2013) Quantitative multi-parameter mapping of R1, PD(*), MT, and R2(*) at 3T: a multi-center validation. Front Neurosci 7:95. https://doi.org/10.3389/fnins.2013.00095
42. Deoni SCL, Dean DC, Remer J, Dirks H, O'Muircheartaigh J (2015) Cortical maturation and myelination in healthy toddlers and young children. NeuroImage 115:147–161. https://doi.org/10.1016/j.neuroimage.2015.04.058
43. Shah M et al (2011) Evaluating intensity normalization on MRIs of human brain with multiple sclerosis. Med Image Anal 15:267–282. https://doi.org/10.1016/j.media.2010.12.003
44. Jiang X, Nardelli J (2016) Cellular and molecular introduction to brain development. Neurobiol Dis 92(Pt A):3–17. https://doi.org/10.1016/j.nbd.2015.07.007
45. Van Essen DC, Maunsell JHR (1980) Two-dimensional maps of the cerebral cortex. J Comp Neurol 191:255–281. https://doi.org/10.1002/cne.901910208
46. Wright R et al (2015) Construction of a fetal spatio-temporal cortical surface atlas from in utero MRI: application of spectral surface matching. NeuroImage 120:467–480. https://doi.org/10.1016/j.neuroimage.2015.05.087
47. Evans AC, Brain Development Cooperative Group (2006) The NIH MRI study of normal brain development. NeuroImage 30:184–202. https://doi.org/10.1016/j.neuroimage.2005.09.068
48. Régis J et al (2005) "Sulcal root"; generic model: a hypothesis to overcome the variability of the human cortex folding patterns. Neurol Med Chir 45:1–17
49. Nie J et al (2012) A computational growth model for measuring dynamic cortical development in the first year of life. Cereb Cortex 22:2272–2284. https://doi.org/10.1093/cercor/bhr293
50. Collins DL et al (1998) Design and construction of a realistic digital brain phantom. IEEE

Trans Med Imaging 17:463–468. https://doi.org/10.1109/42.712135

51. Dahnke R, Yotter RA, Gaser C (2013) Cortical thickness and central surface estimation. NeuroImage 65:336–348. https://doi.org/10.1016/j.neuroimage.2012.09.050
52. Tustison NJ et al (2014) Large-scale evaluation of ANTs and FreeSurfer cortical thickness measurements. NeuroImage 99:166–179. https://doi.org/10.1016/j.neuroimage.2014.05.044
53. Poldrack RA, Gorgolewski KJ (2014) Making big data open: data sharing in neuroimaging. Nat Neurosci 17:1510–1517. https://doi.org/10.1038/nn.3818
54. Ashburner J, Friston KJ (2005) Unified segmentation. NeuroImage 26:839–851. https://doi.org/10.1016/j.neuroimage.2005.02.018
55. Coupé P, Yger P, Barillot C (2006) Fast non local means denoising for 3D MR images. Med Image Comput Comput Assist Interv 9:33–40
56. Shattuck DW, Leahy RM (2002) BrainSuite: an automated cortical surface identification tool. Med Image Anal 6:129–142. https://doi.org/10.1016/S1361-8415(02)00054-3
57. Bazin P-L, Pham DL (2008) Homeomorphic brain image segmentation with topological and statistical atlases. Med Image Anal 12:616–625. https://doi.org/10.1016/j.media.2008.06.008
58. Wang L et al (2013) Longitudinally guided level sets for consistent tissue segmentation of neonates. Hum Brain Mapp 34:956–972. https://doi.org/10.1002/hbm.21486
59. Mendrik AM et al (2015) MRBrainS challenge: online evaluation framework for brain image segmentation in 3T MRI scans. Comput Intell Neurosci 2015:813696–813616. https://doi.org/10.1155/2015/813696
60. Ou Y, Akbari H, Bilello M, Da X, Davatzikos C (2014) Comparative evaluation of registration algorithms in different brain databases with varying difficulty: results and insights. IEEE Trans Med Imaging 33 (10):2039–2065. https://doi.org/10.1109/TMI.2014.2330355
61. Anwander A, Tittgemeyer M, von Cramon DY, Friederici AD, Knösche TR (2007) Connectivity-based Parcellation of Broca's area. Cereb Cortex 17:816–825. https://doi.org/10.1093/cercor/bhk034
62. Schubotz RI, Anwander A, Knösche TR, von Cramon DY, Tittgemeyer M (2010) Anatomical and functional parcellation of the human lateral premotor cortex. NeuroImage 50:396–408. https://doi.org/10.1016/j.neuroimage.2009.12.069
63. MacDonald D, Kabani NJ, Avis D, Evans AC (2000) Automated 3-D extraction of inner and outer surfaces of cerebral cortex from MRI. NeuroImage 12:340–356. https://doi.org/10.1006/nimg.1999.0534
64. Nakamura K, Fox R, Fisher E (2011) CLADA: cortical longitudinal atrophy detection algorithm. NeuroImage 54:278–289. https://doi.org/10.1016/j.neuroimage.2010.07.052
65. Xu C, Pham DL, Rettmann ME, Yu DN, Prince JL (1999) Reconstruction of the human cerebral cortex from magnetic resonance images. IEEE Trans Med Imaging 18:467–480. https://doi.org/10.1109/42.781013
66. Smith SM (2002) Fast robust automated brain extraction. Hum Brain Mapp 17:143–155. https://doi.org/10.1002/hbm.10062
67. Tosun D et al (2004) Cortical surface segmentation and mapping. NeuroImage 23 (Suppl 1):S108–S118. https://doi.org/10.1016/j.neuroimage.2004.07.042
68. Kriegeskorte N, Goebel R (2001) An efficient algorithm for topologically correct segmentation of the cortical sheet in anatomical mr volumes. NeuroImage 14:329–346. https://doi.org/10.1006/nimg.2001.0831
69. Cachia A et al (2003) A generic framework for the parcellation of the cortical surface into gyri using geodesic Voronoï diagrams. Med Image Anal 7:403–416
70. Eskildsen SF, Østergaard LR (2006) Active surface approach for extraction of the human cerebral cortex from MRI. Med Image Comput Comput Assist Interv 9:823–830
71. Fischl BR, Dale AM (2000) Measuring the thickness of the human cerebral cortex from magnetic resonance images. Proc Natl Acad Sci U S A 97:11050–11055. https://doi.org/10.1073/pnas.200033797
72. Paniagua B et al (2013) Lateral ventricle morphology analysis via mean latitude axis. Proc SPIE Int Soc Opt Eng 8672:86720M. https://doi.org/10.1117/12.2006846
73. Qiu A, Miller MI (2008) Multi-structure network shape analysis via normal surface momentum maps. NeuroImage 42:1430–1438. https://doi.org/10.1016/j.neuroimage.2008.04.257

74. Qiu A et al (2010) Surface-based analysis on shape and fractional anisotropy of white matter tracts in Alzheimer's disease. PLoS One 5: e9811. https://doi.org/10.1371/journal.pone.0009811
75. Fischl BR, Sereno MI, Dale AM (1999) Cortical surface-based analysis. II: inflation, flattening, and a surface-based coordinate system. NeuroImage 9:195–207. https://doi.org/10.1006/nimg.1998.0396
76. Frey PJ (2001) Anisotropic surface remeshing. Elsevier
77. Yotter RA, Dahnke R, Thompson PM, Gaser C (2011) Topological correction of brain surface meshes using spherical harmonics. Hum Brain Mapp 32:1109–1124. https://doi.org/10.1002/hbm.21095
78. Germanaud D et al (2012) Larger is twistier: spectral analysis of gyrification (SPANGY) applied to adult brain size polymorphism. NeuroImage 63:1257–1272. https://doi.org/10.1016/j.neuroimage.2012.07.053
79. Kleinnijenhuis M et al (2015) Diffusion tensor characteristics of gyrencephaly using high resolution diffusion MRI in vivo at 7T. NeuroImage 109:378–387. https://doi.org/10.1016/j.neuroimage.2015.01.001
80. Nie J et al (2014) Longitudinal development of cortical thickness, folding, and fiber density networks in the first 2 years of life. Hum Brain Mapp 35:3726–3737. https://doi.org/10.1002/hbm.22432
81. Savadjiev P et al (2014) Fusion of white and gray matter geometry: a framework for investigating brain development. Med Image Anal 18:1349–1360. https://doi.org/10.1016/j.media.2014.06.013
82. Jones SE, Buchbinder BR, Aharon I (2000) Three-dimensional mapping of cortical thickness using Laplace's equation. Hum Brain Mapp 11:12–32
83. Im K et al (2008) Sulcal morphology changes and their relationship with cortical thickness and gyral white matter volume in mild cognitive impairment and Alzheimer's disease. NeuroImage 43:103–113. https://doi.org/10.1016/j.neuroimage.2008.07.016
84. Yotter RA, Nenadic I, Ziegler G, Thompson PM, Gaser C (2011) Local cortical surface complexity maps from spherical harmonic reconstructions. NeuroImage 56:961–973. https://doi.org/10.1016/j.neuroimage.2011.02.007
85. Zilles K, Armstrong E, Schleicher A, Kretschmann H-J (1988) The human pattern of gyrification in the cerebral cortex. Anat Embryol 179:173–179. https://doi.org/10.1007/BF00304699
86. Rodriguez-Carranza C, Mukherjee P, Vigneron DB, Barkovich AJ, Studholme C (2008) A framework for in vivo quantification of regional brain folding in premature neonates. NeuroImage 41:462–478. https://doi.org/10.1016/j.neuroimage.2008.01.008
87. Schaer M et al (2008) A surface-based approach to quantify local cortical gyrification. IEEE Trans Med Imaging 27:161–170. https://doi.org/10.1109/TMI.2007.903576
88. Mandelbrot B (1967) How long is the coast of britain? Statistical self-similarity and fractional dimension. Science 156:636–638. https://doi.org/10.1126/science.156.3775.636
89. Jiang J et al (2008) A robust and accurate algorithm for estimating the complexity of the cortical surface. J Neurosci Methods 172:122–130. https://doi.org/10.1016/j.jneumeth.2008.04.018
90. Pienaar R, Fischl BR, Caviness VS, Makris N, Grant PE (2008) A methodology for analyzing curvature in the developing brain from preterm to adult. Int J Imaging Syst Technol 18(1):42–68
91. Luders E et al (2006) A curvature-based approach to estimate local gyrification on the cortical surface. NeuroImage 29:1224–1230. https://doi.org/10.1016/j.neuroimage.2005.08.049
92. Yun HJ, Im K, Yang J-J, Yoon U, Lee J-M (2013) Automated sulcal depth measurement on cortical surface reflecting geometrical properties of sulci. PLoS One 8:e55977. https://doi.org/10.1371/journal.pone.0055977
93. Kochunov PV et al (2008) Relationship among neuroimaging indices of cerebral health during normal aging. Hum Brain Mapp 29:36–45. https://doi.org/10.1002/hbm.20369
94. Levy-Cooperman N, Ramirez J, Lobaugh NJ, Black SE (2008) Misclassified tissue volumes in Alzheimer disease patients with white matter hyperintensities: importance of lesion segmentation procedures for volumetric analysis. Stroke 39:1134–1141. https://doi.org/10.1161/STROKEAHA.107.498196
95. Shishegar R, Britto JM, Johnston LA (2014) Conference proceedings: Annual International Conference of the IEEE Engineering

in Medicine and Biology Society IEEE Engineering in Medicine and Biology Society Conference, vol 2014. pp 1525–1528

96. Desikan RSR et al (2006) An automated labeling system for subdividing the human cerebral cortex on MRI scans into gyral based regions of interest. NeuroImage 31:968–980. https://doi.org/10.1016/j.neuroimage.2006.01.021
97. Li G, Guo L, Nie J, Liu T (2009) Automatic cortical sulcal parcellation based on surface principal direction flow field tracking. NeuroImage 46:923–937. https://doi.org/10.1016/j.neuroimage.2009.03.039
98. Meng Y, Li G, Lin W, Gilmore JH, Shen D (2014) Spatial distribution and longitudinal development of deep cortical sulcal landmarks in infants. NeuroImage 100:206–218. https://doi.org/10.1016/j.neuroimage.2014.06.004
99. Hopkins WD et al (2014) Evolution of the central sulcus morphology in primates. Brain Behav Evol 84:19–30. https://doi.org/10.1159/000362431
100. Reuter M et al (2015) Head motion during MRI acquisition reduces gray matter volume and thickness estimates. NeuroImage 107:107–115. https://doi.org/10.1016/j.neuroimage.2014.12.006

Chapter 5

The General Linear Model: Theory and Practicalities in Brain Morphometric Analyses

Cyril R. Pernet

Abstract

The general linear model (GLM) is the statistical method of choice used in brain morphometric analyses because of its ability to incorporate a multitude of effects. This chapter starts by presenting the theory, focusing on modeling, and then goes on discussing multiple comparisons issues specific to voxel-based approaches. The end of the chapter discusses practicalities: variable selection and covariates of no interest. Researchers have often a multitude of demographic and behavioral measures they wish to use, and methods to select such variables are presented. We end with a note of caution as the GLM can only reveal covariations between the brain and behavior, and prediction and causation mandate specific designs and analyses.

Key words General linear model, Inference, Variable selection, Covariates

1 Introduction

At the core, the general linear model (GLM) comes from regression and correlational methods, and it can be understood as a general multiple regression model. It derives from the theory of algebraic invariants developed in the nineteenth century along with the development of linear algebra. The theory searches for explicit description of polynomial functions[1] that do not change under transformations from a given linear group. Linear correlations, for example, are invariant, i.e., the linear correlation coefficient does not change after linearly transforming the data. For a thorough explanation of the GLM, see [1]. The GLM has been successfully used in the analysis of brain structures because of its flexibility to handle both categorical (e.g., groups of subjects) and continuous variables (e.g., test scores). It can be used to examine regions of interest (ROIs) from which various morphometric markers can be extracted, but it has been most successful in whole brain analysis

[1] A polynomial function is a function, such as a quadratic, a cubic, a quartic, and so on, involving only nonnegative integer powers of x.

Gianfranco Spalletta et al. (eds.), *Brain Morphometry*, Neuromethods, vol. 136,
https://doi.org/10.1007/978-1-4939-7647-8_5,

using a voxel-based approach for gray matter (voxel-based morphometry or VBM [2]) and white matter (tract-based spatial statistics or TBSS [3]).

2 Theory

The dependent variable (Υ) can be described as a linear combination of independent variables ($X\beta$). Written as an equation, this gives

$$\Upsilon = X\beta + \varepsilon \tag{1}$$

Each term is a matrix of data with Υ of dimensions $n^* m$, X of dimensions $n * p$, β of dimension $p * m$, and ε of dimensions $n * m$ representing the residuals, i.e., the part of the data not explained by $X\beta$. Typically $m = 1$ when performing morphometric analyses (Fig. 1), that is, each region of interest (ROI) or each voxel of the brain is modeled separately (massive univariate approach). When using multiple ROI, it can be useful to have a single model with $m > 1$, taking advantage of correlational structure in the data to find a significant multivariate effect in the absence of significant univariate differences (e.g., patients can differ from controls by exhibiting lower volumes in regions A and B and higher volumes

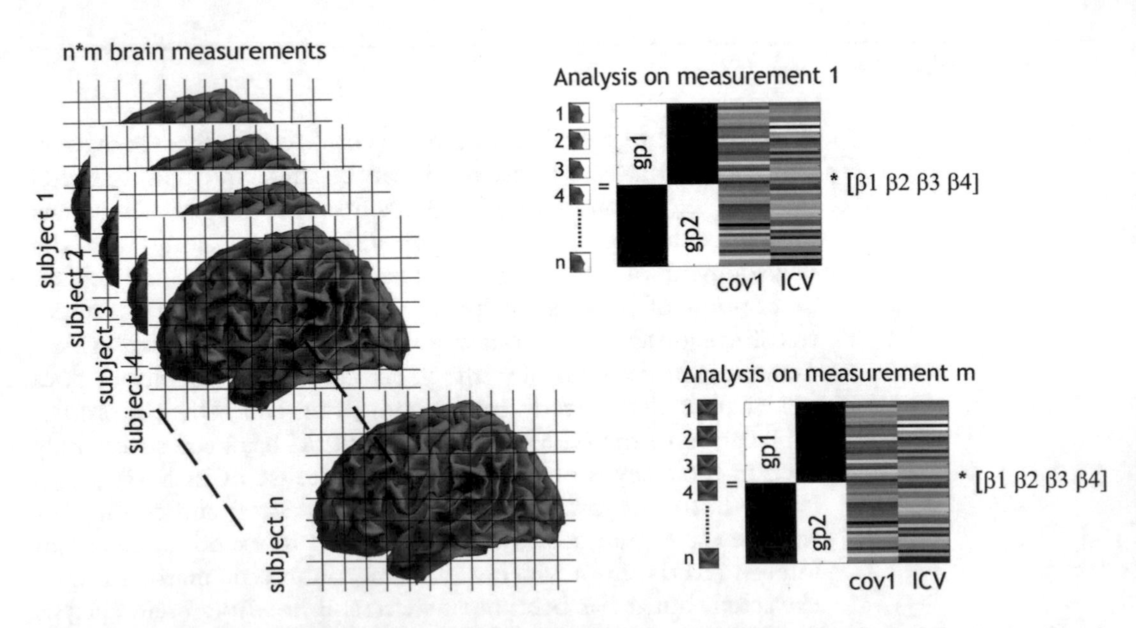

Fig. 1 Illustration of the mass-univariate GLM approach. Data consist in n subjects and the brain of each subject is "cut" into m measurements (voxels or ROI). Each measurement is then submitted to the same analysis, i.e., the same model X (here illustrated with a matrix of four columns, modeling membership of group 1 or 2, a covariate of interest and the intracranial volume (ICV) as a controlling confounding variable). The parameters β, or linear combinations of them, are then tested for significance either for each of these measurements or for topological features associated to these measurements (i.e., cluster's height or size)

in regions C and D, but none of those regions show on their own a significant difference).

3 Linearity

The first aspect to recognize in Eq. 1 is linearity. Linearity refers to a relationship between variables that can be geometrically represented as a straight line (2D), a plane (3D), or a hyperplane (4D and above). A linear function satisfies the properties of additivity and scaling, which means that outputs (Y) are simply an addition of inputs (X: x_1, x_2, x_3, etc.) each one of them multiplied by some constants (β: b_1, b_2, b_3, etc.).

Let us consider the application of such model to hippocampal volume measurements. I use here rounded data from [4] (Fig. 3) and look at the relation between healthy subject ages and normalized volumes of the right hippocampus.[2] The model is a simple linear regression such as $y = x_1 b_1 + b_2$. The equation is linear since it states that y, the hippocampal volume, is equal to age (x_1) scaled by b_1 plus an average volume value (b_2). Geometrically, the relationship is defined as a straight line (Fig. 2).

4 Modeling Data

The second aspect to recognize in Eq. 1 is the modeling. Consider again the regression model for the hippocampus volume, adding now the group of Alzheimer patients. We can either model the data with a single mean or having group-specific means in addition of group-specific age regressors. In the former case, we build a model in which controls and patients have overall the same volumes, but we allow age-related changes to differ between groups (i.e., the hypothesis is that aging does not affect the brain the same way between subjects who have or do not have the disease). In the latter case, we build a model in which controls and patients have different overall volume values (i.e., they come from different distributions, which is a reasonable assumption given what we know about Alzheimer and memory) while also exhibiting differential age-related changes (Fig. 3).

It is important to recognize that any data analysis using the GLM is driven by hypotheses. While this might seem trivial, differences between models can lead to large differences in results, and

[2] In morphometric analyses, the total intracranial volume (or related measurement) is typically accounted for, either in the model or in the data. Here the hippocampal volume is normalized to the total brain volume—this transformation is mandatory as bigger heads give bigger volume and vice versa, and bias in a sample can lead to spurious results.

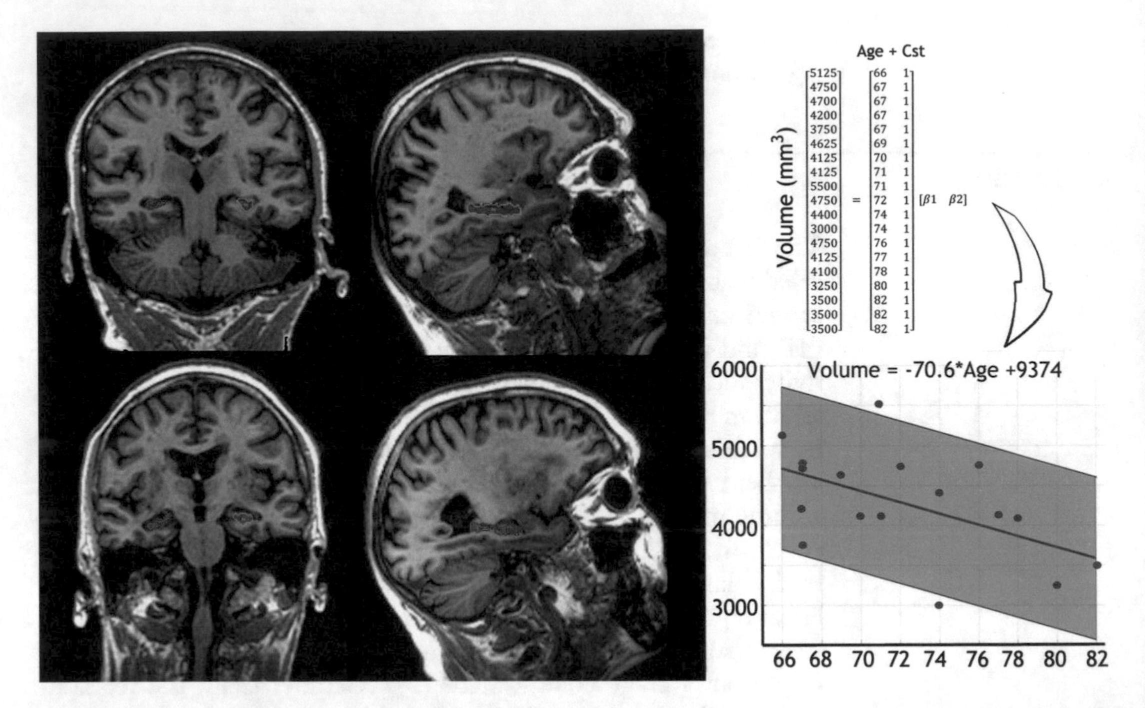

Fig. 2 Simple regression on a morphological feature: here hippocampal volume. After delineating the hippocampus of 19 subjects, (image courtesy from the Centre for Clinical Brain Science, Edinburgh Imaging Library brain-DA0001), Frisoni et al. [4] computed the normalized hippocampal volume and performed a regression analysis of the volume as a function of the subject's age (plus a constant term), showing a significant reduction of hippocampus volume as one gets older

authors and readers of scientific articles should be aware of this. Here both models are statistically significant (the first model has a R^2 of 0.56 and the second model a R^2 of 0.61). The simple inclusion or exclusion of the group variable leads however to opposite conclusions. If we consider that both controls and patients come from the same distributions, patients show a significant correlation with age. If we consider group-specific volumes, the correlation with age is only significant in control subjects (implying the disease "destroyed" or obscured age-related changes).

5 Model Estimation

Once the model is set up, a software solves the equation, i.e., it estimates the model parameters. If we consider a linear system, the solution is found simply by multiplying each side of Eq. 1 by the inverse of X giving the solution in Eq. 2. Most often X is not square and the system is not fully determined (i.e., we don't have an exact set of equations that describes the data). The solution is therefore a set of estimates. Typically, the Moore-Penrose pseudoinverse (Eq. 3) is used, but other solutions exist (e.g., QR decomposition,

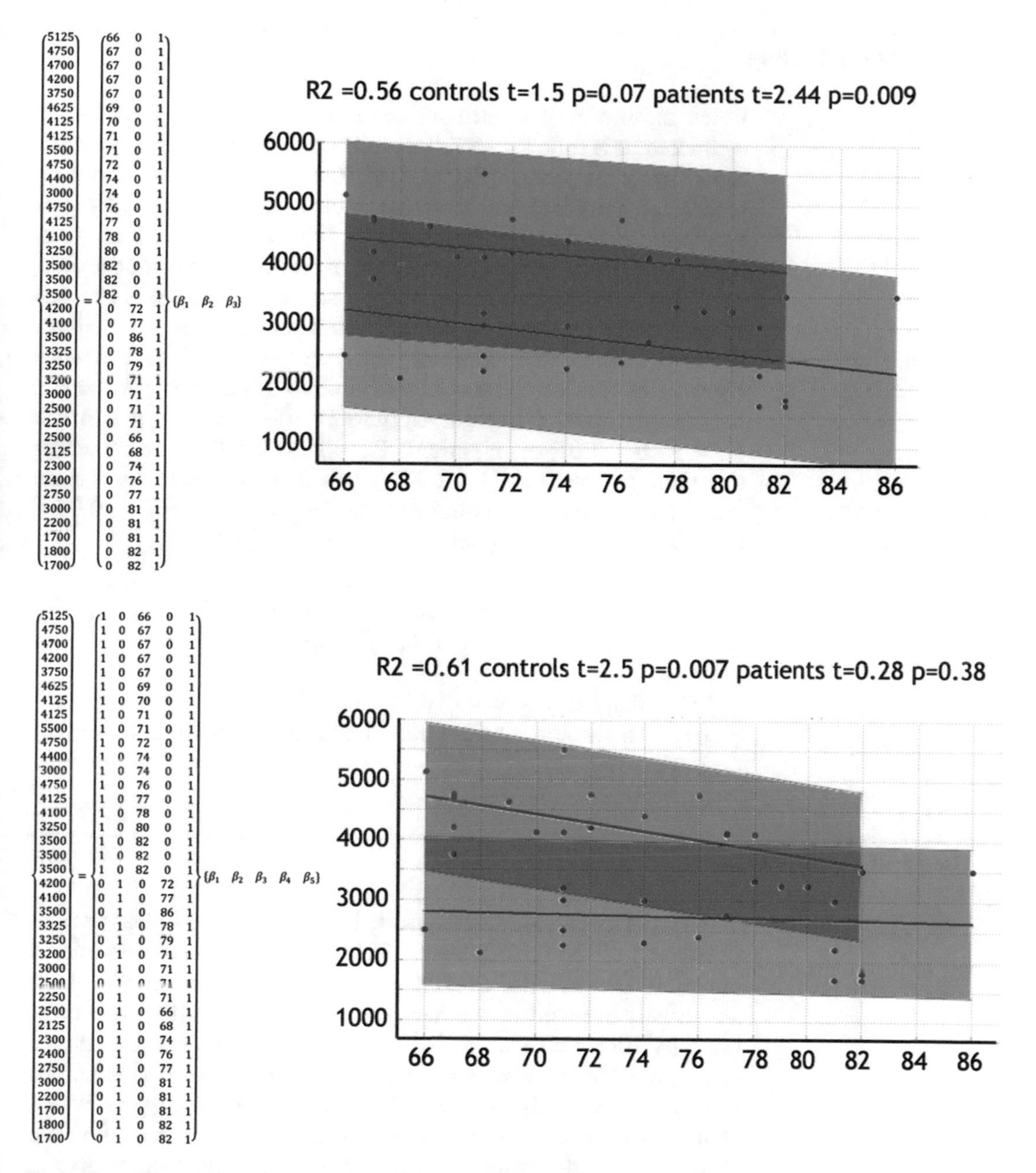

Fig. 3 Two different models comparing the hippocampal volumes of healthy controls and Alzheimer patients. Including the group variable changes the slope of the age regressors, giving different results/conclusions about how gray matter plasticity changes with age

Bayesian estimation), which implies that slightly different results can sometimes be obtained because of the estimation method.

$$\widehat{\beta} = X^{-1}\Upsilon \tag{2}$$

$$\widehat{\beta} = X^{\sim}\Upsilon \tag{3}$$

6 Hypothesis Testing

Once the model has been set up and the parameters have been estimated, it is time for statistical testing. Testing always follow the same rule: effect/error (Eq. 4). One way to think about hypothesis testing is to think about how to combine parts of the model (i.e., the columns of X) to explain the data (Y). One can test if only one regressor explains the data, or if a set of regressors explain the data, or even if one regressor or set of regressors explain more of the data than another regressor or set of regressors. Contrasts are simple linear functions that combine the columns of X to test such hypotheses. Taking the model with five columns in Fig. 3, we can test using Eq. 4 if there is significant effect of age in controls ($C = [0\ 0\ 1\ 0\ 0]$) or in patients ($C = [0\ 0\ 0\ 1\ 0]$) or even across the whole sample ($C = [0\ 0\ 1\ 1\ 0]$). It is also possible to test if there is a difference in overall volumes between groups ($C = [1 {-1}\ 0\ 0\ 0]$) or a difference in the age effect ($C = [0\ 0\ 1 {-1}\ 0]$).

$$t = \frac{C\widehat{\beta}}{\sqrt{\sigma^2 C^T \left(X^T X\right)^{-1} C}} \tag{4}$$

C defined the contrast of interest, $\widehat{\beta}$ are the parameter estimates, σ^2 is the variance obtained from the residuals, and X is the design matrix.

7 Multiple Comparisons Correction

The type I error rate corresponds to probability to observe false positives, i.e., the probability to declare a ROI or a voxel significant while H0 (the absence of effect) is true. Because m statistical tests are usually performed, we have to consider the family-wise error rate (FWER), i.e., the probability to make one or more type I errors, in order to ensure an overall error at the prespecified alpha level. Many methods have been devised to control the type I FWER at the voxel or cluster level: Bonferroni correction, random field theory [5], and randomization tests (permutation or bootstrap [6]). Note that false discovery rate (FDR) procedures do not control the type I error in the classical sense. FDR procedures control for the expected number of false positives among all positive tests.

If testing multiple ROI, one of those techniques must be used: either FDR in the context of data exploration or FWER for hypothesis-driven studies. While both Bonferroni and randomization tests will control the type 1 FWER to a similar level, Bonferroni correction is not optimal because it considers ROI as independent, while ROI data are often correlated. Randomization tests are likely to provide more power, accounting for correlations between ROIs. With full brain analyses, i.e., testing many spatially contiguous voxels (i.e., VBM or TBSS), the choice of the multiple comparison technique has an even greater impact. It has been shown that statistical values are attracted toward regions of low residual variance—which should not happen if the data were stationary. Unfortunately, full brain voxel-based analyses suffer from nonstationarity, and this means that in smooth regions, clusters tend to be large even in the absence of true signals, resulting in increased false positives, while in rough regions, clusters tend to be small, resulting in reduced power. Because of this sensitivity to smoothness, it is not recommended to threshold maps based on cluster size [2] unless using specifically a nonstationary permutation cluster test [6]. An interesting solution is to integrate size and height over all thresholds, a technique known as threshold-free cluster enhancement [7] which has been successfully applied to TBSS and is becoming more popular for VBM when, again, it is adjusted for nonstationarity [8, 9]. Finally, it is also worth considering the number of contrasts performed—the more covariates are tested, the more likely one of them will be significant by chance, and alpha level adjustment could also be performed at that stage [10].

8 Practicalities

8.1 Assumptions

The first job of the experimenter and data analyst is to define the independent variables. Sometimes people worry about the normality assumption. It is important to understand that this only applies to the data (and model residuals) and not to the regressors, which means that as long as subjects are independent and identically distributed (i.i.d.), anything can be entered as regressors, for instance, a quadratic effect of age. It is however worth also considering the relationships between variables as it is often not clear if brain structural changes are explained by the behavior or if this is the behavioral data that are predicted by brain morphology [11].

Normality. When performing the analysis on a few ROI, it is advisable to check the data distributions as long tails will give spurious results. With VBM, gray matter values are bounded, and it is thus recommended to have sufficient spatial smoothing, from 4 to 8 mm [12], to render the distributions normal and thus avoid high false-positive rate. In this context, unbalance between groups is the most problematic with extreme cases such as comparing a single subject to a group leading to high number of false positives [13].

Independence. Lack of independence occurs when the error covariance is not diagonal. There is no good way to fix such problem because it does not relate to the data but to the sampling. The single best advice is thus to have a good sampling strategy for the population of interest.

8.2 Choosing Covariates

To investigate structural brain changes in health and disease, many variables of interest are obtained for each participant. For instance, researchers working on language disabilities measure basic sensory performances along with IQ and auditory and visual cognitive abilities. Similarly, research in aging will try to characterize many cognitive dimensions using multiple tests along with a more general test such as the MoCA [14]. Of particular interest for GLM analyses are (a) the accuracy of these measures and (b) the relationship between these measures.

All sensory and cognitive measures are affected by random measurement errors. This means that measurements fluctuate randomly around their true values because of inherent biological variability, of imprecise tool, or both. The total error in a variable with random measurement error averages out to zero, which implies that on average this is accurate. Many people assume that it implies that it does not affect regression analyses. Since GLM estimates minimize the square distance in X, errors in the regressors (i.e., the sensory or cognitive measurements) lead to poor data fit, biasing the regression coefficient toward the null, a phenomenon known as attenuation or regression dilution bias [15]. When we have multiple variables which measure the same sensory or cognitive dimension, it is best to choose the one with the smallest error, i.e., the one with the smallest standardized standard deviation.

Having multiple variables that measure the same or overlapping dimensions can also be a problem for morphometric analyses. If the goal were to explain the data with a model, collinearity between regressors would not be an issue. Since the goal is usually to investigate how each variable relates to structural changes, this becomes a problem. Multicollinearity leads to a decrease in the unique amount of variance explained by a single regressor, thereby posing difficulty in interpreting their respective contribution. Kraha et al. [16] propose a set of tools that can be useful in this context (e.g., commonality analysis, dominance analysis), examining relationships between the regressors and the variable of interest. These tools are easy to apply for ROI analyses but are intractable for VBM or TBSS analyses. One useful tool is the variance inflation factor, which examines the impact of each regressor on the others, therefore making sure no regressors in the model are too much correlated with others or a combination of them. In many cases, the experimenter or data analyst will still have to select some variables. Without a tractable solution over all voxels, one option is to rely on a decomposition of behavioral data to build a new (smaller) set of regressors that reflect specific dimensions with minimal (factor analysis) to no correlation (principal component analysis) between them. Using such approaches has also the advantage to have predictors with minimized variance, i.e., likely reduced measurement errors.

9 Common Covariates of No Interest

Depending on the research question at hand, several confounding variables can be included in the design matrix. A few other covariates, related to brain size, should also be included: age, sex, and total intracranial volume (TIV). The exact relationships between global and local morphology to age and sex are not known, but these factors are known to affect gray and white matter tissues and thus should be included. Because morphometric analyses are interested in local changes, it is also recommended to control for the TIV and/or related measurements [17]. For instance, we can include TIV to account for total head size or the total gray matter (TGM) or total white matter (TWM) to account for tissue-specific global effects. However, not all of these covariates are needed, and their inclusion/exclusion must be justified. As discussed in [18], adjusting data using TIV eliminates differences due to sex, making that variable irrelevant. Similarly, adjusting for age and TGM approximates TIV because age and TGM are correlated in normal subjects. Such choice is however invalid in neurodegenerative disease since TGM does not follow TIV. In all case, one global measure is likely to be included, and there are different ways to perform this control [19]:

1. *Local covariation*: the model assumes that observed changes are explained by the sum of a global and a local effect. The global effect (TIV, TGM, or TWM) is however scaled differently at each voxel/ROI, which implies that one accepts the hypothesis that the global effect affects different regions differently. For instance, one brain region can be smaller as head size increases, while another region can be bigger. This approach is computed by simply adding the global measure as covariate in the GLM. The results are interpreted as significant changes, in addition of, or despite changes in the global.
2. *Global scaling*: the model assumes changes are explained by a local effect that scales beyond changes in the global measurement. Here, the global effect is accounted for identically for every voxel/ROI by simply dividing them by the global estimate. A significant effect of a regressor indicates changes that are stronger than changes in the global measurement. This approach is particularly useful when changes in TGM or TWM are known and/or compete with the local effect tested like in, e.g., aging.
3. *Local scaling*: the model assumes changes are explained by a local effect that scales beyond changes in the global measurement as for global scaling. This effect however scales differently at each voxel/ROI as in the local covariation approach. This is computed by dividing the data by the local mean values. Compared

to the local covariation approach, this allows looking for effects that scale beyond global changes, rather than in addition of the global effect. Since the adjustment is local, it seems that part of the effects observed (compared to the local covariation) can be attributed to partial volumes rather than uniquely attributed to a global confound.

In general, local covariation has shown to be more effective to control for the global effect of head size (see references in [17]), but scaling methods allow to answer slightly different questions and have therefore merits of their own. No matter what model is used, it is important to check effects with and without covariates of no interest (in particular TIV or related), to better understand observed changes [20]. Good reporting practices are there to (a) say if an effect is there with/without the covariates; (b) report standardized regression coefficient ($\text{std}(X)/\text{std}(y) * \beta$) along with statistical values, allowing quantitative comparison across studies; and (c) also report changes in terms of effect size (e.g., raw gray matter values) to provide a more direct biological interpretation.

10 Discussion

As presented above, the GLM is a mathematical tool of great flexibility allowing one to combine categorical (e.g., patients vs. controls) and continuous variables (e.g., age, TIV), to test multiple hypotheses within the same model, thus accounting for confounding effects. Results from such an approach must however be considered carefully because (a) of the assumptions of the normality, of independence, and of linearity (see, e.g., [21] for nonlinear effects); (b) multiple models are often possible and can give different results, i.e., the model reflects the experimenter's hypotheses and assumptions about the population(s); and (c) significant effects of regressors are not always good biomarkers or predictors. This last point is of utmost clinical importance. The GLM is useful to understand how brain structures linearly covary with behavioral measures, but it does not provide information of causation or diagnosis. Causation can only be established by design, as, for instance, in a randomized controlled interventional study (pre- and post-treatment with placebo), showing that changes in behavior after treatment correspond to changes in brain structure in treated patients but not placebo. Similarly, to establish that a behavioral variable or a brain region is a good biomarker, cross-validation techniques must be used, and a new independent subject sample must be tested to validate the prediction because significant GLM regressors are not necessarily good predictors [22].

Acknowledgments

Thank you to Ged Ridgway for providing useful references related to variables of no interest and reviewing the manuscript and to David Raffelt for pointing out difference between looking at voxel content (VBM/TBSSS) and morphometry per se (looking at shapes).

References

1. Christensen R (2002) Plane answers to complex questions. The theory of linear models, 3rd edn. Springer, New-York
2. Ashburner J, Friston KJ (2000) Voxel-based morphometry—the methods. NeuroImage 11:805–821
3. Smith SM, Jenkinson M, Johansen-Berg H et al (2006) Tract-based spatial statistics: voxelwise analysis of multi-subject diffusion data. NeuroImage 31:1487–1505
4. Frisoni GB, Ganzola R, Canu E et al (2008) Mapping local hippocampal changes in Alzheimer's disease and normal ageing with MRI at 3 Tesla. Brain J Neurol 131:3266–3276
5. Worsley KJ, Marrett S, Neelin P et al (1996) A unified statistical approach for determining significant signals in images of cerebral activation. Hum Brain Mapp 4:58–73
6. Hayasaka S, Phan KL, Liberzon I et al (2004) Nonstationary cluster-size inference with random field and permutation methods. NeuroImage 22:676–687
7. Smith S, Nichols T (2009) Threshold-free cluster enhancement: addressing problems of smoothing, threshold dependence and localisation in cluster inference. NeuroImage 44:83–98
8. Salimi-Khorshidi G, Smith SM, Nichols TE (2009) Adjusting the neuroimaging statistical inferences for nonstationarity. Med Image Comput Comput Assist Interv 12. Springer:992–999
9. Salimi-Khorshidi G, Smith SM, Nichols TE (2011) Adjusting the effect of nonstationarity in cluster-based and TFCE inference. NeuroImage 54:2006–2019
10. Ridgway GR, Henley SMD, Rohrer JD et al (2008) Ten simple rules for reporting voxel-based morphometry studies. NeuroImage 40:1429–1435
11. Huang L, Rattner A, Liu H, Nathans J (2013) How to draw the line in biomedical research. elife 2:e00638
12. Salmond CH, Ashburner J, Vargha-Khadem F et al (2002) Distributional assumptions in voxel-based morphometry. NeuroImage 17:1027–1030
13. Scarpazza C, Sartori G, De Simone MS, Mechelli A (2013) When the single matters more than the group: very high false positive rates in single case voxel based morphometry. NeuroImage 70:175–188
14. Sink KM, Craft S, Smith SC et al (2015) Montreal cognitive assessment and modified mini mental state examination in African Americans. J Aging Res 2015:872018
15. Hutcheon JA, Chiolero A, Hanley JA (2010) Random measurement error and regression dilution bias. BMJ 340:c2289
16. Kraha A, Turner H, Nimon K et al (2012) Tools to support interpreting multiple regression in the face of multicollinearity. Front Psychol 3:44
17. Malone IB, Leung KK, Clegg S et al (2015) Accurate automatic estimation of total intracranial volume: a nuisance variable with less nuisance. NeuroImage 104:366–372
18. Henley SMD, Ridgway GR, Scahill RI et al (2010) Pitfalls in the use of voxel-based morphometry as a biomarker: examples from huntington disease. AJNR Am J Neuroradiol 31:711–719
19. Peelle JE, Cusack R, Henson RNA (2012) Adjusting for global effects in voxel-based morphometry: gray matter decline in normal aging. NeuroImage 60:1503–1516
20. Tu Y-K, Gunnell D, Gilthorpe MS (2008) Simpson's paradox, Lord's paradox, and suppression effects are the same phenomenon--the reversal paradox. Emerg Themes Epidemiol 5:2
21. Pernet CR, Poline J, Demonet J, Rousselet GA (2009) Brain classification reveals the right cerebellum as the best biomarker of dyslexia. BMC Neurosci 10:67
22. Lo A, Chernoff H, Zheng T, Lo S-H (2015) Why significant variables aren't automatically good predictors. Proc Natl Acad Sci 112:13892–13897

Part II

Brain Morphometry: Non Clinical Applications

Check for updates

Chapter 6

Relating High-Dimensional Structural Networks to Resting Functional Connectivity with Sparse Canonical Correlation Analysis for Neuroimaging

Brian B. Avants

Abstract

Human brain mapping is increasingly faced with the need to efficiently interrogate small sample size, but high-dimensional ("short and wide") data-sets. Such data may derive from rare or difficult to identify populations wherein we seek to detect subtle network changes that precede disease. Few prior hypotheses may exist in these cases and yet, due to small sample size, exploratory analysis is power challenged. We overview how to use sparse canonical correlation analysis to produce biologically principled low-dimensional representations before proceeding to hypothesis testing. This strategy conserves power by taking advantage of the underlying neurobiological covariation across modalities to compress large data-sets. We provide an example that maps voxel-wise cortical thickness measurements to resting state network correlations in order to identify structure-function sub-networks with little further supervision. The resulting network-like, sparse basis functions allow one to predict traditional univariate outcomes from multiple neuroimaging modalities even when sample sizes are relatively small. Importantly, these data-driven functions are anatomically and edge-wise specific, allowing a nearly traditional neuroscientific interpretation.

Key words Dimensionality reduction, Sparse canonical correlation analysis, SCCAN, Eigenanatomy, Brain, Neuroimaging, Resting state, Network, Thickness, Hubs, DiReCT

1 Introduction

Sparse canonical correlation analysis is gaining popularity within biomedical analysis as a tool for handling multiview high-dimensional data-sets [1–9]. Perhaps the first effort to extend Hotelling's original work [10] to sparse solutions involved relating specific words within musical annotations directly to specific sounds [11] by adding ℓ_1 regularization. More recently, the method is gaining traction in imaging genomics to relate imaging phenotype

Dr. Avants recently became a Biogen employee.

Gianfranco Spalletta et al. (eds.), *Brain Morphometry*, Neuromethods, vol. 136,
https://doi.org/10.1007/978-1-4939-7647-8_6, © Springer Science+Business Media, LLC 2018

to genotype, potentially by taking advantage of prior knowledge in the form of structured sparsity [9].

Despite these new applications, the advantages of CCA remain the same today as in 1936: CCA allows one to symmetrically relate sets of variables to each other. Specifically, consider two matrices, X (dimension $\times p$) and Y (dimension $n \times q$). If we want a direct univariate comparison between the columns of these matrices, then we would have to perform on the order of $p \times q$ pairwise calculations. Clearly, if p and q are large (e.g. greater than several thousand), then this results in a very large computational burden. Because neuroimaging information is often highly redundant, this univariate comparison forces many more individual comparisons than should be necessary to gain an understanding of the relationship between these measurements. In contrast, CCA can make this comparison with a single multivariate calculation:

$$[x^\star, y^\star] = \text{argmax}_{x,y} \ \text{Corr}(Xx, Yy)$$

where Corr represents Pearson correlation and $x^\star$ and $y^\star$ are the optimal solution vectors (or matrices if we are computing multiple canonical variates.) The variables x and y project X and Y into two $n \times 1$ vectors. While it is clear that the directly multivariate approach gains power by avoiding multiple comparisons, one compromises interpretability. That is, the factors that traditional CCA produces are dense. In analogy to linear regression, the "β weights" are non-zero on both sides of the equation.

The need to increase interpretability motivates sparse CCA, which allows x and y to have entries, which are "mostly zero" and therefore have more specificity. This is a useful property in neuroscience, which seeks to understand the hierarchical and (somewhat) segregated nature of the brain (and focal impact of brain disorders). Sparse CCA allows one to powerfully address, for example, questions such as how patterns of variability in cortical structure relate to patterns of resting activity, as measured by functional magnetic resonance imaging. This type of question is fundamental to understanding not only how structure recapitulates function but also how changes in the organization of the brain may precede the onset of clinical symptoms. Relatively few studies address the question of how functional connectivity directly associates with cortical (or structural) networks [12, 13] where we define a structural network as a set of covarying cortical regions [14]. The relative paucity of studies may relate to the lack of freely available statistical frameworks for addressing this question.

The ability to project large sets of complementary neuroimaging measurements into a low-dimensional and interpretable space is a powerful one for brain mapping. The CCA (and sparse CCA) methodology permits hypotheses about the brain to be framed in new and fundamentally multivariate ways. As we will see below, with proper constraints, sparse CCA can produce solutions that in some sense capture what we think of as networks: sets of measurements that covary, presumably due to underlying neurobiological mechanisms. This mitigates several problems: multiple comparisons correction, perhaps too rigid prior definitions of network characteristics and the need to "translate" artificially independent voxel-based statistical tests into regularized statistical maps (by, for example, cluster constrained post-processing). Below, we will make these statements specific by employing a public data-set to show how we can employ sparse canonical correlation analysis for neuroimaging (SCCAN) [7] to compute a structure-function basis set by associating whole brain cortical thickness directly to resting state network correlation matrices. We use SCCAN to infer the structural patterns that relate to functional connectivity patterns and how these may be used to build an interpretable, predictive regression model (based on multiple neuroimaging modalities) for demographic variables. We detail elements of the software, visualization, interpretation and evaluation with reference to the clinical considerations raised by the freely available Pre-symptomatic Evaluation of Novel or Experimental Treatments for Alzheimer's Disease (PREVENT-AD) data-set [15] on which we focus.

2 Materials

2.1 Imaging Data

We demonstrate our application on the T1-weighted and resting state fMRI neuroimages collected in the PREVENT-AD study described in detail in [15]. The cohort subjects are cognitively normal with average age 65.4 ± 6.3 years (min 55.0, max 84.0); PREVENT-AD includes 58 females and 22 males with documented family history of Alzheimer's disease (AD) but with good general health at enrollment. Subjects will be followed longitudinally to quantify the difference in normal aging and progression to AD in this high-risk group. All 80 subjects are scanned at two different sessions conducted within 111.4 ± 24.3 days of each other. Each session includes 1 mm^3 T1-weighted MRI and resting state functional MRI collected on a 3T Siemens machine as described [15]. The resting state fMRI comprises two 5 min 45 s runs within each session. Due to the relative temporal proximity of the sessions and collection of multiple runs within each session, this data-set presents a valuable opportunity to assess fMRI reliability in this special cohort, as covered in the data source publication. However, we focus only on the first session data, which precedes randomized treatment by Naproxen that could potentially impact the second session data.

2.1.1 Structural Image Processing

We processed each T1 image through the ANTs cortical thickness pipeline [16]. After viewing several example data-sets from the PREVENT-AD cohort, we selected the "Oasis" template from [16] due to its similarity to PREVENT-AD in terms of T1 contrast and defacing protocol. ANTs cortical thickness produces a six tissue segmentation which has been validated with respect to FreeSurfer [16]. This segmentation procedure performs brain extraction followed by tissue segmentation within the brain mask. The brain masking algorithm involves bias correction, two registration steps, segmentation and morphological operations. The probability maps that emerge from this part of the pipeline are passed to a volumetric, voxel-wise cortical thickness measurement algorithm, DiReCT [17]. DiReCT accounts for buried sulci while incorporating tissue-class uncertainties in the thickness estimation step and provides a volumetric alternative that may extract more meaningful variability in comparison to Freesurfer in some cohorts [16]. The thickness data for each subject is transferred to the structural T1 template space. By masking the thickness data with a group-wise cortical mask, we transform the voxel representation to a matrix representation. Below, the thickness data will comprise the "left-hand" view in SCCAN, i.e. the X matrix.

2.1.2 Functional Image Processing

We organize our functional image processing strategy around the Power coordinate system [18], which we will use to generate connectivity matrices of size 270×270. The additional two nodes correspond to the left and right head of the hippocampus. Such matrices allow a compressed representation of whole-brain connectivity patterns [19] indexed at reliable anatomical coordinates derived from meta-analysis in task studies. We motion correct each run for the baseline session in the PREVENT-AD cohort to a subject-specific, iteratively computed times series mean and collect the final transformation parameters for later analysis as part of quality assurance. We also compute nuisance variables within non-cortical tissue via CompCor, keeping ten nuisance predictors as in [20]. We residualize nuisance variables from the BOLD signal within the cerebrum and compute the pairwise correlation between all Power nodes. We subsequently align the mean resting BOLD image to the individual time point brain extracted T1. We then map the resting BOLD data to the group space through a composite transformation. Finally, we generate a vector representation of the connection matrix from the upper triangle. This will, below, enter into Y, i.e. the "right hand" matrix in SCCAN. A subset of individual timepoints of resting state exhibited excessive motion artifact, based on framewise displacement, and were removed from further analysis ($n = 8$). We investigated the reliability of this processing stream and found it to be consistent with results reported in [15].

2.2 Software

We employ R version 3.3.1 ("Bug in Your Hair") for basic statistical processing. We also employ ANTsR version 0.3.3 [21] for core image analysis, data organization and early development efforts. ANTsR, itself, links to ITKR, ANTs and the ITK toolkits. The specific versions of the concomitant libraries are documented within ANTsR software. The only "pure ANTs" tool used in this analysis is the well-validated ants cortical thickness pipeline as above. The remainder of the processing is performed within ANTsR, available both at http://stnava.github.io/ANTsR/ and via Neuroconductor https://neuroconductor.org/. In addition to the PREVENT-AD neuroimaging, experiments in this paper additionally require the boot R package. ANTsR, itself, has several dependencies, which are identified at its webpage.

3 Methods

Sparse canonical correlation analysis for neuroimaging (SCCAN) is a general purpose tool for "two-sided" multiple regression [7]. SCCAN, available within ANTsR [22], exploits covariation across large data-sets while imposing data-set-specific spatial regularization that helps prevent overfitting, especially important when $p \gg n$ i.e. when the number of data measurements is much greater than the number of subjects on which these are measured. SCCAN allows one to symmetrically compare one matrix of data to another and find linear relationships between them in a low-dimensional space, just like CCA, but with additional regularization parameters:

$$x^\star, y^\star = \text{argmax}_{x,y} \frac{xX^T Yy}{\|Xx\|\|Yy\|} - \gamma_x \| G^t_{\sigma x} \star x\|_1^+ - \gamma_y \| G^t_{\sigma y} \star y\|_1^+.$$

Here, $\gamma_\cdot$ is a scalar weighting term for each variate that controls the impact of the ℓ_1^+ penalty on the objective function (here denoted $\| \cdot \|_1^+$). The ℓ_1^+ penalty differs from ℓ_1 in that it sends negative values to zero. The operator $G^t_{\sigma\cdot}$ performs spatial regularization (usually Gaussian or a modified Gaussian) on the solution vectors x or y where the amount of regularization is determined by $\sigma\cdot$. Much like singular value decomposition (SVD), the resulting subspace encodes a basis that may be used to reconstruct the original data-set. Also, like SVD, multiple solution vectors can be obtained from the above optimization, which can be done in parallel or via deflation [1]. In comparison to methods which employ standard ℓ_1 regularization, SCCAN has the benefit of inserting the spatial regularization inside the operator and projecting solutions to the non-negative space. This means that the resulting solution vectors can be interpreted within the original units of the data i.e. similar to regions of interest. *See* **Note 1**, the methods for the optimization of the SCCAN functional are beyond the scope of the current paper. However, they rely on projected gradient descent on the SCCAN objective, as in [23]. See the compressed sensing literature for additional justification of these methods [24–28], which shows conditions under which such methods reach local optima (*see* **Note 2**).

3.1 Primer: Example 2D Data

Before proceeding to a more complex brain mapping study, we introduce a step-by-step example via a structural population data-set available in ANTsR. By normalizing these brain slices to a common template, we can generate intuition about SCCAN covariance patterns via fast and easy to reproduce examples. We employ six subjects and use the first as template.

```
library( ANTsR )
fns = c( "r16", "r27", "r30", "r62", "r64", "r85" )
ref = antsImageRead(getANTsRData( fns[1] ) )
fns = fns[-1] # exclude first image because it is the template
```

We first rigidly register the "moving" images to the template. We then deformably register the rigid results to the template. We also compute the jacobian determinant of the deformation field and the initial difference between the rigidly aligned image and the template. The jacobian features are sent to the X matrix and the difference images to the Y matrix.

```
rX=4  # controls resolution for X matrix
rY=8  # controls resolution for Y matrix
maskX  = getMask( ref )  %>% resampleImage( c( rX, rX ) ) %>% iMath("MD",2)
maskY  = resampleImage( maskX, c( rY, rY ) )
refsub = resampleImageToTarget( ref, maskY )
X = matrix( nrow = length( fns ), ncol = sum( maskX ) )
Y = matrix( nrow = length( fns ), ncol = sum( maskY ) )
for ( i in 1:length( fns ) )
  {
  tar = antsImageRead(  getANTsRData( fns[i] )  )
  reg = antsRegistration( ref, tar, typeofTransform = 'Rigid' )
  tgr = resampleImageToTarget( reg$warpedmovout, maskY )
  reg = antsRegistration( ref, reg$warpedmovout, typeofTransform = 'SyNOnly')
  jac = createJacobianDeterminantImage( ref, reg$fwdtransforms[1], 1 ) %>%
    resampleImageToTarget( maskX )  # the jacobian, mapped to low resolution
  X[ i, ] = jac[ maskX == 1 ]
  reftardif = ( refsub - tgr )  # the initial difference image, low resolution
  Y[ i, ] = reftardif[ maskY == 1 ]
  }
```

The minimal processing above makes clear the construction of the matrix representations and suggests there is no need for X, Y to have the same number of columns. Given a matrix representation of image structure, we can express the relationship between these measurements in a lower dimensional space, via SCCAN. This allows us to assess *the multivariate relationship between the initial difference image and the jacobian determinant*. These should be related because the larger the initial difference, the greater the necessary deformation. Here is the code:

```
ccaMatrixList = list(
  scale(X, center=F, scale=F ),
  scale(Y, center=F, scale=F )) # see Note section for more on these choices
cth = 25 # cluster threshold
sccanToy = sparseDecom2( ccaMatrixList,      # input data
  inmask = c(maskX, maskY ),                 # masks that allow regularization
  sparseness = c(0.25, 0.25),                # 25 percent sparseness
  nvecs = 4,                                 # 4 pseudo-eigenvectors
  smooth = 0.0, cthresh = c(cth, cth),       # spatial regularization
  perms = 50)                                # permutations
```

The parameters passed to the software relate directly to the optimization criterion. More specifically:

- The sparseness entry relates to the terms γ_x, γ_y. However, for user convenience, the units are expressed in terms of target percent sparseness (relative to the size of the mask). In the above example, the pseudo-eigenvectors will be roughly 25% sparse, depending on the optimization outcome.
- The smooth and cthresh entries relate to the *operator* within the ℓ_1 penalty term. Above, we set $G^5_{0.0}$ where the super-scripted t sets isolated voxel clusters below t to zero. Larger settings of t lead to pseudo-eigenvectors that tend toward greater spatial connectivity.
- nvecs sets the number of pseudo-eigenvectors to compute. We also set perms which forces the algorithm to internally permute the data and recompute p times, comparing the original to the permuted correlations. This provides an empirical, if conservative, estimate of significance for each pseudo-eigenvector pair. We can also visualize the first pair of pseudo-eigenvectors on the template. *See* Fig. 1 for an overview.

3.2 Structural Networks and Resting Connectivity in PREVENT-AD

Age and cognition impact structural covariance [14] as well as resting connectivity [29]. If structural networks covary with resting connectivity, then we can exploit SCCAN to identify sparse sub-networks from both types of data that may—in downstream hypothesis testing—relate to age, cognition or, potentially, disease status. To test this question, we follow a "clustering before hypothesis testing" approach similar to that proposed in eigenanatomy [23]. The present works differs in that the low-dimensional projection of the imaging data (dimensionality reduction) is based on

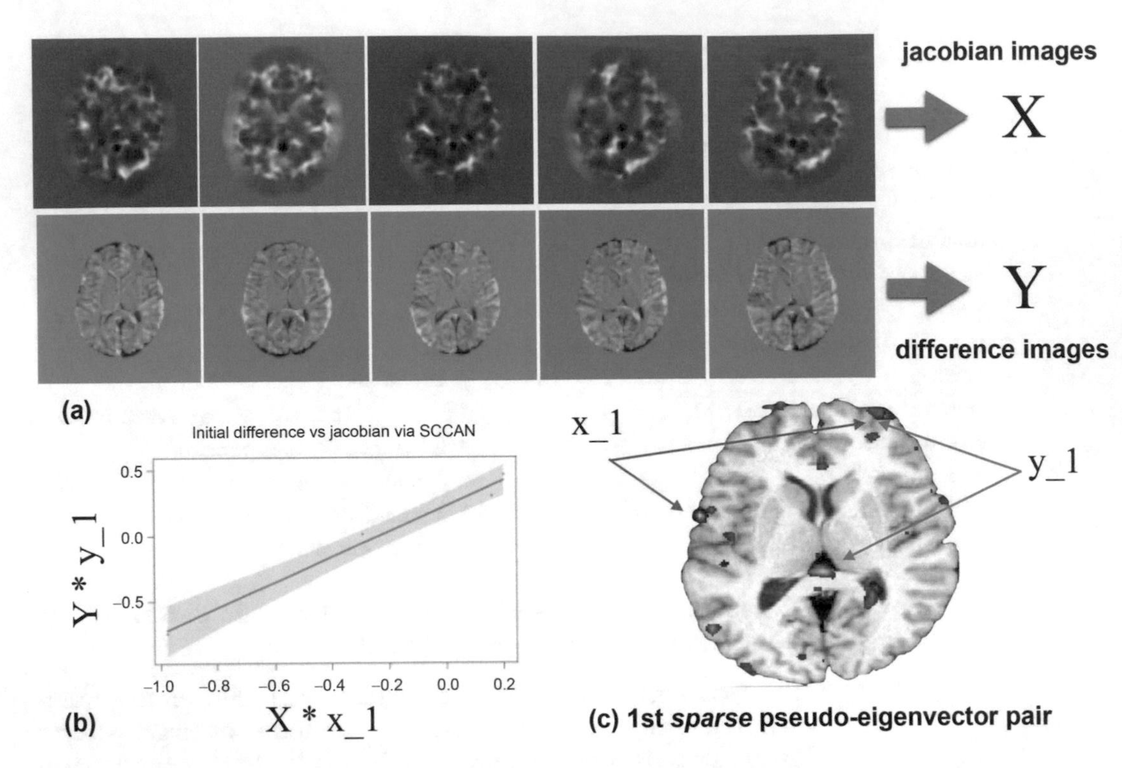

Fig. 1 The input jacobian and difference images are in panel (**a**). The low-dimensional correlation produced by SCCAN is in panel (**b**). Panel (**c**) shows the first pseudo-eigenvector overlaid on the template. The first pseudo-eigenvector pair captures the majority of the covariation and subsequent solutions are less powerful. The overlay on the template identifies the regions (in magma colormap) that maximize the correlation with the intial difference (in viridis colormap). Finally, note that classic CCA would not be able to compute such an interpretable solution. First, it is constrained to the case where $n > p$ and, second, solutions would be both signed and everywhere non-zero (i.e. not sparse)

covariation across modalities rather than within a single modality. Our study design is therefore similar to principal component regression. We first perform dimensionality reduction on the high-dimensional data and subsequently build a general linear model that relates the low-dimensional representation to population variables. Principal component regression uses the SVD to reduce the data and may plug the left eigenvectors into a regression model which may be investigated for significance. The procedure below is similar but uses SCCAN instead of SVD and evaluates the validity of our hypothesis using both model significance as well as cross-validation.

The public version of the PREVENT-AD data-set releases very limited demographic information, primarily age and gender. We use this data to test the hypothesis that multivariate covariation between cortical thickness networks and resting correlation matrices (assessed in the Power nodal system) will identify age-related sub-networks. Furthermore, we will assess—via cross-validation—whether these sub-networks improve the cross-validated prediction of age when used together in the same model. That is, we hope to show that cortical thickness and resting connectivity provide complementary information regarding brain aging in this cohort.

The input data, as in the toy example, comprises two sets of high-dimensional measurements. Given our preprocessed data, we must now consider the necessary steps for running SCCAN and performing hypothesis testing in PREVENT-AD.

3.2.1 Step 1: Define the X, Y Matrix Content

Preprocessing yielded normalized cortical thickness images. Consistent with the goals of our hypothesis, we organize this data into matrix form for use in SCCAN. A key consideration, here, is which voxels to include. This provides the scientist the opportunity to restrict to prior-defined regions and to investigate covariation only within these regions. However, here, we are interested in identifying cortical networks that may cover large spans of the brain and, as such, we choose to be inclusive. We create a mask from the average of the normalized cortical thickness images, thresholded at level 0.5. This results in an X matrix with 784,557 columns and 72 rows (eight subjects were removed from the resting data due to excessive motion). The thickness images are first smoothed with a 6 mm Gaussian kernel before being passed into the matrix format.

The input data for the Υ matrix is the upper triangle of the Power node correlation matrix. Some may recommend passing data through Fisher's z-transformation but we employ generalized scaling procedures described below. This results in $\frac{n^*(n-1)}{2}$ entries (36,315). We do not require a mask for the resting state network correlation data because it is, in effect, already spatially regularized. Note that it may be reasonable to induce a structured sparsity on the matrix, given that nodes are members of suspected functional systems. We forego this option because we are primarily interested in identifying new patterns from whole brain connectivity that may capture cross-system changes relating to cortical structure.

3.2.2 Step 2: Determine SCCAN Hyperparameters

This joint X, Υ matrix pair enables us to use SCCAN directly. However, we must first set a few parameters, as noted in the prior example.

- Matrix scaling: This is an often overlooked component of dimensionality reduction. We recommend the default option of simply centering the data before passing to SCCAN. This avoids forcing all columns to have unit variance, which may inflate the impact of noisy data sources (*see* **Note 3** for a robust alternative).
- Embedded smoothing within the optimization: Regularization acts as a hedge against overfitting. In SCCAN, this involves not only the sparsity levels but also setting the values for G^t_σ, which is the operator within the ℓ_1 norm. In this example, we only need to consider the impact on the thickness pseudo-eigenvectors.

In general, we recommend setting the σ value to be 0.5 for voxel data, which provides sufficiently smooth results in nearly every example we have encountered, even when little to no prior smoothing is employed. We also choose a cluster threshold (the t parameter) of 1000 which focuses results on the scale of roughly a ten voxel cube (a 1 cm^3 volume, given 1 mm input data.)

- Setting the number of pseudo-eigenvectors: As the goal of this study is to ultimately use the computed pseudo-eigenvectors within a general linear model, we elect to compute relatively few pairs. Here, we set nvec = 6 and will only employ the top k with correlations greater than a given threshold in our evaluation below. Computation time is also a concern, here, due to the relatively high dimensional data for which we want a compressed and interpretable representation (SCCAN computation scales roughly linearly with nvec, given fixed matrices.)
- Searching for the optimal sparsity settings: Choosing hyperparameters can be challenging unless one has substantial experience with a methodology or specific prior hypotheses about a dataset. Given the relative novelty of this application, we cannot explicitly recommend a given sparsity level for the computation of regularized sparse canonical correlation. In this case, our best option is to exploit the low cost of modern computing and search over a reasonable range of potential sparsity values. We define "reasonable" a set of sparsity levels for which we expect to interpret the derived networks. In several prior SCCAN studies, we have searched over a sparsity parameter space that allows solutions to range from very sparse (1%) up to 15% of data coverage. In the latter case, we can expect a component to cover roughly 15% of the voxels. Our recommended spatial regularization, above, also will restrict the solution to be spatially contiguous down to the selected scale. In this application, we search regular increments of sparsity from 0.01 to 0.15 as a fraction of the number of columns in X or Y. If we parcellate the search region into K equal segments, this leads to a $K \times K$ parameter search space which we send to a compute cluster.

3.2.3 Step 3: Assess the Validity of the SCCAN Models

Here, we make reference not to the significance of the SCCAN results but to their interaction with the target biological hypothesis. The cross-validation model will compare three linear regression models:

$$\text{Age} = \sum_{i=1}^{k} \beta_i^x \, X x_i + \text{Sex} + \epsilon$$

$$\text{Age} = \sum_{i=1}^{k} \beta_i^y \, \Upsilon y_i + \text{Sex} + \epsilon$$

$$\text{Age} = \sum_{i=1}^{k} (\beta_i^x \, X x_i + \beta_i^y \, \Upsilon y_i) + \text{Sex} + \epsilon$$

We use k-folds cross-validation to assess the relative accuracy of each equation rather than using model-based interpretation of these prediction equations. In short, we ask how well—in left out data—each model can predict age based on a subset of training data. We repeatedly perform 18-fold validation (training on 75% and testing on 25% data) in order to gain a distribution of performance for each model, using the boot package in R. This enables us to find out which set of sparseness parameters leads to the most predictive model. In essence, this searches over the space of possible hypotheses regarding structure-function-age interactions. While it does introduce a multiple comparisons problem, it is only on the order of the number of searched models, rather than the number of voxels or number of components. Thus, it remains fairly conservative. Alternative methods for assessing the validity of hyperparameters and candidate solutions are discussed in **Notes 4** and **5**.

3.2.4 Step 4: Visualize and Interpret the Multivariate Results

The results of our cross-validation parameter search (*see* Fig. 2) show that the best model uses both modalities and yields good predictive value with an average error of between 4.5 and 5.0 years. In comparison, either the thickness or resting correlation model alone yields error of over 5.5 years, a 1 year reduction. While it is unclear if this is a clinically meaningful measurement of "brain age", it does provide evidence that resting state correlations change in concert with structural networks along the aging spectrum. The model, which explains over 50% of the age variance, is not improved by the inclusion of mean or max framewise displacement or brain volume as covariates. The improvement due to both modalities may be best illustrated by the performance histograms shown in Fig. 3. The best model used sparseness 0.075 for thickness and 0.125 for resting bold. Table 1 shows its regression results on the whole dataset.

One key advantage of the regularization operator $\| G_\sigma^t \star \cdot \|_1^+$ is that it leads to unsigned pseudo-eigenvectors that may be displayed and interpreted with anatomical specificity. Figure 2 shows that there are two networks implicated as "significant" in the best linear model. The two networks Thk-e2 (cool) and Thk-e4 (hot) (the second and fourth pseudo-eigenvectors for thickness) have alternating sign indicating that it is a difference of thickness in these networks that is predictive of age. Similarly, resting state pseudo-eigenvectors contribute significantly to the model regression. These also have different sign, again implying that differences in patterns of connectivity relate to the aging process. These patterns are also visualized in Fig. 2, where we highlight (in the brain

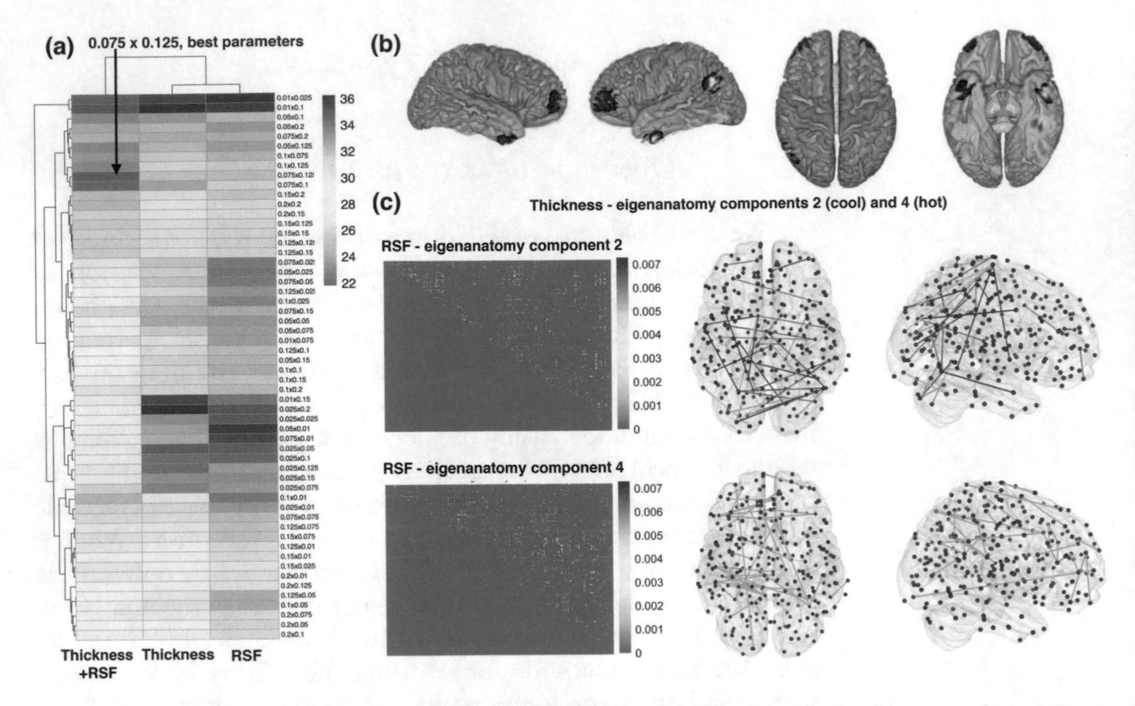

Fig. 2 Panel (**a**) visualizes the results of the hyperparameter search where we display the squared error over a selection of the parameter space and for both modalities (left column), the thickness modality alone (center column) and the resting state network features alone (right column). *Darker blue* indicates better performance. Panel (**b**) shows the best model's two most significant thickness pseudo-eigenvectors overlaid on the template image. Panel (**c**) shows the sub-networks extracted from SCCAN in terms of both a heatmap and within the glass brain space. Power nodes are shown along with the most heavily weighted edges in the SCCAN pseudo-eigenvectors. Again, we display the two most significant resting network pseudo-eigenvectors in the dual modality model

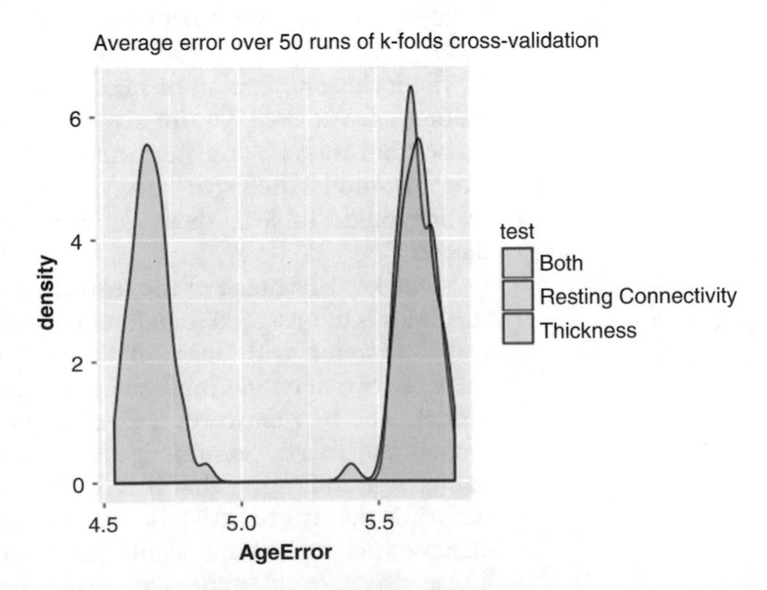

Fig. 3 Cross-validation results establish that the best model's performance shows a clear advantage of using both modalities together to predict subject age

Table 1
The regression results for the best model

	Predictor	Beta	SE	*T*-stat	*P*-value
2	Sex	−1.2	1.3	−0.92	0.36
3	Thk_e1	−3.0	14.8	−0.20	0.84
4	Thk_e2	−53.2	18.6	−2.86	0.01
5	Thk_e3	18.9	14.9	1.27	0.21
6	Thk_e4	55.3	12.9	4.28	0.00
7	RSF_e1	−36.0	16.4	−2.19	0.03
8	RSF_e2	56.9	13.0	4.38	0.00
9	RSF_e3	−3.7	14.8	−0.25	0.80
10	RSF_e4	−54.7	13.7	−4.00	0.00

Thk indicates thickness predictors; *RSF* indicates the resting correlation matrix predictors

Table 2
The anatomical coordinates for the two most important thickness pseudo-eigenvectors in the best model

NetworkID	*x*	*y*	*z*	Brodmann	AAL	pval
MTL	−30	−41	−22	37	Fusiform_L	0.01
MTL	37	−33	−23	20	Fusiform_R	NA
MTL	4	−44	48	0	Precuneus_R	NA
Anterior-angular	36	47	8	10	Frontal_Mid_R	0.00
Anterior-angular	−34	52	2	47	Frontal_Mid_L	NA
Anterior-angular	50	−64	29	39	Angular_R	NA
Anterior-angular	47	−2	−37	20	Temporal_Inf_R	NA
Anterior-angular	−41	2	−37	20	Temporal_Inf_L	NA

Recall that these are multivariate regions of interest. This table identifies the key loci detailing the spatial extent of the pseudo-eigenvector (referred to in the table as *_node). The NetworkID column shows an overall descriptive name differentiating each pseudo-eigenvector. The *x*, *y*, *z* coordinate columns show the MNI space coordinates. The Brodmann and AAL columns give an approximate anatomical region/label for each node. The pval column shows the pval for the overall eigenvector and shows NA for the sub-nodes

rendering) only the strongest of the sparse correlations shown in the heatmap.

Finally, these results indicate that aging impacts both structural hubs and resting connectivity in PREVENT-AD. Table 2 highlights regions and coordinates of key thickness components. In cortical thickness, one bilateral sub-component of Thk-e4 is likely to be part of a network convergence site (hub) in the inferior parietal lobe, the angular gyrus. This component's bilateral anterior and inferior temporal region may be a semantic hub. The anterior portion of the middle frontal gyrus may relate to working memory and perhaps be a hub for attention networks. Thk-e2, on the other hand, focuses more on the medial temporal lobe network. Das et al. describe the MTL network as "an area with extensive bi-directional connections to the rest of the brain, and thus often thought to integrate complex sets of information from multiple sensory modalities" [30] but also highlights a second network hub, the precuneus. Visualization of the resting state components reveals that changes in superior and anterior resting connectivity are relevant in one component. In the other component, we see a pattern highlighting posterior connectivity. Overall, we conclude that age-related structural changes in hub regions may relate to age-related changes in network connectivity in this PREVENT-AD data-set, a cohort at high risk for imminent AD. Whether these findings would be replicated in a purely control cohort is not yet known. Further work is therefore needed in larger data-sets to confirm the relevance of these findings and the value of this methodology, more generally.

4 Notes

1. Consider that a set of regions of interest may be considered as two-level functions in a common voxel space. Then, we represent k regions of interest as an "ROI matrix" μ of dimension $k \times p$. If population data is normalized to a common space (an assumption behind our X and Y matrices), then the region of interest totals may be computed via $r_X = X\mu^T$. This operation is analogous to what we achieve with SCCAN via data-driven dual image decomposition.
2. Iterative thresholding methods are known to be sensitive to initialization. ANTsR therefore provides several initialization options. See initializeEigenanatomy and options that allow SCCAN to be guided by spatial priors.
3. Robust data transformations can address the sensitivity of correlation-based analyses to outliers [31]. ANTsR provides a function robustMatrixTransform which transforms each datapoint by a rank transform and thus acts as a general purpose and easy to use safeguard against such issues (although we note that careful data inspection is always a superior approach). Referring

back to our example data-set, we can see how this option can be used to create a zero-centered rank transform:

```
mat  = replicate( 2, rnorm( 5 ) )
rmat = robustMatrixTransform( mat )
```

4. Permutation is, at times, recommended for evaluating significance [1]. However, in our experience this may be anti-conservative in $p \gg n$ data-sets. An effective alternative is to employ cross-validation of the correlation with permutation. That is, rather than asking "how well do these correlations hold up under permutation?", one might ask "how does cross-validation performance change between real and permuted data?" This latter strategy may be more informative in some data-sets.
5. At times, it may be useful to visually investigate candidate solutions generated by machine learning algorithms before passing them on to a biological model. The researcher might treat the SCCAN output as "suggested hypotheses." Candidate solutions may be rejected before passing them down to the stage wherein they are tested against variables that will compromise power (age in the example above).

References

1. Witten DM, Tibshirani R, Hastie T (2009) A penalized matrix decomposition, with applications to sparse principal components and canonical correlation analysis. Biostatistics (Oxford, England) 10:515–534
2. Avants BB, Cook PA, Ungar L et al (2010) Dementia induces correlated reductions in white matter integrity and cortical thickness: a multivariate neuroimaging study with sparse canonical correlation analysis. NeuroImage 50:1004–1016
3. Avants B, Cook PA, McMillan C et al (2010) Sparse unbiased analysis of anatomical variance in longitudinal imaging. In: Medical image computing and computer-assisted intervention: MICCAI, International Conference on Medical Image Computing and Computer-Assisted Intervention, vol 13. Springer, Berlin, Heidelberg, pp 324–331
4. Chalise P, Batzler A, Abo R et al (2012) Simultaneous analysis of multiple data types in pharmacogenomic studies using weighted sparse canonical correlation analysis. Omics 16:363–373
5. Duda JT, Detre JA, Kim J et al (2013) Fusing functional signals by sparse canonical correlation analysis improves network reproducibility. In: Medical image computing and computer-assisted intervention: MICCAI, International Conference on Medical Image Computing and Computer-Assisted Intervention, vol 16. Springer, Berlin, Heidelberg, pp 635–642
6. Lin D, Calhoun VD, Wang Y-P (2014) Correspondence between fMRI and snp data by group sparse canonical correlation analysis. Med Image Anal 18:891–902
7. Avants BB, Libon DJ, Rascovsky K et al (2014) Sparse canonical correlation analysis relates network-level atrophy to multivariate cognitive measures in a neurodegenerative population. NeuroImage 84:698–711
8. Fang J, Lin D, Schulz SC et al (2016) Joint sparse canonical correlation analysis for detecting differential imaging genetics modules. Bioinformatics (Oxford, England). 32:3480–3488
9. Du L, Huang H, Yan J et al (2016) Structured sparse canonical correlation analysis for brain imaging genetics: an improved graphnet method. Bioinformatics (Oxford, England) 32:1544–1551
10. Hotelling H (1936) Relations between two sets of variates. Biometrika 28(377):321
11. Torres DA, Turnbull D, Sriperumbudur BK et al (2007) Finding musically meaningful

words by sparse CCA. In: Neural information processing systems (nips) workshop on music, the brain and cognition

12. Romero-Garcia R, Atienza M, Cantero JL (2014) Predictors of coupling between structural and functional cortical networks in normal aging. Hum Brain Mapp 35:2724–2740
13. Marstaller L, Williams M, Rich A et al (2015) Aging and large-scale functional networks: white matter integrity, gray matter volume, and functional connectivity in the resting state. Neuroscience 290:369–378
14. Khundrakpam BS, Lewis JD, Reid A et al (2017) Imaging structural covariance in the development of intelligence. NeuroImage 144:227–240
15. Orban P, Madjar C, Savard M et al (2015) Test-retest resting-state fMRI in healthy elderly persons with a family history of Alzheimer's disease. Sci Data 2:150043
16. Tustison NJ, Cook PA, Klein A et al (2014) Large-scale evaluation of ants and freesurfer cortical thickness measurements. NeuroImage 99:166–179
17. Das SR, Avants BB, Grossman M et al (2009) Registration based cortical thickness measurement. NeuroImage 45:867–879
18. Power JD, Cohen AL, Nelson SM et al (2011) Functional network organization of the human brain. Neuron 72:665–678
19. Shirer WR, Ryali S, Rykhlevskaia E et al (2012) Decoding subject-driven cognitive states with whole-brain connectivity patterns. Cereb Cortex 1991(22):158–165
20. Shirer WR, Jiang H, Price CM et al (2015) Optimization of rs-fMRI pre-processing for enhanced signal-noise separation, test-retest reliability, and group discrimination. NeuroImage 117:67–79
21. Tustison NJ, Shrinidhi KL, Wintermark M et al (2015) Optimal symmetric multimodal templates and concatenated random forests for supervised brain tumor segmentation (simplified) with ANTsR. Neuroinformatics 13:209–225
22. Avants BB, Duda JT, Kilroy E et al (2015) The pediatric template of brain perfusion. Sci Data 2:150003
23. Kandel BM, Wang DJJ, Gee JC et al (2015) Eigenanatomy: sparse dimensionality reduction for multi-modal medical image analysis. Methods 73:43–53
24. Donoho DL, Tsaig Y, Drori I et al (2012) Sparse solution of underdetermined systems of linear equations by stagewise orthogonal matching pursuit. IEEE Trans Inf Theory 58:1094–1121
25. Beck A, Teboulle M (2009) A fast iterative shrinkage-thresholding algorithm for linear inverse problems. SIAM J Imaging Sci 2:183–202
26. Bredies K, Lorenz DA (2008) Linear convergence of iterative soft-thresholding. J Fourier Anal Appl 14:813–837
27. Blumensath T, Davies ME (2008) Iterative thresholding for sparse approximations. J Fourier Anal Appl 14:629–654
28. Herrity KK, Gilbert AC, Tropp JA (2006) Sparse approximation via iterative thresholding. In: Acoustics, speech and signal processing. icassp 2006 proceedings. 2006 ieee international conference on, pp III–III IEEE
29. Franzmeier N, Buerger K, Teipel S et al (2017) Cognitive reserve moderates the association between functional network anti-correlations and memory in mci. Neurobiol Aging 50:152–162
30. Das SR, Pluta J, Mancuso L et al (2015) Anterior and posterior MTL networks in aging and MCI. Neurobiol Aging 36(Suppl 1):S141.e1–S150.e1
31. Wilms I, Croux C (2016) Robust sparse canonical correlation analysis. BMC Syst Biol 10:72

Check for updates

Chapter 7

Source-Based Morphometry: Data-Driven Multivariate Analysis of Structural Brain Imaging Data

Cota Navin Gupta, Jessica A. Turner, and Vince D. Calhoun

Abstract

This chapter discusses a now established linear multivariate technique called source-based morphometry (SBM), a data-driven multivariate approach for decomposing structural brain imaging data into commonly covarying components and subject-specific loading parameters. It has been used to study neuroanatomic differences between healthy controls and patients with neuropsychiatric diseases. We start by discussing the advantages of data-driven multivariate techniques over univariate analysis for imaging studies. We then discuss results from a range of recent imaging studies which have successfully applied this linear technique. We also present extensions of this framework such as nonlinear SBM, morphometric analysis using independent vector analysis (IVA), and related approaches such as parallel independent component analysis with reference (pICA-R). This chapter thus reviews a wide range of multivariate, data-driven approaches which have been successfully applied to brain imaging studies.

Key words Independent component analysis (ICA), Source-based morphometry (SBM), Multivariate analysis, Genome-wide association, Voxel-based morphometry (VBM), Univariate analysis, Nonlinear independent component analysis (NICE), Independent vector analysis (IVA)

1 Source-Based Morphometry for Neuroimaging

Source-based morphometry (SBM) [1] is a data-driven algorithm which provides a multivariate extension to voxel-based morphometry [2] using independent component analysis (ICA) [3]. SBM provides patterns of common variation (e.g., gray or white matter patterns) among participants [1]. SBM is an attractive multivariate alternative to univariate voxel-based morphometry (VBM) [2] as it combines information across different voxels using ICA and performs testing on the subject covariation of these patterns rather than testing each voxel separately. As such, it preserves spatial correlation between different brain regions [4, 5]. For neuroimaging it does not require a priori selection of regions to analyze and

Jessica A. Turner and Vince D. Calhoun contributed equally to this work.

Gianfranco Spalletta et al. (eds.), *Brain Morphometry*, Neuromethods, vol. 136,
https://doi.org/10.1007/978-1-4939-7647-8_7,

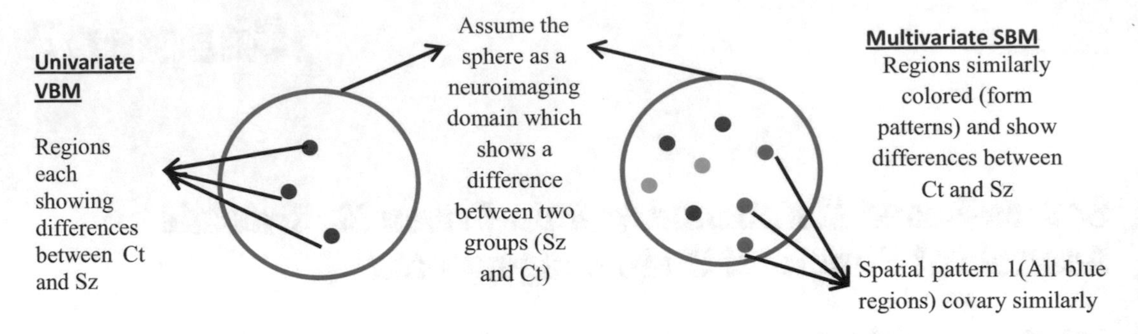

Fig. 1 Comparison of univariate and multivariate analysis [5] in a single neuroimaging modality

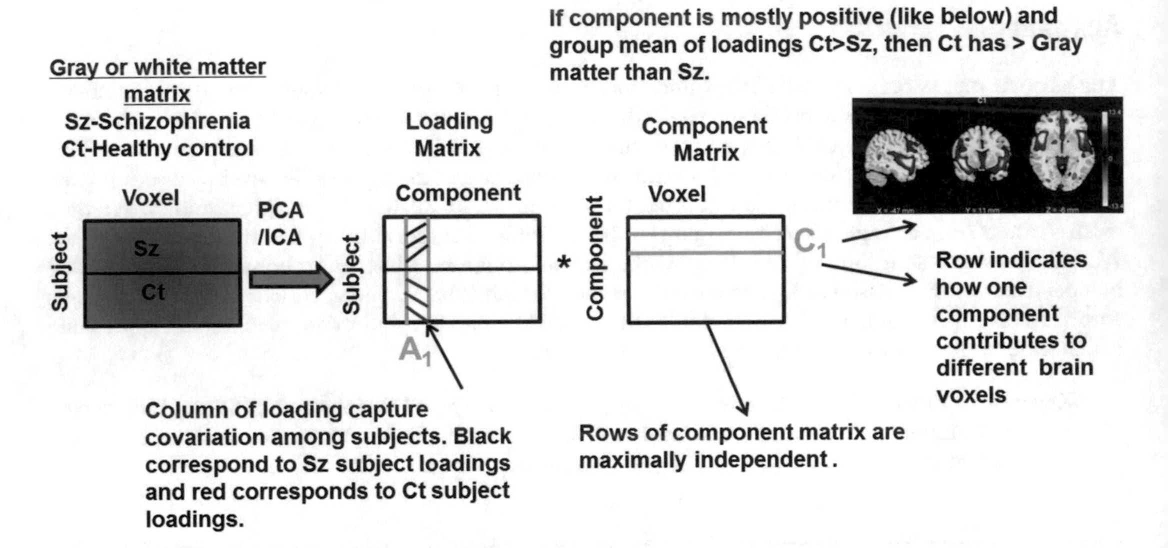

Fig. 2 Source-based morphometry algorithm [1] illustrated. A two-sample t-test is done on each column of loading coefficients using a corrected threshold of $p < 0.05$ (which controls for the false discovery rate [FDR]), thereby indicating the corresponding components which show group difference. Associations with medication and symptoms can then be performed on A_1 using correlation. The map C_1 is common for all subjects with column A_1 having weights

acts as a spatial filter. The difference between these two established methods when applied on the same dataset is pictorially depicted in Fig. 1 for a single modality (e.g., imaging) when considering two groups (e.g., Sz (patients with schizophrenia) and Ct (healthy controls)) [5].

A high-level summary of the SBM algorithm for structural imaging is depicted in Fig. 2. The assumption is that the structural imaging data have been preprocessed using VBM to provide a voxel-wise map of gray matter concentration or volume or a white matter measure at each voxel in the image for each subject. Each row of the matrix is a flattened image from one subject, with the voxels from the 3D image unwrapped into a single row. Sz refer to patients with schizophrenia, while Ct refers to healthy controls. ICA is applied to a gray or white matter subject-by-voxel matrix, resulting in a loading matrix and component matrix [1]. The columns (in the loading matrix) and rows (in the component matrix) are usually considered in pairs (in green). The row C_1 translates into the component map and is an indication of how one component contributes to different brain voxels. A two-sample t-test can then be performed on each column of loading coefficients, typically using a corrected threshold of $p < 0.05$ (which controls for the false discovery rate [FDR]), thereby indicating the components which show group differences in the loading components [1].

ICA belongs to a class of blind source separation (BSS) methods used to recover underlying source signals from linearly mixed signals without any prior information about the source signals or the mixing process [6, 7]. ICA is a subset of BSS that assumes the underlying sources are independent. The instantaneous (linear) noise-free mixture model [6, 7] is given in Eq. 1.

$$X = AS; \ \widehat{S} = WX, \ W = \text{inv}(A) \tag{1}$$

where X stands for linearly recorded sources, S represents the underlying independent components, A represents the mixing matrix or loadings, and W denotes the unmixing matrix, which is the (pseudo)inverse of the mixing matrix A. The task of any ICA algorithm is to obtain the matrix W as in Eq. 1, which enables estimating the underlying independent sources $\widehat{S}$, up to a permutation and scaling factor. A good review of ICA methods for multimodal fusion and biomedical applications is available in these articles [8, 9].

2 Neuroimaging Studies Using Source-Based Morphometry

Source-based morphometry was initially proposed to extract maximally spatially independent sources in structural gray matter MRI images and to identify structural magnetic resonance imaging (sMRI) differences between Ct and Sz [1]. Reliable replication of gray matter concentration (GMC) components showing (Ct/Sz) diagnostic differences were assessed recently in the largest aggregated structural imaging dataset to date in schizophrenia consisting of 784 Sz and 936 Ct across 23 scanning sites [10]. The study reported nine components having diagnostic differences which comprised of separate cortical, subcortical, and cerebellar regions. Seven components showed greater GMC in Ct than Sz, while two (brainstem and cerebellum) showed greater GMC for Sz.

The component with the greatest GMC decrease in Sz and the largest effect size was a pattern comprising regions of superior temporal gyrus, inferior frontal gyrus, and medial frontal cortex,

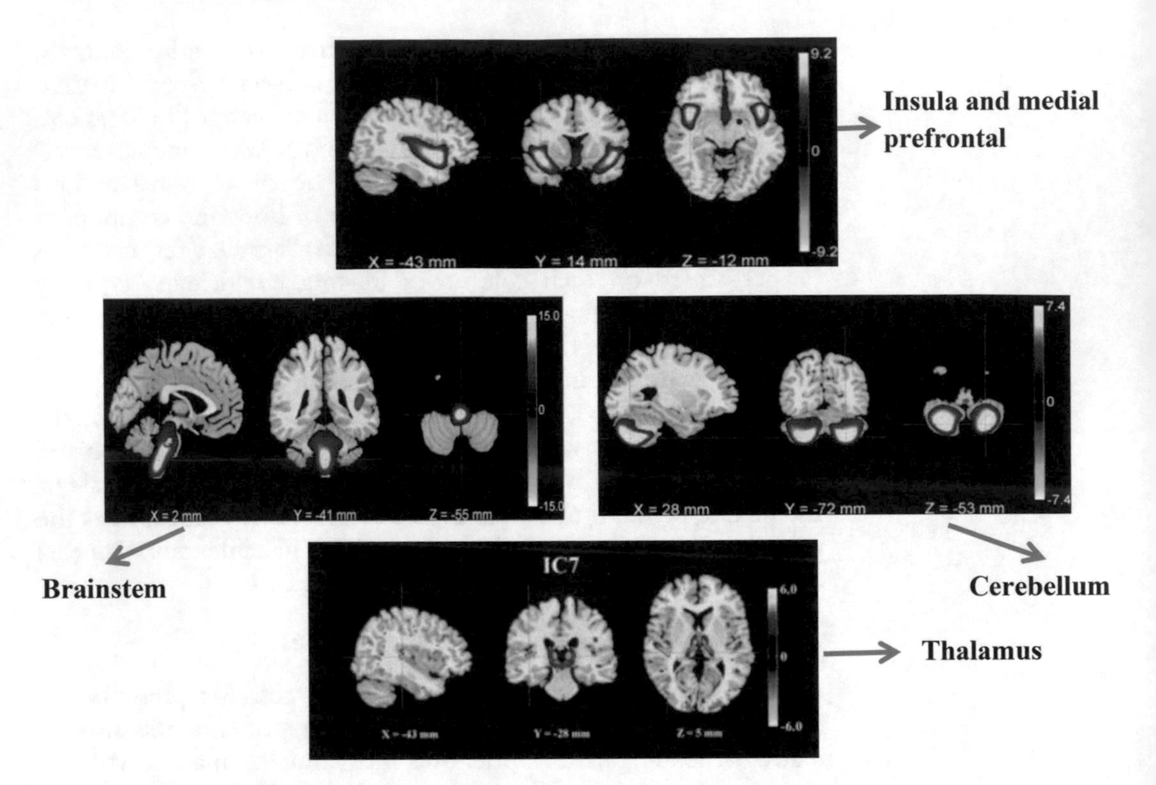

Fig. 3 Significant SBM components from various studies showing differences between Sz and Ct. Top: insula and medial prefrontal component [1, 10]. Middle left: brainstem [10, 15]. Middle right: cerebellum [10]. Bottom: thalamus [15]

and it replicated reliably across different datasets [10]. In family studies, this component has also exhibited heritability estimates from 0.49 to 0.59 [11, 12].

Two other components (brainstem and cerebellum) also depicted in Fig. 3 (middle) showed increased loading coefficients in Sz when compared to Ct [10]. We speculated that increased GMC in brainstem component for Sz could have resulted from chronic exposure to dopamine D2-blocking [10] therapeutic agents.

Reduction of medial and inferior frontal, insular, and bilateral temporal gray matter volumes between patients with persistent auditory verbal hallucinations and without persistent auditory verbal hallucinations [13] proved that SBM can also perform challenging clinical subtyping. Multivariate morphometric patterns have suggested that concomitant increase and decrease in gray matter occur in association with persistent negative thought disorder in clinically stable individuals with schizophrenia [14].

2.1 SBM for Artifact and Motion Detection

Aggregation of multisite data helps increase statistical power in neuroimaging analysis. However different scanning platforms used can typically introduce systematic differences [15]. Using

SBM, independent components are extracted from the data and assessed for associations with scanning parameters [15]. Identified scanning-related components were then eliminated from the original data for correction. SBM can therefore improve sensitivity in neuroimaging by finding artifact components associated with motion and scanner effects which are eventually removed, thereby reducing the number of corrections for multiple statistical tests [15]. Such an approach can also be useful in identifying motion-related effects in structural brain imaging data.

2.2 SBM Applied to Fractional Anisotropy Data

The importance of utilizing multivariate approaches in fractional anisotropy-based morphometric studies was first presented for schizophrenia [16]. This approach did not limit the results to a priori regions or examine each voxel separately; rather it identified patterns from the entire fractional anisotropy map (FA). SBM on the FA data identified several components comprising global disjoint brain regions and multiple white matter tracts [16] as reported in other univariate analysis, but without relying on averaging with tracts or on pre-constraints using specific atlas.

2.3 SBM Applied to Gray and White Matter

Independent SBMs on gray matter volume (GMV) and white matter volume (WMV) also revealed two patterns having significantly lower gray and higher white matter volume in delusional patients compared to healthy controls [17]. Lower GMV occurred in the regions of medial prefrontal cortex, anterior cingulate cortex, medial temporal lobe structures (parahippocampus and hippocampus), sensorimotor cortices, bilateral insula and thalamus, and inferior parietal regions, while higher WMV was found in medial and middle frontal and temporal cortices, left insula, and lentiform nucleus [17].

The combination of tissue types can also be analyzed through the use of a more complex SBM approach. The relationship between gray and white matters is not straightforward, and it is reasonable to expect that morphometric changes in one tissue may result in disturbance in the other [18]. SBM fusion approaches for group-level analysis were also applied to angle and power images sensitive to gray and white matter interrelationship [18]. The angle image reflects the gray-to-white matter ratio and is sensitive to small changes in regions where gray matter is increasing and white matter is decreasing or vice versa [18], while the power image indicates overall tissue concentration and highlights tissue presence in each voxel, especially in regions where both gray and white matter concentrations are low [18]. Subject differences in angle and power images are shown in Fig. 4. SBM approach on the angle and power matrixes separately showed six networks having lower white-to-gray matter ratio in schizophrenia. It included thalamus, right precentral-postcentral, left pre-/postcentral, parietal, right cuneus-frontal, and left cuneus-frontal sources. A few of the networks looked similar to sensorimotor and vision functional patterns [18].

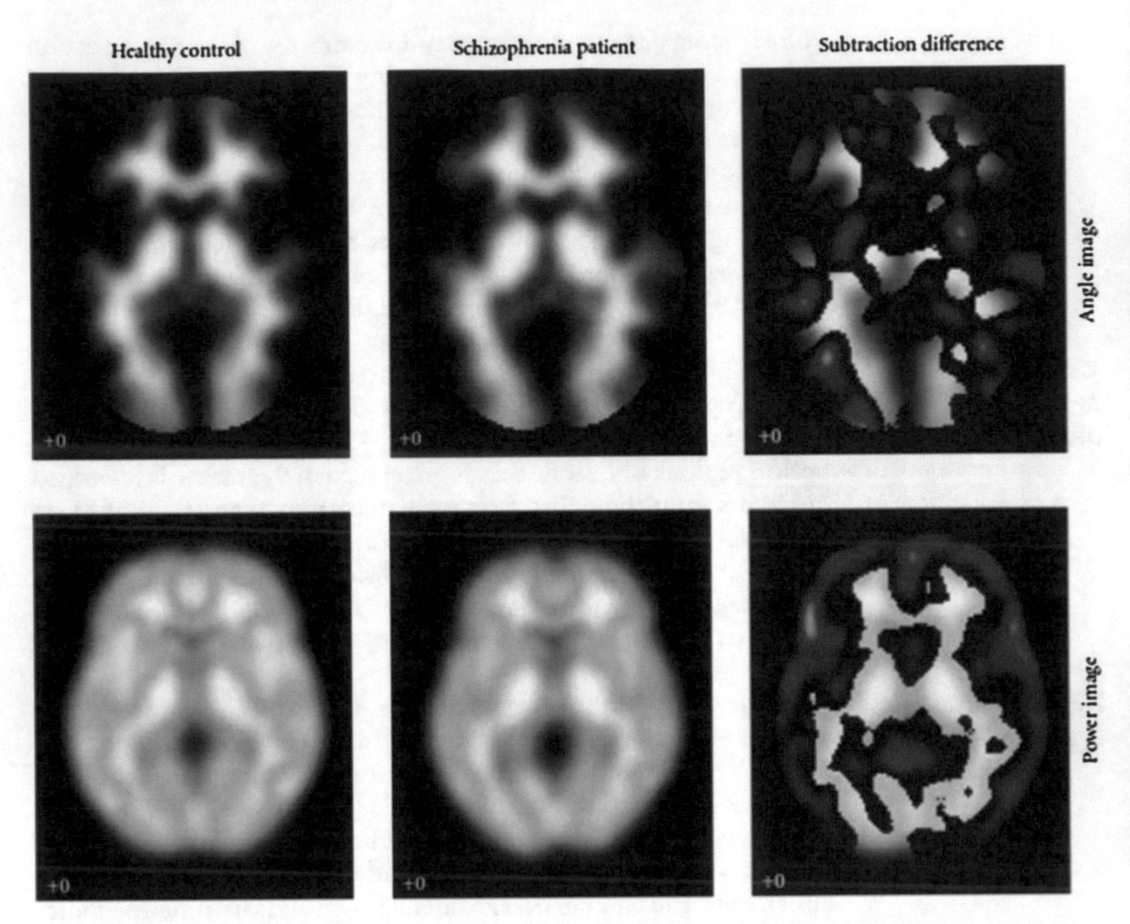

Fig. 4 Subject differences in angle and power [18]. The upper row consists of angle images, and the bottom consists of power images. The first column is an image from healthy control; the second is the image from the schizophrenia patient; the third is the subtraction showing the subject differences [18]

Joint source-based morphometry (jSBM) [19] is a multimodal multivariate neuroimaging algorithm which identifies links between multiple imaging modalities, revealing networks of brain regions showing diagnostic difference between groups. This data fusion framework using both white and gray matter together identified four joint sources as significantly associated with schizophrenia, namely, (1) temporal, corpus callosum; (2) occipital/frontal, inferior fronto-occipital fasciculus; (3) frontal/parietal/occipital/temporal, superior longitudinal fasciculus; and (4) parietal/frontal, thalamus.

2.4 SBM Patterns are Similar to Resting State Functional Networks

The spatial relationship between structure and function was also assessed by comparing SBMs from sMRI images with group ICA maps estimated from resting-state functional magnetic resonance imaging (rs-fMRI). Several GMC structural components that spatially corresponded to resting-state functional components were

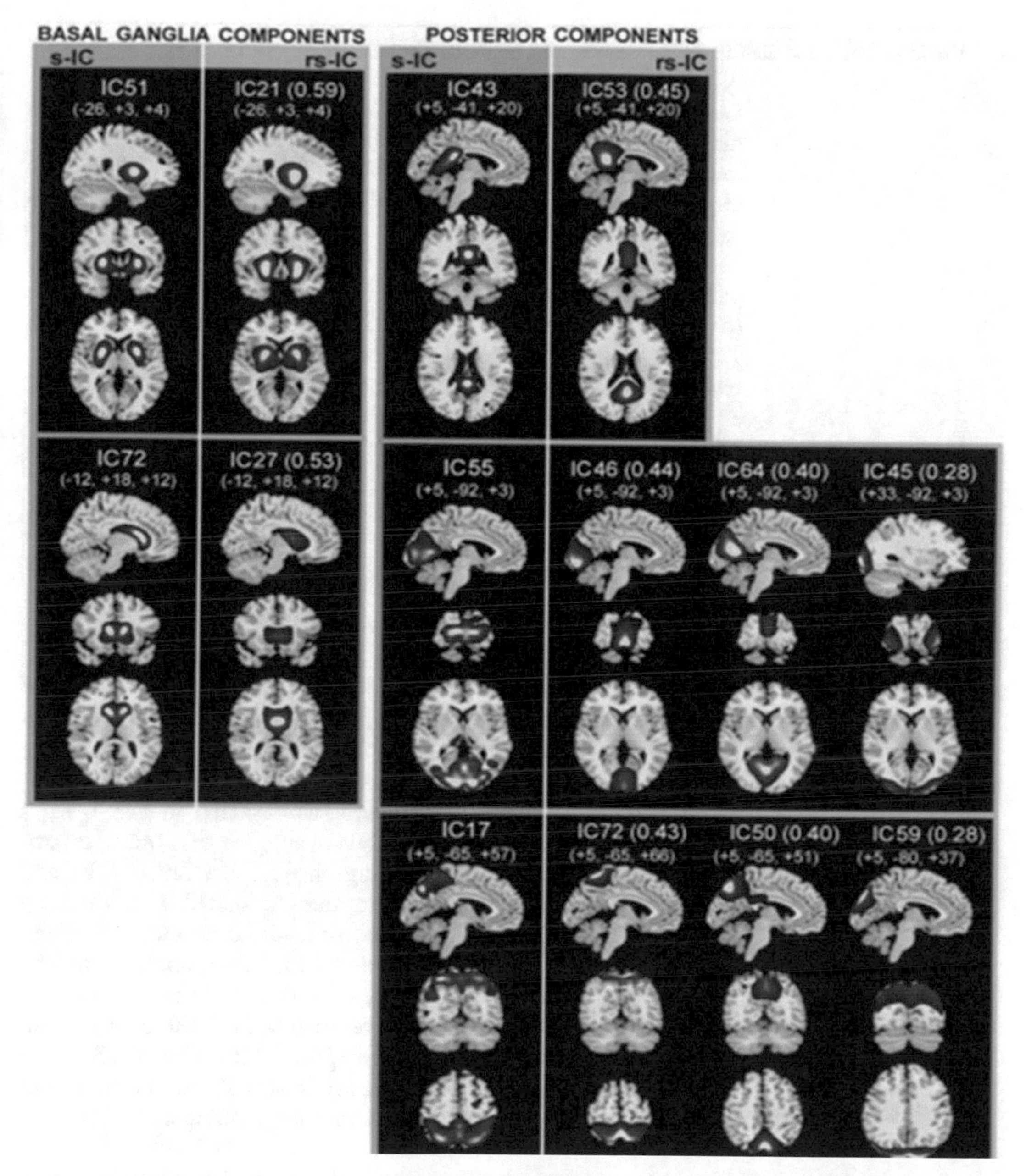

Fig. 5 Examples of identified spatial relationships between gray matter covariation in structural (sMRI) components estimated using SBM (*red*) and corresponding group ICA rs-fMRI components (*blue*) [20]

identified [20] as depicted in Fig. 5. Basal ganglia components showed the strongest structural to resting-state functional correlation, while cortical components displayed correspondence between a single structural component and several resting-state functional components [20]. This work also identified precuneus as a hub using structural network correlation analysis [20].

3 Classification Studies and Extensions of Source-Based Morphometry

Classification of images as being from individuals with schizophrenia can be done using a variety of machine learning techniques and imaging modalities, including SBM as a feature identification technique [21]. In the case of SBM, the classification system does not use the raw image values but the loading coefficients on the spatial patterns of the image to identify diagnostic groups. Using GMC patterns obtained from SBM with a bagged support vector machine (SVM) and resampling framework [22] achieved an accuracy of 73% distinguishing patients with schizophrenia from healthy controls. This machine learning approach avoids overfitting and trains an ensemble of classifiers with different data samples to give a prediction based on the combination of them. Applications of these classification techniques could be used to distinguish prodromal individuals who will convert to full psychosis versus those who will not, for example, as has been initially done in other studies [23].

3.1 Nonlinear SBM

Nonlinear ICA or nonlinear SBM using a deep learning architecture to capture nonlinearities in structural MRI data has been successfully implemented recently [24]. This work compared several methods of image decomposition including linear SBM, nonlinear ICA (NICE), and a nonlinear variational autoencoder algorithm (VAE) on the same dataset. It was observed that nonlinear ICA detected more components, as depicted in Fig. 6 for a given dataset. Initial results from NICE suggest that this approach can detect components resembling those from linear ICA and biologically relevant nonlinear components missed by assuming a linear mixing. The discovered additional components covered regions that were known to be implicated in schizophrenia, thereby giving a more comprehensive view of the underlying anatomic changes. Since it is unknown if the mixing is linear or nonlinear for a given dataset, we recommend applying both linear SBM and NICE frameworks and then draw inferences about the nature of mixing from the obtained components and loadings.

3.2 Deep Learning Applied to Gray Matter Data

Application of deep learning methods in neuroimaging has been gaining popularity [25]. In recent work, we used deep belief networks (DBN) and its building block, the restricted Boltzmann machine, to learn physiologically important representations and detect latent relations from sMRI data in schizophrenia and Huntington disease [25]. This work showed that the depth of DBN improves classification and increases group separation. The group separation map between various classes in Huntington disease is shown in Fig. 7.

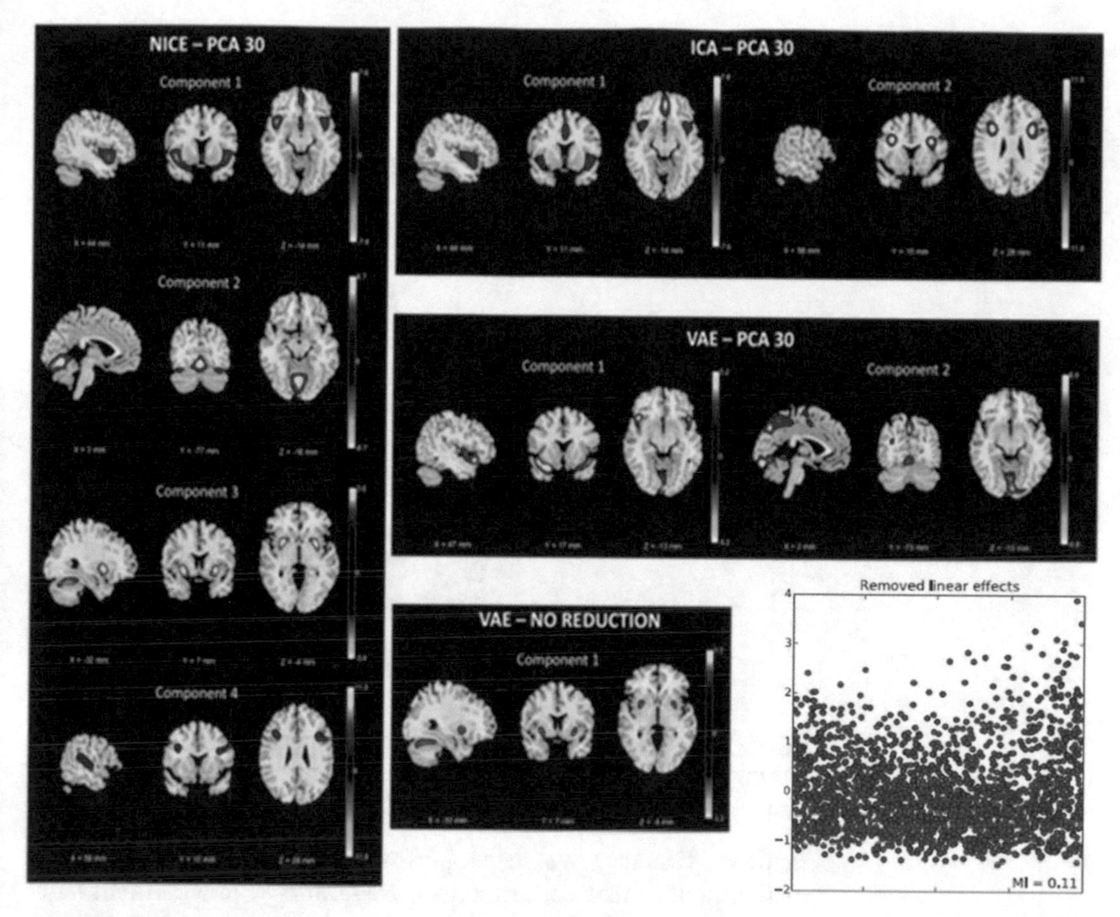

Fig. 6 Significant SBM components estimated by nonlinear ICA (NICE), ICA, and VAE showing (Ct/Sz) differences after data dimensionality reduction to 30 [24]. Included is also VAE without prior dimensionality reduction [24]. The identified nonlinear relationship in loading coefficients for Component 3 (NICE-PCA30) after removing linear effects

Deep learning approaches can be challenging due to the amount of training data needed. This has been addressed by our recent work which developed an independent component analysis methodology [26] using real sMRI to generate an extensive synthetic structural magnetic resonance imaging (sMRI) dataset. This simulated but realistic data used at the pretraining stage of deep learning application for neuroimaging enables us to train the algorithm on a limited amount of data. As a result it achieved a 5% increase in classification accuracy compared to regular approaches.

3.3 Flexible SBM via Independent Vector Analysis Applied to Multisite/Scanner Data

SBM is a data-driven technique, and applying it to different datasets that are collected at different research labs using different equipment can lead to slightly different components or patterns in the different datasets, though as noted above there is some replication [10]. The ability to identify components across datasets while simultaneously allowing dataset-specific variability requires a new

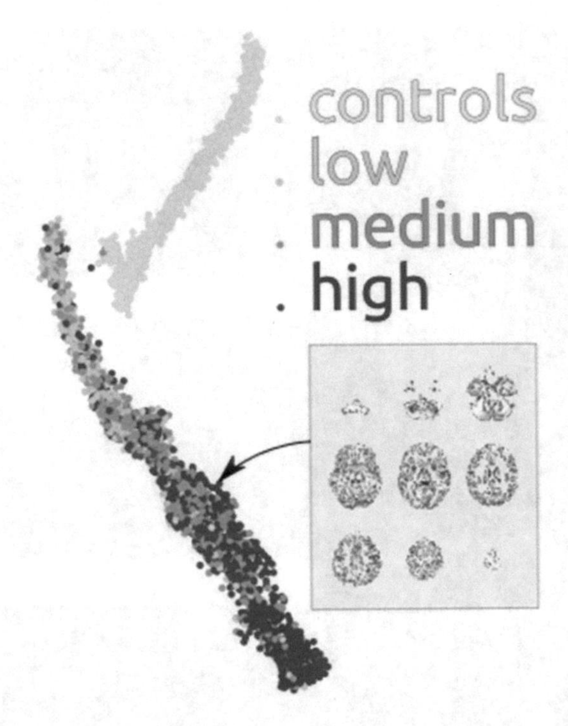

Fig. 7 Patient and control group separation map of sMRI scans by Huntington disease severity identified using deep learning [25]

method. Recently we developed a methodology for automatic matching of similar components across multiple independent datasets, without manual intervention, using independent vector analysis (IVA) [27]. IVA is a generalization of ICA framework for analysis of multiple datasets [28]. IVA allows maintaining the independence among components within a dataset while increasing dependency of components between datasets [28]. Comparing IVA with an SBM-based methodology in a multi-scanner dataset, we observed that additional spatial patterns were recovered by IVA. The generalization of SBM to IVA will be of great use in analyzing aggregated datasets from different scanners and study; IVA is the next toward allowing dataset-specific artifacts while identifying common components across datasets.

4 Source-Based Morphometry and Genetics

Schizophrenia is a heritable disorder with both common and rare genetic variants thought to contribute toward genetic risk [29]. *Imaging genetics* refers to combining genetic information and neuroimaging data from the same set of subjects to discover neuromechanisms linked to diseases. Measurable components called endophenotypes along the pathway between disease and distal genotype are identified in imaging genetic studies [30]. An endophenotype shows association with an illness, is heritable, and is associated with a gene region. It could be neurophysiological, biochemical, endocrinological, neuroanatomical, cognitive, or neuropsychological in nature [30]. The goal of imaging genetics in psychiatry is to identify genetic effects on neuroimaging measures, in order to clarify the link between genetic effects and the eventual syndromes and disorders related to those brain measures. A common approach is to examine each genetic locus individually against a brain imaging measure [31, 32], but SBM and its extensions can do more.

In fusing high-dimensional neuroimaging and genetic data, the aim is to identify aggregated genetic effects on covarying imaging patterns; all of the genetic loci are included in the analysis at once rather than individually, reducing the dimensionality of the genome-wide data while leveraging similarities in gray matter or functional brain networks. While it potentially helps identify associations between two completely different modalities, it is considered a huge challenge from an algorithmic viewpoint [33]. A comprehensive review of the various algorithms and the factors to be considered for various imaging genetic scenarios is discussed in our recent review paper [33]. Using SBM and related algorithms, we can use SBM on the imaging data to find GMC patterns to use as phenotypes, or we can examine both the genetics and the imaging together. We present examples of both below.

The combination of imaging and genome-wide scan data in a family study has been used to identify the heritability and genetic influences on the SBM components. The SBM component comprising the superior temporal gyrus, inferior frontal gyrus, and medial frontal cortex (which showed greatest GMC deficit in patients with schizophrenia) was used as a phenotype in a family study of multiple multigenerational families, to identify how the measure varied within and across families and what part of the genome was associated with the phenotype across families [12]. It showed a significant linkage peak at a specific genomic region 12q24 (LOD = 3.76). This region contained the Darier's disease locus and has been of significant interest in psychiatric genetics. It also contained other susceptibility genes (e.g., DAO, NOS1) and has been linked to schizophrenia in multiple populations before [12]. Significant heritability of this SBM component was also found with healthy sibling data [11].

Also using multivariate analysis in a separate case-control study, the effects of rs1625579 genotype with the genetic risk score (GRS) of miR-137-regulated genes (TCF4, PTGS2, MAPK1, and MAPK3) were assessed in a three-way interaction with diagnosis on GMC patterns [34]. Schizophrenia subjects homozygous for the MIR137HG risk allele showed significant decreases in the occipital, parietal, and temporal lobe GMC (depicted in Fig. 8) with

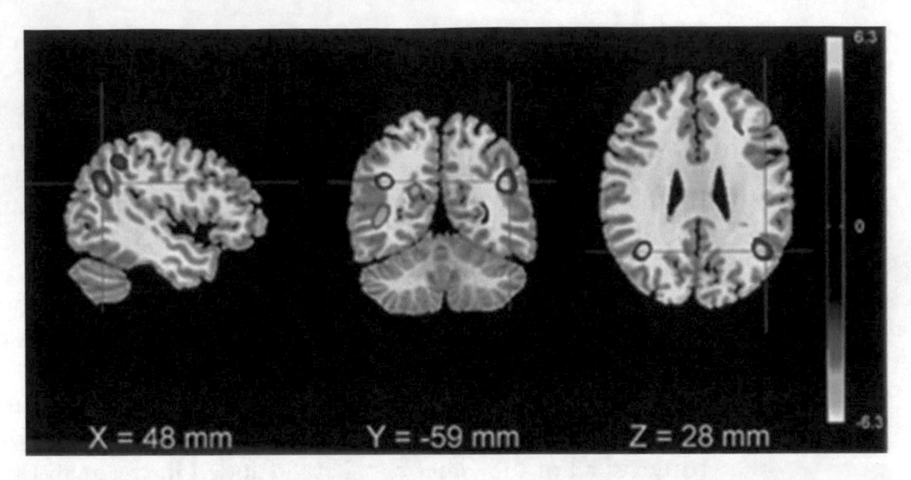

Fig. 8 SBM component showing a genetic and diagnosis interactive effect in a study involving MIR-137 [34]

increasing miR-137-regulated genetic risk scores (GRS), whereas those carrying the protective minor allele showed significant increases in GMC with GRS [34].

The studies reported above used SBM on the gray matter images and used the loading coefficients as a phenotype in regressions against genetic data. However, the method can be applied to the genetic data as well. An extension of ICA applied to imaging genetics is parallel ICA (pICA) [5]. pICA is a multivariate multimodal algorithm which simultaneously maximizes independence within modalities and correlations across the two modalities; it is thus an example of symmetric data fusion in which both modalities are informing the solution jointly [35]. When the input is imaging data and single nucleotide polymorphism (SNP) arrays, the output is correlated imaging components and genetic patterns; individuals with larger loading coefficients on the imaging component would have larger loadings on the genetic pattern, identifying an aggregated genetic effect on the brain measures. It was first applied to a genetic study of schizophrenia using a SNP array and auditory oddball fMRI data [36]. A correlation of 0.38 between one fMRI component and one genetic component was observed, and both the linked component loadings showed (Ct/Sz) group difference [36]. The fMRI component in Fig. 9 consisted mainly of parietal lobe activations, while the single nucleotide polymorphism (SNP) component was contributed by SNPs located in genes, including those coding for nicotinic a-7 cholinergic receptor, aromatic amino acid decarboxylase, and disrupted in schizophrenia 1[36] as depicted in Fig. 9. The algorithm was also successfully applied to investigate the genetic underpinnings of white matter abnormalities in schizophrenia [37]. A significant correlation ($r = -0.37$) was identified between one genetic factor and one fractional anisotropy (FA) component after controlling for scanning site, ethnicity, age, and sex. The FA component reflected decreased white matter

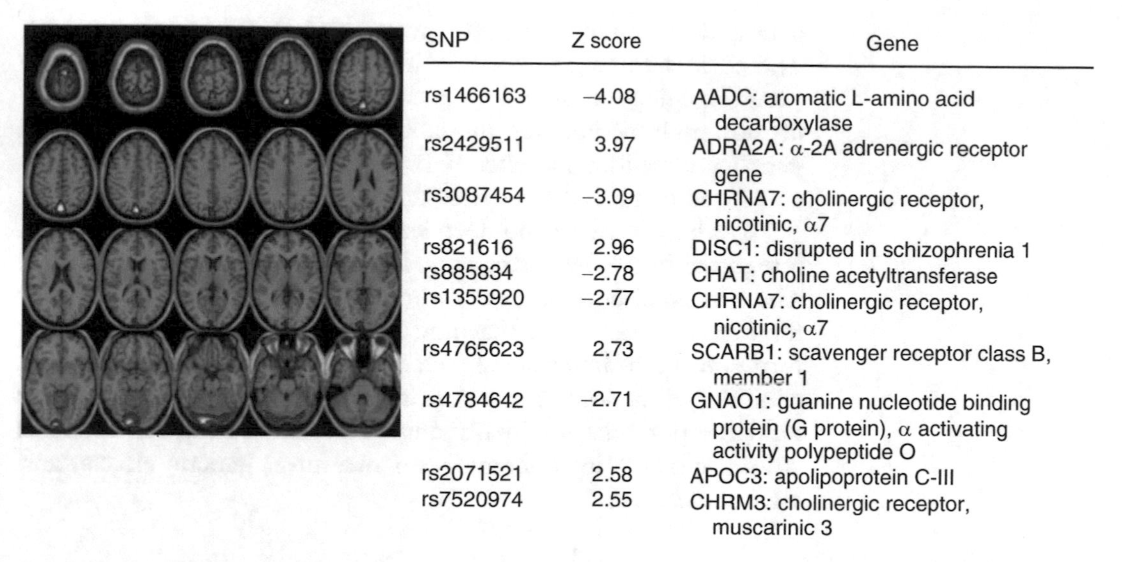

SNP	Z score	Gene
rs1466163	−4.08	AADC: aromatic L-amino acid decarboxylase
rs2429511	3.97	ADRA2A: α-2A adrenergic receptor gene
rs3087454	−3.09	CHRNA7: cholinergic receptor, nicotinic, α7
rs821616	2.96	DISC1: disrupted in schizophrenia 1
rs885834	−2.78	CHAT: choline acetyltransferase
rs1355920	−2.77	CHRNA7: cholinergic receptor, nicotinic, α7
rs4765623	2.73	SCARB1: scavenger receptor class B, member 1
rs4784642	−2.71	GNAO1: guanine nucleotide binding protein (G protein), α activating activity polypeptide O
rs2071521	2.58	APOC3: apolipoprotein C-III
rs7520974	2.55	CHRM3: cholinergic receptor, muscarinic 3

Fig. 9 The linked fMRI component map (on left) and the SNP components (on right) detected by a parallel ICA study on genetic and fMRI data [36]

integrity in the forceps major for patients with schizophrenia [37]. The SNP component was overrepresented in genes whose products are involved in corpus callosum morphology (e.g., CNTNAP2, NPAS3, and NFIB) as well as canonical pathways of synaptic long-term depression and protein kinase A signaling [37].

It is also important to note that parallel ICA has been applied to other types of data in addition to genetics and neuroimaging datasets [38–40]. For a relatively small sample size with large genetic array data, extensions of parallel ICA have been developed to improve the performance by incorporating prior information about genetic or imaging data called parallel ICA with reference [41] and parallel ICA with multiple references [42]. These two approaches help identify particular genetic components using prior knowledge, which may not be detected otherwise [33].

5 Conclusion

This chapter discusses a wealth of imaging, genetic, and machine learning studies which are based on source-based morphometry and independent component analysis frameworks. Multivariate approaches such as SBM are ideally suited for identifying complex and weak effects from high-dimensional datasets while simultaneously performing data reduction procedures [1]. The imaging data can be reduced from 20,000 voxels to a handful of networks or patterns and the genetic data from several million common polymorphisms to a few genetic structures. The loading coefficients then capture the contribution of these patterns to each individual's data and can be examined for group differences or other effects. The SBM framework was first implemented for single modality (neuroimaging or genetics) and then successfully expanded to include multiple imaging modalities (joint SBM) or imaging and genetic modalities (parallel ICA and its extensions), to identify correlated patterns across these different data types.

Machine learning and deep learning approaches have also been integrated into SBM framework for various application scenarios. The successful development of nonlinear SBM recently paves way for further genetic and machine learning research in the coming future. As high-dimensional data becomes the norm for examining clinical dysfunctions, these data reduction and fusion techniques hold the potential for identifying and summarizing relationships among patterns in different brain measures, genetic effects, and clinical measures.

Acknowledgments

This work was supported by NIH 1R01MH094524 (to JT and VDC) as well as P20GM103472, 1R01EB006841, and R01EB005846 (to VDC).

References

1. Xu L, Groth KM, Pearlson G, Schretlen DJ, Calhoun VD (2009) Source-based morphometry: the use of independent component analysis to identify gray matter differences with application to schizophrenia. Hum Brain Mapp 30(3):711–724
2. Ashburner J, Friston KJ (2000) Voxel-based morphometry—the methods. NeuroImage 11 (6):805–821
3. McKeown MJ, Sejnowski TJ (1998) Independent component analysis of fMRI data: examining the assumptions. Hum Brain Mapp 6 (5–6):368–372
4. Sui J, Adali T, Yu Q, Chen J, Calhoun VD (2012) A review of multivariate methods for multimodal fusion of brain imaging data. J Neurosci Methods 204(1):68–81
5. Pearlson GD, Liu J, Calhoun VD (2015) An introductory review of parallel independent component analysis (p-ICA) and a guide to applying p-ICA to genetic data and imaging phenotypes to identify disease-associated biological pathways and systems in common complex disorders. Front Genet 6:276
6. Hyvärinen A, Oja E (2000) Independent component analysis: algorithms and applications. Neural Netw 13(4):411–430
7. Lee T-W (1998) Independent component analysis: theory and applications. Springer, New York, London, pp 27–66
8. Comon P, Jutten C (2010) Handbook of blind source separation: independent component analysis and applications. Academic press
9. Calhoun VD, Liu J, Adali T (2009) A review of group ICA for fMRI data and ICA for joint inference of imaging, genetic, and ERP data. NeuroImage 45(1):S163–S172
10. Gupta CN, Calhoun VD, Rachakonda S et al (2015) Patterns of gray matter abnormalities in schizophrenia based on an international mega-analysis. Schizophr Bull 41(5):1133–1142
11. Turner JA, Calhoun VD, Michael A et al (2012) Heritability of multivariate gray matter measures in schizophrenia. Twin Res Hum Genet 15(03):324–335
12. Sprooten E, Gupta CN, Knowles EE et al (2015) Genome-wide significant linkage of schizophrenia-related neuroanatomical trait to 12q24. Am J Med Genet B Neuropsychiatr Genet 168(8):678–686
13. Kubera KM, Sambataro F, Vasic N et al (2014) Source-based morphometry of gray matter volume in patients with schizophrenia who have persistent auditory verbal hallucinations. Prog

Neuro-Psychopharmacol Biol Psychiatry 50:102–109

14. Palaniyappan L, Mahmood J, Balain V, Mougin O, Gowland PA, Liddle PF (2015) Structural correlates of formal thought disorder in schizophrenia: an ultra-high field multivariate morphometry study. Schizophr Res 168 (1):305–312
15. Chen J, Liu J, Calhoun VD et al (2014) Exploration of scanning effects in multi-site structural MRI studies. J Neurosci Methods 230:37–50
16. Caprihan A, Abbott C, Yamamoto J et al (2011) Source based morphometry analysis of group differences in fractional anisotropy in schizophrenia. Brain Connect 1(2):133–145
17. Wolf RC, Huber M, Lepping P et al (2014) Source-based morphometry reveals distinct patterns of aberrant brain volume in delusional infestation. Prog Neuro-Psychopharmacol Biol Psychiatry 48:112–116
18. Xu L, Adali T, Schretlen D, Pearlson G, Calhoun VD (2011) Structural angle and power images reveal interrelated gray and white matter abnormalities in schizophrenia. Neurol Res Int 2012:735249
19. Xu L, Pearlson G, Calhoun VD (2009) Joint source based morphometry identifies linked gray and white matter group differences. NeuroImage 44(3):777–789
20. Segall J, Allen EA, Jung RE, Erhardt E, Arja S, Kiehl KA, Calhoun VD (2012) Correspondence between structure and function in the human brain at rest. Front Neuroinform 6:10
21. Arbabshirani MR, Plis S, Sui J, Calhoun VD (2016) Single subject prediction of brain disorders in neuroimaging: promises and pitfalls. NeuroImage 145(Pt B):137–165
22. Castro E, Gupta CN, Martínez-Ramón M, Calhoun VD, Arbabshirani MR, Turner J (2014) Identification of patterns of gray matter abnormalities in schizophrenia using source-based morphometry and bagging. Paper presented at: 2014 36th Annual International Conference of the IEEE Engineering in Medicine and Biology Society
23. Koutsouleris N, Meisenzahl EM, Davatzikos C et al (2009) Use of neuroanatomical pattern classification to identify subjects in at-risk mental states of psychosis and predict disease transition. Arch Gen Psychiatry 66(7):700–712
24. Castro E, Hjelm RD, Plis SM, Dinh L, Turner JA, Calhoun VD (2016) Deep independence network analysis of structural brain imaging: application to schizophrenia. IEEE Trans Med Imaging 35:1729–1740
25. Plis SM, Hjelm DR, Salakhutdinov R, Allen EA, Bockholt HJ, Long JD, Johnson HJ, Paulsen JS, Turner JA, Calhoun VD (2014) Deep learning for neuroimaging: a validation study. Front Neurosci 8:229
26. Castro E, Ulloa A, Plis SM, Turner JA, Calhoun VD (2015) Generation of synthetic structural magnetic resonance images for deep learning pre-training. Paper presented at: 2015 I.E. 12th International Symposium on Biomedical Imaging (ISBI)
27. Gupta CN, Arias-Vasquez A, Liu J, Andreassen O, Agartz I, Calhoun VD (2016) Canonicality of structural patterns compared using source based morphometry and independent vector analysis. Organization for Human Brain Mapping Conference, June 2016
28. Kim T, Lee I, Lee T-W (2006) Independent vector analysis: definition and algorithms. Paper presented at: 2006 Fortieth Asilomar Conference on Signals, Systems and Computers
29. Sullivan PF, Kendler KS, Neale MC (2003) Schizophrenia as a complex trait: evidence from a meta-analysis of twin studies. Arch Gen Psychiatry 60(12):1187–1192
30. Gottesman II, Gould TD (2003) The endophenotype concept in psychiatry: etymology and strategic intentions. Am J Psychiatr 160 (4):636–645
31. Stein JL, Medland SE, Vasquez AA et al (2012) Identification of common variants associated with human hippocampal and intracranial volumes. Nat Genet 44(5):552–561
32. Thompson PM, Stein JL, Medland SE et al (2014) The ENIGMA Consortium: large-scale collaborative analyses of neuroimaging and genetic data. Brain Imaging Behav 8 (2):153–182
33. Liu J, Calhoun VD (2014) A review of multivariate analyses in imaging genetics. Front Neuroinform 8:29
34. Wright C, Gupta C, Chen J et al (2016) Polymorphisms in MIR137HG and microRNA-137-regulated genes influence gray matter structure in schizophrenia. Transl Psychiatry 6 (2):e724
35. Calhoun VD, Sui J (2016) Multimodal fusion of brain imaging data: a key to finding the missing link(s) in complex mental illness. Biol Psychiatry Cogn Neurosci Neuroimaging 1 (3):230–244
36. Liu J, Pearlson G, Windemuth A, Ruano G, Perrone-Bizzozero NI, Calhoun V (2009) Combining fMRI and SNP data to investigate connections between brain function and

genetics using parallel ICA. Hum Brain Mapp 30(1):241–255

37. Gupta CN, Chen J, Liu J et al (2014) Genetic markers of white matter integrity in schizophrenia revealed by parallel ICA. Front Hum Neurosci 9:100–100
38. Yarosh HL, Meda SA, De Wit H, Hart AB, Pearlson GD (2015) Multivariate analysis of subjective responses to d-amphetamine in healthy volunteers finds novel genetic pathway associations. Psychopharmacology 232 (15):2781–2794
39. Narayanan B, Soh P, Calhoun V et al (2015) Multivariate genetic determinants of EEG oscillations in schizophrenia and psychotic bipolar disorder from the BSNIP study. Transl Psychiatry 5(6):e588
40. Meier T, Wildenberg J, Liu J et al (2012) Parallel ICA identifies sub-components of resting state networks that covary with behavioral indices. Front Hum Neurosci 6:281
41. Chen J, Calhoun VD, Pearlson GD et al (2013) Guided exploration of genomic risk for gray matter abnormalities in schizophrenia using parallel independent component analysis with reference. NeuroImage 83:384–396
42. Chen J, Calhoun VD, Ulloa AE, Liu J (2014) Parallel ICA with multiple references: A semi-blind multivariate approach. Paper presented at: 2014 36th Annual International Conference of the IEEE Engineering in Medicine and Biology Society

Check for updates

Chapter 8

Integrating Cytoarchitectonic Probabilities with MRI-Based Signal Intensities to Calculate Regional Volumes of Interest

Florian Kurth, Lutz Jancke, and Eileen Luders

Abstract

Hypotheses on specific brain structures are frequently assessed using anatomically defined regions of interest (ROIs) in structural T1-weighted images. The definition of these ROIs is often based on macroscopic landmarks (sulci, gyri, etc.) that do not necessarily coincide with the functional architecture of the brain. Microscopic labeling, on the other hand, enables a precise localization of anatomical boundaries. Thus, defining ROIs using cytoarchitectonic information might be a suitable alternative to the traditional landmark-based approach. In this chapter, we describe in detail how to perform such cytoarchitectonic ROI analyses by integrating voxel-wise cytoarchitectonic probabilities (using freely available maps created *post mortem*) with MR-based signal intensities (using standard T1-weighted data obtained in vivo). After elucidating common techniques to create ROIs (*see* Sect. 1), detailed information is provided with respect to the cytoarchitectonically defined probabilistic maps, i.e., the foundation for the proposed cytoarchitectonic ROI approach (*see* Sect. 2). The methodological aspects pertaining to this approach constitute the heart of the chapter and comprise three main steps: probability map selection, preprocessing, and integration (*see* Sect. 3). The chapter concludes with practical tips and helpful pointers for conducting a cytoarchitectonic ROI analysis (*see* Sect. 4).

Key words Atlas, Cytoarchitecture, Gray Matter, Morphometry, MRI, Neuroimaging, Normalization, Segmentation, Structural, Tissue Classification, ROI

1 Introduction

1.1 Overview

Anatomically defined regions of interest (ROIs) are frequently used in basic and clinical research to assess specific hypotheses pertaining to specific brain structures (on correlations, group differences, changes over time, etc.). However, given the large interindividual variability in the brain's anatomy, many structures vary in their shape, size, and exact location. Moreover, visible macroanatomy rarely matches underlying microanatomy and thus lacks precise functional correspondence. Therefore, the classic landmark-based approach for ROI analyses, as commonly applied in neuroimaging studies, comes with various limitations. In this chapter, we describe an alternative

Gianfranco Spalletta et al. (eds.), *Brain Morphometry*, Neuromethods, vol. 136,
https://doi.org/10.1007/978-1-4939-7647-8_8, © Springer Science+Business Media, LLC 2018

approach that uses cytoarchitectonically defined probability maps rather than macroscopic landmarks to define ROIs.

1.2 Brain Segregation and Region-of-Interest (ROI) Analyses

Broca's discovery of a circumscribed speech region [1] and the publication of Brodmann's atlas [2] were important milestones in the history of neuroscience that profoundly shaped our understanding of the brain's architecture. Today, the view of the brain as segregated into functionally and anatomically different regions has been widely accepted and provides a common ground for clinicians, scientists, and researchers around the globe. Specifically, it serves as a framework to develop and assess hypotheses with regard to specific brain regions and also provides a reference to relate the multitude of observations in neuroimaging research. One approach employing our knowledge of the brain's structural and functional segregation directly is the so-called region-of-interest (ROI) analysis. As the name implies, an ROI analysis allows for the testing of hypotheses with particular respect to a specific brain region or structure.

In conventional analyses, ROIs are usually created (e.g., as 2D areas or 3D volumes) by either employing automated algorithms or manual tracings that follow defined structural borders (ideally based on standardized protocols). While manually created ROIs are considered more precise than automatically created ones, the tracing procedure is time-consuming and also prone to user bias. However, both manual and automatic techniques rely on visible and/or detectable landmarks, which poses a major drawback as macroanatomic landmarks (sulci, gyri, etc.) rarely match actual cytoarchitectonic boundaries and thus lack precise functional correspondence [3–13]. Moreover, for large parts of the brain, it is extremely difficult to precisely define (or identify) unambiguous boundaries due to the limited resolution of MRI scans as well as large interindividual variabilities in terms of cortical folding patterns, where tertiary and sometimes even secondary sulci may be missing entirely.

This is where the integration of information on cytoarchitectonic probabilities and image intensities comes into play, as this approach combines the advantages of automated and manual techniques while avoiding their aforementioned limitations [14, 15]. More specifically, the concept is based on multiplying MRI-based signals (e.g., indicative of gray matter; also *see* **Note 1**) obtained in vivo with voxel-wise structural probabilities obtained *post mortem*. The latter probabilities indicate the likelihood of each voxel to belong to a specific cytoarchitectonically distinct structure or substructure (e.g., the hippocampus or hippocampal subiculum). The main advantage of this proposed cytoarchitectonic ROI approach compared to conventional ROI analyses lies in its observer independence and the ability to capture location, extent, and interindividual variability across subjects within a 3D reference frame in MNI space. This reference frame allows for a seamless integration of signal-based information obtained in vivo (e.g., using MRI) enabling the definition/creation of ROIs even in the absence of any macroanatomic landmarks.

2 Materials

The required cytoarchitectonic probability maps are available as part of the Anatomy Toolbox for SPM [16] as well as from the Institute of Neuroscience and Medicine at the Research Center Jülich (www.fz-juelich.de/inm/inm-1/EN/Forschung/_docs/Gehirnkarten/gehirnkarten_node.html).[1] The maps are derived from cytoarchitectonic studies using cell-body-stained histological sections of *post mortem* brains, as detailed elsewhere [8, 9, 17]. Briefly, those studies were each based on five men and five women without any clinical history of neurological or psychiatric diseases, whose brains were removed from the skull within 24 h after death and fixated. After fixation, each brain underwent structural MRI scanning and was embedded in paraffin, cut into 20 μm serial sections, and stained for cell bodies. Using the cell-body-stained sections, the borders between distinct cortical areas were detected by quantifying the changes of the laminar pattern and validated by statistical testing, as described elsewhere [18, 19]. Effectively this defines the location and extent of distinct cytoarchitectonic areas for each subject. Subsequently, these areas were digitized, warped into MNI single subject space (*see* **Note 3**), and converted into probabilities translating the information on location and extent from each subject into one single 3D probability map per structure. At the end, each voxel within a given 3D probability map contains a count of how many brains (out of ten) have that voxel labeled as the respective brain structure, therefore coding interindividual cytoarchitectonic variability in MNI space (*see* **Note 3**). Figure 1 illustrates the probability maps (in selected sagittal planes) for three randomly chosen subregions of the hippocampal complex: cornu ammonis, fascia dentata, and subiculum. The entire list of available probability maps can be found at http://www.fz-juelich.de/inm/inm-1/EN/Forschung/_docs/SPMAnatomyToolbox/SPMAnatomyToolbox_node.html.

[1] As earlier versions of the maps may have been calculated differently, we recommend to always use the most recent maps to ensure that resulting volumes are as accurate as possible.

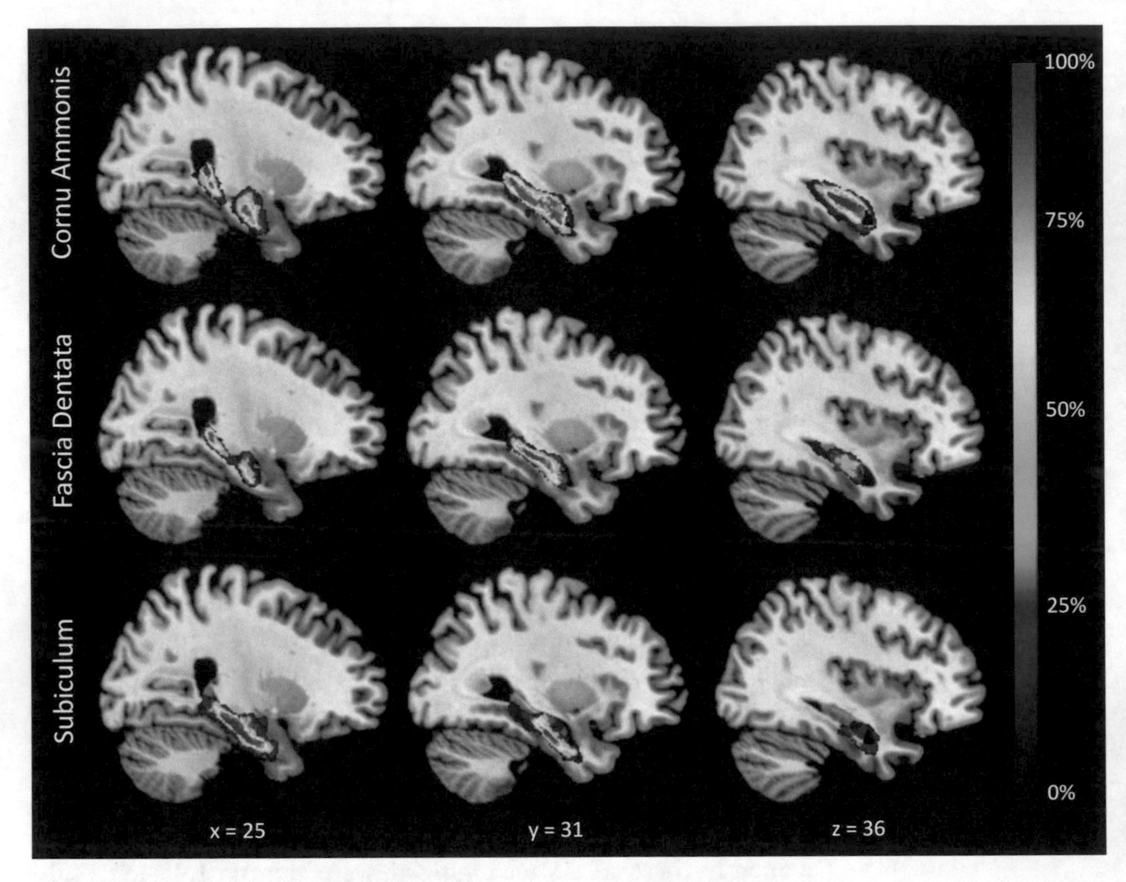

Fig. 1 Probabilistic maps of the cornu ammonis (top row), fascia dentata (middle row), and subiculum (bottom row) overlaid onto the ICBM single subject brain template. For each region, the voxel-wise probability is coded in color corresponding to the bar on the right (0–100%)

3 Methods

The proposed cytoarchitectonic ROI approach comprises three basic steps: [1] selecting the desired cytoarchitectonic probability maps, [2] preprocessing the T1-weighted images, and [3] integrating the cytoarchitectonic tissue probabilities with the MR-based signal intensities, as further detailed below and illustrated in Fig. 2.

3.1 Selection of the Cytoarchitectonic Probability Maps

Obviously, the selection of the particular region(s) of interest and thus the respective probability map(s) for the analysis should be hypothesis-driven. For example, in a study on speech production, it seems reasonable to focus on Broca's region, which consists of Areas 44 and 45 in the left hemisphere [20], rather than, for example, including primary somatosensory or visual areas. However, it is important to know that not the entire brain has been mapped yet as creating cytoarchitectonic probabilistic labels is

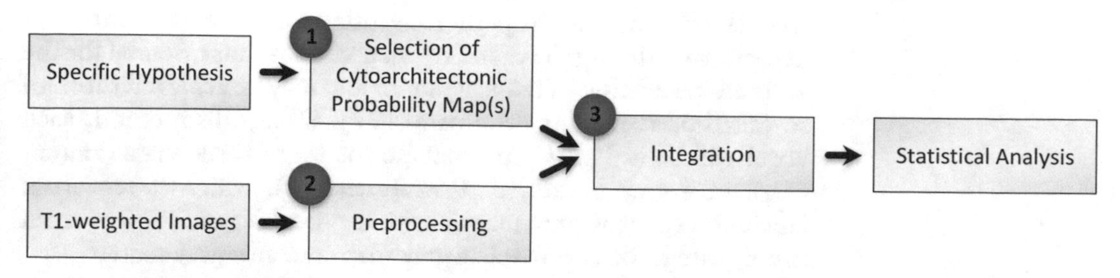

Fig. 2 The three main steps of the cytoarchitectonic ROI approach are numbered in red. Each analysis starts out with a region-specific hypothesis and a set of T1-weighted brain images. The region-specific hypothesis determines the (1) selection of the respective cytoarchitectonic probability map(s). Then, the T1-weighted images need to undergo adequate (2) preprocessing, comprising of tissue classification and spatial normalization. Finally, during the (3) integration, the regional volumes are calculated by multiplying the voxel-wise gray matter (using the preprocessed T1-weighted images) with the voxel-wise probability (using the cytoarchitectonic probability maps). The resulting regional volumes constitute the input for the statistical analysis

extremely labor-intensive. In other words, while maps for many brain regions are already available, there are still some gaps which hopefully will be successively closed over the coming years. Finally, one should be aware that the proposed cytoarchitectonic ROI analysis (or any other ROI analysis for that matter) might require a correction for multiple comparisons if more than just one region is included. Consequently, in the case of open regional hypotheses (or an interest in large parts of the brain), it might be prudent to consider a whole-brain approach, such as voxel-based morphometry (VBM), rather than an ROI approach.

3.2 Preprocessing of the T1-weighted Images: Tissue Classification and Spatial Normalization

First, all T1-weighted images have to undergo a tissue segmentation in order to isolate the gray matter. Then, the isolated gray matter needs to be spatially normalized in order to ensure spatial correspondence across subjects as well as to facilitate integration with the cytoarchitectonically defined ROIs. These steps, detailed below, are similar to the ones applied in standard VBM analyses [21–23] and can be conducted using the CAT12 toolbox (*see* **Note 2**).

Tissue segmentation is based on intensity values and basically serves to classify the T1-weighted image as gray matter, white matter, cerebrospinal fluid, and background [24–26]. The segmentation process is usually preceded or accompanied by a correction for image intensity inhomogeneities (also known as bias correction) that were caused by the magnetic field of the scanner. Also note that, given the common voxel size of $1 \times 1 \times 1$ mm^3, any given voxel may contain more than one tissue, which leads to ambiguous signal intensities. This is generally the case at the border between brain parenchyma and cerebrospinal fluid, at boundaries between gray and white matter, and in structures where white matter fibers cross the gray matter. Properly accounting for those tissue mixtures in each voxel during the tissue segmentation is thus essential for the accurate calculation of tissue volumes and may be achieved through so-called partial volume estimates [27]. Using this method, each voxel will be assigned a percentage for every tissue class content (e.g., 70% gray matter, 30% white matter), which differs from labeling a voxel as gray matter with a probability of 70% or from simply categorizing a voxel as gray matter or not gray matter.

After the tissue segmentation, the resulting gray matter segments in native space must be spatially normalized to correspond to a common reference space (also *see* **Note 3**), so that a voxel-wise comparability is guaranteed—both across subjects as well as between gray matter segments and cytoarchitectonic ROIs [24, 25, 28, 29]. While brains vary greatly in their local anatomy across individuals, modern normalization techniques make it possible to achieve a reasonable local comparability across brains [28] and also between gray matter segments and cytoarchitectonic ROIs. However, some possible issues need consideration: first, spatial normalization alters the brain size globally and locally which, in turn, will affect the quantification of tissue volumes in normalized images. This can be amended by modulating the normalized gray matter segments (i.e., applying a correction for the local and global volume changes occurring during normalization), as further described elsewhere [22, 23, 30]. If one is interested in brain-size-independent effects, it is recommended to enter brain volume as a covariate in the statistical model. Last but not least, it needs to be pointed out that a registration algorithm, such as DARTEL [28], works using a study-specific reference space. This space is slightly different from the anatomical MNI space of the cytoarchitectonic probability map (*see* **Note 3**). Thus, to properly match the two spaces, an additional spatial transformation must be calculated and applied to bring the cytoarchitectonic probability map into the study-specific reference space.

3.3 Integration of Cytoarchitectonic Probabilities to Calculate ROI Volumes

The preprocessing step leaves the gray matter isolated and eventually in the same space as the cytoarchitectonic probability map. Since partial volumes were estimated during the segmentation step and preserved during the normalization procedure by modulation, the local gray matter volume within each voxel can be accurately quantified. That is, with the voxel size known (e.g., $1 \times 1 \times 1$ mm^3) as well as the relative amount of gray matter (e.g., 30%), both values can simply be multiplied to obtain the voxel-wise gray matter volume (e.g., 0.3 mm^3). This means further that the gray matter volume within any region of interest can be calculated as the sum of all its voxel-wise volumes. These considerations apply without restriction when using binary labels, where a voxel is either included/counted or not.

Things are slightly more complex when using non-binary labels, such as the cytoarchitectonic probability maps. As described above (*see* Section 2), cytoarchitectonic probability maps are unique in that they reflect the location, extent, and interindividual variability across the ten *post mortem* brains labeled cytoarchitectonically. Taken the cytoarchitectonic map of the subiculum as example, this means that a specific voxel has been assigned a 20% probability if it was apparent as subiculum in two brains (out of the ten brains). Thus, when calculating the voxel-wise volume using the aforementioned formula (e.g., $1 \times 1 \times 1\ \text{mm}^3 \times 30\% = 0.3\ \text{mm}^3$), one has to weigh the resulting voxel-wise volume by 20% (e.g., $0.3\ \text{mm}^3 \times 20\% = 0.06\ \text{mm}^3$). In other words, one needs to integrate the voxel-wise gray matter volume with the voxel-wise probability to belong to the structure that one is interested in.

Integrating the voxel-wise gray matter volumes with the voxel-wise probabilities is achieved by multiplying both sets of data at each voxel. This is possible because gray matter segments and cytoarchitectonic probability maps are in the same space—a consequence of the spatial normalization(s). The resulting gray matter volume of the ROI is then calculated as the sum of all resulting voxel-wise volumes and entered as dependent variable in the statistical analysis (there will be one value per ROI per subject).

4 Notes

1. Throughout the chapter, we have focused on the ROI analysis of *gray matter* volumes, as this was the emphasis of previous studies and will likely be the main interest in future analyses. However, the principles of this approach may also be extended to examine *white matter* or non-volumetric properties (e.g., diffusion or perfusion), although adaptations will be necessary in terms of data preprocessing. While those extended applications are beyond the scope of this chapter, it seems crucial to point out that all analyses—regardless whether based on T1-weighted, T2-weighted, or diffusion-weighted data—ultimately depend on (and thus are limited by) the available cytoarchitectonic probability maps. A list of all brain regions that have been mapped cytoarchitectonically is provided here: http://www.fz-juelich.de/inm/inm-1/EN/Forschung/_docs/SPMAnatomyToolbox/SPMAnatomyToolbox_node.html. We highly recommend to always use the latest version of these cytoarchitectonic probability maps.
2. The CAT12 toolbox (http://www.neuro.uni-jena.de/cat/) is a free and useful tool to perform both the tissue segmentation and the spatial normalization in one combined step. This toolbox is an extension to the spm12 software (http://www.fil.ion.

ucl.ac.uk/spm). Since the implemented tissue segmentation algorithm in this toolbox uses a partial volume estimation [27], it already satisfies all requirements for a cytoarchitectonic ROI analysis. A manual for the CAT12 toolbox is freely available (http://www.neuro.uni-jena.de/cat/).

3. Note that the cytoarchitectonic probability maps distributed with the Anatomy Toolbox [16] are provided in "anatomical MNI space" (aMNI). This space resembles the common MNI space with a slight spatial shift as the origin was set to the anterior commissure [31]. Consequently, all cytoarchitectonic probability maps need to be transformed from the aMNI space to the study-specific space, so probability maps and brain images can be properly multiplied (integrated). The spatial transformation can be easily achieved by calculating (and applying) the required parameters using the ICBM single subject brain in aMNI space ("colin27T1_seg.nii"), which is provided with the Anatomy Toolbox.

References

1. Broca P. (1861) Sur le Siége de la faculté du langage articulé, avec deux observations d'aphémie (perte de la parole), par le Dr Paul Broca,... Texte imprimé. V. Masson et fils, Paris
2. Brodmann K (1909) Vergleichende Lokalisationslehre der Grosshirnrinde. Verlag von Johann Ambrosius Barth, Leipzig
3. Amunts K, Schleicher A, Burgel U, Mohlberg H, Uylings HB, Zilles K (1999) Broca's region revisited: cytoarchitecture and intersubject variability. J Comp Neurol 412 (2):319–341
4. Caspers S, Geyer S, Schleicher A, Mohlberg H, Amunts K, Zilles K (2006) The human inferior parietal cortex: cytoarchitectonic parcellation and interindividual variability. NeuroImage 33 (2):430–448
5. Choi HJ, Zilles K, Mohlberg H, Schleicher A, Fink GR, Armstrong E, Amunts K (2006) Cytoarchitectonic identification and probabilistic mapping of two distinct areas within the anterior ventral bank of the human intraparietal sulcus. J Comp Neurol 495(1):53–69
6. Eickhoff SB, Schleicher A, Zilles K, Amunts K (2006) The human parietal operculum. I. Cytoarchitectonic mapping of subdivisions. Cereb Cortex 16(2):254–267
7. Kujovic M, Zilles K, Malikovic A, Schleicher A, Mohlberg H, Rottschy C, Eickhoff SB, Amunts K (2013) Cytoarchitectonic mapping of the human dorsal extrastriate cortex. Brain Struct Funct 218(1):157–172
8. Amunts K, Schleicher A, Zilles K (2007) Cytoarchitecture of the cerebral cortex--more than localization. NeuroImage 37 (4):1061–1065
9. Zilles K, Amunts K (2010) Centenary of Brodmann's map--conception and fate. Nat Rev Neurosci 11(2):139–145
10. Kurth F, Eickhoff SB, Schleicher A, Hoemke L, Zilles K, Amunts K (2010) Cytoarchitecture and probabilistic maps of the human posterior insular cortex. Cereb Cortex 20 (6):1448–1461
11. Scheperjans F, Eickhoff SB, Homke L, Mohlberg H, Hermann K, Amunts K, Zilles K (2008) Probabilistic maps, morphometry, and variability of cytoarchitectonic areas in the human superior parietal cortex. Cereb Cortex 18(9):2141–2157
12. Palomero-Gallagher N, Eickhoff SB, Hoffstaedter F, Schleicher A, Mohlberg H, Vogt BA, Amunts K, Zilles K (2015) Functional organization of human subgenual cortical areas: relationship between architectonical segregation and connectional heterogeneity. NeuroImage 115:177–190
13. Rottschy C, Eickhoff SB, Schleicher A, Mohlberg H, Kujovic M, Zilles K, Amunts K (2007) Ventral visual cortex in humans: cytoarchitectonic mapping of two extrastriate areas. Hum Brain Mapp 28(10):1045–1059
14. Kurth F, Cherbuin N, Luders E (2015) Reduced age-related degeneration of the

hippocampal subiculum in long-term meditators. Psychiatry Res 232(3):214–218
15. Luders E, Kurth F, Toga AW, Narr KL, Gaser C (2013) Meditation effects within the hippocampal complex revealed by voxel-based morphometry and cytoarchitectonic probabilistic mapping. Front Psychol 4:398
16. Eickhoff SB, Stephan KE, Mohlberg H, Grefkes C, Fink GR, Amunts K, Zilles K (2005) A new SPM toolbox for combining probabilistic cytoarchitectonic maps and functional imaging data. NeuroImage 25 (4):1325–1335
17. Zilles K, Schleicher A, Palomero-Gallagher N, Amunts K (2002) Quantitative analysis of cyto- and receptor architecture of the human brain. In: Mazziotta J, Toga A (eds) Brain mapping: the methods. Elsevier, USA, pp 573–602
18. Schleicher A, Amunts K, Geyer S, Kowalski T, Schormann T, Palomero-Gallagher N, Zilles K (2000) A stereological approach to human cortical architecture: identification and delineation of cortical areas. J Chem Neuroanat 20 (1):31–47
19. Schleicher A, Palomero-Gallagher N, Morosan P, Eickhoff SB, Kowalski T, de Vos K, Amunts K, Zilles K (2005) Quantitative architectural analysis: a new approach to cortical mapping. Anat Embryol (Berl) 210 (5–6):373–386
20. Kurth F, Jancke L, Luders E (2017) Sexual dimorphism of Broca's region: more gray matter in female brains in Brodmann areas 44 and 45. J Neurosci Res 95(1–2):626–632
21. Ashburner J, Friston K (2007) Voxel-based Morphometry. In: Friston K, Ashburner J, Kiebel S, Nichols TE, Penny WD (eds) Statistical parametric mapping: the analysis of functional brain images. Elsevier, London, pp 92–100
22. Ashburner J, Friston KJ (2000) Voxel-based morphometry--the methods. NeuroImage 11 (6 Pt 1):805–821
23. Kurth F, Luders E, Gaser C (2015) Voxel-based Morphometry. In: Toga A (ed) Brain mapping: an encyclopedic reference. Academic Press, London, pp 345–349
24. Ashburner J, Friston KJ (2005) Unified segmentation. NeuroImage 26(3):839–851
25. Ashburner J, Friston K (1997) Multimodal image coregistration and partitioning--a unified framework. NeuroImage 6(3):209–217
26. Rajapakse JC, Giedd JN, Rapoport JL (1997) Statistical approach to segmentation of single-channel cerebral MR images. IEEE Trans Med Imaging 16(2):176–186
27. Tohka J, Zijdenbos A, Evans A (2004) Fast and robust parameter estimation for statistical partial volume models in brain MRI. NeuroImage 23(1):84–97
28. Ashburner J (2007) A fast diffeomorphic image registration algorithm. NeuroImage 38 (1):95–113
29. Ashburner J, Friston KJ (1999) Nonlinear spatial normalization using basis functions. Hum Brain Mapp 7(4):254–266
30. Good CD, Johnsrude IS, Ashburner J, Henson RN, Friston KJ, Frackowiak RS (2001) A voxel-based morphometric study of ageing in 465 normal adult human brains. NeuroImage 14(1 Pt 1):21–36
31. Amunts K, Kedo O, Kindler M, Pieperhoff P, Mohlberg H, Shah NJ, Habel U, Schneider F, Zilles K (2005) Cytoarchitectonic mapping of the human amygdala, hippocampal region and entorhinal cortex: intersubject variability and probability maps. Anat Embryol 210 (5–6):343–352

Check for updates

Chapter 9

Morphometry of the Corpus Callosum

Eileen Luders, Paul M. Thompson, and Florian Kurth

Abstract

The corpus callosum is the largest commissure in the human brain and the principal connection transmitting information between the two hemispheres. Importantly, midsagittal callosal size scales with the number of small diameter fibers crossing through, suggesting that a larger corpus callosum relays more connections between the hemispheres than a small one. Moreover, fibers of the corpus callosum are arranged following a specific organization implying that regional callosal size is functionally significant. This has sparked considerable interest in the neuroscience community to develop and refine approaches to analyze callosal morphology not only in terms of its overall size but also with respect to its local dimensions. In this chapter, we first briefly review traditional parcellation schemes allowing for local measures of callosal morphology. Subsequently, we detail a newer computational technique, commonly referred to as "callosal thickness" approach, measuring callosal distances at 100 equally spaced nodes. To demonstrate an application of this approach, we examined callosal thickness in relation to chronological age in 72 healthy subjects (36 men, 36 women) aged between 30 and 69 years. The chapter then progresses by summarizing advantages and potential drawbacks of this approach and concludes with practical tips and helpful pointers for conducting a callosal thickness analysis.

Key words Brain, Corpus Callosum, Imaging, Morphology, Morphometry, MRI, Thickness

1 Introduction

Rapid technological developments in the fields of neuroimaging and brain mapping have been accompanied by an exponential increase in the number of investigations exploring the anatomy of the human brain, with many studies focusing on the corpus callosum. This prominent white matter fiber tract is located about 10 cm deep within the interhemispheric fissure and interconnects the two hemispheres via over 200 million fibers. The majority of callosal fibers are homotopic in nature connecting corresponding cortical areas, although some heterotopic callosal connections exist as well. The corpus callosum is topographically organized, where fibers connecting anterior brain regions travel primarily through rostral

Gianfranco Spalletta et al. (eds.), *Brain Morphometry*, Neuromethods, vol. 136,
https://doi.org/10.1007/978-1-4939-7647-8_9,

callosal sections and fibers connecting posterior regions travel through caudal callosal sections. There are regional differences in the diameter of callosal fibers and their density across the corpus callosum along the rostro-caudal axis. However, perhaps surprisingly, there is no significant link between fiber density and midsagittal callosal area size, and a larger area is assumed to reflect a larger number of fibers crossing through. This link was originally established, at least for small diameter fibers, using *post mortem* data [1]. Today, this important relationship is the foundation for interpreting callosal findings in most imaging studies, where midsagittal callosal area size can be easily determined.

2 Callosal Parcellation

As different sections of the corpus callosum contain fibers that connect different parts of the brain, it seems appropriate to measure the size of callosal subsections, rather than merely the size of the corpus callosum as a whole. For this purpose, various approaches have subdivided the corpus callosum into different segments along the rostro-caudal axis at the midsagittal section of the brain (*see* Fig. 1), which is followed by calculating the resulting segment-specific midsagittal areas.

For example, as illustrated in **panel A**, after rotating the corpus callosum to maximize its anterior-posterior length, it may be divided into vertical partitions based on equal fractions of its entire length. Alternatively, equiangular radial partitions may be created based on rays emanating either from the callosal centroid or the midpoint of an artificial line joining the most inferior points within the anterior and posterior section. As illustrated in **panel B**, other parcellation schemes derive a curvilinear reference and place a set of equally spaced nodes on that line. Callosal segments are defined by connecting upper and lower callosal boundaries through these nodes, ranging from only a few to up to 100.

Out of all traditional parcellations, however, the one most frequently used is the so-called Witelson scheme [2]. As illustrated in **panel C**, this scheme separates the corpus callosum into five or seven vertical partitions based on defined fractions of its maximum anterior-posterior length. The Witelson partition has generated some controversy [3, 4], but it is still widely used in the neuroscience community, either to analyze callosal morphology or as a frame of reference[1] when describing the location of analysis outcomes or when relating findings across studies.

[1] For the sake of clarity, when describing the tracing procedure (Sect. 3) and the location of effects (Sect. 4), we will also refer to the callosal segments based on the Witelson scheme.

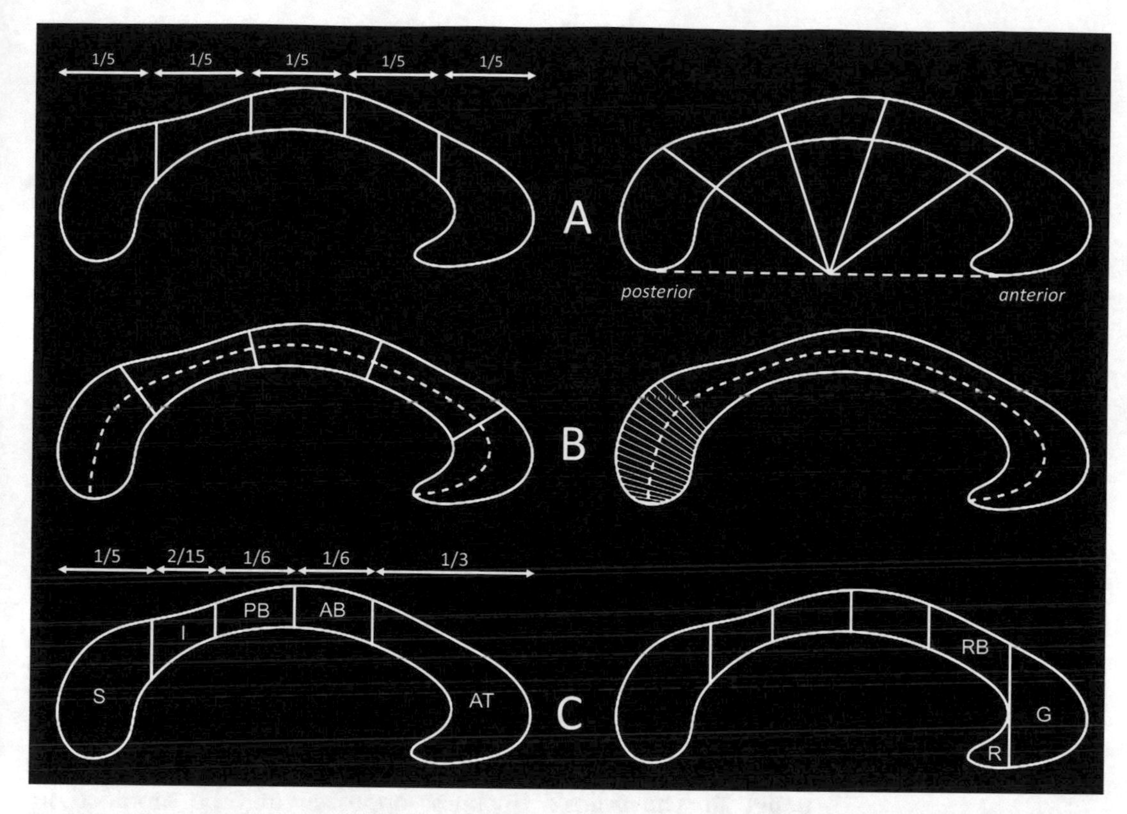

Fig. 1 Common callosal parcellation schemes. **Panel A**: The midsagittal surface of the corpus callosum may be divided into vertical partitions based on equal fractions of its maximum length (left) or equiangular sectors relative to the midpoint of an artificial line joining the two most inferior points posteriorly and anteriorly (right). **Panel B**: Subdivisions may also be generated using a curvilinear reference with equally spaced nodes at which upper and lower callosal boundaries are connected to define partitions. The number of resulting partitions depends on the number of nodes, ranging between a few (left) and up to 100 (right). Note that callosal parcellations shown in panels A and B—reflecting either true or adapted schemes [47–52]—are not exhaustive but merely provide examples of how to dissect the corpus callosum. **Panel C**: The well known Witelson scheme [2] divides the corpus callosum into five (left) or seven (right) vertical partitions based on fractions of its maximum length, such as the splenium [S] representing the posterior fifth, the isthmus [I] occupying two fifteenths, the posterior midbody [PB] and the anterior midbody [AB] each one sixth, and the anterior third [AT], which itself may be further subdivided into rostral body [RB], genu [G], and rostrum [R], by placing a vertical line through the most anterior point on the inner convexity of the anterior corpus callosum

3 Callosal Thickness

In the following, we will detail a newer computational approach measuring callosal distances at 100 equally spaced nodes. The measurement is widely known under the term "callosal thickness" and has been applied in a wide range of studies investigating, for example, links between callosal thickness and biological sex [5, 6], developmental stages [7, 8], mindfulness practices [9], intelligence and cognitive performance [10–12], handedness and hand motor performance [13, 14], as well as in association with several disorders and diseases [15–27]. Moreover, the callosal thickness approach has been applied to investigate the relationship between callosal size and brain asymmetry [28] and even to determine the asymmetry of the corpus callosum itself [29].

The callosal thickness approach, illustrated in Fig. 2, requires that two separate callosal outlines are generated at each brain's midsagittal section (*see* **Note 1**), one following the upper and one following the lower callosal boundary. Both outlines, either created manually or automatically, start at the tip of the callosal rostrum and converge at the most ventral part of the splenium. An example of the two resulting outlines is shown in **panel A**. To obtain the highly localized measures, the upper and lower callosal boundaries are first separated into 100 nodes and re-sampled at regular intervals rendering the discrete points comprising the two boundaries spatially uniform (**panel B**). Then, a new midline curve is created by calculating the 2D average from the 100 equidistant nodes representing the upper and the lower callosal boundaries (**panel C**). Finally, the distances between the 100 nodes of the upper as well as the lower callosal boundaries to the 100 nodes of the midline curve are calculated (**panel D**). These distances (in mm) indicate callosal thickness at 100 locations distributed evenly over the callosal surface. An example of a color-coded distance map is shown in **panel E**. The callosal thickness approach may be modified by increasing or decreasing the number of nodes or by varying the way distances are measured, such as perpendicular to the midline curve, following the shortest distance between the upper and lower callosal boundaries, etc. Numerous refined and sophisticated variations of this approach have been developed and used across studies (for selected applications and comprehensive summaries, please refer to [30–36]).

To obtain statistics from the callosal thickness approach, a mass univariate analysis can be performed. More specifically, the resulting point-wise distances are analyzed simultaneously via t-tests, analyses of variance, or regression analyses, with or without accounting for possible confounds (*see* **Note 2**). The resulting point-specific statistics are then mapped back (and color-coded) onto their respective locations, in the same way as shown for the point-wise distance measures (**panel E**). Given the large number of statistical tests, it is essential to apply corrections for multiple comparisons. While Bonferroni procedures would be overly conservative (as the measurements are not independent), controlling the false discovery rate [37] or permutation testing is suitable.

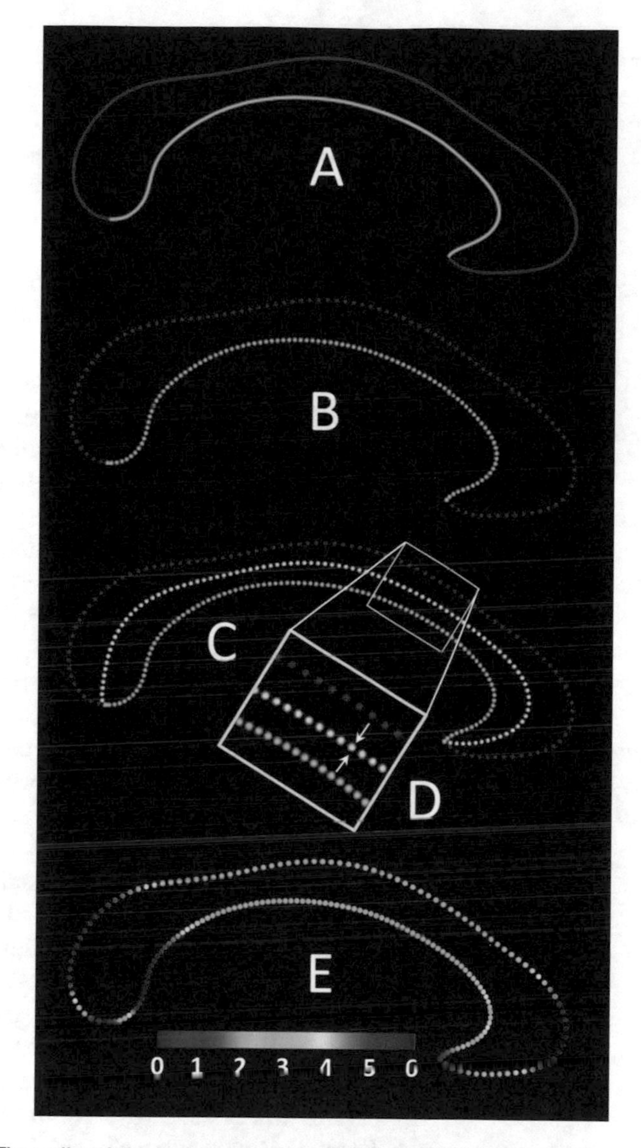

Fig. 2 The callosal thickness approach. **Panel A**: The corpus callosum is traced at the midsagittal section, generating two separate outlines by following the upper (red) and lower (cyan) callosal boundaries. **Panel B**: The two separate outlines are each redigitized into 100 equidistant nodes. **Panel C**: A new midline curve (yellow) is created by calculating the 2D average from the 100 equidistant nodes representing the upper and the lower callosal boundaries. **Panel D**: The distances between the 100 nodes of the upper as well as the lower callosal boundaries to the 100 nodes of the midline curve are calculated. **Panel E**: The resulting point-wise distances (in mm) may be color-coded (*see* color bar) and mapped back onto the callosal outlines

4 Example Application

To demonstrate an application of the callosal thickness approach, including data processing and interpretation of findings, we will provide an example analysis investigating the link between aging and callosal morphology in healthy subjects.

4.1 Subjects and Imaging Data

Seventy-two subjects (36 men, 36 women) were selected from the ICBM database of normal adults (http://www.loni.usc.edu/ICBM/Databases/). As we aimed to illustrate significant effects of age, we deliberately excluded subjects younger than 30 years. We were also interested in possible interactions between age and sex as revealed for other brain structures [38–41], so we pair-wise matched men and women as closely as possible for age, with similar resulting mean ages and standard deviations in males (46.77 ± 10.69 years) and in females (48.31 ± 10.91 years). The age range for both men and women was 30–69 years. All brain data was collected on a 1.5 Tesla scanner (Siemens Sonata) using an 8-channel head coil and a T1-weighted sequence, as detailed elsewhere [42, 43]. The image resolution was $1 \times 1 \times 1$ mm^3.

4.2 Image Processing and Callosal Tracing

Although the corpus callosum is usually easy to identify in a standard T1-weighted brain image, correcting intensity drifts caused by magnetic field inhomogeneities can further enhance the gray/white contrast. Thus, automated bias field corrections were first applied to all image volumes making the course of the callosal boundaries clearer. In addition, all image volumes were spatially normalized to a template in MNI standard space using automated six-parameter (rigid-body) transformations ensuring that all brains have the same orientation and alignment (*see* **Note 3**). Using the bias-corrected and normalized image volumes, the corpus callosum was traced in each brain as described above, and the resulting upper and lower callosal boundaries were processed accordingly (*see* Sect. 3). Eventually, the thickness of the corpus callosum was calculated at 100 locations distributed evenly over the callosal surface. In addition, a smooth callosal contour was derived as the spatial average from all upper and lower callosal boundaries ($n = 72$, each). This study-specific template will serve to visualize the outcomes (*see* Fig. 3).

4.3 Descriptive Statistics

As shown in Fig. 3, when calculating the mean thickness, the largest measures are evident at the anterior and posterior callosal bend within the genu and splenium (**panel A**; *top*). Other descriptors of interest, such as the standard deviation, variance, median, mode, minimum, maximum, etc., may also be calculated (**panel A**; *bottom*). This shows one of the many benefits of this approach: by color-coding the respective values and mapping them back onto

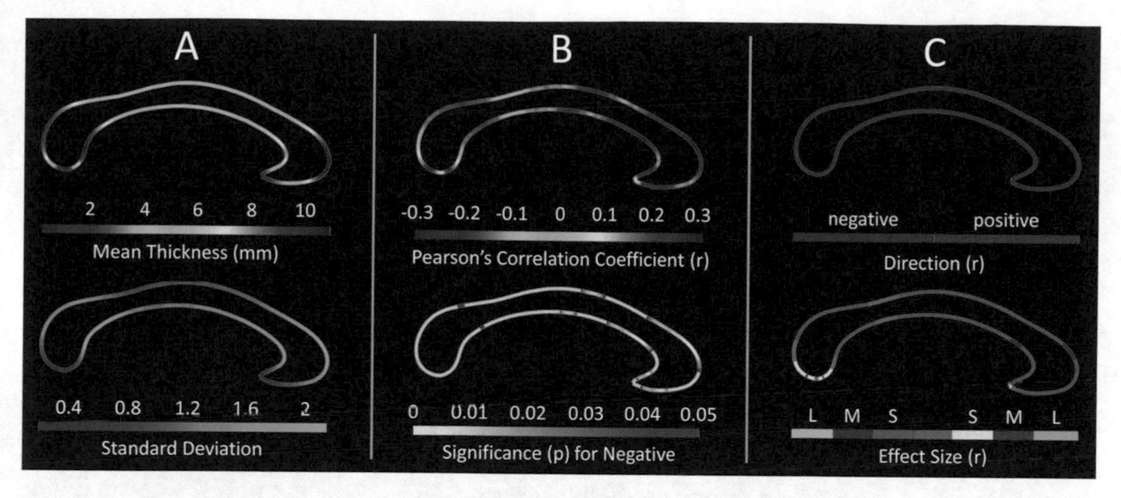

Fig. 3 Correlations between callosal thickness and chronological age. Shown are the descriptive and inferential statistics in a sample of 72 subjects (36 men, 36 women) ranging between 30 and 69 years. **Panel A**: Mean thickness and standard deviation. **Panel B**: Pearson's correlation coefficients (*r*) and significance values (*p*) for negative correlations; nonsignificant effects are shown in gray. **Panel C**: Direction of the correlation and effect sizes, encoding small [S], medium [M], and large [L] effects, according to $|r| \geq 0.1$, $|r| \geq 0.3$, and $|r| \geq 0.5$; negligible effects ($-0.1 < r < 0.1$) are shown in gray

their original location, one is provided with a graphical representation of the data, rather than just numeric values as in the case of applying standard callosal parcellations and obtaining area measures (*see* Sect. 2).

4.4 Inferential Statistics

To investigate the effect of aging on callosal morphology, we computed Pearson correlations between the point-wise callosal distances and chronological age. The analysis was performed using a general linear model (GLM), where age was included as the independent variable, total intracranial volume[2] as a nuisance variable, and the point-wise callosal distances as the dependent variable.[3] Correlation coefficients (r) can range from -1 to $+1$ indicating the magnitude and direction of the effect. However, as shown in Fig. 3, our r-scale was truncated at -0.3 and $+0.3$, based on the actual range of the r-values (**panel B**, *top*).

To establish significance, alpha was set at 0.05, and the resulting point-wise significance (p) values were used to create the significance profile (**panel B**; *bottom*) for negative correlations (there were no significant positive correlations). Permutation testing with 10,000 iterations was employed as previously detailed [19]

[2] Total intracranial volume was calculated automatically using the VBM8 toolbox (http://dbm.neuro.uni-jena.de/vbm/download/) in SPM8 (http://www.fil.ion.ucl.ac.uk/spm/software/spm8/).

[3] A second GLM was estimated testing for possible interactions between age and sex (also while removing the variance associated with total intracranial volume). However, there were no significant effects. Thus, all subsequent descriptions refer to outcomes of the first GLM only.

to confirm significance. Altogether, our analyses indicate that the corpus callosum is thinner in older people, with pronounced age effects within the rostrum, rostral body, and anterior midbody and more subtle effects within the isthmus (*see* **Note 4**).

As further illustrated in **panel C,** other descriptors of interest, such as the simple direction of the correlation (*top*) or effect sizes, coded either on a continuous scale or as discrete indices (*bottom*) may be generated. Of note, regardless of significance, callosal thickness and age are almost exclusively negatively correlated (only the posterior splenium shows a positive link). Large effects ($r \leq -0.5$ and ≥ 0.5) are entirely absent, and medium effects ($r \leq -0.3$ and ≥ 0.3) only occur within the rostral body and tip of the splenium, while small effects ($r \leq -0.1$ and ≥ 0.1) prevail in almost all remaining areas. Two sections located within splenium and posterior midbody show negligible effects ($-0.1 < r < 0.1$).

5 Advantages and Drawbacks

In addition to the aforementioned benefit—a graphical representation of the data that is easy to grasp and comprehend—the callosal thickness approach offers several other advantages. The regional specificity is much higher (100 data points) in comparison to most of the segment-specific approaches, which usually yield between five and seven data points. This increases the detail of scientific observation and may, under some circumstances, also be more sensitive in detecting significant effects. For example, when applying traditional parcellation schemes, effects may be missed if significant correlations only exist in a small portion of a callosal segment or right at the border between two segments. Similarly, if both positive and negative correlations occurred in a single callosal segment, they may cancel each other out. Thus, the high regional specificity of the callosal thickness approach is clearly a benefit. However, it also comes with a drawback. Given the required corrections for multiple comparisons (*see* Sect. 4), effects hovering at the brink of significance may not survive and thus remain undetected. Notwithstanding, altogether, the callosal thickness approach should be considered a suitable and attractive technique, either as alternative or complimentary to traditional area measurements. Importantly, its outcomes are easy to relate to other findings based on any parcellation scheme (as also done in this chapter when referring to the Witelson scheme). Moreover, such comparisons are not only possible within the same imaging modality but also across modalities. For example, according to schemes derived using DTI-based fiber tractography [4, 44], the negative correlations occurring in anterior callosal regions, as observed in our MRI-based approach, may indicate age effects in fibers connecting predominantly prefrontal, premotor, and supplementary motor regions.

6 Notes

1. Callosal fibers fan out differently away from the brain's midline, and significant parasagittal asymmetries are already evident only 4–6 mm into each hemisphere [29]. Thus, it is essential to trace the corpus callosum in the sagittal section as closely to midline as possible (preferably midsagittal) and to be consistent across subjects. Standard visualization programs usually offer simultaneous views of sagittal, coronal, and axial sections, and the brain's midline can be located by scrolling through the image volume and identifying the interhemispheric fissure in the coronal or axial plane. Then, within the sagittal plane, the midsection (midsagittal) may also be confirmed by the presence of the septum pellucidum, the double membrane separating the anterior horns of the two lateral ventricles.
2. The inclusion of nuisance variables strongly depends on the individual dataset and research question, but total intracranial volume (or similar indicators of brain size) should be considered as a possible confound as brain dimensions and callosal dimensions (including callosal thickness) are closely related [5, 45, 46].
3. Outlining callosal boundaries in native space is not recommended. Due to different head positions in the scanner, there are large variations in the orientation of the brains. This may become a problem during the callosal tracing procedure: different brain tilts across subjects along the rostro-caudal axis would lead to variable end points at the inferior tip of the splenium; different brain tilts across subjects along the other two axes may result in oblique/slanted slices and affect the general (midsagittal) outline of the corpus callosum. Thus, it is imperative to standardize all brain volumes in their orientation and alignment via applying six-parameter (rigid-body) transformations to a common space.
4. Importantly, given the highly selective sample (*see* Sect. 4.1), the derived callosal maps should be interpreted with caution and not be used as a frame of reference in terms of absolute numbers. Nevertheless, our current findings seem to indicate that anterior callosal sections—especially rostrum, rostral body, and anterior midbody—show pronounced and stronger age-related correlations than posterior callosal regions, at least between the ages of 30 and 69 years. While these significant negative correlations observed cross-sectionally may reflect actual age-related decreases over time, longitudinal studies are needed to confirm this.

References

1. Aboitiz F, Scheibel AB, Fisher RS, Zaidel E (1992) Fiber composition of the human corpus callosum. Brain Res 598(1–2):143–153
2. Witelson SF (1989) Hand and sex differences in the isthmus and genu of the human corpus callosum. A postmortem morphological study. Brain 112(Pt 3):799–835
3. Tomaiuolo F, Scapin M, Di Paola M, Le Nezet P, Fadda L, Musicco M, Caltagirone C, Collins DL (2007) Gross anatomy of the corpus callosum in Alzheimer's disease: regions of degeneration and their neuropsychological correlates. Dement Geriatr Cogn Disord 23 (2):96–103
4. Hofer S, Frahm J (2006) Topography of the human corpus callosum revisited--comprehensive fiber tractography using diffusion tensor magnetic resonance imaging. NeuroImage 32 (3):989–994
5. Luders E, Toga AW, Thompson PM (2014) Why size matters: differences in brain volume account for apparent sex differences in callosal anatomy: the sexual dimorphism of the corpus callosum. NeuroImage 84:820–824
6. Luders E, Narr KL, Zaidel E, Thompson PM, Toga AW (2006) Gender effects on callosal thickness in scaled and unscaled space. Neuroreport 17(11):1103–1106
7. Chavarria MC, Sanchez FJ, Chou YY, Thompson PM, Luders E (2014) Puberty in the corpus callosum. Neuroscience 265:1–8
8. Luders E, Thompson PM, Toga AW (2010) The development of the corpus callosum in the healthy human brain. J Neurosci 30 (33):10985–10990
9. Luders E, Phillips OR, Clark K, Kurth F, Toga AW, Narr KL (2012) Bridging the hemispheres in meditation: thicker callosal regions and enhanced fractional anisotropy (FA) in long-term practitioners. NeuroImage 61 (1):181–187
10. Luders E, Thompson PM, Narr KL, Zamanyan A, Chou YY, Gutman B, Dinov ID, Toga AW (2011) The link between callosal thickness and intelligence in healthy children and adolescents. NeuroImage 54 (3):1823–1830
11. Luders E, Narr KL, Bilder RM, Thompson PM, Szeszko PR, Hamilton L, Toga AW (2007) Positive correlations between corpus callosum thickness and intelligence. NeuroImage 37(4):1457–1464
12. Westerhausen R, Luders E, Specht K, Ofte SH, Toga AW, Thompson PM, Helland T, Hugdahl K (2011) Structural and functional reorganization of the corpus callosum between the age of 6 and 8 years. Cereb Cortex 21 (5):1012–1017
13. Kurth F, Mayer EA, Toga AW, Thompson PM, Luders E (2013) The right inhibition? Callosal correlates of hand performance in healthy children and adolescents callosal correlates of hand performance. Hum Brain Mapp 34 (9):2259–2265
14. Luders E, Cherbuin N, Thompson PM, Gutman B, Anstey KJ, Sachdev P, Toga AW (2010) When more is less: associations between corpus callosum size and handedness lateralization. NeuroImage 52(1):43–49
15. Di Paola M, Luders E, Di IF, Cherubini A, Passafiume D, Thompson PM, Caltagirone C, Toga AW, Spalletta G (2010) Callosal atrophy in mild cognitive impairment and Alzheimer's disease: different effects in different stages. NeuroImage 49(1):141–149
16. Luders E, Di Paola M, Tomaiuolo F, Thompson PM, Toga AW, Vicari S, Petrides M, Caltagirone C (2007) Callosal morphology in Williams syndrome: a new evaluation of shape and thickness. Neuroreport 18(3):203–207
17. Minnerop M, Luders E, Specht K, Ruhlmann J, Schimke N, Thompson PM, Chou YY, Toga AW, Abele M, Wullner U, Klockgether T (2010) Callosal tissue loss in multiple system atrophy--a one-year follow-up study. Mov Disord 25(15):2613–2620
18. Rusch N, Luders E, Lieb K, Zahn R, Ebert D, Thompson PM, Toga AW, van Elst LT (2007) Corpus callosum abnormalities in women with borderline personality disorder and comorbid attention-deficit hyperactivity disorder. J Psychiatry Neurosci 32(6):417–422
19. Luders E, Narr KL, Hamilton LS, Phillips OR, Thompson PM, Valle JS, Del'Homme M, Strickland T, McCracken JT, Toga AW, Levitt JG (2009) Decreased callosal thickness in attention-deficit/hyperactivity disorder. Biol Psychiatry 65(1):84–88
20. Weber B, Luders E, Faber J, Richter S, Quesada CM, Urbach H, Thompson PM, Toga AW, Elger CE, Helmstaedter C (2007) Distinct regional atrophy in the corpus callosum of patients with temporal lobe epilepsy. Brain 130(Pt 12):3149–3154
21. Minnerop M, Luders E, Specht K, Ruhlmann J, Schneider-Gold C, Schroder R, Thompson PM, Toga AW, Klockgether T, Kornblum C (2008) Grey and white matter loss along cerebral midline structures in myotonic dystrophy type 2. J Neurol 255(12):1904–1909

22. Zito G, Luders E, Tomasevic L, Lupoi D, Toga AW, Thompson PM, Rossini PM, Filippi MM, Tecchio F (2014) Inter-hemispheric functional connectivity changes with corpus callosum morphology in multiple sclerosis. Neuroscience 266:47–55
23. Schneider C, Helmstaedter C, Luders E, Thompson PM, Toga AW, Elger C, Weber B (2014) Relation of callosal structure to cognitive abilities in temporal lobe epilepsy. Front Neurol 5:16
24. Walterfang M, Luders E, Looi JC, Rajagopalan P, Velakoulis D, Thompson PM, Lindberg O, Ostberg P, Nordin LE, Svensson L, Wahlund LO (2014) Shape analysis of the corpus callosum in Alzheimer's disease and frontotemporal lobar degeneration subtypes. J Alzheimers Dis 40(4):897–906
25. Freitag CM, Luders E, Hulst HE, Narr KL, Thompson PM, Toga AW, Krick C, Konrad C (2009) Total brain volume and corpus callosum size in medication-naive adolescents and young adults with autism spectrum disorder. Biol Psychiatry 66(4):316–319
26. Bearden CE, van Erp TG, Dutton RA, Boyle C, Madsen S, Luders E, Kieseppa T, Tuulio-Henriksson A, Huttunen M, Partonen T, Kaprio J, Lonnqvist J, Thompson PM, Cannon TD (2011) Mapping corpus callosum morphology in twin pairs discordant for bipolar disorder. Cereb Cortex 21(10):2415–2424
27. Anastasopoulou S, Kurth F, Luders E, Savic I (2016) Generalized epilepsy syndromes and callosal thickness: differential effects between patients with juvenile myoclonic epilepsy and those with generalized tonic-clonic seizures alone. Epilepsy Res 129:74–78
28. Cherbuin N, Luders E, Chou YY, Thompson PM, Toga AW, Anstey KJ (2013) Right, left, and center: how does cerebral asymmetry mix with callosal connectivity? Hum Brain Mapp 34 (7):1728–1736
29. Luders E, Narr KL, Zaidel E, Thompson PM, Jancke L, Toga AW (2006) Parasagittal asymmetries of the corpus callosum. Cereb Cortex 16(3):346–354
30. Men W, Falk D, Sun T, Chen W, Li J, Yin D, Zang L, Fan M (2014) The corpus callosum of Albert Einstein's brain: another clue to his high intelligence? Brain J Neurol 137(Pt 4):e268
31. Joshi SH, Narr KL, Philips OR, Nuechterlein KH, Asarnow RF, Toga AW, Woods RP (2013) Statistical shape analysis of the corpus callosum in schizophrenia. NeuroImage 64:547–559
32. Herron TJ, Kang X, Woods DL (2012) Automated measurement of the human corpus callosum using MRI. Front Neuroinform 6:25
33. Adamson CL, Wood AG, Chen J, Barton S, Reutens DC, Pantelis C, Velakoulis D, Walterfang M (2011) Thickness profile generation for the corpus callosum using Laplace's equation. Hum Brain Mapp 32(12):2131–2140
34. Adamson C, Beare R, Walterfang M, Seal M (2014) Software pipeline for midsagittal corpus callosum thickness profile processing: automated segmentation, manual editor, thickness profile generator, group-wise statistical comparison and results display. Neuroinformatics 12(4):595–614
35. Walterfang M, Yucel M, Barton S, Reutens DC, Wood AG, Chen J, Lorenzetti V, Velakoulis D, Pantelis C, Allen NB (2009) Corpus callosum size and shape in individuals with current and past depression. J Affect Disord 115(3):411–420
36. Clarke S, Kraftsik R, Van d LH, Innocenti GM (1989) Forms and measures of adult and developing human corpus callosum: is there sexual dimorphism? J Comp Neurol 280(2):213–230
37. Hochberg Y, Benjamini Y (1990) More powerful procedures for multiple significance testing. Stat Med 9(7):811–818
38. Coffey CE, Lucke JF, Saxton JA, Ratcliff G, Unitas LJ, Billig B, Bryan RN (1998) Sex differences in brain aging: a quantitative magnetic resonance imaging study. Arch Neurol 55 (2):169–179
39. Xu J, Kobayashi S, Yamaguchi S, Iijima K, Okada K, Yamashita K (2000) Gender effects on age-related changes in brain structure. AJNR Am J Neuroradiol 21(1):112–118
40. Raz N, Gunning Dixon F, Head D, Rodrigue KM, Williamson A, Acker JD (2004) Aging, sexual dimorphism, and hemispheric asymmetry of the cerebral cortex: replicability of regional differences in volume. Neurobiol Aging 25(3):377–396
41. Luders E, Cherbuin N, Gaser C (2016) Estimating brain age using high-resolution pattern recognition: younger brains in long-term meditation practitioners. NeuroImage 134:508–513
42. Luders E, Gaser C, Narr KL, Toga AW (2009) Why sex matters: brain size independent differences in gray matter distributions between men and women. J Neurosci 29(45):14265–14270
43. Luders E, Toga AW, Thompson PM (2014) Why size matters: differences in brain volume account for apparent sex differences in callosal anatomy. NeuroImage 84:820–824

44. Zarei M, Johansen-Berg H, Smith S, Ciccarelli O, Thompson AJ, Matthews PM (2006) Functional anatomy of interhemispheric cortical connections in the human brain. J Anat 209(3):311–320
45. Jancke L, Staiger JF, Schlaug G, Huang Y, Steinmetz H (1997) The relationship between corpus callosum size and forebrain volume. Cereb Cortex 7(1):48–56
46. Jancke L, Preis S, Steinmetz H (1999) The relation between forebrain volume and midsagittal size of the corpus callosum in children. Neuroreport 10(14):2981–2985
47. Clarke JM, Zaidel E (1994) Anatomical-behavioral relationships: corpus callosum morphometry and hemispheric specialization. Behav Brain Res 64(1–2):185–202
48. Denenberg VH, Kertesz A, Cowell PE (1991) A factor analysis of the human's corpus callosum. Brain Res 548(1–2):126–132
49. Duara R, Kushch A, Gross-Glenn K, Barker WW, Jallad B, Pascal S, Loewenstein DA, Sheldon J, Rabin M, Levin B et al (1991) Neuroanatomic differences between dyslexic and normal readers on magnetic resonance imaging scans. Arch Neurol 48(4):410–416
50. Larsen JP, Hoien T, Odegaard H (1992) Magnetic-resonance-imaging of the corpus-callosum in developmental dyslexia. Cognitive Neuropsych 9(2):123–134
51. Ganjavi H, Lewis JD, Bellec P, MacDonald PA, Waber DP, Evans AC, Karama S (2011) Negative associations between corpus callosum midsagittal area and IQ in a representative sample of healthy children and adolescents. PLoS One 6(5):e19698
52. Weis S, Kimbacher M, Wenger E, Neuhold A (1993) Morphometric analysis of the corpus callosum using MR: correlation of measurements with aging in healthy individuals. AJNR Am J Neuroradiol 14(3):637–645

Check for updates

Chapter 10

Morphometry and Development: Changes in Brain Structure from Birth to Adult Age

Christian K. Tamnes and Ylva Østby

Abstract

This chapter gives an overview of the field of brain morphometry and development from birth to adult age, including selected methodological considerations and fields of application. Brain development is an area of research where morphometry studies have greatly increased our knowledge, revealing organized patterns where regional differences in cortical, subcortical, and white matter structural maturation play a role for cognitive development. Studies show that early rapid increases in gray matter structures are generally followed by decreases, whereas white matter continues to increase throughout childhood and adolescence. The chapter also highlights the importance of developmental perspectives in structural neuroimaging studies for our understanding of clinical conditions such as schizophrenia, autism spectrum disorders, and epilepsy.

Key words Adolescence, Autism spectrum disorder, Brain structure, Childhood, Cognition, Epilepsy, Infancy, Maturation, MRI, Schizophrenia

1 Introduction

Genes, maturation, and experience continuously interact to shape who we are, the brain, and our cognitive abilities at any point in time. Knowledge of human brain development was initially based in large part on postmortem studies and histological studies (examination of tissue under the microscope). While such studies provide invaluable information regarding the basic processes underlying brain development, they are unable to inform us fully about how different brain structures change over time within and between individuals and how these changes relate to behavioral and cognitive changes. Over the last couple of decades, however, the use of neuroimaging techniques, especially magnetic resonance imaging (MRI), has given us a much better understanding of how the brain changes during development [1] as well as throughout life [2].

Structural MRI provides high-quality, detailed images of brain anatomy. Using sophisticated analysis software, we can perform

Gianfranco Spalletta et al. (eds.), *Brain Morphometry*, Neuromethods, vol. 136,
https://doi.org/10.1007/978-1-4939-7647-8_10,

quantitative measurements of a range of different aspects of brain morphology for different types of brain tissue, such as gray matter, which includes both the cerebral cortex and a number of subcortical structures, and white matter, as well as for specific structures and regions. Morphometry, measurements of form, including size and shape, has been used to study both age-related differences across individuals (cross-sectional studies) and developmental changes by following the same individuals over time (longitudinal studies). The earliest longitudinal structural MRI project originated at the Child Psychiatry Branch of the National Institute of Mental Health [3]. Other MRI techniques can be used to examine, e.g., microstructural properties of fiber tracts in the brain (diffusion tensor imaging, DTI), brain activity during the performance of various tasks (functional MRI, fMRI), and intrinsic activity patterns in brain networks (resting state fMRI: rs-fMRI). Together, these techniques have provided new and exciting insights into the extensive and complex changes that occur in the brain from birth to adult age.

Studying brain development is a window into understanding characteristic features of child and adolescent behavior [4, 5], psychological and cognitive development [6], as well as emerging sex differences and the possible role of puberty-related hormonal changes, which differ dramatically for girls and boys [7, 8]. Also, many mental illnesses have their onset in adolescence or early adulthood, and the developing adolescent brain might, in combination with new social demands and stressors in this period of life, partly explain that [9, 10].

This chapter will in Sect. 2 begin with a brief discussion of a few selected methodological considerations that are of particular relevance for neuroimaging studies of brain development. For broader and more in-depth coverage of such issues, we refer the reader elsewhere [1, 11]. Section 3 will give an overview of typical development of brain morphology in infancy and throughout childhood and adolescence and how this relates to lifespan changes. We will also introduce some factors influencing individual differences in brain development. Section 4 will focus on atypical development of brain morphology in selected clinical populations and discuss the relationship between typical and atypical brain development from a dimensional perspective. Finally, Sect. 5 will consider the behavioral and cognitive relevance of structural brain development.

2 Methodological Considerations for Studies of Brain Development

To date there have been few studies of typical brain development in infancy and almost none that investigate the period between infancy and school age. The main reason for this is that young children are more likely to move while inside the MRI scanner, which results in lower image quality. While anesthesia is an option in clinical examinations, it is generally not used in research involving volunteers. MRI examinations of infants can be performed during natural sleep after feeding, but scanning of young children is more challenging, and studies of this age group are therefore particularly rare. Planning and preparation using instructional videos and mock scanner visits and friendly scanning operators can however help to alleviate the anxieties of young participants and improve image quality. It is sometimes also helpful to let young participants come to the scanner facilities a few days prior to their appointment, to see it for themselves. That way, there is less pressure mixed with apprehension on the day of their scan.

Quality control procedures, both pre- and post-processing, are of great importance in brain morphometry studies to reduce noise in the data and guard against spurious findings. In the context of development, this was clearly shown in a study by Ducharme and colleagues [12]. Their results showed that post-processing quality control, in the form of exclusion of scans defined as quality control failures on the basis of visual inspection and review of extreme values, had a large impact on identified developmental trajectories for cortical thickness from mid-childhood to early adulthood, with a shift toward more complex trajectories when including scans of lower quality. An image quality issue of particular importance is motion-related artifacts, which may greatly affect results in all types of imaging studies, especially in studies of development. It is reasonable to assume that such artifacts are more common in younger participants and thus can be confounded with age or time-point effects. While this issue has received increasing attention in fMRI studies in the past years, structural MRI studies would also benefit greatly from an increased focus on quantitative motion detection and measurement, as well as increased use of both prospective and retrospective motion compensation procedures, and the inclusion of such procedures in commonly used software packages [13, 14].

Other conceptual and methodological considerations that are of importance for developmental brain morphometric studies, discussed further elsewhere [1], include the use of appropriate terminology when describing results from cross-sectional studies (e.g., to refer to "age-related differences"), the modeling of nonlinear longitudinal data and the interpretation of the resulting trajectories [15], and whether and how to correct for global quantities, such as the intracranial volume or the total brain size measures, which might also change during development [16]. Studies of infants and young children have additional major challenges, including image registration; use of atlases, which are often based on adult brains; the large scale of anatomical changes; and the change of image intensity contrasts [11]. Despite these methodological issues, great advances in the understanding of normal brain development have resulted from brain morphometric studies.

3 Typical Development of Brain Morphology

3.1 Brain Development in Infancy

We now know that the brain and our cognitive abilities change continuously throughout our lives. Nevertheless, some periods of life are marked by larger and more extensive changes in the brain than others, typically the periods early and late in life: development and aging. Furthermore, it is clear that the development of the brain that occurs prior to birth (prenatal) and during the first few years of life (postnatal) qualitatively and quantitatively far exceeds the changes seen in the rest of the life cycle. With careful preparation, it is often possible to perform high-quality MRI studies of infants that provide invaluable information on early brain development. While the majority of our neurons are present at birth, the brain also continues to develop very rapidly after birth. Neuronal processes (axons and dendrites) mature and the density of contact points (synapses) between neurons increases, axons undergo myelination, and those that are already myelinated get increased thickness of their insulating myelin sheaths.

The extremely rapid changes in brain morphology in early infancy were mapped in a longitudinal MRI study, which examined a large group of newborns aged 2–90 days old [17]. The results showed that the brain's total volume increased by approximately 1% per day in the period immediately after birth (Fig. 1). In a series of studies in infants, John H. Gilmore and colleagues have shown how brain structures undergo major changes during the first 2 years of life, reflecting the enormous psychological and behavioral development seen in this period. For example, they found that the volume of the cerebral cortex more than doubled (+108%) during the first year of life and showed a further, but markedly smaller, increase (+18%) in the second year [18]. To put these huge changes in perspective, we know that the cerebral cortex of elderly patients with Alzheimer's disease decreases by 1–3%, dependent on region, per year [19].

3.2 Brain Development in Childhood and Adolescence

As discussed above, there are scarcely any brain imaging studies of the age period between infancy and school age. This is primarily because it is challenging to perform high-quality MRI examinations of participants in this age range due to movement. However, we know that the brain continues to develop throughout childhood and adolescence via processes including further myelination and elimination of synapses [1]. In common with the prenatal overproduction and elimination of neurons, after birth we see an initial overproduction of synapses followed by an extensive pruning, which in certain areas of cortex probably continues well into the

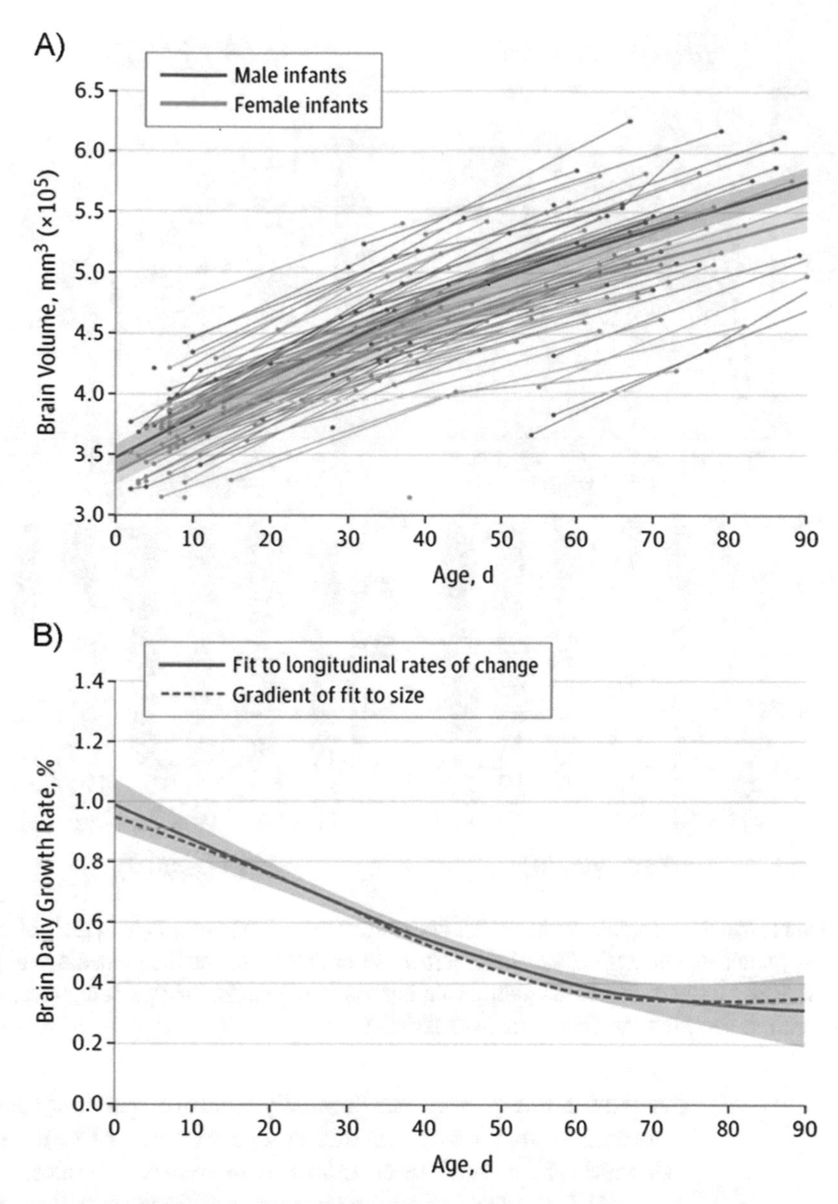

Fig. 1 Brain development in early infancy. (**a**) Spaghetti plot showing whole-brain volume across the first 90 days after birth, along with generalized additive mixed model fits to the data, and 95% confidence intervals. (**b**) Daily growth rate estimates for whole-brain across the first 90 days after birth (for male and female infants combined). Reproduced from [17] with permission from the American Medical Association

teens. These processes contribute to increasing efficiency and further specialization of information processing, at the expense of the possibility for change. Interestingly, the human brain seems to undergo a slow and especially protracted development compared to other species. For example, myelination of the human brain continues beyond adolescence, whereas degree of myelination in the chimpanzee brain reaches adult levels at roughly the same time

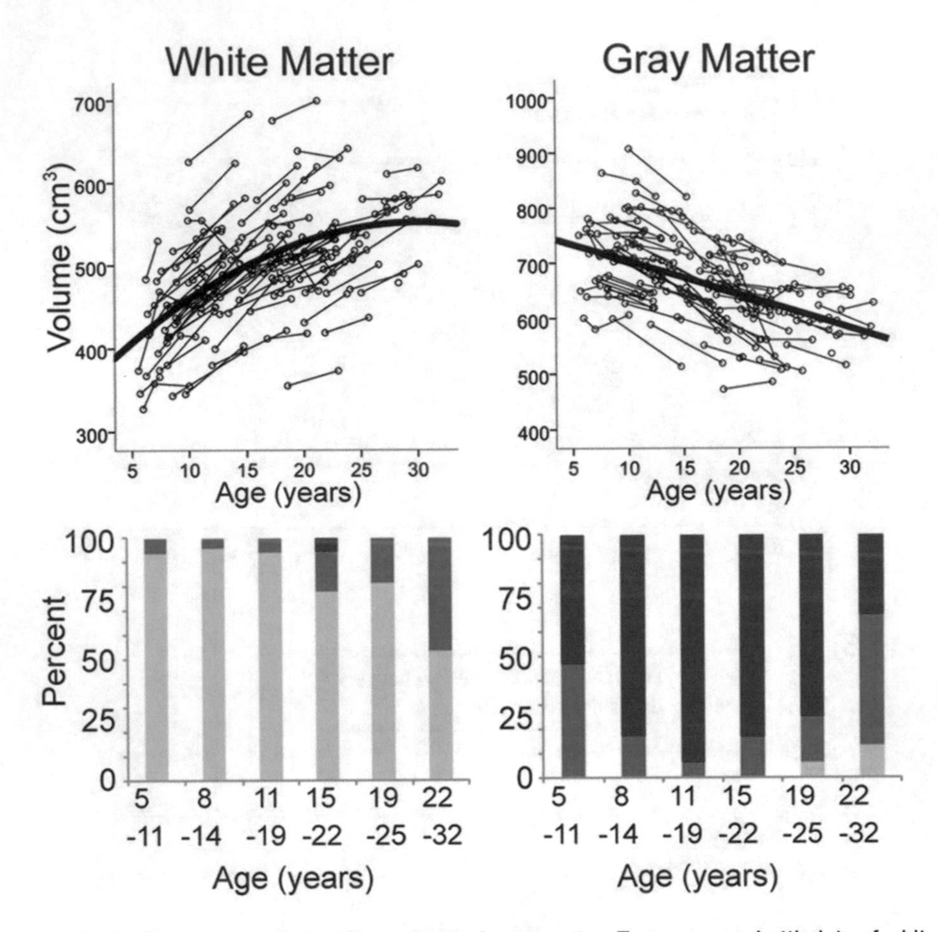

Fig. 2 Longitudinal volume changes of white matter and gray matter. Top row: spaghetti plots of white matter volume and gray matter volume against age. Bottom row: bar graphs reflecting the percentage of participants with volume increases (green), decreases (red), or no change (blue) within six age categories. Reproduced from [21] with permission from the Society for Neuroscience

as the animal becomes sexually mature [20]. Much of the potential—and many of the vulnerabilities—of our brains will depend on this prolonged maturation and experiences.

MRI studies of children and adolescents show that early increases in the volume of cortex and subcortical structures are followed by reductions, whereas white matter continues to increase in volume [16, 21]. Morphometric changes in the brain from late childhood to adulthood are thus tissue specific: the amount of white matter increases, whereas that of gray matter decreases (Fig. 2). Unfortunately, our knowledge about the underlying neurobiological processes largely relies upon extrapolation from very limited postmortem material and from data acquired in other species, and both of these approaches have limitations. The postmortem material is rare, there is concern that these brains are not representative of healthy brains, and we obviously cannot do longitudinal studies. With data from other species, we cannot be sure that the processes are the same or happened at the same rate as in humans. Nonetheless, it is generally thought that both gray matter decreases and white matter increases observed in morphometric studies, from late childhood to adulthood, are partly caused by increasing caliber and myelination of axons [22–24], which also involves sub- and intracortical myelination and white matter encroachment into the lower cortical layers. The gray matter reductions can partly also be explained by regressive changes in the form of simplification or elimination of neuronal processes and synapses and associated processes [25–27].

In addition to measures such as cortical volume and density, it is clear that more specific and distinct components of cortical morphology, thickness and surface area (and also other morphometric features [28, 29]), in many contexts should be investigated separately, as these are influenced by different evolutionary [30], genetic [31], and cellular [32] processes. In the first 2 years of life, both cortical thickness and surface area increase over time [33], but from mid-childhood to adulthood, these distinct components show very different developmental patterns. Available data suggest a monotonic decline for cortical thickness, albeit faster at younger ages, while surface area increases until early adolescence and then slightly decreases [34–37], and both measures show decreases across the adult lifespan [38].

Importantly, neurodevelopmental processes in childhood and adolescence occur to differing degrees and at different times in different parts of the brain. Brain development is thus characterized by marked regional differences. A number of studies have described sequences of development of various brain regions. In the cerebral cortex [39, 40] and for white matter fiber tracts [21, 41], development appears to generally follow a posterior–anterior pattern, with relatively late development of prefrontal brain regions (Fig. 3) and connections. Regional developmental patterns are also clearly seen in the heterogeneous changes in subcortical gray matter volumes (Fig. 4). In particular, the medial temporal lobe structures, hippocampus and amygdala, appear to follow different developmental patterns across adolescence than the basal ganglia structures, with the former showing volume increases or little or no change and the latter showing volume decreases in most studies [42–45].

The brain has a modular organization—delimited regions have specific functions. Examples include the role of visual cortex in visual perception and that of the hippocampus in certain forms of memory. Despite this specialization, a single brain region is never solely responsible for a specific function—even a relatively "simple" task such as recognizing a family member or a friend requires communication between a large number of dispersed brain regions. Brain development therefore involves maturation of circuits and entire systems—and the development of different regions must be organized and coordinated. It has for instance been found that the

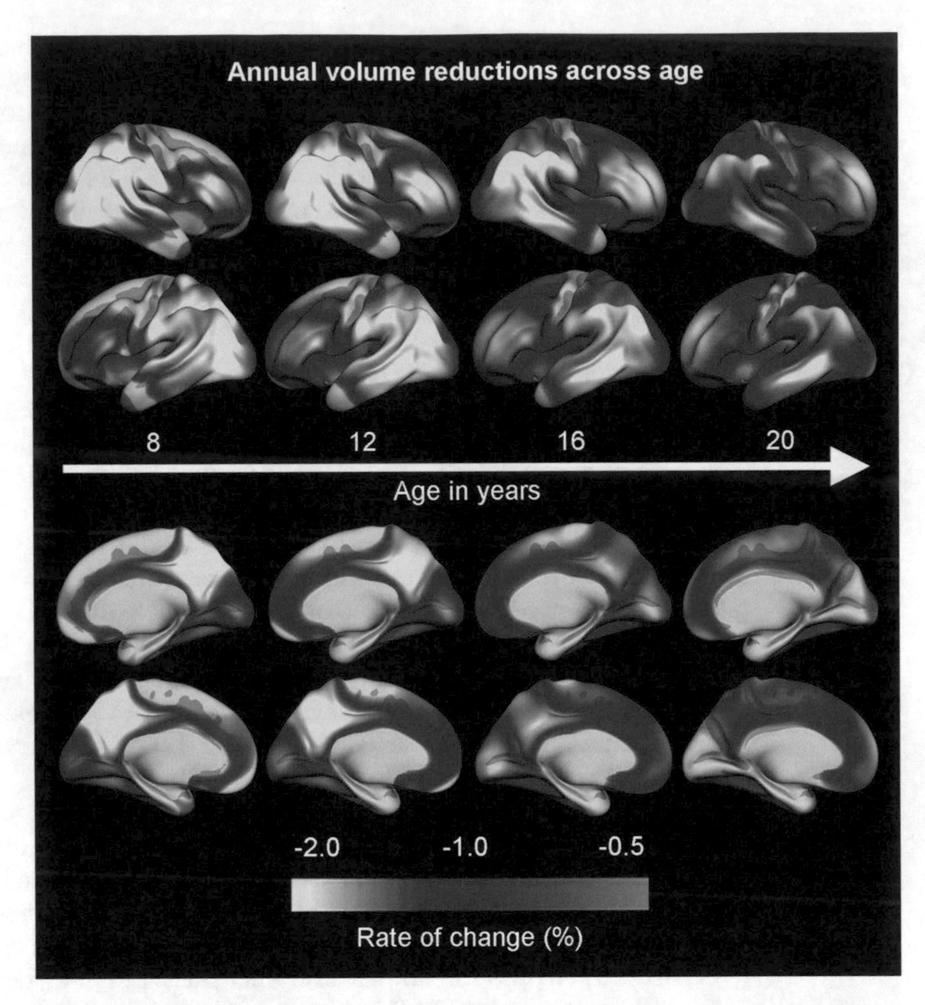

Fig. 3 Development of the cerebral cortex in childhood and adolescence. The color scale shows annual percentage change in cortical volume. The upper two rows show the two hemispheres of the brain as seen from the side (lateral view), while the bottom two rows show the brain as seen from the middle (medial view). A sequence of development can be seen across different brain regions, with cortical development generally following a posterior–anterior pattern. Modified from [39] with permission from Elsevier

rates of developmental change in different cortical regions are organized with respect to one another [46]. The results of this study showed that the degree to which change in any one region was coordinated with the change in other regions varied systematically. The rate of change in areas of association cortex in the frontal and temporal lobes showed the strongest correlations with change in other cortical areas, whereas the pace of development of primary sensory and motor areas was less closely related to development in the rest of the cortex. It was speculated that this might be because association areas in the frontal and temporal lobes are particularly important for integrative cognitive processes that require a high degree of functional coordination with other brain regions.

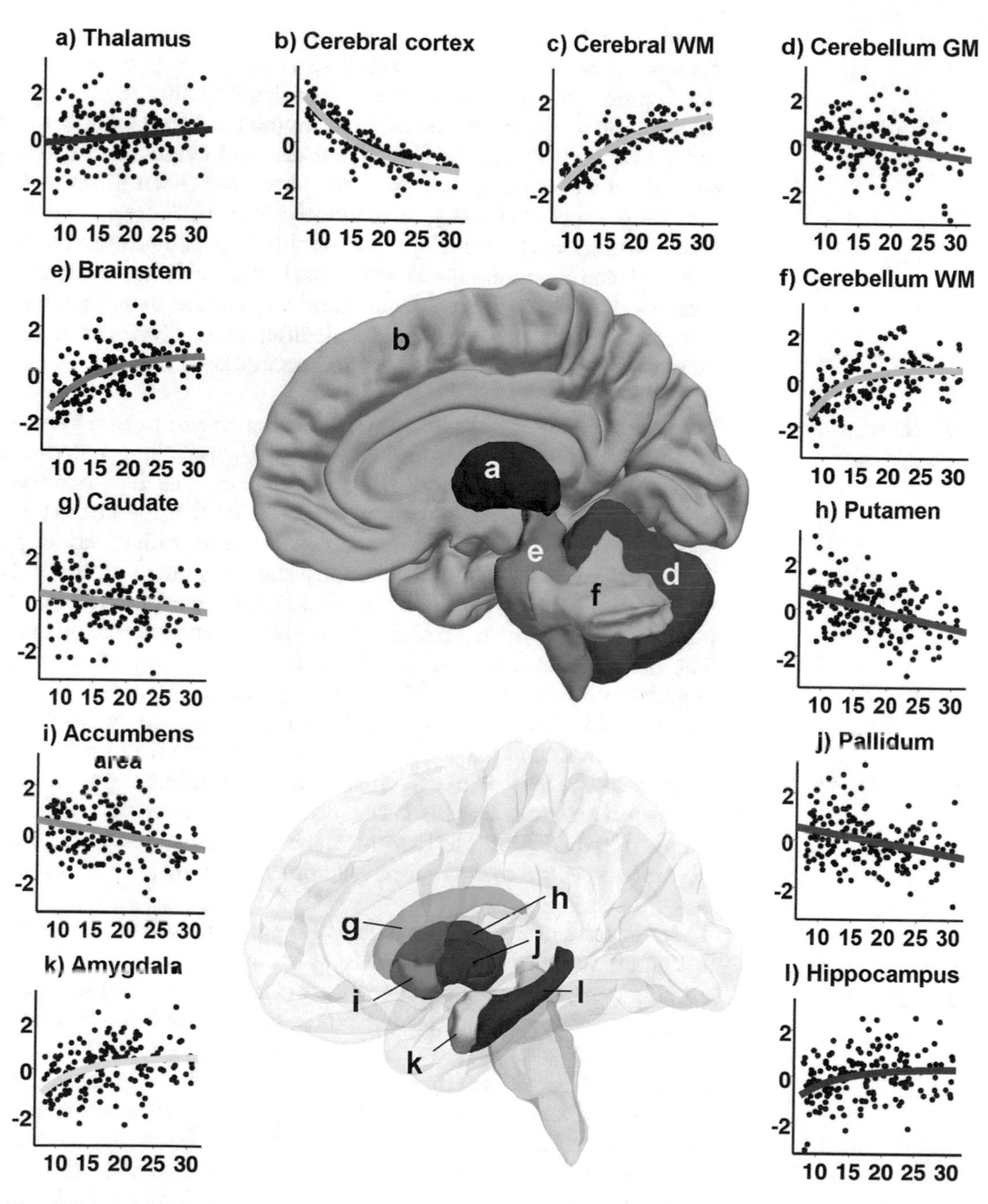

Fig. 4 Development of regional brain volumes in childhood and adolescence. Regression plots showing the relationship between age and bilateral volumes of the (**a**) thalamus, (**b**) cerebral cortex, (**c**) cerebral white matter, (**d**) cerebellum gray matter, (**e**) brainstem, (**f**) cerebellum white matter, (**g**) caudate, (**h**) putamen, (**i**) accumbens area, (**j**) pallidum, (**k**) amygdala, and (**l**) hippocampus. The volumes are corrected for total brain volume and shown in *z*-scores. Also displayed are samples of the segmentation (cerebral white matter not shown), based on the group mean. Reproduced from [42] with permission from the Society for Neuroscience

Another study showed coordinated patterns of cortical–subcortical change within known neurocognitive systems [47]. Specifically, development of the hippocampus was related mainly to development of the temporal lobe, while development of basal ganglia was related to development of frontal, insular, and cingulate cortices. Several other organization systems have also been proposed, including functional [40], cytoarchitectonic [48], topographic [49], evolutionary [50–52], genetic [36], and network-based [53, 54], and these systems likely coexist but possibly have different roles for different structural parameters [55]. Although brain development is characterized by regional differences, diverse changes take place in a carefully organized and coordinated fashion.

3.3 Lifespan Perspectives

When does the human brain stop developing? Neuroimaging studies clearly show that it is not possible to specify an age when development stops or when aging begins, and that it is not the case that the brain is static at any age, but rather that it is characterized by life-long experience-dependent neurocognitive plasticity. Although a certain degree of stability can be seen in adulthood relative to the earlier development, there does not appear to be any period of life in which brain structure and function remain fixed. The brain continues to change throughout life, with positive and negative processes presumably occurring alongside one another as we grow older [56]. Interestingly, there is increasing recognition of how early life influences on brain and cognition can affect the whole lifespan and of how neurocognitive changes in different periods of life may be related. For instance, recent studies provide evidence supporting the hypothesis that normal brain degeneration in aging in some respects mirrors brain development [39, 57].

3.4 Individual Difference in Brain Development

There is great individual variability in brain structure at any given age and in brain development. And as for most aspects of a child's development, the development of the brain is the result of longitudinally ongoing and interrelated influence (transactions) between multiple endogenous (internal) and exogenous (external) factors [58]. Here, we will briefly discuss how some selected factors can affect individual differences in brain morphology and development in children and adolescents. Specifically, we will discuss heritability, as well as how early nutrition, substance use, the mental status of pregnant women, and birth-related factors could affect children's structural brain development.

Twin studies have found substantial heritability estimates not just for the volume of specific brain structure, cortical thickness, and surface area in adults [31] but also for developmental changes in these aspects of brain morphology in childhood and adolescence [43, 59]. So far, we have limited knowledge about the role of specific genes. But recently reported results suggest that polymorphisms in several putative risk genes for mental illnesses or dementias may have similar effects on brain structure early in life as in adult age [60], suggesting that the influences of these genes may represent stable life-long effects. An increasing number of neuroimaging studies also show that influences from the environment both prenatally and in infancy can have effects on later brain structure. Women's nutrition during pregnancy is critical for the initial development of the central nervous system of the fetus and may, together with the infant's early diet, also have effects on later brain structure and development [61]. And it is well established that women's use of a range of different substances during pregnancy, including alcohol, tobacco, cocaine, methamphetamine, marihuana, and opioids, can have long-lasting negative effects on the brain of the fetus [62, 63]. Intriguingly, pregnant women's mental health and emotional state may also possibly have long-term, although much more subtle, effects on the brain of the fetus. For instance, a recent study showed weak but relatively consistent negative associations between maternal depressive symptoms at 19, 25, and 31 weeks' gestation and cortical thickness in children when aged 6–9 years old [64].

Early influences such as those briefly discussed above may also influence birth-related (perinatal) factors, which in turn may have consequences for the child's further development. It is for instance well established that prenatal stress can cause lower birthweight for gestational age and earlier delivery [65], and we know that these factors, on a group level, are associated with a range of adverse neurodevelopmental outcomes for the children. While it has long been known that premature delivery and low birth weight, typically defined as less than 2.500 g, can influence brain development, it has now also been shown that normal variation in birth weight is positively associated with cortical surface area in several regions and brain volumes many years later [66, 67].

In addition to the importance of genetic factors and the long-lasting and in some cases possibly life-long effects of early environmental influences on brain morphology, it is also important to remember that the brain, in interaction with the physical, social, and cultural environment, continues to change throughout life. A large number of animal studies, both studies comparing animals reared in so-called enriched environments with animals in standard experimental environments and studies comparing animals in the wild and in captivity, have documented a range of different neural changes in response to complex stimulation and experiences [68]. In humans, an increasing number of controlled training studies, both with adults and with young participants, support the conclusion that experiences and learning can have substantial influences on brain structure [69, 70]. This general conclusion is further supported by naturalistic studies of early negative experiences, e.g., neglect, abuse, and stress [71, 72]. A particularly interesting study compared brain structure in a relatively large sample of young adolescents who were internationally adopted from institutional care of variable quality as young children to a comparison group reared with their biological families [73]. The results showed smaller volume of frontal cortical regions, driven primarily by differences in cortical surface area, in post-institutionalized youth, and also that longer duration of institutional care was associated with smaller hippocampal volumes. It should be stressed that such effects generally are subtle and that the causal relationships between early negative experiences and later brain structure likely are complex and may also involve multiple other factors. Further investigations, both controlled experiments and naturalistic studies, are needed to examine if and how the potential for change in brain structure as a result of experiences and learning differs quantitatively and/or qualitatively with age.

4 Atypical Development of Brain Morphology

4.1 Brain Development in Clinical Populations

For many research groups, a strong motivation for studying typical development has been to eventually identify and understand development when it has gone awry. It is essential to know what is normal to be able to say what is abnormal or pathological. Below, we will briefly discuss schizophrenia, autism spectrum disorder, and epilepsy as examples of clinical populations where aspects of brain morphology may show atypical development.

MRI studies of children and adolescents with early-onset schizophrenia overall implicate similar brain regions as those delineated in adult samples, including, but not limited to, prefrontal, medial temporal, and superior temporal regions [74, 75]. It is however not known whether the brain abnormalities are more or less severe than those observed in adult-onset schizophrenia. Of great interest, studies of brain structure in children and adolescents with schizophrenia also indicate altered developmental trajectories of gray matter volumes and regional cortical thickness [76, 77]. Moreover, a recent study found high spatial overlap between a widespread network of mainly transmodal gray matter regions, which show prolonged development during adolescence and regions showing atypical development in adolescents with schizophrenia [57]. This might indicate that the pattern of brain structure alterations in schizophrenia is influenced and to some extent determined by the timing of the pathological processes in relation to typical brain development patterns.

Group-level variations in brain morphology have also been found to be associated with autism spectrum disorder, and these appear to change across the lifespan so that case-control differences, e.g., in young children, may differ from those observed in other stages of life [78]. Early brain development in autism spectrum disorder seems to be characterized by accelerated volume increases, but less is known about the development during adolescence. New longitudinal studies do however suggest accelerated developmental decreases in regional brain volumes and cortical thickness across adolescence [79–81].

Two cases of epilepsy syndromes illustrate how developmental perspectives may inform us of pathological brain functioning, in these cases in the form of epileptic seizures and decreased cognitive functioning. In juvenile myoclonic epilepsy (JME), seizures in the form of myoclonic jerks and generalized tonic–clonic seizures appear during adolescence or young adulthood. The seizures originate in frontal networks, bilaterally, and the syndrome is often accompanied by mild executive dysfunction. Several voxel-based morphometry studies have shown increased gray matter volume or cortical thickness associated with this syndrome, possibly suggesting a deficit in the cortical thinning process [82, 83]. This may help explain the adolescent onset of the disorder. However, another study actually found decreased cortical thickness in similar brain regions [84]. Yet another study, using measures of thickness, surface area, and curvature from FreeSurfer analyses, found evidence of cortical morphology abnormalities indicating early-onset disruption in cortical folding [85]. As no studies have followed the developmental trajectory of cortical thickness or curvature in new-onset JME, it remains to be seen whether there is an element of early-onset developmental abnormality, a developmental deficit in cortical pruning, or atrophy related to the seizure activity. A developmental perspective is therefore paramount.

A related problem in the field of epilepsy is temporal lobe epilepsy (TLE) and the question of progressive disease or developmental vulnerability. TLE often presents during adolescence or early adulthood but may arise at almost any age (rarely in old age, though). For a subgroup of patients, antiepileptic treatment is unsuccessful, leading to chronic TLE, sometimes with epilepsy surgery as a last resort treatment. One hypothesis is that TLE follows a progressively deteriorating course with increased hippocampal sclerosis and seizure frequency, evident as presenting with more serious memory deficits in middle-aged to older patient groups [86]. A cross-sectional cortical thickness study of TLE patients aged 14–60 years showed subtle widespread brain morphometric alterations compared to controls, i.e., alterations not limited to the temporal lobes [87]. The study also suggested a rather fixed discrepancy between the patient group and the controls throughout the studied age span, although ventricular size was increasing with age to a greater degree in patients than in controls. A developmental origin has been proposed by Helmstaedter and Elger [88], based on a cross-sectional study of memory function in TLE patients aged 6–68 years. They noted an early point at which the trajectory of memory performance in the TLE group diverged from the controls, then keeping a steady distance to the controls throughout the rest of the studied age span. This could indicate a developmental disturbance in brain development that ultimately may lead to TLE and memory dysfunction. A developmental perspective and the use of brain morphometric methods are needed to find this out.

Importantly, it should be stressed that further longitudinal research is generally needed to establish whether and how brain developmental trajectories are altered in different specific disorders. In the case of schizophrenia for instance, popular models suggest that abnormal neurodevelopmental processes [89, 90] and brain connectivity [91] play pivotal roles, but although DTI studies of white matter microstructure consistently find lower regional fractional anisotropy (FA) in children and adolescents with early-onset schizophrenia compared with healthy control participants, only three studies have investigated case-control developmental differences, and the results are highly mixed [92]. One study concluded that adolescents with schizophrenia and controls show diverging white matter developmental trajectories, a second study concluded with converging trajectories, and a recent study found parallel trajectories.

4.2 Dimensional Perspectives

An important question is whether the traits that underlie pathology are always unique to disease states. New evidence suggests that this is not the case: often it is not a matter of categorical distinctions but of variation along continuous dimensions. This has been clearly demonstrated in studies of children with diagnosed attention deficit hyperactivity disorder (ADHD) and children with subclinical ADHD symptoms. First, it has been found that maturation of the cerebral cortex is delayed in children with an ADHD diagnosis compared with a control group, especially in frontal areas [93]. Next, the same research group showed that degree of hyperactivity and impulsivity in typically developing healthy children was also associated with the rate of cortical development in some of the same cortical regions [94]. Similar results have been obtained for conduct disorder and related subclinical symptoms. While some studies indicate that children diagnosed with conduct disorder have reduced gray matter volume in specific brain regions, including prefrontal cortices [95], a recent study found that also symptoms of conduct problems within the normal range were associated with thinner left hemisphere prefrontal and supramarginal cortices [96]. Results such as these suggest that abnormal cortical development in certain brain regions is characteristic of both ADHD and "normal" hyperactivity and impulsivity, as well as conduct disorder and minor symptoms of behavioral problems, just to differing degrees in each case. The findings therefore suggest that there may be neuroanatomical continuity between subclinical symptoms and at least certain clinical disorders.

For psychosis spectrum disorders, there is evidence from studies of individuals at increased risk for developing such disorders, either individuals with diagnosed relatives (genetic high-risk) or individuals showing specific symptoms or functional decline (clinical high-risk), that structural and microstructural brain changes might precede the disorder [97, 98]. However, in both these cases, a large proportion of these individuals will likely not develop clinical-level psychotic disorders, and these findings, together with studies of population-based samples [99], support a dimensional perspective also on the brain morphology phenotypes associated with psychosis spectrum disorders and psychotic experiences (see also [100]).

5 Behavioral and Cognitive Relevance of Brain Development

Developmental differences across brain regions are relatively consistent with behavioral studies, which show, for example, rapid development of visual acuity and perception in the first year of life. More complex functions, such as the abilities to plan and to inhibit responses, which are dependent on regions including prefrontal cortices, develop over a longer period extending well into the teens. Brain regions involved in more basic abilities and skills thus appear to develop earlier than those supporting more complex cognition. However, there is a major need for longitudinal studies that directly examine the relationship between brain development and cognitive development, as only a few studies exist. In fact, in the light of the increasing number of studies examining brain–behavior relationship in various clinical conditions, there are surprisingly few studies documenting such links in typically developing children and adolescents. It is important to establish these principles in normal development, in order to make a framework for understanding abnormal development. A reason for the lack of studies could be related to the methodological issues mentioned earlier. In addition, associations between changes in brain morphometric variables and changes in behavioral measures are likely quite subtle. Thus, a large study population is needed, as well as time and patience.

As histological studies have shown, the neuronal processes and synapses in the cerebral cortex are gradually pruned, and this likely contributes, together with associated changes and a range of other processes, to making the cortex thinner as development progresses. A natural assumption then is that cognitive functions, as they also improve with increasing age, are related to cortical reduction and thinning. In line with this, studies show moderate associations between age-related or longitudinal cortical volume reductions or thinning and improvements in general intellectual abilities [101], memory [102], and working memory and executive functioning [103–105].

The heterogeneity of brain development is not only seen in gray matter structures. As previously mentioned, white matter tracts also undergo refinement during childhood and adolescence. This refinement, through processes such as increased myelination and alignment of axon fibers, is important for increasing speed and consistency of signal transmission necessary for network communication. An integration of morphometric and DTI studies is needed for understanding the dynamic interplay between these developmental processes in underlying cognitive development. In one such study, we investigated the simultaneous contribution of cortical thinning and developing white matter tract microstructure on the development of working memory [105]. We found unique effects of morphometric and DTI measures on digit span performance in children and adolescents. Furthermore, the effects varied dynamically during development, with white matter measures having the largest effect during early/middle adolescence and cortical thickness having the largest effect during late adolescence. To speculate, this might for instance mean that the integrity of the whole network during development is first dependent on the development of the communicating white matter tracts.

In a prospective multimodal imaging study of normal development, Ullman et al. [106] investigated the correlations between DTI, gray matter volume/density, and functional MRI activation, on the one hand, and measures of working memory, on the other hand. As expected, there were cross-sectional relationships between frontal and parietal cortical regions and visuospatial working memory performance. However, even more interestingly, when they used the MRI-derived variables to predict working memory performance as measured 2 years later, a new pattern emerged. Here, functional MRI and DTI in and around thalamus and the caudate nucleus best predicted future performance, even when current performance on the same task was included in the regression model. Morphometric measures also predicted future performance when entered by itself in a regression model; however, when entered together with the other two modalities of MRI data, morphometry did not remain statistically significant. The authors speculate that a network including thalamus and caudate might be especially involved in working memory training, facilitating developmental improvement as well. This study shows that multimodal imaging and longitudinal research designs have great potential for revealing new insight into brain development and development of cognition and behavior. As brain imaging acquisition and analysis methods become increasingly sensitive to new aspects of brain morphology and function, the diversity of developmental processes may be mapped in greater detail.

6 Conclusion

Morphometry studies have documented that although the most dramatic developmental changes in brain morphology take place before birth and during the first few years of life, the human brain also continues to undergo substantial structural remodeling throughout childhood and adolescence and into adulthood. Different tissue classes, brain regions, and structural features develop differently, and there are large interindividual differences in brain morphology at any given age and in its development. Available data suggest that aspects of structural brain development may be altered in certain clinical conditions, including schizophrenia spectrum disorders, autism spectrum disorder, and epilepsy, but also neuroanatomical continuity between subclinical symptoms and certain clinical disorders. Developmental perspectives may shed new light on the natural course of clinical disorders and subclinical symptoms. Morphometric studies, especially in combination with other neuroimaging modalities, are currently our best tool for capturing the complex and multifaceted nature of brain development in healthy children and adolescents, although special care must be taken regarding methodological challenges when studying children.

Acknowledgments

This work was supported by the Research Council of Norway and the University of Oslo (to CKT) and the South-Eastern Norway Regional Health Authority (to YØ).

References

1. Mills KL, Tamnes CK (2014) Methods and considerations for longitudinal structural brain imaging analysis across development. Dev Cogn Neurosci 9:172–190
2. Hedman AM, van Haren NE, Schnack HG, Kahn RS, Hulshoff Pol HE (2012) Human brain changes across the life span: a review of 56 longitudinal magnetic resonance imaging studies. Hum Brain Mapp 33:1987–2002
3. Giedd JN, Raznahan A, Alexander-Bloch A, Schmitt E, Gogtay N, Rapoport JL (2015) Child psychiatry branch of the National Institute of Mental Health longitudinal structural magnetic resonance imaging study of human brain development. Neuropsychopharmacology 40:43–49
4. Blakemore SJ, Mills KL (2014) Is adolescence a sensitive period for sociocultural processing? Annu Rev Psychol 65:187–207
5. Crone EA, Dahl RE (2012) Understanding adolescence as a period of social-affective engagement and goal flexibility. Nat Rev Neurosci 13:636–650
6. Walhovd KB, Tamnes CK, Fjell AM (2014) Brain structural maturation and the foundations of cognitive behavioral development. Curr Opin Neurol 27:176–184
7. Blakemore SJ, Burnett S, Dahl RE (2010) The role of puberty in the developing adolescent brain. Hum Brain Mapp 31:926–933
8. Peper JS, Dahl RE (2013) Surging hormones: brain-behavior interactions during puberty. Curr Dir Psychol Sci 22:134–139
9. Paus T, Keshavan M, Giedd JN (2008) Why do many psychiatric disorders emerge during adolescence? Nat Rev Neurosci 9:947–957
10. Keshavan MS, Giedd J, Lau JY, Lewis DA, Paus T (2014) Changes in the adolescent

brain and the pathophysiology of psychotic disorders. Lancet Psychiatry 1:549–558

11. Sled JG, Nossin-Manor R (2013) Quantitative MRI for studying neonatal brain development. Neuroradiology 55(Suppl 2):97–104
12. Ducharme S, Albaugh MD, Nguyen TV, Hudziak JJ, Mateos-Perez JM, Labbe A, Evans AC, Karama S, Brain Development Cooperative Group (2016) Trajectories of cortical thickness maturation in normal brain development - the importance of quality control procedures. NeuroImage 125:267–279
13. Yendiki A, Koldewyn K, Kakunoori S, Kanwisher N, Fischl B (2013) Spurious group differences due to head motion in a diffusion MRI study. NeuroImage 88C:79–90
14. Brown TT, Kuperman JM, Erhart M, White NS, Roddey JC, Shankaranarayanan A, Han ET, Rettmann D, Dale AM (2010) Prospective motion correction of high-resolution magnetic resonance imaging data in children. NeuroImage 53:139–145
15. Fjell AM, Walhovd KB, Westlye LT, Østby Y, Tamnes CK, Jernigan TL, Gamst A, Dale AM (2010) When does brain aging accelerate? Dangers of quadratic fits in cross-sectional studies. NeuroImage 50:1376–1383
16. Mills KL, Goddings AL, Herting MM, Meuwese R, Blakemore SJ, Crone EA, Dahl RE, Guroglu B, Raznahan A, Sowell ER, Tamnes CK (2016) Structural brain development between childhood and adulthood: convergence across four longitudinal samples. NeuroImage 141:273–281
17. Holland D, Chang L, Ernst TM, Curran M, Buchthal SD, Alicata D, Skranes J, Johansen H, Hernandez A, Yamakawa R, Kuperman JM, Dale AM (2014) Structural growth trajectories and rates of change in the first 3 months of infant brain development. JAMA Neurol 71:1266–1274
18. Gilmore JH, Shi F, Woolson SL, Knickmeyer RC, Short SJ, Lin W, Zhu H, Hamer RM, Styner M, Shen D (2012) Longitudinal development of cortical and subcortical gray matter from birth to 2 years. Cereb Cortex 22:2478–2485
19. Fjell AM, Walhovd KB, Fennema-Notestine-C, McEvoy LK, Hagler DJ, Holland D, Brewer JB, Dale AM (2009) One-year brain atrophy evident in healthy aging. J Neurosci 29:15223–15231
20. Miller DJ, Duka T, Stimpson CD, Schapiro SJ, Baze WB, McArthur MJ, Fobbs AJ, Sousa AM, Sestan N, Wildman DE, Lipovich L, Kuzawa CW, Hof PR, Sherwood CC (2012) Prolonged myelination in human neocortical evolution. Proc Natl Acad Sci U S A 109:16480–16485
21. Lebel C, Beaulieu C (2011) Longitudinal development of human brain wiring continues from childhood into adulthood. J Neurosci 31:10937–10947
22. Benes FM (1989) Myelination of cortical-hippocampal relays during late adolescence. Schizophr Bull 15:585–593
23. Benes FM, Turtle M, Khan Y, Farol P (1994) Myelination of a key relay zone in the hippocampal formation occurs in the human brain during childhood, adolescence, and adulthood. Arch Gen Psychiatry 51:477–484
24. Yakovlev PA, Lecours IR (1967) The myelogenetic cycles of regional maturation of the brain. In: Minkowski A (ed) Regional development of the brain in early life. Blackwell, Oxford
25. Bourgeois JP, Rakic P (1993) Changes of synaptic density in the primary visual cortex of the macaque monkey from fetal to adult stage. J Neurosci 13:2801–2820
26. Huttenlocher PR, Dabholkar AS (1997) Regional differences in synaptogenesis in human cerebral cortex. J Comp Neurol 387:167–178
27. Petanjek Z, Judas M, Simic G, Rasin MR, Uylings HB, Rakic P, Kostovic I (2011) Extraordinary neoteny of synaptic spines in the human prefrontal cortex. Proc Natl Acad Sci U S A 108:13281–13286
28. Aleman-Gomez Y, Janssen J, Schnack H, Balaban E, Pina-Camacho L, Alfaro-Almagro-F, Castro-Fornieles J, Otero S, Baeza I, Moreno D, Bargallo N, Parellada M, Arango C, Desco M (2013) The human cerebral cortex flattens during adolescence. J Neurosci 33:15004–15010
29. Mutlu AK, Schneider M, Debbane M, Badoud D, Eliez S, Schaer M (2013) Sex differences in thickness, and folding developments throughout the cortex. NeuroImage 82:200–207
30. Geschwind DH, Rakic P (2013) Cortical evolution: judge the brain by its cover. Neuron 80:633–647
31. Kremen WS, Fennema-Notestine C, Eyler LT, Panizzon MS, Chen CH, Franz CE, Lyons MJ, Thompson WK, Dale AM (2013) Genetics of brain structure: contributions from the Vietnam era twin study of aging. Am J Med Genet B Neuropsychiatr Genet 162B:751–761
32. Chenn A, Walsh CA (2002) Regulation of cerebral cortical size by control of cell cycle

exit in neural precursors. Science 297:365–569

33. Lyall AE, Shi F, Geng X, Woolson S, Li G, Wang L, Hamer RM, Shen D, Gilmore JH (2015) Dynamic development of regional cortical thickness and surface area in early childhood. Cereb Cortex 25:2204–2212
34. Tamnes CK, Herting MM, Goddings AL, Meuwese R, Blakemore SJ, Dahl RE, Güroğlu B, Raznahan A, Sowell ER, Crone EA, Mills KL (2017) Development of the cerebral cortex across adolescence: a multisample study of interrelated longitudinal changes in cortical volume, surface area and thickness. J Neurosci 37(12):3402–3412
35. Wierenga LM, Langen M, Oranje B, Durston S (2014) Unique developmental trajectories of cortical thickness and surface area. NeuroImage 87:120–126
36. Fjell AM, Grydeland H, Krogsrud SK, Amlien I, Rohani DA, Ferschmann L, Storsve AB, Tamnes CK, Sala-Llonch R, Due-Tønnessen P, Bjornerud A, Sølsnes AE, Håberg AK, Skranes J, Bartsch H, Chen CH, Thompson WK, Panizzon MS, Kremen WS, Dale AM, Walhovd KB (2015) Development and aging of cortical thickness correspond to genetic organization patterns. Proc Natl Acad Sci U S A 112:15462–15467
37. Vijayakumar N, Allen NB, Youssef G, Dennison M, Yucel M, Simmons JG, Whittle S (2016) Brain development during adolescence: a mixed-longitudinal investigation of cortical thickness, surface area, and volume. Hum Brain Mapp 37:2027–2038
38. Storsve AB, Fjell AM, Tamnes CK, Westlye LT, Øverbye K, Aasland HW, Walhovd KB (2014) Differential longitudinal changes in cortical thickness, surface area and volume across the adult life span: regions of accelerating and decelerating change. J Neurosci 34:8488–8498
39. Tamnes CK, Walhovd KB, Dale AM, Østby Y, Grydeland H, Richardson G, Westlye LT, Roddey JC, Hagler DJ Jr, Due-Tønnessen P, Holland D, Fjell AM, Alzheimer's Disease Neuroimaging Initiative (2013) Brain development and aging: overlapping and unique patterns of change. NeuroImage 68:63–74
40. Gogtay N, Giedd JN, Lusk L, Hayashi KM, Greenstein D, Vaituzis AC, Nugent TF 3rd, Herman DH, Clasen LS, Toga AW, Rapoport JL, Thompson PM (2004) Dynamic mapping of human cortical development during childhood through early adulthood. Proc Natl Acad Sci U S A 101:8174–8179
41. Westlye LT, Walhovd KB, Dale AM, Bjørnerud A, Due-Tønnessen P, Engvig A, Grydeland H, Tamnes CK, Østby Y, Fjell AM (2010) Life-span changes of the human brain white matter: diffusion tensor imaging (DTI) and volumetry. Cereb Cortex 20:2055–2068
42. Østby Y, Tamnes CK, Fjell AM, Westlye LT, Due-Tønnessen P, Walhovd KB (2009) Heterogeneity in subcortical brain development: a structural magnetic resonance imaging study of brain maturation from 8 to 30 years. J Neurosci 29:11772–11782
43. Swagerman SC, Brouwer RM, de Geus EJ, Hulshoff Pol HE, Boomsma DI (2014) Development and heritability of subcortical brain volumes at ages 9 and 12. Genes Brain Behav 13:733–742
44. Goddings AL, Mills KL, Clasen LS, Giedd JN, Viner RM, Blakemore SJ (2014) The influence of puberty on subcortical brain development. NeuroImage 88:242–251
45. Dennison M, Whittle S, Yucel M, Vijayakumar N, Kline A, Simmons J, Allen NB (2013) Mapping subcortical brain maturation during adolescence: evidence of hemisphere- and sex-specific longitudinal changes. Dev Sci 16:772–791
46. Raznahan A, Lerch JP, Lee N, Greenstein D, Wallace GL, Stockman M, Clasen L, Shaw PW, Giedd JN (2011) Patterns of coordinated anatomical change in human cortical development: a longitudinal neuroimaging study of maturational coupling. Neuron 72:873–884
47. Walhovd KB, Tamnes CK, Bjørnerud A, Due-Tønnessen P, Holland D, Dale AM, Fjell AM (2015) Maturation of cortico-subcortical structural networks - segregation and overlap of medial temporal and fronto-striatal systems in development. Cereb Cortex 25:1835–1841
48. Shaw P, Kabani NJ, Lerch JP, Eckstrand K, Lenroot R, Gogtay N, Greenstein D, Clasen L, Evans A, Rapoport JL, Giedd JN, Wise SP (2008) Neurodevelopmental trajectories of the human cerebral cortex. J Neurosci 28:3586–3594
49. Vandekar SN, Shinohara RT, Raznahan A, Roalf DR, Ross M, DeLeo N, Ruparel K, Verma R, Wolf DH, Gur RC, Gur RE, Satterthwaite TD (2015) Topologically dissociable patterns of development of the human cerebral cortex. J Neurosci 35:599–609
50. Amlien IK, Fjell AM, Tamnes CK, Grydeland H, Krogsrud SK, Chaplin TA, Rosa MG, Walhovd KB (2016) Organizing principles of human cortical development - thickness and area from 4 to 30 years: insights from comparative primate neuroanatomy. Cereb Cortex 26:257–267

51. Hill J, Inder T, Neil J, Dierker D, Harwell J, Van Essen D (2010) Similar patterns of cortical expansion during human development and evolution. Proc Natl Acad Sci U S A 107:13135–13140
52. Fjell AM, Westlye LT, Amlien I, Tamnes CK, Grydeland H, Engvig A, Espeseth T, Reinvang I, Lundervold AJ, Lundervold A, Walhovd KB (2015) High-expanding cortical regions in human development and evolution are related to higher intellectual abilities. Cereb Cortex 25:26–34
53. Alexander-Bloch A, Giedd JN, Bullmore E (2013) Imaging structural co-variance between human brain regions. Nat Rev Neurosci 14:322–336
54. Alexander-Bloch A, Raznahan A, Bullmore E, Giedd J (2013) The convergence of maturational change and structural covariance in human cortical networks. J Neurosci 33:2889–2899
55. Krongold M, Cooper C, Bray S (2017) Modular development of cortical gray matter across childhood and adolescence. Cereb Cortex 27(2):1125–1136
56. Lindenberger U (2014) Human cognitive aging: corriger la fortune? Science 346:572–578
57. Douaud G, Groves AR, Tamnes CK, Westlye LT, Duff EP, Engvig A, Walhovd KB, James A, Gass A, Monsch AU, Matthews PM, Fjell AM, Smith SM, Johansen-Berg H (2014) A common brain network links development, aging, and vulnerability to disease. Proc Natl Acad Sci U S A 111:17648–17653
58. Sameroff A (2010) A unified theory of development: a dialectic integration of nature and nurture. Child Dev 81:6–22
59. van Soelen IL, Brouwer RM, van Baal GC, Schnack HG, Peper JS, Collins DL, Evans AC, Kahn RS, Boomsma DI, Hulshoff Pol HE (2012) Genetic influences on thinning of the cerebral cortex during development. NeuroImage 59:3871–3880
60. Knickmeyer RC, Wang J, Zhu H, Geng X, Woolson S, Hamer RM, Konneker T, Lin W, Styner M, Gilmore JH (2014) Common variants in psychiatric risk genes predict brain structure at birth. Cereb Cortex 24:1230–1246
61. Isaacs EB (2013) Neuroimaging, a new tool for investigating the effects of early diet on cognitive and brain development. Front Hum Neurosci 7:445
62. Derauf C, Kekatpure M, Neyzi N, Lester B, Kosofsky B (2009) Neuroimaging of children following prenatal drug exposure. Semin Cell Dev Biol 20:441–454
63. Roussotte F, Soderberg L, Sowell E (2010) Structural, metabolic, and functional brain abnormalities as a result of prenatal exposure to drugs of abuse: evidence from neuroimaging. Neuropsychol Rev 20:376–397
64. Sandman CA, Buss C, Head K, Davis EP (2015) Fetal exposure to maternal depressive symptoms is associated with cortical thickness in late childhood. Biol Psychiatry 77:324–334
65. Glover V (2014) Maternal depression, anxiety and stress during pregnancy and child outcome; what needs to be done. Best Pract Res Clin Obstet Gynaecol 28:25–35
66. Walhovd KB, Fjell AM, Brown TT, Kuperman JM, Chung Y, Hagler DJ Jr, Roddey JC, Erhart M, McCabe C, Akshoomoff N, Amaral DG, Bloss CS, Libiger O, Schork NJ, Darst BF, Casey BJ, Chang L, Ernst TM, Frazier J, Gruen JR, Kaufmann WE, Murray SS, van Zijl P, Mostofsky S, Dale AM, Pediatric Imaging Neurocognition Genetics Study (2012) Long-term influence of normal variation in neonatal characteristics on human brain development. Proc Natl Acad Sci U S A 109:20089–20094
67. Raznahan A, Greenstein D, Lee NR, Clasen LS, Giedd JN (2012) Prenatal growth in humans and postnatal brain maturation into late adolescence. Proc Natl Acad Sci U S A 109:11366–11371
68. van Praag H, Kempermann G, Gage FH (2000) Neural consequences of environmental enrichment. Nat Rev Neurosci 1:191–198
69. Zatorre RJ, Fields RD, Johansen-Berg H (2012) Plasticity in gray and white: neuroimaging changes in brain structure during learning. Nat Neurosci 15:528–536
70. Jolles DD, Crone EA (2012) Training the developing brain: a neurocognitive perspective. Front Hum Neurosci 6:76
71. Lim L, Radua J, Rubia K (2014) Gray matter abnormalities in childhood maltreatment: a voxel-wise meta-analysis. Am J Psychiatry 171:854–863
72. McCrory E, De Brito SA, Viding E (2011) The impact of childhood maltreatment: a review of neurobiological and genetic factors. Front Psych 2:48
73. Hodel AS, Hunt RH, Cowell RA, Van Den Heuvel SE, Gunnar MR, Thomas KM (2015) Duration of early adversity and structural brain development in post-institutionalized adolescents. NeuroImage 105:112–119
74. Brent BK, Thermenos HW, Keshavan MS, Seidman LJ (2013) Gray matter alterations in schizophrenia high-risk youth and early-onset schizophrenia: a review of structural

MRI findings. Child Adolesc Psychiatr Clin N Am 22:689–714

75. Janssen J, Aleman-Gomez Y, Schnack H, Balaban E, Pina-Camacho L, Alfaro-Almagro-F, Castro-Fornieles J, Otero S, Baeza I, Moreno D, Bargallo N, Parellada M, Arango C, Desco M (2014) Cortical morphology of adolescents with bipolar disorder and with schizophrenia. Schizophr Res 158:91–99
76. Fraguas D, Diaz-Caneja CM, Pina-Camacho-L, Janssen J, Arango C (2016) Progressive brain changes in children and adolescents with early-onset psychosis: a meta-analysis of longitudinal MRI studies. Schizophr Res 173:132–139
77. Thormodsen R, Rimol LM, Tamnes CK, Juuhl-Langseth M, Holmen A, Emblem KE, Rund BR, Agartz I (2013) Age-related cortical thickness differences in adolescents with early-onset schizophrenia compared with healthy adolescents. Psychiatry Res 214:190–196
78. Ecker C, Bookheimer SY, Murphy DG (2015) Neuroimaging in autism spectrum disorder: brain structure and function across the lifespan. Lancet Neurol 14:1121–1134
79. Zielinski BA, Prigge MB, Nielsen JA, Froehlich AL, Abildskov TJ, Anderson JS, Fletcher PT, Zygmunt KM, Travers BG, Lange N, Alexander AL, Bigler ED, Lainhart JE (2014) Longitudinal changes in cortical thickness in autism and typical development. Brain 137:1799–1812
80. Lange N, Travers BG, Bigler ED, Prigge MB, Froehlich AL, Nielsen JA, Cariello AN, Zielinski BA, Anderson JS, Fletcher PT, Alexander AA, Lainhart JE (2015) Longitudinal volumetric brain changes in autism spectrum disorder ages 6-35 years. Autism Res 8:82–93
81. Wallace GL, Eisenberg IW, Robustelli B, Dankner N, Kenworthy L, Giedd JN, Martin A (2015) Longitudinal cortical development during adolescence and young adulthood in autism spectrum disorder: increased cortical thinning but comparable surface area changes. J Am Acad Child Adolesc Psychiatry 54:464–469
82. Cao B, Tang Y, Li J, Zhang X, Shang HF, Zhou D (2013) A meta-analysis of voxel-based morphometry studies on gray matter volume alteration in juvenile myoclonic epilepsy. Epilepsy Res 106:370–377
83. Alhusaini S, Ronan L, Scanlon C, Whelan CD, Doherty CP, Delanty N, Fitzsimons M (2013) Regional increase of cerebral cortex thickness in juvenile myoclonic epilepsy. Epilepsia 54:e138–e141
84. Tae WS, Kim SH, Joo EY, Han SJ, Kim IY, Kim SI, Lee JM, Hong SB (2008) Cortical thickness abnormality in juvenile myoclonic epilepsy. J Neurol 255:561–566
85. Ronan L, Alhusaini S, Scanlon C, Doherty CP, Delanty N, Fitzsimons M (2012) Widespread cortical morphologic changes in juvenile myoclonic epilepsy: evidence from structural MRI. Epilepsia 53:651–658
86. Engman E, Malmgren K (2012) Long-term follow-up of memory in patients with epilepsy. In: Zeman A, Kapur N, Jones-Gotman M (eds) Epilepsy & memory. Oxford Scholarship Online, Oxford
87. Dabbs K, Becker T, Jones J, Rutecki P, Seidenberg M, Hermann B (2012) Brain structure and aging in chronic temporal lobe epilepsy. Epilepsia 53:1033–1043
88. Helmstaedter C, Elger CE (2009) Chronic temporal lobe epilepsy: a neurodevelopmental or progressively dementing disease? Brain 132:2822–2830
89. Rapoport JL, Giedd JN, Gogtay N (2012) Neurodevelopmental model of schizophrenia: update 2012. Mol Psychiatry 17:1228–1238
90. Fatemi SH, Folsom TD (2009) The neurodevelopmental hypothesis of schizophrenia, revisited. Schizophr Bull 35:528–548
91. Pettersson-Yeo W, Allen P, Benetti S, McGuire P, Mechelli A (2011) Dysconnectivity in schizophrenia: where are we now? Neurosci Biobehav Rev 35:1110–1124
92. Tamnes CK, Agartz I (2016) White matter microstructure in early-onset schizophrenia: a systematic review of diffusion tensor imaging studies. J Am Acad Child Adolesc Psychiatry 55:269–279
93. Shaw P, Eckstrand K, Sharp W, Blumenthal J, Lerch JP, Greenstein D, Clasen L, Evans A, Giedd J, Rapoport JL (2007) Attention-deficit/hyperactivity disorder is characterized by a delay in cortical maturation. Proc Natl Acad Sci U S A 104:19649–19654
94. Shaw P, Gilliam M, Liverpool M, Weddle C, Malek M, Sharp W, Greenstein D, Evans A, Rapoport J, Giedd J (2011) Cortical development in typically developing children with symptoms of hyperactivity and impulsivity: support for a dimensional view of attention deficit hyperactivity disorder. Am J Psychiatry 168:143–151
95. Huebner T, Vloet TD, Marx I, Konrad K, Fink GR, Herpertz SC, Herpertz-Dahlmann B (2008) Morphometric brain abnormalities in boys with conduct disorder. J Am Acad Child Adolesc Psychiatry 47:540–547

96. Walhovd KB, Tamnes CK, Østby Y, Due-Tønnessen P, Fjell AM (2012) Normal variation in behavioral adjustment relates to regional differences in cortical thickness in children. Eur Child Adolesc Psychiatry 21:133–140
97. Moran ME, Hulshoff Pol H, Gogtay N (2013) A family affair: brain abnormalities in siblings of patients with schizophrenia. Brain 136:3215–3226
98. Peters BD, Karlsgodt KH (2015) White matter development in the early stages of psychosis. Schizophr Res 161:61–69
99. Jacobson S, Kelleher I, Harley M, Murtagh A, Clarke M, Blanchard M, Connolly C, O'Hanlon E, Garavan H, Cannon M (2010) Structural and functional brain correlates of subclinical psychotic symptoms in 11-13 year old schoolchildren. NeuroImage 49:1875–1885
100. Satterthwaite TD, Vandekar SN, Wolf DH, Bassett DS, Ruparel K, Shehzad Z, Craddock RC, Shinohara RT, Moore TM, Gennatas ED, Jackson C, Roalf DR, Milham MP, Calkins ME, Hakonarson H, Gur RC, Gur RE (2015) Connectome-wide network analysis of youth with psychosis-Spectrum symptoms. Mol Psychiatry 20:1508–1515
101. Goh S, Bansal R, Xu D, Hao X, Liu J, Peterson BS (2011) Neuroanatomical correlates of intellectual ability across the life span. Dev Cogn Neurosci 1:305–312
102. Østby Y, Tamnes CK, Fjell AM, Walhovd KB (2012) Dissociating memory processes in the developing brain: the role of hippocampal volume and cortical thickness in recall after minutes versus days. Cereb Cortex 22:381–390
103. Faradi N, Karama S, Burgaleta M, White MT, Evans AC, Fonov V, Collins DL, Waber DP (2015) Neuroanatomical correlates of behavioral rating versus performance measures of working memory in typically developing children and adolescents. Neuropsychologia 29:82–91
104. Tamnes CK, Walhovd KB, Grydeland H, Holland D, Østby Y, Dale AM, Fjell AM (2013) Longitudinal working memory development is related to structural maturation of frontal and parietal cortices. J Cogn Neurosci 25:1611–1623
105. Østby Y, Tamnes CK, Fjell AM, Walhovd KB (2011) Morphometry and connectivity of the fronto-parietal verbal working memory network in development. Neuropsychologia 49:3854–3862
106. Ullman H, Almeida R, Klingberg T (2014) Structural maturation and brain activity predict future working memory capacity during childhood development. J Neurosci 34:1592–1598

Check for updates

Chapter 11

Morphometry in Normal Aging

Hiroshi Matsuda

Abstract

Magnetic resonance imaging (MRI)-based evaluation of brain anatomy is regarded as a well-validated method for assessing the age-related trajectories of human regional brain volumes over the life span. Automated softwares such as FreeSurfer have enabled automated quantification of regional and even subregional volumes. Vowel-based morphometry is easily applicable with a short scanning time. Age-related volume reductions are observed in the perisylvian, pericentral, and cingulate cortex as well as the thalamic radiations, internal capsule, corpus callosum, cerebellum, and deep white matter. Hippocampal subfields show age-related volume decline prominently in CA1, the dentate gyrus, and the perirhinal cortex. Recent application of graph theory to structural connectivity has revealed significant network differences between young and old age groups, particularly in the default mode network.

Key words MRI, Aging, Voxel-based morphometry, Hippocampal subfields, Structural connectivity

1 Introduction

Studies of brain morphometry using magnetic resonance imaging (MRI) have been conducted by many researchers along with the increasing resolution of anatomical scans of a whole brain and decreasing acquisition times. MRI-based evaluation is regarded as a well-validated method for assessing age-related and pathological changes in brain volume [1]. However, differences in the shape and neuroanatomical configuration of individual brains may hinder the visual identification of important structural alterations. Moreover, visual inspection alone cannot detect subtle cross-sectional and longitudinal volume changes. To overcome the drawbacks of visual inspection, manual outlining has conventionally been performed for the detection of volume changes [2]. Although manual outlining by an expert human rater is a validated procedure used to estimate volume changes, it may give rise to undesired variations in volumetric measures [3]. To minimize these variations by human raters, there has been an ongoing international effort to harmonize existing protocols for manual volumetry of brain regions such as

Gianfranco Spalletta et al. (eds.), *Brain Morphometry*, Neuromethods, vol. 136,
https://doi.org/10.1007/978-1-4939-7647-8_11, © Springer Science+Business Media, LLC 2018

the hippocampus [4]. Nevertheless, manual volumetry is still time-consuming, precluding its becoming a routine procedure.

Meanwhile, surface-based morphometry tools developed over the last decade have made it possible to quantify gray matter volumes from human MR images in a more automated fashion. Software packages such as FreeSurfer [5] (http://surfer.nmr.mgh.harvard.edu/) provide measurements of cortical and subcortical gray matter features based on MRI data. This surface-based morphometry can measure not only volume but also cortical thickness. Using this technique, we recently revealed that poor episodic memory and depressive state in patients with mild cognitive impairment were associated with thinner cortices in the left entorhinal region, anterior medial temporal lobe, and gyrus adjacent to the amygdala [6]. Moreover, a recent advance in MRI analysis has made it possible to automatically measure hippocampal subfields and adjacent cortical subregions [7]. However, these software tools are not in routine use because the execution time of the volumetric workflow, required for a single MRI analysis, is more than 10 h. At present, voxel-based morphometry (VBM) is the most frequently used method for routine volumetry, although VBM provides only statistical indices, not absolute volume measures. The advantage of the VBM approach [8] is that it is not biased to one particular structure and gives an impartial and comprehensive assessment of anatomical differences throughout the brain [9].

This chapter describes common VBM techniques using MRI incorporated in statistical parametric mapping (SPM) software (http://www.fil.ion.ucl.ac.uk/spm/) running on the MATLAB platform and the application of the software to the assessment of age-related volume changes in hippocampal subfields. Recent applications of graph theory to the study of connectivity changes in normal aging are also described.

2 Voxel-Based Morphometry

2.1 General Concept of VBM

Data for VBM are acquired as three-dimensional (3D) volumetric T1-weighted MR images with 1–1.5 mm slice thickness without interslice gaps. The image matrix is usually 256×256. VBM of MRI data involves segmentation into gray matter, white matter, and cerebrospinal fluid (CSF) partitions, anatomical standardization of all images to the same stereotactic space by linear affine transformation, and further nonlinear warping, smoothing, and finally statistical analysis. The output is a statistical parametric map.

2.2 Optimized VBM

Optimized VBM has been developed for removing areas of mis-segmented non-gray matter voxels by the introduction of additional preprocessing steps prior to anatomical standardization and subsequent segmentation [10, 11]. Initial steps involve segmentation of the original structural MR images in native space into gray and white matter images, followed by a series of fully automated morphological operations for removing unconnected non-brain voxels of the skull and dural venous sinus from the segmented images. The resulting images are extracted gray and white matter partitions in native space. The extracted segmented gray and white matter images are then standardized to the gray/white matter templates, thereby preventing any contribution of non-brain voxels and affording optimal anatomical standardization. To facilitate optimal segmentation, the optimized standardization parameters are reapplied to the original whole brain structural images in native space. The optimally standardized whole brain structural images in stereotactic space are segmented into gray matter, white matter, and CSF partitions and then subject to a second extraction of standardized segmented gray/white matter images. The brain extraction step is repeated at this stage because some non-brain voxels from the scalp, skull, or venous sinuses in the optimally normalized whole brain images may still remain outside the brain margins on segmented gray/white matter images.

As a result of nonlinear spatial normalization based on discrete cosine transforms, certain brain regions may grow, whereas others may shrink. To preserve the volume of a particular tissue within a voxel, a further processing step is incorporated. This involves multiplying (or modulating) voxel values in the segmented images by the Jacobian determinants derived from the anatomical standardization step. In effect, an analysis of modulated data tests for regional differences in the absolute amount (volume) of gray matter, whereas an analysis of unmodulated data tests for regional differences in the concentration of gray matter. Finally, each optimally standardized, segmented, modulated image is smoothed by convolving with an isotropic Gaussian kernel with 12-mm full width at half maximum (FWHM). The smoothing step helps to compensate for the inexact nature of the anatomical standardization. Moreover, it has the effect of rendering the data more normally distributed, thereby increasing the validity of parametric statistical tests.

2.3 SPM8 Plus Diffeomorphic Anatomical Registration Using Exponentiated Lie Algebra (DARTEL)

DARTEL incorporated in SPM8 is an algorithm implemented for diffeomorphic image registration [12]. The software includes an option for estimating inverse consistent deformations. This nonlinear registration is considered a local optimization problem, which is solved using a Levenberg–Marquardt strategy. A constant Eulerian velocity framework is used, which allows a rapid scaling and squaring method to be used in the computations. This technique improves intersubject registration. The VBM8 toolbox (http://dbm.neuro.uni-jena.de/vbm8/) using SPM8 plus DARTEL is available for both cross-sectional and longitudinal VBM studies.

In the SPM8 plus DARTEL procedure, T1-weighted images are classified into gray matter, white matter, and CSF using the segmentation routine implemented in SPM8, which yields both the native space and DARTEL imported versions of the tissues. The VBM8 toolbox using SPM8 provides better segmentation accuracy and reliability than the FMRIB Software Library or FreeSurfer software program [13]. The DARTEL imported versions of gray and white matter are used to generate the flow fields encoding the shapes and a series of template images by running the "DARTEL (create templates)" routine. In this step, DARTEL increases the accuracy of intersubject alignment by modeling the shape of each brain using millions of parameter values. DARTEL works by aligning gray matter among the images while simultaneously aligning the white matter. This is achieved by generating increasingly precise average template data, to which the data are iteratively aligned. The flow fields and final template image created in the previous step are used to generate smoothed (8-mm FWHM), modulated, spatially normalized, and Jacobian-scaled gray and white matter images resliced to isotropic voxel size in MNI (Montreal Neurological Institute) space. DARTEL provides better or equivalent registration accuracy as compared to other widely used nonlinear deformation algorithms that include optimized VBM [14].

2.4 VBM in Normal Aging

Studying the morphometric changes in the brain during normal aging is important for understanding the mechanisms underlying age-related neurological disorders. The advent of VBM techniques for analysis of MRI structural data has facilitated sensitive detection of regional patterns of gray matter and white matter volume changes. Most studies have employed a cross-sectional design in which correlations of volume with age at specific time points are used to make inferences about how aging affects brain structure. However, the changes found in cross-sectional studies inevitably include overlap in changes associated with normal aging and those associated with disorders such as Alzheimer's disease [15]. Thus, large longitudinal VBM studies, involving the acquisition of serial MRI measurements over time in the same subjects, are desirable [16].

There have been numerous VBM investigations of gray matter volume changes with advancing age. Many cortical regions, prominently in frontal and insular areas, have been reported to show linear negative associations between volume and age [10, 17–24] (Fig. 1a). In contrast, many reports have found preservation of gray matter volume in specific structures such as the amygdala, hippocampus, and thalamus [10, 18, 20, 23, 24]. Relative preservation of these subcortical limbic or paralimbic regions is consistent with the functional importance of the thalamo-limbic circuits in sensory integration, arousal, emotion, and memory. This preservation in areas with the early maturation may lend credence to the idea that late-maturing cortical regions are more vulnerable to age-related morphologic changes. These age-related features may be already

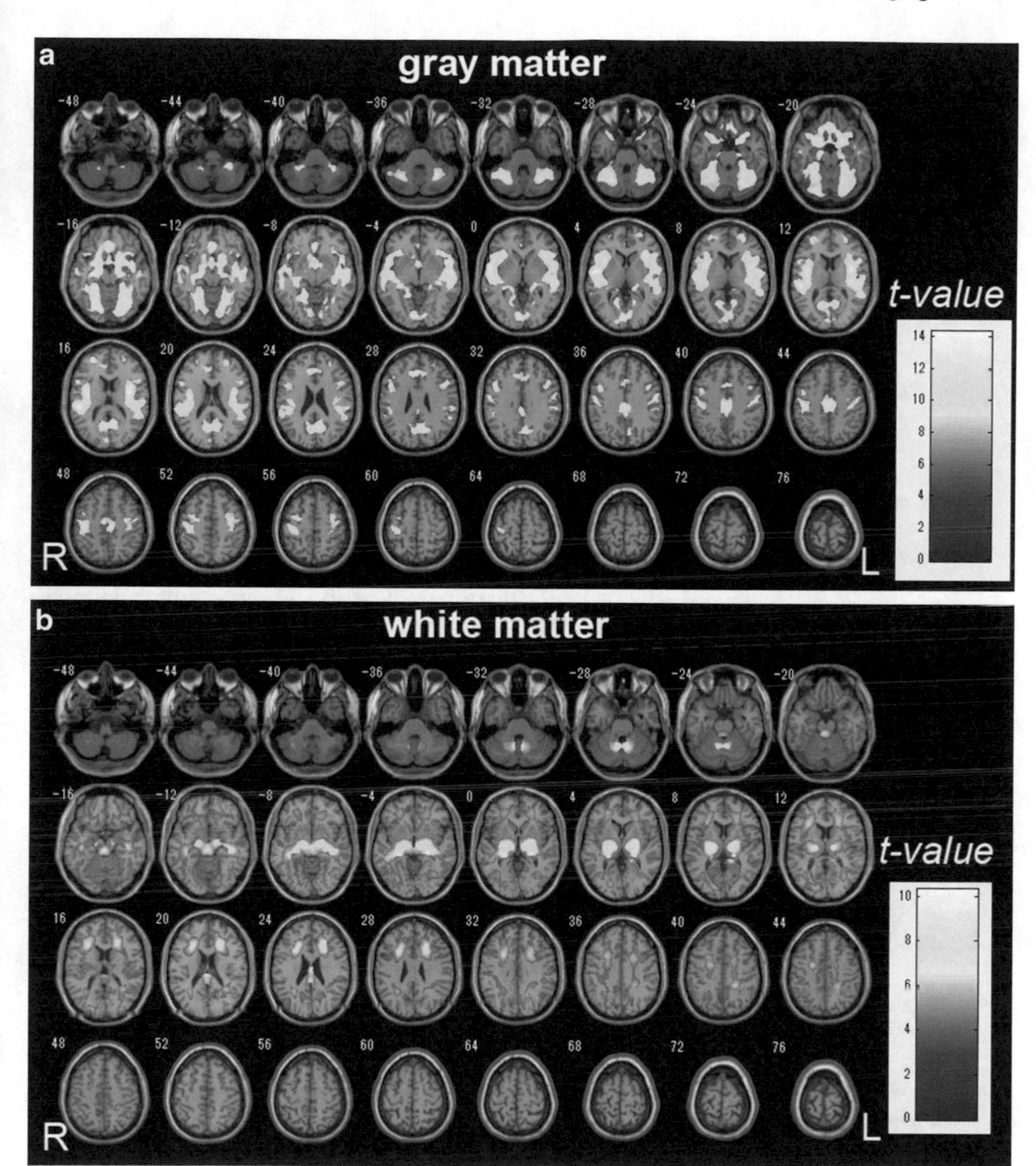

Fig. 1 Age-related atrophy of gray and white matter in healthy controls. Voxel-based morphometry analysis of 101 healthy volunteers age 20–89 years reveals significant decreases in regional gray matter (**a**) and white matter volumes (**b**) with advancing age. Age-related gray matter volume reductions are prominent in the perisylvian, pericentral, and cingulate cortex, while white matter reductions are observed in the thalamic radiation, internal capsule, midbrain, corpus callosum, cerebellum, and deep white matter

present during early adulthood. Indeed, prefrontal areas showed progressive linear volume reductions even before old age, while medial temporal regions showed volume preservation throughout adulthood [24]. Furthermore, some regional cortical gray matter volumes exhibited age-related reductions even during adolescence [25].

White matter structures such as the anterior thalamic radiation, internal capsule, cerebral peduncle, cerebellar white matter, and external capsule also showed linear negative associations between volume and age [26] (Fig. 1b). In contrast, nonlinear relationships between white matter volume and age were found in the superior longitudinal fascicle and superior corona radiata. Volumes gradually increased before 40 years of age, peaked around 50 years of age, and rapidly declined after 60 years of age [27]. This nonlinear inverted U-shaped relationship with age is consistent with the notion of ongoing maturation of the white matter beyond adolescence, reaching its peak in the fourth decade of life. A longitudinal study conducted at 2.5-year intervals during adolescence reported a significant increase in frontal white matter bilaterally connected via the body of the corpus callosum [25]. It should be noted, however, that individual regional volumes are prone to acute changes due to physiological status. Dehydration may induce significant widespread loss of white matter volume in the temporoparietal areas [28]. Consequently, VBM studies investigating white matter changes should consider controlling subject hydration state to avoid this potential confounder.

The morphology of cortical gray matter is assessed not only by volumetry but also by cortical thickness measures. Hutton and colleagues examined how gray matter changes identified using voxel-based cortical thickness (VBCT) measures compare with local gray matter volume changes identified using VBM during normal aging [29]. VBCT and VBM yielded generally consistent results, but VBCT provided a more sensitive measure of age-associated decline in gray matter compared with VBM. Voxel-based cortical thickness specifically measures cortical thickness, while VBM provides a mixed measure of gray matter including cortical surface area or cortical folding as well as cortical thickness.

3 Morphometry of Hippocampal Subfields

3.1 General Concept of Subfield Volumetry

Aging can be associated with cognitive decline even in the absence of disease. One of the most prominently affected cognitive domains is declarative memory. This suggests that the hippocampal formation, which is crucial for storage and retrieval of declarative information, may be particularly vulnerable to aging. Until recently, most MRI studies evaluated the hippocampal formation as a singular structure. However, the hippocampus is not a homogeneous structure; it is divided into several subfields with distinct histological characteristics and functions: the subiculum with subdivisions presubiculum, parasubiculum, and subiculum proper, the four fields of cornu ammonis (CA1–CA4), and the dentate gyrus.

Further, these subfields have different vulnerabilities. For example, the CA1 subfield is known to be vulnerable to vascular disease [30]. Recent developments in high-resolution submillimeter MRI scanners with wider fields as well as advances in manual and automated segmentation have allowed for assessment of human hippocampal subfields in vivo. However, it is not possible to directly visualize the boundaries between two contiguous subiculum–CA subfields. Subfields are therefore distinguished by landmarks derived from anatomical atlases and geometric rules. Unfortunately, previous studies of hippocampal subfield volumetry have shown marked variability in the number of segmented subfields, which are segmented separately or grouped together, in the subfield borders, and whether subfield segmentation is performed on the whole hippocampus or on only the hippocampal body.

3.2 Automatic Volumetry of Hippocampal Subfields

Automatic methods have been developed for subfield assessment in research and for potential clinical applications without requiring as much time, labor, and anatomical expertise as manual delineation. A group from the University of Pennsylvania has developed an open-source, semiautomatic method called automatic segmentation of hippocampal subfields (ASHS) designed for use on high-resolution anisotropic T2-weighted images [7]. This algorithm initially relied on the segmentation protocol developed by Mueller et al. [31]. ASHS was recently improved to identify more numerous hippocampal subfields and parahippocampal subregions, to segment these along the full length of the hippocampus, and to allow both volume and thickness analyses [32] (Fig. 2). This updated, fully automated version of ASHS requires classical three-dimensional T1-weighted images in addition to the high-resolution (e.g., 0.4 × 0.4 × 2-mm) oblique coronal T2-weighted images, oriented perpendicular to the longitudinal axis of the hippocampus. Using this technique, CA1, CA2, CA3, the dentate gyrus, subiculum, the entorhinal cortex, Brodmann area 35, Brodmann area 36, the collateral sulcus, and miscellaneous are labeled.

Automatic methods to measure the volume of hippocampal subfields using standard (approximately 1 mm^3) isotropic T1-weighted images have also been developed. The most widely used is the one implemented in FreeSurfer 5.3. Using this technique, CA1, combined CA2/CA3, combined CA4/dentate gyrus, fimbria, subiculum, presubiculum, whole hippocampus, and hippocampal fissure are labeled. This method is user-friendly and directly applicable to standard 1.5-T or 3-T T1-weighted images and so has been used extensively over the last several years to study hippocampal changes related to development, aging, and various pathological conditions. However, some authors have recently expressed concern about the subfield segmentation tool implemented in FreeSurfer 5.3 [32–34], arguing that it has not been validated on the standard images on which it is commonly used. Doubts are cast not

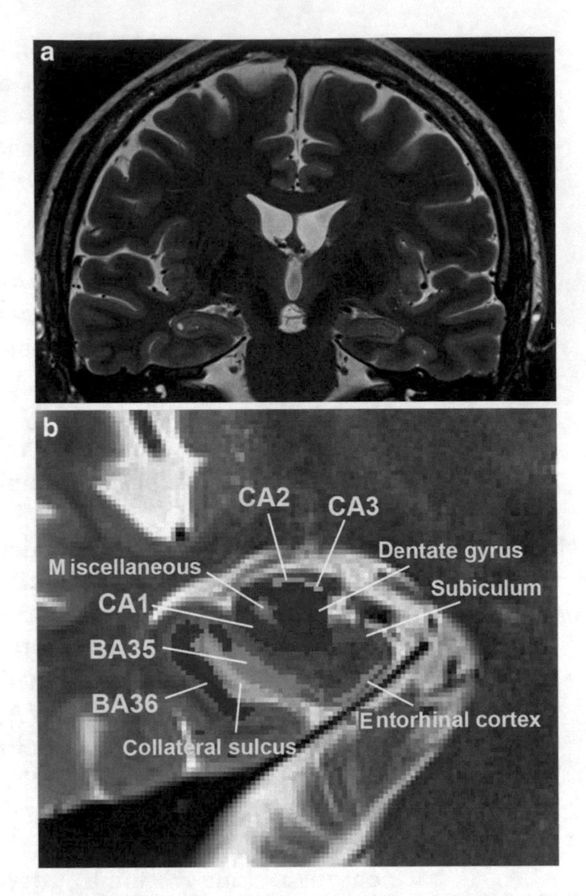

Fig. 2 Automated volumetry of hippocampal subfields. (**a**) A high-field T2-weighted magnetic resonance imaging with 0.4 × 0.4 × 2 mm voxel resolution. (**b**) Segmentation of the hippocampal subfields (CA1, CA2, CA3, subiculum), dentate gyrus, entorhinal cortex, and perirhinal cortex (Brodmann areas 35 and 36) using the automatic segmentation of hippocampal subfields (ASHS) program. *BA* Brodmann area, *CA* cornu ammonis

only on the ability to distinguish subfields in low-resolution images but also on the boundaries of the parcellation scheme, which strongly differ from the majority of in vivo and imaging atlases. For instance, the FreeSurfer 5.3 CA1 is the smallest subfield, while the combined CA2/CA3 is the largest, in contrast to histological data showing the opposite. Very recently, the creators of the FreeSurfer 5.3 subfield segmentation tool acknowledged the flaws of their initial method and developed an alternative tool implemented in FreeSurfer 6.0 [35]. Briefly, this new method uses a specifically developed atlas derived from ex vivo 7-T MRI scans of 15 subjects and comprised 13 different labels. FreeSurfer 6.0 can be used for analysis of classical 3D T1-weighted images, specific T2-weighted images, or both. As a consequence of these major changes, results are likely to be very different from those obtained with FreeSurfer 5.3. Using this newest FreeSurfer version, alveus, parasubiculum, presubiculum, subiculum, CA1, combined CA2/CA3, CA4, granule cell layer of the dentate gyrus, hippocampus–amygdala transition area, fimbria, molecular layer, hippocampal fissure, and tail are labeled. Results obtained with FreeSurfer 6.0 should thus be more accurate than those from FreeSurfer 5.3. However, because this update was released only recently, it has not been widely used and tested yet.

3.3 Age-Related Volumetric Changes of Hippocampal Subfields

Several studies on brain structure investigating age-related changes in the absence of disease and significant cognitive decline have indicated that hippocampal subfield volumes shrink unevenly with age. However, the specific pattern has been inconsistent across studies. Mueller et al. [31] found a linear effect of age in CA1. The 60s and older age group had a significantly smaller CA1 than all other age groups. In a subsequent study including more individuals, they also found an additional age effect on the combined CA3/dentate gyrus [36]. Another group found age-related linear atrophy in the combined CA1/CA2 [37, 38]. Kerchner et al. [39] described a diminution of the entorhinal cortex and CA1 stratum radiatum and stratum lacunosum moleculare widths in older controls compared to the younger comparison group using a 7-T MRI scanner. La Joie et al. [40] described a linear effect of age on the subiculum with relative preservation of CA1 and other subfields in a group of individuals between 18 and 68 years old. In a follow-up studies including the same 50 adults together with another 48 new individuals, de Flores et al. [34] observed a linear volume decrease of the subiculum and a quadratic decrease of CA1 volume with an inflection point around 50 years but no significant changes in the other subfields. These trajectories of subfield atrophy over the adult life span were almost identical to those reported by Ziegler et al. [41] in a cohort ranging from 19 to 86 years. In an independent sample of 29 elderly subjects between 65 and 80 years old, Wisse et al. [33] reported a significant age-related volumetric decrease in CA1 and the combined CA4/dentate gyrus, with annual atrophy rates of 1.4 and 2.4%, respectively. Using the automated FreeSurfer 5.3 method on a cohort of individuals between 50 and 75 years old, Pereira et al. [42] reported a linear effect of age on the combined CA2/CA3 and combined CA4/dentate gyrus. Voineskos et al. [43] reported that all right and left hippocampal subfield volumes were inversely related to age except for the right and left CA1.

Daugherty et al. [44] reported age differences in hippocampal subfield volumes over almost the entire life span from 8 to 82 years. The magnitude of these differences and the pattern of age–volume associations varied across the regions examined. From childhood to late adulthood, the volumes of combined CA1/CA2, combined CA3/dentate gyrus, and entorhinal cortex but not subiculum showed negative associations with age. The

Table 1
Subfield volumes of medial temporal structures measured by the ASHS program in 107 healthy controls ranging from 21 to 82 years

Region	Left		Right	
	Volume (μL)	Correlation with age (ICV corrected)	Volume (μL)	Correlation with age (ICV corrected)
Hippocampal formation				
Whole	2680.1±318.8	−0.415 (−0.223)	2732.2±303.8	−0.354 (−0.157)
CA1	1321.8±187.2	−0.496 (−0.328)	1333.6±166.6	−0.448 (−0.236)
CA2	20.8±5.7	−0.406 (−0.322)	23.6±4.9	−0.252 (−0.134)
CA3	62.0±18.8	0.303 (0.375)	72.6±23.8	0.030 (0.079)
Dentate gyrus	775.3±109.7	−0.454 (−0.309)	796.3±109.1	−0.383 (−0.221)
Miscellaneous	124.1±39.6	0.643 (0.685)	126.5±41.3	0.538 (0.593)
Subiculum	376.1±52.4	−0.353 (−0.181)	379.3±53.9	−0.354 (−0.167)
Entorhinal cortex	526.7±70.4	−0.332 (−0.158)	526.6±71.6	−0.303 (−0.130)
Brodmann area 35	422.2±78.4	−0.308 (−0.192)	436.9±74.7	−0.509 (−0.402)
Brodmann area 36	1574.2±322.7	−0.412 (−0.350)	1505.1±328.2	0.461 (−0.443)
Collateral sulcus	250.1±75.6	0.340 (0.421)	202.4±79.5	0.056 (0.128)

ASHS automatic segmentation of hippocampal subfields; *CA* cornu ammonis; *ICV* intracranial volume

association between younger age and larger combined CA1/CA2 volume was linear across the life span, whereas nonlinear functions described age-dependent shrinkage during later adulthood in combined CA3/dentate gyrus volume and entorhinal cortex volume.

Our study using ASHS on 107 healthy controls between 21 and 82 years old revealed age-related volume declines prominently in CA1, the dentate gyrus, and Brodmann areas 35 and 36 (Table 1). Inconsistent results among these cited studies may reflect inherent flaws in the cross-sectional design, including potential cohort bias and the influence of elderly individuals at the presymptomatic stage of a neurodegenerative disorder (which may cause an overestimation of the normal age effect), in addition to the different methods used for segmentation. Longitudinal studies in which the effect of age is examined within subjects over time may be optimal.

A few groups have assessed memory–structure correlations in healthy controls. In healthy young adults, Chadwick et al. [45] found that CA3 volume predicted the precision of memory recall as assessed by the ability to distinguish memories with a high degree of similarity. This effect was not associated with volume changes in other hippocampal subfields. In healthy adults between 52 and 82 years old, Bender et al. [46] found that larger combined CA3/CA4/dentate gyrus volume was associated with better associative memory over repeated tests regardless of vascular risk, whereas combined CA1/CA2 volume was specifically associated with free recall of common nouns but only in hypertensive individuals. These findings suggest that relatively small regions of the hippocampus may play a role in age-related memory decline and that vascular risk factors associated with advanced age may modify this relationship.

4 Structural Connectivity Analyzed by Graph Theory

4.1 General Concept of Graph Theory

For conventional MRI data, anatomical connectivity has been inferred by thresholding a matrix of interregional covariation in cortical volume or thickness. Strong between-subject covariation in local gray matter measurements has been interpreted as indicative of greater axonal connectivity between covarying regions on the grounds that connectivity has mutually trophic effects on the growth of connected neurons or regions when measured after a period of development. An early attempt to investigate human structural connectivity using morphometric correlations of VBM-derived gray matter density [47] extracted the regional gray matter densities from multiple regions of interest as seed regions and revealed patterns of positive and negative covariance that provided insight into the topographical organization of multiple cortical regions. A similar connectivity analysis using cortical thickness measures took Brodmann area 44 as the seed region [48]. This analysis revealed patterns of anatomical correlations across the cerebral cortex strikingly similar to tractography maps obtained from diffusion tensor imaging.

Analysis of structural connectivity using a seed region has recently been replaced by an approach using graph theory, a branch of mathematics that offers a diverse range of quantitative measures for characterizing the topology of structural networks for the whole brain without an a priori hypothesis. These measures largely relate to three broad properties: topological integration, topological segregation, and hub dominance. These graph-theoretical studies suggest that the human brain can be modeled as a complex network and may have a small-world structure at the levels of both anatomical and functional connectivity. This small-world structure is hypothesized to facilitate rapid synchronization and efficient information transfer with minimal wiring costs through an optimal balance between local processing and global interaction [49]. In this framework of a brain graph, nodes represent regions and edges the network connections between them [50, 51]. Based on the number and distribution of edges, a variety of measures can be computed to describe global and local connectivity properties. A brain graph can be subdivided or partitioned into subsets or modules of nodes. The mathematically optimal modular decomposition for a brain network is to find the partition that maximizes the ratio of intramodular to intermodular edges. Resolving the modular structure of a brain graph is likely to add important information about which anatomical regions have the most critical topological roles in transmission of information across brain networks. The term structural connectivity in the context of neuroscience refers to the physical network structure, particularly the long-range connections formed by neurons via white matter tracts, which subserves brain function. This is distinct from functional connectivity, which describes statistically associated functional signals observed through various functional imaging methods.

Little is currently known about the biological underpinnings of structural correlations across subjects or within subjects. It is therefore still unclear how different methods of graph construction map onto different biological processes that influence anatomical correlations, such as the presence of white matter tracts, and direct and indirect functional coupling. One of the few studies that has examined this issue [52] showed that when structural graphs are constructed based on Pearson correlations, about 40% of the positive correlations but only 10% of negative correlations map onto fiber connections as determined by diffusion tensor imaging. Another study [53] also found a relatively high similarity of 60% between the structural network and resting state fMRI network based on positive correlations, as well as a low to moderate similarity of 10–40% for negative correlations. These findings highlight the potential for morphometric correlational analyses to characterize structural connectivity in vivo, its longitudinal changes over the course of a neurodegenerative process, and its association with behavioral performance metrics. However, one drawback of the morphological correlative approach is that it describes only group-wise networks, from which only limited conclusions can be drawn at the individual level. On the other hand, Tijms et al. [54, 55] proposed a new method to represent the cortical morphology of individual subjects as networks using information about the similarity of gray matter structure within the cortex. This similarity-based extraction technique is now being investigated for the evaluation of individual networks.

4.2 Structural Connectivity in Aging

Organizational alterations of the structural brain network at the system level have been reported in normal aging. Chen et al. [56] reported that aging adults showed significantly reduced modularity compared with young adults, which might be indicative of reduced functional segregation in the aging brain. The aging brain network exhibited reduced intra-/inter-module connectivity among modules corresponding to the executive function and default mode networks of young adults, which may be associated with age-related cognitive decline. Wu et al. [57] investigated structural brain networks in 350 healthy subjects divided into young, middle-aged, and old age groups. The small-world efficiency and node betweenness varied significantly and revealed a U- or inverted U-curve model tendency among the age groups. The modular organization of structural brain networks was similar between the young and middle-aged groups but quite different for the old age group. The brain network changed slightly from young to middle age, developing into a more distributed organization, but then shifted to a more localized organization in old age. Zhu et al. [58] found that an older cohort had lower global efficiency due to higher local clustering in the brain structural networks compared with the younger cohort. The older cohort had lower centrality of certain brain regions, such as the bilateral hippocampus, bilateral insula, and left posterior cingulate gyrus. In our study on structural connectivity in healthy subjects age 20 to 89 years, the older age group (57 subjects older than 50 years) showed significantly lower global and local efficiency and clustering coefficients than the younger age group (44 subjects below 50 years old). The older age group showed significantly lower clustering and higher betweenness centrality in the main nodes of the default mode network, namely, the posterior cingulate gyrus, precuneus, and medial prefrontal areas, than the younger age group (Fig. 3).

5 Conclusions

The histological phenomena driving macroscopic volume reductions in normal aging are still uncertain. Data from premortem, in vivo MRI combined with postmortem histological analysis could clarify these mechanisms. Tau PET imaging could also allow us to investigate the links between subfield atrophy and the development of tau pathology in aging. Further investigations are needed on the relationship between structural connectivity and cognitive performance.

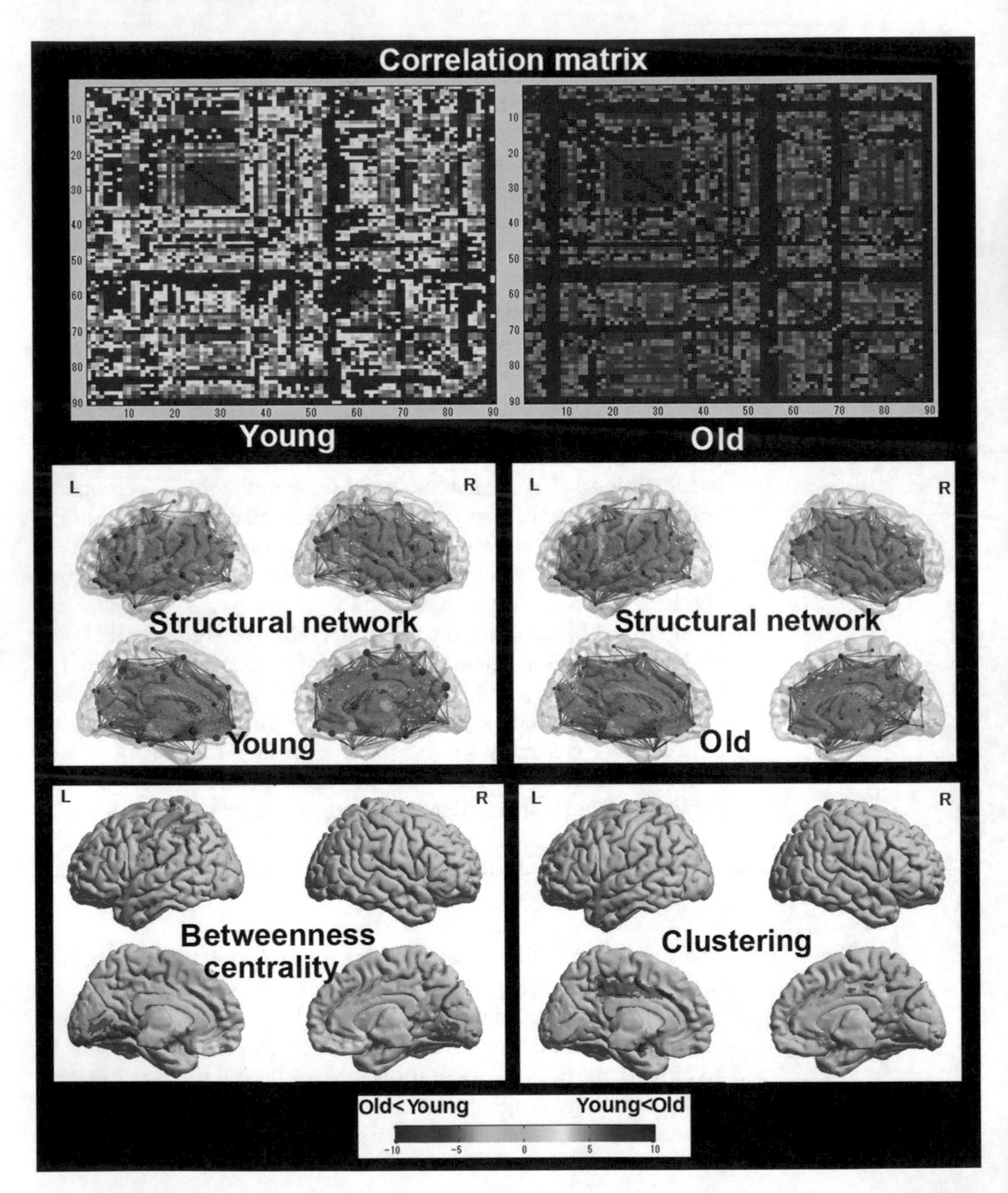

Fig. 3 Group differences in regional topography of structural networks between young and old age groups of healthy volunteers. Structural connectivity analysis based on graph theory reveals differences in the association matrix (upper row) and in the network comprising nodes and edges (middle row; hubs are colored green). The old age group shows significant regional increases and decreases (false discovery rate of 0.05) in betweenness centrality and clustering, respectively, in main part of default mode network, the posterior cingulate gyrus, precuneus, and medial prefrontal areas (lower row)

Acknowledgments

This work was carried out under the Brain Mapping by Integrated Neurotechnologies for Disease Studies (Brain/MINDS) project, funded by the Japan Agency for Medical Research and Development (AMED).

References

1. Frisoni GB, Fox NC, Jack CR Jr, Scheltens P, Thompson PM (2010) The clinical use of structural MRI in Alzheimer disease. Nat Rev Neurol 6:67–77
2. Jack CR Jr, Barkhof F, Bernstein MA et al (2011) Steps to standardization and validation of hippocampal volumetry as a biomarker in clinical trials and diagnostic criterion for Alzheimer's disease. Alzheimers Dement 7:474–485
3. Geuze E, Vermetten E, Bremner JD (2005) MR-based in vivo hippocampal volumetrics: 1. Review of methodologies currently employed. Mol Psychiatry 10:147–159
4. Frisoni GB, Jack CR Jr, Bocchetta M et al (2015) The EADC-ADNI harmonized protocol for manual hippocampal segmentation on magnetic resonance: evidence of validity. Alzheimers Dement 11:111–125
5. Fischl B, Salat DH, Busa E et al (2002) Whole brain segmentation: automated labeling of neuroanatomical structures in the human brain. Neuron 33:341–355
6. Fujishima M, Maikusa N, Nakamura K et al (2014) Mild cognitive impairment, poor episodic memory, and late-life depression are associated with cerebral cortical thinning and increased white matter hyperintensities. Front Aging Neurosci 6:306
7. Yushkevich PA, Wang H, Pluta J et al (2010) Nearly automatic segmentation of hippocampal subfields in in vivo focal T2-weighted MRI. NeuroImage 53:1208–1224
8. Ashburner J, Friston KJ (2000) Voxel-based morphometry--the methods. NeuroImage 11:805–821
9. Ashburner J, Friston KJ (2001) Why voxel-based morphometry should be used. NeuroImage 14:1238–1243
10. Good CD, Johnsrude IS, Ashburner J et al (2001) A voxel-based morphometric study of ageing in 465 normal adult human brains. NeuroImage 14:21–36
11. Karas GB, Burton EJ, Rombouts SA et al (2003) A comprehensive study of gray matter loss in patients with Alzheimer's disease using optimized voxel-based morphometry. NeuroImage 18:895–907
12. Ashburner J (2007) A fast diffeomorphic image registration algorithm. NeuroImage 38:95–113
13. Eggert LD, Sommer J, Jansen A, Kircher T, Konrad C (2012) Accuracy and reliability of automated gray matter segmentation pathways on real and simulated structural magnetic resonance images of the human brain. PLoS One 7: e45081
14. Klein A, Andersson J, Ardekani BA et al (2009) Evaluation of 14 nonlinear deformation algorithms applied to human brain MRI registration. NeuroImage 46:786–802
15. Raji CA, Lopez OL, Kuller LH, Carmichael OT, Becker JT (2009) Age, Alzheimer disease, and brain structure. Neurology 73:1899–1905
16. Raz N, Lindenberger U, Rodrigue KM et al (2005) Regional brain changes in aging healthy adults: general trends, individual differences and modifiers. Cereb Cortex 15:1676–1689
17. Resnick SM, Pham DL, Kraut MA, Zonderman AB, Davatzikos C (2003) Longitudinal magnetic resonance imaging studies of older adults: a shrinking brain. J Neurosci 23:3295–3301
18. Matsuda H, Ohnishi T, Asada T et al (2003) Correction for partial-volume effects on brain perfusion SPECT in healthy men. J Nucl Med 44:1243–1252
19. Tisserand DJ, van Boxtel MP, Pruessner JC, Hofman P, Evans AC, Jolles J (2004) A voxel-based morphometric study to determine individual differences in gray matter density associated with age and cognitive change over time. Cereb Cortex 14:966–973
20. Grieve SM, Clark CR, Williams LM, Peduto AJ, Gordon E (2005) Preservation of limbic and paralimbic structures in aging. Hum Brain Mapp 25:391–401
21. Smith CD, Chebrolu H, Wekstein DR, Schmitt FA, Markesbery WR (2007) Age and gender effects on human brain anatomy: a voxel-based morphometric study in healthy elderly. Neurobiol Aging 28:1075–1087

22. Curiati PK, Tamashiro JH, Squarzoni P et al (2009) Brain structural variability due to aging and gender in cognitively healthy elders: results from the Sao Paulo ageing and health study. Am J Neuroradiol 30:1850–1856
23. Kalpouzos G, Chételat G, Baron JC et al (2009) Voxel-based mapping of brain gray matter volume and glucose metabolism profiles in normal aging. Neurobiol Aging 30:112–124
24. Terribilli D, Schaufelberger MS, Duran FL et al (2011) Age-related gray matter volume changes in the brain during non-elderly adulthood. Neurobiol Aging 32:354–368
25. Giorgio A, Watkins KE, Chadwick M et al (2010) Longitudinal changes in grey and white matter during adolescence. NeuroImage 49:94–103
26. Giorgio A, Santelli L, Tomassini V et al (2010) Age-related changes in grey and white matter structure throughout adulthood. NeuroImage 51:943–951
27. Liu H, Wang L, Geng Z et al (2016) A voxel-based morphometric study of age- and sex-related changes in white matter volume in the normal aging brain. Neuropsychiatr Dis Treat 12:453–465
28. Streitbürger DP, Möller HE, Tittgemeyer M, Hund-Georgiadis M, Schroeter ML, Mueller K (2012) Investigating structural brain changes of dehydration using voxel-based morphometry. PLoS One 7:e44195
29. Hutton C, Draganski B, Ashburner J, Weiskopf N (2009) A comparison between voxel-based cortical thickness and voxel-based morphometry in normal aging. NeuroImage 48:371–380
30. Wu W, Brickman AM, Luchsinger J et al (2008) The brain in the age of old: the hippocampal formation is targeted differentially by diseases of late life. Ann Neurol 64:698–706
31. Mueller SG, Stables L, AT D et al (2007) Measurement of hippocampal subfields and age-related changes with high resolution MRI at 4T. Neurobiol Aging 28:719–726
32. Yushkevich PA, Pluta JB, Wang H et al (2015) Automated volumetry and regional thickness analysis of hippocampal subfields and medial temporal cortical structures in mild cognitive impairment. Hum Brain Mapp 36:258–287
33. Wisse LE, Biessels GJ, Geerlings MI (2014) Critical appraisal of the hippocampal subfield segmentation package in FreeSurfer. Front Aging Neurosci 6:261
34. de Flores R, La Joie R, Landeau B et al (2015) Effects of age and Alzheimer's disease on hippocampal subfields: comparison between manual and FreeSurfer volumetry. Hum Brain Mapp 36:463–474
35. Iglesias JE, Augustinack JC, Nguyen K et al (2015) A computational atlas of the hippocampal formation using ex vivo, ultra-high resolution MRI: application to adaptive segmentation of in vivo MRI. NeuroImage 115:117–137
36. Mueller SG, Weiner MW (2009) Selective effect of age, Apo e4, and Alzheimer's disease on hippocampal subfields. Hippocampus 19:558–564
37. Shing YL, Rodrigue KM, Kennedy KM et al (2011) Hippocampal subfield volumes: age, vascular risk, and correlation with associative memory. Front Aging Neurosci 3:2
38. Raz N, Daugherty AM, Bender AR, Dahle CL, Land S (2015) Volume of the hippocampal subfields in healthy adults: differential associations with age and a pro-inflammatory genetic variant. Brain Struct Funct 220:2663–2674
39. Kerchner GA, Bernstein JD, Fenesy MC et al (2013) Shared vulnerability of two synaptically-connected medial temporal lobe areas to age and cognitive decline: a seven tesla magnetic resonance imaging study. J Neurosci 33:16666–16672
40. La Joie R, Fouquet M, Mézenge F et al (2010) Differential effect of age on hippocampal subfields assessed using a new high-resolution 3T MR sequence. NeuroImage 53:506–514
41. Ziegler G, Dahnke R, Jäncke L, Yotter RA, May A, Gaser C (2012) Brain structural trajectories over the adult lifespan. Hum Brain Mapp 33:2377–2389
42. Pereira JB, Valls-Pedret C, Ros E et al (2014) Regional vulnerability of hippocampal subfields to aging measured by structural and diffusion MRI. Hippocampus 24:403–414
43. Voineskos AN, Winterburn JL, Felsky D et al (2015) Hippocampal (subfield) volume and shape in relation to cognitive performance across the adult lifespan. Hum Brain Mapp 36:3020–3037
44. Daugherty AM, Bender AR, Raz N, Ofen N (2016) Age differences in hippocampal subfield volumes from childhood to late adulthood. Hippocampus 26:220–228
45. Chadwick MJ, Bonnici HM, Maguire EA (2014) CA3 size predicts the precision of memory recall. Proc Natl Acad Sci U S A 111:10720–10725
46. Bender AR, Daugherty AM, Raz N (2013) Vascular risk moderates associations between hippocampal subfield volumes and memory. J Cogn Neurosci 25:1851–1862
47. Mechelli A, Friston KJ, Frackowiak RS, Price CJ (2005) Structural covariance in the human cortex. J Neurosci 25:8303–8310

48. Lerch JP, Worsley K, Shaw WP et al (2006) Mapping anatomical correlations across cerebral cortex (MACACC) using cortical thickness from MRI. NeuroImage 31:993–1003
49. Stam CJ, Reijneveld JC (2007) Graph theoretical analysis of complex networks in the brain. Nonlinear Biomed Phys 1:3
50. Bullmore ET, Bassert DS (2011) Brain graphs: graphical models of the human brain connectome. Annu Rev Clin Psychol 7:113–140
51. Hosseini SM, Hoeft F, Kesler SR (2012) GAT: a graph-theoretical analysis toolbox for analyzing between-group differences in large-scale structural and functional brain networks. PLoS One 7:e40709
52. Gong G, He Y, Chen ZJ, Evans AC (2012) Convergence and divergence of thickness correlations with diffusion connections across the human cerebral cortex. NeuroImage 59:1239–1248
53. Hosseini SM, Kesler SR (2013) Comparing connectivity pattern and small-world organization between structural correlation and resting-state networks in healthy adults. NeuroImage 78:402–414
54. Tijms BM, Seriès P, Willshaw DJ, Lawrie SM (2012) Similarity-based extraction of individual networks from gray matter MRI scans. Cereb Cortex 22:1530–1541
55. Tijms BM, Kate MT, Wink AM et al (2016) Gray matter network disruptions and amyloid beta in cognitively normal adults. Neurobiol Aging 37:154–160
56. Chen ZJ, He Y, Rosa-Neto P, Gong G, Evans AC (2011) Age-related alterations in the modular organization of structural cortical network by using cortical thickness from MRI. NeuroImage 56:235–245
57. Wu K, Taki Y, Sato K et al (2012) Age-related changes in topological organization of structural brain networks in healthy individuals. Hum Brain Mapp 33:552–568
58. Zhu W, Wen W, He Y, Xia A, Anstey KJ, Sachdev P (2012) Changing topological patterns in normal aging using large-scale structural networks. Neurobiol Aging 33:899–913

Check for updates

Chapter 12

Morphometry and Genetics

Ali Bani-Fatemi, Samia Tasmim, Tayna Santos, Jose Araujo, and Vincenzo De Luca

Abstract

Several twin and family studies have shown that the influence of genes on brain volume is already evident in childhood. However, the influence of those genes on brain development and structure remains unclear. Most current research was done on candidate gene polymorphisms and their possible associations with brain structural abnormalities in psychiatric diseases such as schizophrenia, major depressive disorder, and bipolar disorder. The polymorphisms are often studied through genome-wide association studies (GWAS), and the brain imaging is often done by magnetic resonance imaging (MRI) techniques, including diffusion tensor imaging (DTI). Although there are many studies on the effects of gene polymorphisms and structural neuroimaging, a comprehensive review is lacking. Thus, the scope of this chapter is to review the structural imaging genetics studies across several neuropsychiatric disorders. This chapter will review the current literature on imaging genetics studies, provide additional considerations on the imaging genetics studies of suicidal behaviors and childhood onset disorders, and also discuss studies that have investigated rare genetic variants.

Key words Neuroimaging, Genetics, Polymorphisms, Neuropsychiatric disorders, Structural MRI, Brain morphometry

1 Imaging Genetics in the General Population

1.1 Genetic Influence on Brain Structure

The degree to which the environment and genetics influence the brain's susceptibility to psychiatric disorder development is still unclear. Several twin and family studies have shown that the influence of genes on both global and gray matter volumes of the human brain is already evident in childhood. In a comparison between groups of dizygotic (DZ) and monozygotic (MZ) twins, it was found that the global gray matter volume was similar in genetically alike individuals, suggesting that brain volume in adulthood is highly heritable [1]. Moreover, a recent meta-analysis suggests that cerebellar, subcortical, ventricular, and corpus callosum volumes seen in adults are also highly heritable [2]. However, because the findings are still inconsistent, further

Gianfranco Spalletta et al. (eds.), *Brain Morphometry*, Neuromethods, vol. 136,
https://doi.org/10.1007/978-1-4939-7647-8_12,

investigations are needed in order to understand the impact of genetics on brain imaging.

1.2 Studies of the Candidate Genes Influencing Human Brain Volume in Healthy Individuals

The influence of genetics on the human brain has been studied in healthy individuals, and some genes have become prominent as candidate genes that could be associated with a particular phenotype of interest based on a priori hypotheses. The three most widely studied candidate genes with regard to brain volume are *BDNF* [3–5], *COMT* [6–8], and *DISC1* [9–14].

1.2.1 BDNF *Gene*

Brain-derived neurotrophic factor (*BDNF*) is found in the brain and spinal cord and belongs to a family of proteins called neurotrophins. It is encoded by the *BDNF* gene. *BDNF* is involved in the survival, development, and maintenance of neurons in the central and peripheral nervous systems [3]. A recent meta-analysis reported increased bilateral hippocampal volumes in healthy Val/Val individuals when compared to Met allele carriers of the *BDNF* gene *Val66Met* single nucleotide polymorphism (SNP) [4]. Another study [5] showed that Met carriers have reduced hippocampal volumes, independently of gender and age, suggesting that these changes occurred before adulthood. In addition, it was found that the *Val/Met* heterozygotes show a reduction in the lateral convexity of the frontal cortex gray matter volumes [5].

1.2.2 COMT *Gene*

Catechol-O-methyltransferase (*COMT*), encoded by the *COMT* gene, is an enzyme that catalyzes the transfer of a methyl group from S-adenosylmethionine to catecholamines such as dopamine, epinephrine, and norepinephrine [6]. *COMT* is also important in the metabolism of various drugs with catechol structures. Zinkstok et al. [7] reported an increased gray matter volume with increasing age in female Val carriers of *COMT* Val158Met polymorphism. They also found decreased gray matter volumes in the right parietal lobe and left parahippocampal gyrus, and decreased white matter volumes in the right frontal lobe, left parahippocampal gyrus, right inferior parietal lobe, and left corpus callosum, with increasing age in female Met homozygotes. However, these effects of the *COMT* genotype were not found in males. One possible explanation for this gender difference may be related to estrogen. Recent findings [8] show that individuals who carry both *COMT*-Met and *5-HTTLPR* (serotonin transporter length polymorphic region) *short* alleles have decreased gray matter volumes in parahippocampal gyrus (bilaterally), cerebellum, right putamen/insula, amygdala, and hippocampus when compared with homozygotes for either *COMT*-Val or *5-HTTLPR long* allele.

1.2.3 DISC1 *Gene*

The gene encoding for the disrupted in schizophrenia-1 (*DISC1*) protein has been reported as a risk factor for several psychiatric disorders [9]. This protein influences neural growth and synaptic modulation [10]. DISC1 has been associated with hippocampal abnormalities in healthy individuals, as Callicott et al. [11] showed that Ser704 homozygotes had a significant reduction in hippocampal gray matter volume. In addition, recent studies suggest that the effect of the Ser704Cys polymorphism on the hippocampus takes place before the onset of mid-childhood, with Ser704 homozygotes having smaller hippocampi than Cys704 carriers [12].

Nevertheless, Raznahan et al. [12] have also shown that Phe607 carriers (rs6675281) exhibited a decreased thickness in temporal-parietal areas. Moreover, it was found that Phe607 carriers had greater striatal volume than Leu607 homozygotes [13].

We have unpublished data from healthy subjects genotyped for the Ser704Cys (rs821616) and Arg264Gln (rs3738401) SNPs finding no significant difference in either left or right hippocampal volumes.

Another study by Trost et al. [14] on the T allele carriers of rs821616 from different diagnostic groups showed a reduction in gray matter volumes in the frontal brain: the right superior gyrus and bilaterally the middle gyri and, on the left side, the subgenual anterior cingulate cortex; in addition, the T carriers had a reduction of gray matter volumes in the right medial temporal cortex and right inferior temporal gyrus. On the other hand, the A/A homozygotes had reduced gray matter volumes in the left superior gyrus.

1.3 GWAS

GWAS is a relatively new approach to identify genes involved in human disease. In this hypothesis-free method, the whole genome is searched for genetic variations (typically single-nucleotide polymorphisms) that arise more frequently in individuals with a particular disease compared to people without the disease.

A genome-wide association meta-analysis done by Stein et al. [15] reported some associations of brain volumes with the common SNP rs7294919 (on chromosome 12q24.22), and hippocampal volumes with rs10784502 (on chromosome 12q14.3). However, a study done with two independent samples of healthy German individuals [16] showed significant evidence of association between the T-allele of rs7294919 and a reduction of bilateral gray matter volumes in the hippocampus. However, no association was found with the white matter volumes.

A cross-sectional genome-wide association analysis [17] suggested that the G allele for rs17178006 in the region 12q14 is associated with decreased hippocampal volumes. In addition, the G allele of the SNP rs6741941 (on chromosome 12q24) and the C allele of the SNP rs7852872 (on chromosome 9) showed suggestive evidence of a relationship with smaller hippocampal volumes but did not reach genome-wide significance. A study discussing the effect of common schizophrenia risk SNPs (MIR137, TCF4, and ZNF804A) on healthy subjects [18] showed no effect on white and gray matter, total brain, or hippocampal volumes. However, in schizophrenic patients, the SNP rs1625579 on MIR137 was associated with decreased hippocampal volumes [19].

1.4 Copy Number Variation Studies

Copy number variations (CNVs) have been identified as possible contributors to neurodevelopmental disorders, as some studies have found that at certain spots in the genome, CNVs are more frequently present in neuropsychiatric patients than in controls.

Copy number changes at 16p11.2 have been strongly associated with increased risk for some psychiatric disorders including autism spectrum disorder (ASD), schizophrenia, developmental delay, and epilepsy. The deletion in this genomic region represents an increased risk for ASD, while its duplication increases risk for both ASD and schizophrenia [20]. According to a study done by Stein et al. [20] about the effects of CNVs at 16p11.2 on brain structure, deletion carriers seem to have smaller global brain volumes, while duplication carriers seem to have smaller volumes of total white and gray matter, when compared to controls. In addition, thalamic and cerebellar volumes also seem to be associated with CNVs at this locus, as deletion carriers have shown greater volumes in those specific brain regions. Another study [21] on CNVs at the same region showed a cortical thickness reduction in both deletion and duplication carriers. This study also suggested that, in the absence of diagnostic criteria for ASD and schizophrenia, brain anatomy changes in regions related to both disorders, such as in fronto-subcortical regions, could act as endophenotypes for those disorders. However, more studies are needed to elucidate the effects of 16p11.2 CNVs on psychiatric disorders and brain morphometry.

The deletion syndrome at 22q11.2 is a common contiguous gene deletion syndrome associated with a wide range of neuropsychiatric disorders. One third to one half of the children with this deletion are diagnosed with attention-deficit/hyperactivity disorder (ADHD), anxiety disorder, mood disorder, or ASD. Some investigators also suggest that deletions at 22q11.2 are associated with psychosis; for example, Vorstman et al. [22] suggest that ASD and schizophrenia could be different pleiotropic manifestations of a 22q11.2 deletion. A study done by Bearden et al. [23] showed a reduction in total brain volume in 22q11.2 deletion syndrome individuals, particularly in the parietal lobes, as well as in the occipital lobe and cerebellar midline brain regions, although the frontal lobe volume seemed to be preserved. Even though further studies are needed to understand the effects of CNVs at 22q11.2 and schizophrenia, several studies [24, 25] showed an association between functional variants in genes within the 22q11.2 locus and an increased risk for the development of schizophrenia, as well as significant brain structural abnormalities, such as a decrease in fronto-striatal volume.

1.5 Imaging Genetics of Rare Variants

Although there are not many studies that have investigated rare variants, recent studies in youth with Down syndrome (DS) report reductions in total brain volume and surface area and increased cortical, precentral gyrus, and caudal anterior cingulate thickness, compared with their healthy siblings [26]. Another research study comparing the MRI scans of DS patients with healthy controls showed a significant reduction in the total brain volume and cerebellum in the DS patients. However, specific atrophic changes were not found in DS brains [27]. Nevertheless, a study done through MRI scans in non-demented adults with Down syndrome [28] found a reduction in total gray matter volume with increasing age, particularly in the bilateral parietal, left prefrontal, left occipital, and left temporal cortices. Controversially, this study found that gray matter volume in the cerebellum was preserved. Several neuroimaging studies have investigated structural differences between brains of XXY males (Klinefelter syndrome) and controls. A recent voxel-based analysis [29] found significant atrophy in the insula, temporal gyri, amygdala, hippocampus, cingulate, occipital gyri, and parietal lobe white matter when compared to controls. Previous studies by Patwardhan et al. [30] have also reported significantly reduced amygdala volumes in individuals with Klinefelter syndrome, as well as significantly reduced whole brain volumes in females with the XXX karyotype.

There are few studies on the brain morphometry of pre-mutation carriers of Fragile X (FraX) syndrome (individuals who have between 50 and 200 CGG repeats in the FMR1 gene). A study among a group of pre-mutation FraX females suggests that FraX pre-mutation carriers do not show significant brain volume differences when compared to controls [31]. A study done with male pre-mutation carriers of FraX reported a significant reduction in both gray and white matter volumes in the left thalamus and amygdalo-hippocampal complex [32]. On the other hand, a study including both male and female subjects with FraX syndrome and healthy controls found a significant enlargement of the periventricular structures in the FraX subjects, without any sex-related differences [33].

Spinocerebellar ataxia (SCA) type 2 is associated with a CAG expansion on chromosome 12q23-24.1 [34]. Recent studies toward SCA2 have shown that there is a region-specific (anterior lobe, lobule VI, Crus I, Crus II, lobule VIII, uvula, corpus medullare, and pons) atrophy in the pontocerebellar structure [35].

Several studies have shown cortical atrophy in late-treated patients with phenylketonuria (PKU), suggesting that hyperphenylalaninemia has a toxic effect on brain development, leading to defective brain growth [36–38]. A recent study [39] has shown no significant differences between patients with early-treated PKU and healthy control subjects. However, the results among late-treated patients are consistent with previous studies.

Huntington's disease (HD) is an autosomal-dominant disorder often characterized by neuropathology in the striatum and globus pallidus. Investigations using Voxel-based morphometry suggest significant reductions in striatal and caudate gray matter volumes [40]. In addition, previous studies have shown that the putaminal volume loss is greater than the caudate nucleus loss [41, 42].

2 Imaging Genetics of Psychiatric Disorders

2.1 Imaging Genetics of Major Depressive Disorder

A recent study done by Seok et al. [43] investigated the relationship between the *COMT* Val158Met (rs4680) polymorphism and brain structure, in both healthy subjects and individuals with major depressive disorder (MDD). It has been shown that fractional anisotropy (using DTI) in white matter regions connecting prefrontal areas to other cortical and subcortical structures was reduced only for the valine homozygote individuals with MDD. In addition, it was found that treatment-naïve patients with MDD have smaller caudate volumes than healthy controls, and this reduction is greater in MDD patients with Val/Met genotype compared to controls with the same genotype.

The *5-HTTLPR* seems to be associated with a greater risk of MDD, but its effect on brain morphometry is controversial. Individuals with the *short* functional variant of the *5-HTTLPR* gene seem to be at greater risk for developing depression than the individuals with the *long* allele [44]. A study conducted by Frodl et al. [45] showed that MDD patients with the *long-long (l/l)* genotype have decreased gray matter in the amygdala, hippocampus, anterior cingulate cortex (ACC), dorsomedial prefrontal cortex (PFC), and dorsolateral PFC, while in healthy individuals, the opposite was found. Another study showed reduced gray matter volumes in the inferior frontal gyrus, ACC, and superior temporal gyrus in non-depressed subjects bearing *short* alleles. A study by Jaworska and colleagues [46] showed no *5-HTTLPR* genotype effect on cortical thickness, both on MDD patients and healthy controls. However, increased volume in the left thalamus and putamen were found in *l/l* homozygotes, which may indicate an association between the polymorphism and the regions involved with emotional and reward processing.

Recent studies have shown a significant association between the pathogenesis of MDD and the norepinephrine transporter gene (*SLC6A2*) polymorphisms T-182T/C (rs2242446) and *G1287A* (rs5569) [47]. The *SLC6A2* gene is involved in the selective NE reuptake by the presynaptic terminal, which is a target for several antidepressants. A study done by Ueda et al. [48] showed a significant enlargement of the left dorsolateral PFC in MDD patients with the G/G genotype for G1287A compared with healthy controls with the same genotype. However, no association between the

T-182C polymorphism and structural brain abnormalities in MDD patients was found. More studies with larger samples must be done in order to elucidate potential association between polymorphisms and morphometric changes in patients with MDD.

2.2 Imaging Genetics of Bipolar Disorder

Bipolar disorder (BD) has been associated with neuroanatomical changes in the fronto-limbic regions, hippocampus, basal ganglia, and corpus callosum [49–52].

Recent studies have associated the *BDNF* Val66Met polymorphism (rs6265) (located on chromosome 11p13, a substitution of guanine for adenine at codon 66, resulting in Val being replaced by Met) with a significant reduction in *BDNF* traffic to secretory granules and therefore to a reduced BDNF production by secretory granules [53, 54]. This polymorphism has also been linked to abnormal hippocampus volumes across the mood disorder spectrum in populations with BD [5, 53, 55–57]. However, these findings are still controversial, since recent studies have shown reduced hippocampal volumes in patients with BD compared to healthy controls that were unrelated to the *BDNF* polymorphism, or even no difference on hippocampal volumes compared to healthy controls [4]. Post hoc analyses showed that within *BDNF* Met carriers BD presented smaller hippocampal volumes compared to healthy controls, while MDD patients had hippocampal volumes similar to healthy patients. *BDNF* Val/Val BD and MDD patients had hippocampal sizes similar to controls. Healthy *BDNF* Met carriers were not found to have abnormal hippocampal volumes.

Two GWAS found significant associations between BD type I and the SNP on the *CACNA1C* gene (rs1006737), which codes for the major L-type voltage-dependent calcium channel, alpha 1C subunit [58, 59]. A study done by Perrier and colleagues [60], with 41 euthymic BD type I patients and 50 healthy controls, showed that risk allele carriers had increased gray matter volume in the right amygdala and right hypothalamus in both healthy controls and BD patients. The only significant interaction between diagnosis and genotype was that BD risk allele carriers showed smaller volumes in the left putamen when compared to risk allele carriers in the control group. Nevertheless, other studies [61] on the same gene have shown no association between subcortical structure volumes and risk allele carriers.

Recent studies suggest that the unregulated expression of *DISC1* impacts neurodevelopment and gray and white matter structures, as well as acts as a predisposing factor for the development of mood disorders [62]. According to a recent study [63] on *DISC1*, Cys704-carriers with affective disorders had larger hippocampal volumes in comparison to Ser-homozygotes, while in healthy controls the Cys704 allele was associated with lower hippocampal volumes. Further studies are necessary to understand whether the *DISC1* gene can influence normal neurodevelopment and brain structure in patients with affective disorders. Nevertheless, these measures in BD patients may be affected by psychotropic medication use, familial load, illness progression, age, heterogeneity of subject groups, and variability in imaging methodology [64–66].

2.3 Imaging Genetics of Schizophrenia

Schizophrenia is a complex disorder affected by multiple genes; therefore, it is likely that multiple genes contribute to the structural abnormalities in gray matter volumes in schizophrenic patients [67]. The findings on the *COMT* Val158Met genotype effect in schizophrenia are contradictory, as two investigations reported no association with *COMT* Val158Met genotype in schizophrenic patients [68, 69] and other two reported genotype effects in these patients [70, 71]. Other evidence [72] suggests an association between smaller temporal areas and the Val158 carriers in patients with schizophrenia.

Ho et al. [73], in a study on the effects of *BDNF* Val66Met genotype on brain structure in schizophrenic patients, found that Met carriers had greater reductions in frontal gray matter volume and increases in the ventricular and sulcal (frontal and temporal) volumes, compared to Val homozygotes. In another study, Ho et al. [74] found that Met allele carriers, independently from their diagnosis, had smaller occipital and temporal lobar gray matter volumes when compared to Val homozygous subjects.

Several studies reported an association between candidate genes such as *DISC1* and hippocampal [11, 75] and insular gray matter volume [76] in schizophrenic patients.

Studies also report an association between *RGS4* gene polymorphisms and dorsolateral PFC volumes in patients with schizophrenia compared to healthy controls [77]. Donohoe et al. [78] reported that homozygous schizophrenia carriers of the ZNF804A rs1344706 risk allele had relatively larger gray matter volumes than heterozygous or homozygous noncarrier schizophrenic patients with AC or CC genotypes, specially for hippocampal volumes, although they could not replicate this finding in healthy controls.

2.4 Imaging Genetics of Suicidal Behavior

Although there are no available studies combining genetic factors and brain morphometry in suicide attempters, there are several studies regarding brain morphometry in this population.

Most of the studies on suicide attempt were done on mood disorder patients, where suicidal behavior is often prominent. A study done by Budisic et al. [79] with 17 patients with MDD including 14 with MDD and present suicidal ideation and 40 healthy controls showed a significant reduction in raphe nuclei echogenicity only in the patients with suicidal ideation. In 2001, Ahearn et al. [80] compared MRI data from a sample of unipolar depressed patients with and without suicide attempt history. They found that the patients with a history of suicide attempt presented significantly more subcortical gray matter hyperintensities. Rüsch et al. [81] reported that schizophrenic patients with a history of suicide attempts have larger inferior frontal white matter volumes bilaterally compared to schizophrenic patients without a history of suicide attempts. A recent MRI study by Ding et al. [82] with 67 mood disorder patients with a history of suicidal behavior, 82 mood disorder patients with no history of suicidal behavior, and 82 healthy controls showed that there are significant reductions in the left ventrolateral PFC gray matter volumes in the suicidal behavior group compared to the other two groups. The study also found differences (that did not survive multiple corrections) in the dorsal PFC and orbitofrontal cortex (OFC) between the suicidal behavior group and controls.

Besteher et al. [83] studied a group of schizophrenia patients, with 14 suicide attempters and 23 non-suicidal people, and 50 controls. They found cortical thinning in the right dorsolateral PFC, superior temporal lobe, middle temporal lobe, and temporolabar lobe and the insular cortex in the suicidal patients compared to the non-suicidal patients.

A recent review by Jollant et al. [84] concluded that there are volumetric reductions in the OFC, superior temporal gyrus, and caudate nucleus in the suicidal brain.

In summary, most suicide attempt studies investigating the brain structure in the frontal, parietal, temporal, occipital, and insular lobes and the ACC, basal nuclei, putamen, corpus callosum, midbrain, and cerebellum of suicide attempters showed a reduction in white and gray matter volume. The only exceptions were the white matter volumes of the inferior frontal gyrus and posterior orbital gyri [81] and the gray matter volumes in the amygdala, which seem to be increased [78, 85]. However, no differences in gray or white matter volumes were found in the thalamus, globus pallidus, nucleus accumbens, and caudate [78, 86].In addition to these studies, structural neuroimaging studies have also reported alterations in neuronal or glial cell density. A DTI study done by Jia and colleagues [87] showed that depressed patients with a history of suicide attempts exhibit a reduction in OFC, dorsomedial PFC, frontal cortex, and right lentiform nucleus white matter volumes.

Regarding candidate gene association studies of suicide attempt, there are several studies that have investigated suicide attempt and the genetic polymorphisms significantly associated with brain morphometry variation.

Recent research done by Antypa et al. in BD patients [88] failed to show an association between suicide attempt and polymorphisms within *COMT* and *BDNF*, which are often associated with structural brain abnormalities in mood disorders [89, 90]. However, they suggested an association of *MAPK1* rs13515 and *CREB1* rs6740584 polymorphisms with suicide attempt. Nevertheless, these results do not show a strong association, and further investigations are needed.

We created a table (Table 1) of the SNPs that have been investigated in both morphometry studies and suicide genetic studies. We have also tried to identify if the same allele is conferring risk for suicide attempt and reduced volume in gray matter. Although the combination of imaging and genetics has been the focus of most psychiatric disorders, this approach was not applied to suicide phenotypes. In conclusion, we propose future imaging genetics studies of suicide which can enable us to identify the genes that are conferring risk for both suicide attempt and volume reduction of brain structures.

2.5 Imaging Genetics of Autism Spectrum Disorder

Several studies investigated imaging genetics among common genetic variants associated with ASD. Variations in the *CNTNAP2* gene are known to contribute to ASD vulnerability and to alterations in frontal white matter volumes [91]. Tan et al. [92] found a reduction in cerebellum, fusiform gyrus, cingulum, and posterior occipital lobe volumes in subjects homozygous for the *CNTNAP2* rs779475 risk allele (T). Also, they found sex-specific reductions in fractional anisotropy of normal control subjects homozygous for the same *CNTNAP2* risk allele. Tost et al. [93] found an association between the A allele of the rs53576 polymorphism in the *OXTR* gene with a reduction in hypothalamus volume and increase in amygdala volume.

A study done by Voineskos et al. [91] also showed that healthy subjects homozygous for the rs1045881 C risk allele in the Neurexin-1 gene showed reduced frontal lobe white matter and thalamic volumes. Findings toward the effects of polymorphisms in the monoamine oxidase A (MAOA) promoter are still controversial. However, Davis [94] showed that among a population of male children with autism, the *MAOA* promoter VNTR polymorphism is associated with increased cortical volume.

2.6 Imaging Genetics of Obsessive Compulsive Disorder

In the first meta-analysis of imaging genetics GWAS on candidate genes related to obsessive compulsive disorders (OCD) [15], any strong associations between hippocampal volumes and predicted SNPs were not found. Current research, however, has been focusing on polymorphisms in genes related to the serotoninergic system and imaging in OCD. Several studies [95–97] found a reduction in the OFC volumes in OCD patients with the *short* allele of the 5-HTTLPR. Studies done with children and adolescents with OCD have found an increase in the volumes of the left OFC and right anterior cingulate cortex in risk allele carriers of the *GRIN2B* polymorphism [98, 99], as well as increased total ACC volume and

Table 1
Convergent evidence of SNP effect on suicide and brain morphometry

SNPs	Effects on suicide	Effects on morphometry
BDNF val66met	Study done by Pregelj et al. [132] showed a significant association between the combined met/met and met/val genotypes versus the val/val genotype in female completed suicide victims (especially those who committed suicide violently) when compared to healthy female controls. However, this study did not show an association between *BDNF* val66met and suicidal behavior. A recent meta-analysis [133] reported that the met-carrying genotypes and the met allele confer risk for suicide attempt in various psychiatric disorders	In healthy subjects, the met-*BDNF* allele is associated with decreased hippocampus volume. There were differences found in hippocampal volume between val/val and val/met healthy individuals. There was also a reduction found in the gray matter density of the frontal and prefrontal cortex of *BDNF* 66met allele carriers [57, 134] In subjects with schizophrenia homozygous for the val allele, larger hippocampal volumes were found [135], and among schizophrenic met carriers, gray matter seemed to be decreased in both occipital and temporal lobes when compared to val-homozygous counterparts [74]
COMT	Studies toward the association between *COMT* val108/158met and suicide attempters have shown some inconsistencies, as suicidal attempts have been associated with both met allele [136, 137] and val allele [138], or even no significant relationship between the frequencies of these polymorphisms was found between suicidal attempters and controls [89]	A recent study showed no statistically significant effect of genotype on frontal lobe morphometry in schizophrenic patients and in healthy populations [70]. However, there are current findings that demonstrated an association between the *COMT* val158met polymorphism and a gray matter density reduction in anterior cingulate [71] and an increase in amygdala and hippocampus volume [139] in both schizophrenic and healthy individuals. The *COMT* met allele was also associated with higher medial temporal lobe volume and a significantly higher gray matter density in the hippocampus and parahippocampal gyrus in healthy controls [139]
5-HTTLPR	A recent meta-analysis showed that the *short* allele of *5-HTTLPR* was more frequent in suicide attempters than non-attempters with the same psychiatric disorders, but the polymorphism was not associated with suicidal behavior when compared to healthy controls [95]	Although controversial, *short* allele carriers showed significantly reduced volumes of the perigenual anterior cingulate cortex and amygdala in healthy individuals [140]
DISC1	A study including Japanese suicide completers reported a mild association between the Ser704 allele (major allele) and suicide [141]	Findings from Takahashi et al. [76] show an association between the Ser704 allele (recessive effect) and a reduced volume of the medial superior frontal gyrus (bilaterally) in healthy subjects. The same study also shows that healthy Ser704 homozygotes have decreased right and left short insular cortex volumes

decreased OFC white matter [99]. Atmaca et al. (2010) found a total white matter increase in OCD patients heterozygous for the Val142 allele of the *MOG* gene. In addition, several regions of the cortico-striato-thalamic loops appear to be associated with glutamatergic genes in OCD [100, 101].

2.7 Imaging Genetics of Personality and Dimensions

A study to investigate the potential interaction between the 5-hydroxytryptamine 1A receptor (*5-HTR1A*) C1019G polymorphism and amygdala volume in patients with borderline personality disorder (BPD) [102] suggests an association of the polymorphism and structural changes in the limbic system of BPD patients. This research showed a reduction in amygdala volumes of G allele carriers with BPD, and this association was significantly higher in BPD with comorbid major depressive episode (MDE) compared to the healthy controls with the same genotype. There are several studies on the possible brain structure abnormalities among BPD patients. In a recent literature review, it was shown that MRI studies with BPD patients have found reduced volumes in the frontal lobe, bilateral hippocampus and amygdala, left frontal cortex, right anterior cingulate cortex, and right parietal cortex, and increased volume in the putamen [103]. This review also considered possible genetic markers of BPD, and the *short* allele of the *5-HTTLPR* appeared to be associated with BPD, especially in patients with comorbid eating disorders [103]. Another study, done by Perroud et al. [104] suggests that *BDNF* methylation at CpG exons I and IV is significantly higher in BPD patients than in healthy controls. Previous studies found that BPD patients present higher methylation status in some candidate genes such as *NR3C1*, *HTR2A*, *MAOA*, *MAOB*, and *COMT* [105, 106].

A recent study has shown evidence of increased striatal volume, especially in the ventral putamen, in patients with schizophrenia spectrum personality disorder (SPD) (Cluster A personality disorders). A study done by Rosmond et al. [107] found that schizotypal personality disorder may be linked with dopaminergic dysfunction and with a polymorphism in the gene that codes for the dopamine 2 receptor (*DRD2*). There were no available studies toward schizotypal personality disorders and imaging genetics; on the other hand, imaging studies have shown thalamic and fusiform gyrus abnormalities in schizotypal personality disorder subjects [108, 109]. Recent findings suggest that SPD patients have increased PFC and anterior cingulum white matter volume when compared to healthy controls [110].

3 Imaging Genetics of Neurological Disorders

3.1 Imaging Genetics of Parkinson's Disease

Parkinson's disease (PD) with dementia is characterized by atrophy in the hippocampus, amygdala, entorhinal cortex, medial temporal lobe, and the caudate.

A recent meta-analysis [111] showed bilateral reduction of gray matter volumes in the medial temporal lobe and basal ganglia of demented PD patients compared to healthy controls. A recent study, investigating a possible association between gray matter volume and the Val158Met *COMT* (cathecol-O-methyl-transferase) polymorphism in PD patients, [112] showed that the global gray matter volume was greater in young Val/Val individuals, suggesting that age-related changes can be influenced by the Val158Met *COMT* polymorphism. In this study, any association between the risk of idiopathic PD and *COMT* Val158Met polymorphism was not found.

3.2 Imaging Genetics of Alzheimer's Disease

Alzheimer's disease (AD) is the most common form of dementia in the world, mainly affecting the elderly. It is characterized by a progressive and irreversible deterioration of the brain that impairs memory and cognitive functions [113].

Among all the risk factors, genetics plays a major role. The most well-established genetic risk factor for AD is the apolipoprotein E (*APOE*) gene that exists in three different isoforms (alleles): epsilon 2 (ε2), epsilon 3 (ε3), and epsilon 4 (ε4), with the allele ε2 being protective and ε4 being an AD risk factor [114, 115]. There are many genes with a known correlation with AD onset and progression, such as *ABCA7, BIN1, CD33, CLU, CR1, CD2AP, EPHA1, MS4A6A-MS4A4E*, and *PICALM*. Other genes with significant association with AD include *HLA-DRB5-DRB1* and *SORL1* [116]. Studies suggest that the disease process in autosomal dominant patients, including the appearance of biomarkers in cerebrospinal fluid (CSF) and amyloid-beta (Aβ) deposition, begins decades before the onset of dementia [117, 118]. Therefore, several studies focused on the identification of biomarkers and instrumental indicators to improve diagnostic accuracy of AD [119], suggesting neuroimaging is a valid tool for in vivo evaluation, because it facilitates the measurement of brain structure and function [119].

The evolution of neuroimaging techniques contributed to the characterization of specific structural, functional, and metabolic features of the AD brain [119, 120].

MRI is the most used method of neuroimaging to study AD. It is utilized to assess brain structure and can be crucial for AD diagnosis and prediction of dementia onset. The most common findings include significantly reduced hippocampal and entorhinal cortex volumes, as well as increased ventricular and sulcal volumes, and reduced gray matter or cortical thickness in regions such as the precuneus, posterior cingulate, parietal, and temporal cortex [120].

The presence of the *APOE* alleles is known to affect the function of the human brain throughout the lifespan. The ε4 allele is confirmed to be the strongest genetic risk factor for AD [121]. Recent studies reported no differences in hippocampal volume in young and mid-age healthy adults, between *APOE*-ε4 carriers and noncarriers. However, there was a significant reduction in right hippocampal volume in healthy *APOE*-ε4 carriers aged 49–79 years [115, 122]. The cuneus region, assumed to be affected by Aβ deposition before any other region, presented an increased white matter volume in young *APOE*-ε4 carriers, and when comparing *APOE*-ε4 individuals with *APOE*-ε3/ε3 individuals, the cortical thickness of the parahippocampal region and the white matter volume of the left anterior cingulate were increased in ε4 individuals, formulating the hypothesis that the greater engagement with these regions in youth may lead to observable structural changes at mid-age [115].

In conclusion, healthy individuals that carry genetic risk factors for AD offer a great opportunity to further understand AD at the presymptomatic phase, but the relationship between genetics and neuroimaging in AD is still unclear, and a consensus among authors has not yet been reached.

3.3 Imaging Genetics of Multiple Sclerosis

The ε4 allele of *APOE* is linked to disease severity in multiple sclerosis (MS). The neuroimaging findings in MS patients carrying the ε4 allele have been contradictory, as some research suggests that carriers have more brain atrophy than their counterparts without the allele [123, 124], while a study with a large sample size found no difference in brain volumes between ε4 allele carriers and noncarriers among MS patients [125]. A recent study has also found no significant association between the ε4 allele and brain abnormalities in MS patients [126].

3.4 Morphometry and Genetics in Non-Alzheimer Dementia

Dementia with Lewy bodies (DLB), the second most common neurodegenerative dementia, is associated with candidate genes such as *GBA*, *LRRK2*, *MAPT*, *SCARB2*, *SNCA*, and *APOE* [127]. However, there are no confirmed significant morphological characteristics using MRI that can indicate DLB, although preservation of medial-temporal lobe structures [128] and structural damage in posterior brain regions are commonly present [129].

Frontotemporal dementia (FTD) has an insidious onset and 30–50% heritability, and it is often associated with the *MAPT*, *GRN*, *C9orf72*, *CHMP2B*, *VCP*, and *UBQLN2* genes [130]. Abnormalities detected by MRI are associated with the diagnosis of FTD, and the main findings are frontal and anterior temporal lobe atrophy, while the parietal lobe is affected only in some cases, and the occipital lobe is almost always spared [131].

4 Conclusion

The development of structural imaging genetics studies of healthy and diseased human brains is a promising advance in the neuropsychiatric research field. The candidate gene approach has been the most common strategy, and the most studied genes have been *BDNF*, *DISC1*, *COMT*, and *SLC6A4* in psychiatric disorders, and *APOE* in neurological disorders. Better understanding of brain structural abnormalities and their underlying genetics can lead to the discovery of novel endophenotypes for neuropsychiatric disorders, which could help to develop better treatment for CNS disorders.

References

1. Thompson PM (2001) Genetic influences on brain structure. Nat Neurosci 4 (12):1253–1258
2. Blokland GA (2012) Genetic and environmental influences on neuroimaging phenotypes: a meta-analytical perspective on twin imaging studies. Twin Res Hum Genet 15 (03):351–371
3. Dwivedi Y (2009) Brain-derived neurotrophic factor: role in depression and suicide. Neuropsychiatr Dis Treat 5:433–449
4. Harrisberger FS (2014) The association of the BDNF Val66Met polymorphism and the hippocampal volumes in healthy humans: a joint meta-analysis of published and new data. Neurosci Biobehav Rev 42:267–278
5. Pezawas LV (2004) The brain-derived neurotrophic factor val66met polymorphism and variation in human cortical morphology. J Neurosci 24(15):10099–10102
6. Zammit S (2011) Cannabis, COMT and psychotic experiences. Br J Psychiatry 199:380–385
7. Zinkstok J, Schmitz N (2006) The COMT val158met polymorphism and brain morphometry in healthy young adults. Neurosci Lett 405:34–39
8. Radua JEH (2013) COMT Val158Met× SLC6A4 5-HTTLPR interaction impacts on gray matter volume of regions supporting emotion processing. Soc Cogn Affect Neurosci 9(8):1232–1238
9. Millar J (2000) Disruption of two novel genes by a translocation co-segregating with schizophrenia. Hum Mol Genet 9(9):1415–1423
10. Kamiya A (2006) DISC1-NDEL1/NUDEL protein interaction, an essential component for neurite outgrowth, is modulated by genetic variations of DISC1. Hum Mol Genet 15(22):3313–3323
11. Callicott JH (2005) Variation in DISC1 affects hippocampal structure and function and increases risk for schizophrenia. Proc Natl Acad Sci U S A 102(24):8627–8632
12. Raznahan AL (2011) Common functional polymorphisms of DISC1 and cortical maturation in typically developing children and adolescents. Mol Psychiatry 16(9):917–926
13. Chakravarty MM (2012) DISC1 and striatal volume: a potential risk phenotype for mental illness. Front Psych 3:57
14. Trost SP (2013) DISC1 (disrupted-in-schizophrenia 1) is associated with cortical grey matter volumes in the human brain: a voxel-based morphometry (VBM) study. J Psychiatr Res:188–196
15. Stein JL (2012) Identification of common variants associated with human hippocampal and intracranial volumes. Nat Genet 44 (5):552–561
16. Dannlowski UG (2012) Multimodal imaging of a tescalcin (TESC)-regulating polymorphism (rs7294919)-specific effects on hippocampal gray matter structure. Mol Psychiatry 20(3):398–404
17. Donohoe G, Corvin A (2012) Common variants at 12q14 and 12q24 are associated with hippocampal volume. Nat Genet 44 (5):545–551
18. Cousijn HE-V (2014) No effect of schizophrenia risk genes MIR137,TCF4, and ZNF804A on macroscopic brain structure. Schizophr Res 159(2–3):329–332
19. Lett TA, Chakravarty MM (2013) The genome-wide supported microRNA-137 variant predicts phenotypic heterogeneity within

schizophrenia. Mol Psychiatry 18 (4):443–450
20. Stein JL (2015) Copy number variation and brain structure: lessons learned from chromosome 16p11.2. Genome Med 7(1):1
21. Maillard AM (2015) The 16p11.2 locus modulates brain structures common to autism, schizophrenia and obesity. Mol Psychiatry 20 (1):140–147
22. Vorstman JA (2006) The 22q11. 2 deletion in children: high rate of autistic disorders and early onset of psychotic symptoms. J Am Acad Child Adolesc Psychiatry 45 (9):1104–1113
23. Bearden CE, van Erp TG (2008) Alterations in midline cortical thickness and gyrification patterns mapped in children with 22q11.2 deletion. Cereb Cortex:115–126
24. Kempf LN-L (2008) Functional polymorphisms in PRODH are associated with risk and protection for schizophrenia and fronto-striatal structure and function. PLoS Genet 4 (11):e1000252
25. Williams HJ (2007) Is COMT a susceptibility gene for schizophrenia? Schizophr Bull 33:635–641
26. Adeyemi EI, Giedd JN (2015) A case study of brain morphometry in triples discordant for down syndrome. Am J Med Genet Part A 167A:1007–1110
27. Weis SW (1991) Down syndrome: MR quantification of brain structures and comparison with normal control subjects. Am J Neuroradiol 12(6):1207–1211
28. Teipel SJ (2004) Age-related cortical grey matter reductions in non-demented Down's syndrome adults determined by MRI with voxel-based morphometry. Brain 127 (4):811–882
29. Shen DL (2004) Automated morphometric study of brain variation in XXY males. NeuroImage 23(2):648–653
30. Patwardhan AJ (2002) Reduced size of the amygdala in individuals with 47, XXY and 47, XXX karyotypes. Am J Med Genet 114 (1):93–98
31. Murphy DG (1999) Premutation female carriers of fragile X syndrome: a pilot study on brain anatomy and metabolism. J Am Acad Child Adolesc Psychiatry 38(10):1294–1301
32. Moore CJ (2004) The effect of pre-mutation of X chromosome CGG trinucleotide repeats on brain anatomy. Brain 127(12):2672–2681
33. Lee AD-C (2007) 3D pattern of brain abnormalities in fragile X syndrome visualized using tensor-based Morphometry. NeuroImage 34(3):924–938
34. Manto M-U, Pandolfo M (2002) The cerebellum and its disorders. Cambridge University Press, Cambridge
35. Jung BC (2012) MRI shows a region-specific pattern of atrophy in spinocerebellar ataxia type 2. Cerebellum 11(1):272–279
36. Leuzzi V, Trasimeni G (1995) Biochemical, clinical and neuroradiological (MRI) correlations in late-detected PKU patients. J Inherit Metab Dis 18:624–634
37. Pearson KD, Gean-Marton AD (1990) Phenylketonuria: MR-imaging of the brain with clinical correlation. Radiology 177:437–440
38. Poser CM, van Bogaert L (1959) Neuropathologic observations in phenylketonuria. Brain 82:1–9
39. Pérez-Dueñas B (2006) Global and regional volume changes in the brains of patients with phenylketonuria. Neurology 66 (7):1074–1078
40. Kassubek JJ (2004) Topography of cerebral atrophy in early Huntington's disease: a voxel based morphometric MRI study. J Neurol Neurosurg Psychiatry 75(2):213–220
41. Harris GJ (1999) Reduced basal ganglia blood flow and volume in pre-symptomatic, gene-tested persons at-risk for Huntington's disease. Brain 122(9):1667–1678
42. Rosas HD (2001) Striatal volume loss in HD as measured by MRI and the influence of CAG repeat. Neurology 57(6):1025–1028
43. Seok JH (2013) Effect of the COMT val158-met polymorphism on white matter connectivity in patients with major depressive disorder. Neurosci Lett 545:35–39
44. Caspi AS (2003) Influence of life stress on depression: moderation by a polymorphism in the 5-HTT gene. Science 301:386–389
45. Frodl T, Zill P (2008) Reduced hippocampal volumes associated with the long variant of the tri and diallelic serotonin transporter polymorphism in major depression. Am J Med Genet B Neuropsychiatr Genet 147B (7):1003–1007
46. Jaworska NM (2016) The influence of 5-HTTLPR and Val66Met polymorphisms on cortical thickness and volume in limbic and paralimbic regions in depression: a preliminary study. BMC Psychiatry 16(1):1
47. Klimek V, Stockmeier C (2016) Reduced levels of norepinephrine transporters in the locus coeruleus in major depression. J Neurosci 16(1):1
48. Ueda IK (2016) Relationship between G1287A of the NET gene polymorphisms and brain volume in major depressive

disorder: a voxel-based MRI study. PLoS One 11(3):e0150712
49. Chepenik LG (2009) Effects of the brain-derived neurotrophic growth factor val66met variation on hippocampus morphology in bipolar disorder. Neuropsychopharmacology 34(4):944–951
50. Lavagnino L (2015) Changes in the corpus callosum in women with late-stage bipolar disorder. Acta Psychiatr Scand 131 (6):458–464
51. Radaelli D (2015) Fronto-limbic disconnection in bipolar disorder. Eur Psychiatry 30 (1):82–88
52. Selek S (2013) A longitudinal study of fronto-limbic brain structures in patients with bipolar I disorder during lithium treatment. J Affect Disord 150(2):629–633
53. Benjamin S (2010) The brain-derived neurotrophic factor Val66Met polymorphism, hippocampal volume, and cognitive function in geriatric depression. Am J Geriatr Psychiatry 18(4):323–331
54. Egan MF (2003) The BDNF val66met polymorphism affects activity-dependent secretion of BDNF and human memory and hippocampal function. Cell 112(2):257–269
55. Chepenik LG (2008) Effects of the brain-derived neurotrophic growth factor val66met variation on hippocampus morphology in bipolar disorder. Neuropsychopharmacology 34(4):944–951
56. Gatt JM (2009) Interactions between BDNF Val66Met polymorphism and early life stress predict brain and arousal pathways to syndromal depression and anxiety. Mol Psychiatry 14 (7):681–695
57. Montag C, Weber B (2009) The BDNF Val66Met polymorphism impacts parahippocampal and amygdala volume in healthy humans: incremental support for a genetic risk factor for depression. Psychol Med 39:1831–1839
58. Ferreira MA (2008) Collaborative genome-wide association analysis supports a role for ANK3 and CACNA1C in bipolar disorder. Nat Genet 40(9):1056–1058
59. Sklar P (2008) Whole-genome association study of bipolar disorder. Mol Psychiatry 13 (6):558–569
60. Perrier E (2011) Initial evidence for the role of CACNA1C on subcortical brain morphology in patients with bipolar disorder. Eur Psychiatry 26(3):135–137
61. Franke B (2010) Genetic variation in CACNA1C, a gene associated with bipolar disorder, influences brainstem rather than gray matter volume in healthy individuals. Biol Psychiatry 68(6):586–588
62. Insel T (2010) Research domain criteria (RDoC): toward a new classification framework for research on mental disorders. Am J Psychiatry 167(7):748–751
63. Opmeer EM (2015) DISC1 gene and affective psychopathology: a combined structural and functional MRI study. J Psychiatr Res 61:150–157
64. Frey SH (2006) Modulation of neural activity during observational learning of actions and their sequential orders. J Neurosci 26 (51):13194–13201
65. Foland LC (2008) Increased volume of the amygdala and hippocampus in bipolar patients treated with lithium. Neuroreport 19(2):221–224
66. Yucel K (2007) Bilateral hippocampal volume increases after long-term lithium treatment in patients with bipolar disorder: a longitudinal MRI study. Psychopharmacology 195 (3):357–367
67. Jagannathan K (2010) Genetic associations of brain structural networks in schizophrenia: a preliminary study. Biol Psychiatry 68 (7):657–666
68. Zinkstok J (2006) The COMT val158met polymorphism and brain morphometry in healthy young adults. Neurosci Lett 405:34–39
69. Ohnishi T, Hashimoto R (2006) The association between the Val158Met polymorphism of the catechol-O-methyl transferase gene and morphological abnormalities of the brain in chronic schizophrenia. Brain 129 (Pt 2):399–410
70. Ho BC (2005) Catechol-O-methyl transferase Val158Met gene polymorphism in schizophrenia: working memory, frontal lobe MRI morphology and frontal cerebral blood flow. Mol Psychiatry 10(3):287–298
71. McIntosh AM (2007) Relationship of catechol-O-methyltransferase variants to brain structure and function in a population at high risk of psychosis. Biol Psychiatry 61 (10):1127–1134
72. Ira EZ (2013) COMT, neuropsychological function and brain structure in schizophrenia: a systematic review and neurobiological interpretation. J Psychiatry Neurosci 38 (6):366–380
73. Ho BC (2007) Association between brain-derived neurotrophic factor Val66Met gene polymorphism and progressive brain volume changes in schizophrenia. Am J Psychiatr 164 (12):1890–1899

74. Ho BC (2006) Cognitive and magnetic resonance imaging brain morphometric correlates of brain-derived neurotrophic factor Val66-Met gene polymorphism in patients with schizophrenia and healthy volunteers. Arch Gen Psychiatry 63(7):731–740
75. Di Giorgio A, Blasi G (2008) Association of the Ser704Cys DISC1 polymorphism with human hippocampal formation gray matter and function during memory encoding. Eur J Neurosci 28(10):2129–2136
76. Takahashi TS (2009) The disrupted-in-Schizophrenia-1 Ser704Cys polymorphism and brain morphology in schizophrenia. *Psychiatry Res Neuroimaging* 172(2):128–135
77. Prasad KM (2005) Genetic polymorphisms of the RGS4 and dorsolateral prefrontal cortex morphometry among first episode schizophrenia patients. Mol Psychiatry 10 (2):213–219
78. Donohoe GR (2011) ZNF804A risk allele is associated with relatively intact gray matter volume in patients with schizophrenia. NeuroImage 54(3):2132–2137
79. Budisic M (2010) Brainstem raphe lesion in patients with major depressive disorder and in patients with suicidal ideation recorded on transcranial sonography. Eur Arch Psychiatry Clin Neurosci 260(3):203–208
80. Ahearn EP, Jamison KR (2001) MRI correlates of suicide attempt history in unipolar depression. Biol Psychiatry 50(4):266–270
81. Rüsch N, Spoletini I (2008) Inferior frontal white matter volume and suicidality in schizophrenia. Psychiatry Res 164(3):206–214
82. Ding Y, Lawrence N, Olié E et al (2015) Prefrontal cortex markers of suicidal vulnerability in mood disorders: a model-based structural neuroimaging study with a translational perspective. Transl Psychiatry 5(2):e516. https://doi.org/10.1038/tp.2015.1
83. Besteher B, Wagner G, Koch K, Schachtzabel C, Reichenbach JR, Schlösser R, Sauer H, Schultz CC (2016) Pronounced prefronto-temporal cortical thinning in schizophrenia: neuroanatomical correlate of suicidal behavior? Schizophr Res 176 (2–3):151–157. https://doi.org/10.1016/j.schres.2016.08.010. PubMed PMID: 27567290
84. Jollant F (2016) Neuroimaging of suicidal behavior. In: Kaschka WP, Rujescu D (eds) Biological aspects of suicidal behavior, Adv biol psychiatry, vol 30. Karger, Basel, pp 110–122. https://doi.org/10.1159/000434744. PubMed PMID: 27567290
85. Monkul ES (2007) Fronto-limbic brain structures in suicidal and non-suicidal female patients with major depressive disorder. Mol Psychiatry 12(4):360–366
86. Dombrovski AY (2012) The temptation of suicide: striatal gray matter, discounting of delayed rewards, and suicide attempts in late-life depression. Psychol Med 42 (06):1203–1215
87. Jia ZH (2010) High-field magnetic resonance imaging of suicidality in patients with major depressive disorder. Am J Psychiatr 167 (11):1381–1390
88. Antypa NS (2015) Clinical and genetic factors associated with suicide in mood disorderpatients. Eur Arch Psychiatry Clin Neurosci:1–13
89. Calati RP (2011) Catechol-o-methyltransferase gene modulation on suicidal behavior and personality traits: review, meta-analysis and association study. J Psychiatr Res 45 (3):309–321
90. De Luca V (2008) Power based association analysis (PBAT) of serotonergic and noradrenergic polymorphisms in bipolar patients with suicidal behaviour. Prog Neuro-Psychopharmacol Biol Psychiatry 32 (1):197–203
91. Voineskos AN (2011) Neurexin-1 and frontal lobe white matter: an overlapping intermediate phenotype for schizophrenia and autism spectrum disorders. PLoS One 6(6):e20982
92. Tan GC (2010) Normal variation in fronto-occipital circuitry and cerebellar structure with an autism-associated polymorphism of CNTNAP2. NeuroImage 53(3):1030–1042
93. Tost HK–L (2010) A common allele in the oxytocin receptor gene (OXTR) impacts prosocial temperament and human hypothalamic-limbic structure and function. Proc Natl Acad Sci 107(31):13936–13941
94. Davis LK (2008) Cortical enlargement in autism is associated with a functional VNTR in the monoamine oxidase a gene. Am J Med Genet B Neuropsychiatr Genet 147 (7):1145–1151
95. Lin PY (2004) Association between serotonin transporter gene promoter polymorphism and suicide: results of a meta-analysis. Biol Psychiatry 55(10):1023–1030
96. Taylor S (2013) Molecular genetics of obsessive-compulsive disorder: a comprehensive meta-analysis of genetic association studies. Mol Psychiatry 18(7):799–805
97. Walitza S (2014) Trio study and meta-analysis support the association of genetic variation at the serotonin transporter with early-onset

obsessive-compulsive disorder. Neurosci Lett 580:100–103

98. Arnold PD (2009) Glutamate receptor gene (GRIN2B) associated with reduced anterior cingulate glutamatergic concentration in pediatric obsessive-compulsive disorder. Psychiatry Res 172(2):136–139
99. Wu K (2012) Glutamate system genes and brain volume alterations in pediatric obsessive-compulsive disorder: a preliminary study. Psychiatry Res 211(3):214–220
100. Brem SH (2012) Neuroimaging of cognitive brain function in paediatric obsessive compulsive disorder: a review of literature and preliminary meta-analysis. J Neural Transm 119:1425–1448
101. Huyser C (2009) Paediatric obsessive-compulsive disorder, a neurodevelopmental disorder? Evidence from neuroimaging. Neurosci Biobehav Rev 33(6):818–830
102. Zetzsche TP-J (2008) 5-HT1A receptor gene C −1019 G polymorphism and amygdala volume in borderline personality disorder. Genes Brain Behav 7:306–313
103. Lis EG (2007) Neuroimaging and genetics of borderline personality disorder: a review. J Psychiatry Neurosci 32(3):162–173
104. Perroud NS (2013) Response to psychotherapy in borderline personality disorder and methylation status of the BDNF gene. Transl Psychiatry 3(1):e207
105. Perroud NP-G (2011) Increased methylation of glucocorticoid receptor gene (NR3C1) in adults with a history of childhood maltreatment: a link with the severity and type of trauma. Transl Psychiatry 1(12):e59
106. Dammann GT (2011) Increased DNA methylation of neuropsychiatric genes occurs in borderline personality disorder. Epigenetics 6(12):1454–1462
107. Rosmond R, Rankinen T (2001) Polymorphism in exon 6 of the dopamine D-2 receptor gene (DRD2) is associated with elevated blood pressure and personality disorders in men. J Hum Hypertens 15:553–558
108. Hazlett EA-C (1999) Three-dimensional analysis with MRI and PET of the size, shape, and function of the thalamus in the schizophrenia spectrum. Am J Psychiatr 156 (8):1190–1199
109. McDonald BH (2000) Anomalous asymmetry of fusiform and parahippocampal gyrus gray matter in schizophrenia: a postmortem study. Am J Psychiatry 157:40–47
110. Hazlett EA (2012) A review of structural MRI and diffusion tensor imaging in schizotypal personality disorder. Curr Psychiatry Rep 14(1):70–78
111. Pan PL (2013) Gray matter atrophy in Parkinson's disease with dementia: evidence from meta-analysis of voxel-based morphometry studies. Neurol Sci 34(5):613–619
112. Rowe JB-G (2010) The val158met COMT polymorphism's effect on atrophy in healthy aging and Parkinson's disease. Neurobiol Aging 31(6):1064–1068
113. Campos CR-C-C (2016) Treatment of cognitive deficits in Alzheimer's disease: a psychopharmacological review. Psychiatr Danub 28(1):2–12
114. Kim J (2009) The role of apolipoprotein E in Alzheimer's disease. Neuron 63(3):287–303
115. Dowell NG (2016) Structural and resting-state MRI detects regional brain differences in young and mid-age healthy APOE-e4 carriers compared with non-APOE-e4. NMR Biomed 29(5):614–624
116. Lambert JI-V-B (2013) Meta-analysis of 74,046 individuals identifies 11 new susceptibility loci for Alzheimer's disease. Nat Genet 45(12):1452–1458
117. Bateman R, Xiong C, Benzinger T, Goate A, Fox N, Marcus D et al (2012) Clinical and biomarker changes in dominantly inherited Alzheimer's disease. N Engl J Med 367 (9):795–804
118. Buchhave PM (2012) Cerebrospinal fluid levels of β-amyloid 1-42, but not of Tau, are fully changed already 5 to 10 years before the onset of Alzheimer dementia. Arch Gen Psychiatry 69(1):98–106
119. Bagnoli SP (2014) Advances in imaging-genetic relationships for Alzheimer's disease: clinical implications. Neurodegener Dis Manag 4:73–81
120. Reiman E a (2012) Brain imaging in the study of Alzheimer's disease. NeuroImage 61 (2):505–516. https://doi.org/10.1016/j.neuroimage.2011.11.075
121. Liu C-CK, Kanekiyo T, Xu H, Bu G (2013) Apolipoprotein E and Alzheimer disease: risk, mechanisms, and therapy. Nat Rev Neurol 9 (2):106–118
122. Lind J (2006) Reduced hippocampal volume in non-demented carriers of the apolipoprotein E epsilon4: relation to chronological age and recognition memory. Neurosci Lett 396 (1):23–27
123. De Stefano NB (2004) Influence of Apolipoprotein E ϵ4 genotype on brain tissue integrity in relapsing-remitting multiple sclerosis. Arch Neurol 61(4):536–540

124. Enzinger CRF (2004) Accelerated evolution of brain atrophy and "black holes" in MS patients with APOE- ε4. Ann Neurol 55 (4):563–569
125. van der Walt A (2009) Apolipoprotein genotype does not influence MS severity, cognition, or brain atrophy. Neurology 73 (13):1018–1025
126. Ghaffar OL (2011) Imaging genetics in multiple sclerosis: a volumetric and diffusion tensor MRI study of APOE ε4. NeuroImage 58 (3):724–731
127. Geiger JD, Troncoso (2016) Next-generation sequencing reveals substantial genetic contribution to dementia with Lewy bodies. Neurobiol Dis 94:55–62
128. Tateno MK (2009) Imaging improves diagnosis of dementia with Lewy bodies. Psychiatry investig 6(4):233–240
129. Borroni BP (2015) Structural and functional imaging study in dementia with Lewy bodies and Parkinson's disease dementia. Parkinsonism Relat Disord 21(9):1049–1055
130. Ferrari RH (2014) Frontotemporal dementia and its subtypes: a genome-wide association study. Lancet Neurol 13(7):686–699
131. Josephs KA (2007) Frontotemporal Lobar Degeneration. Neurol Clin 25(3):683–6vi
132. Pregelj PN (2011) The association between brain-derived neurotrophic factor polymorphism (BDNF Val66Met) and suicide. J Affect Disord 128(3):287–290
133. Zai CC (2012) The brain-derived neurotrophic factor gene in suicidal behaviour: a meta-analysis. Int J Neuropsychopharmacol 15(8):1037
134. Bueller JA-H (2006) BDNF Val66Met allele is associated with reduced hippocampal volume in healthy subjects. Biol Psychiatry 59(9):812–815
135. Szeszko PR (2005) Brain-derived neurotrophic factor val66met polymorphism and volume of the hippocampal formation. Mol Psychiatry 10(7):631–636
136. Nedic GN-S (2010) Association study of a functional catechol-O-methyltransferase polymorphism and smoking in healthy Caucasian subjects. Neurosci Lett 473 (3):216–219
137. Nedic GN-S (2011) Association study of a functional catechol-O-methyltransferase (COMT) Val 108/158 met polymorphism and suicide attempts in patients with alcohol dependence. Int J Neuropsychopharmacol 14 (03):377–388
138. Baud PC (2007) Catechol-O-methyltransferase polymorphism (COMT) in suicide attempters: a possible gender effect on anger traits. Am J Med Genet B Neuropsychiatr Genet 144((8):1042–1047
139. Ehrlich SM (2010) The COMT Val108/158Met polymorphism and medial temporal lobe volumetry in patients with schizophrenia and healthy adults. NeuroImage 53 (3):992–1000
140. Pezawas LM-L (2005) 5-HTTLPR polymorphism impacts human cingulate-amygdala interactions: a genetic susceptibility mechanism for depression. Nat Neurosci 8 (6):828–834
141. Ratta-Apha WH (2014) Haplotype analysis of the DISC1 Ser704Cys variant in Japanese suicide completers. Psychiatry Res 215 (1):249–251

Chapter 13

Multicenter Studies of Brain Morphometry

Fabrizio Piras, Mariangela Iorio, Daniela Vecchio, Tommaso Gili, Federica Piras, and Gianfranco Spalletta

Abstract

Methods enabling the characterization of brain morphometry have evolved exponentially in the last decades and have been shown important clinical applications. First MRI studies usually included a low number of subjects, but later on, it has become recognized that efforts from different research centers can be pooled by combining data acquisition and analyses toward large-scale collaborative projects. Large-scale studies have a number of advantages, with the most intuitive being the increase in the number of enrolled subjects and, consequently, statistical power. However, there are also a number of caveats including (but not limited to) the high cost in terms of human and economic resources and the intrinsic variation of acquired data linked to the different MRI technologies.

This chapter depicts the different categories of large-scale studies, according to the level of control and planning on data acquisition and analyses. It also describes the most relevant multicenter studies, from the first precursor consortia in the early 1990s to the most recent worldwide multimodal collaborative efforts.

Key words Mega-analysis, Meta-analysis, Brain morphology, MRI, Multicenter studies, Neuroinformatics, Data sharing

1 History of Large-Scale Neuroimaging Studies

Methods that enable the characterization of human brain morphometry from magnetic resonance imaging (MRI) data are demonstrating important applications in neuroscience. With the advent of modern neuroimaging technologies in the early 1990s, the proliferation of MRI scanners in research institutes around the world has made it possible to investigate different pathophysiological mechanisms in a variety of populations, including, but not limited to, normal and pathological development [1], neurodegeneration [2–4], traumatic brain injury [5], cerebrovascular diseases [6], autism [7], mood disorders [8], obsessive-compulsive disorder [9], epilepsy [10], and schizophrenia [11].

First publications using MRI included single or a handful of subjects, until Desmond and Glover [12], using simulations in

Gianfranco Spalletta et al. (eds.), *Brain Morphometry*, Neuromethods, vol. 136,
https://doi.org/10.1007/978-1-4939-7647-8_13, © Springer Science+Business Media, LLC 2018

order to generate power curves, claimed that, at least in functional MRI experiments, for a liberal threshold of 0.05, about 12 subjects were required to achieve 80% power at the single voxel level for typical activations. At more realistic thresholds that approach those used after correcting for multiple comparisons, the number of subjects doubled to maintain this level of power. Thus, the problem of sample size in neuroimaging studies carried out at individual institutions may be difficult to solve, also because, as it will be clarified further, they are expensive and time-consuming.

Although studies using less than 20–30 subjects are still quite common, it has become recognized that researchers from different centers can pool efforts by combining data acquisition and analyses toward large-scale collaborative projects with the aim of investigating brain structural alterations in diseases characterized by great clinical heterogeneity. In parallel, a number of tools to analyze brain images have been developed, freely distributed, and continuatively updated by the scientific community. Softwares such as Statistical Parametric Mapping (SPM; [13]), FMRIB Software Library (FSL; [14]), and FreeSurfer [15], among others, made it possible to analyze imaging data and obtain relatively standardized measures in a reliable and shared manner.

Studies involving different research sites are usually organized such that raw and/or processed datasets are stored and may be examined collectively by associated partners or by researchers who are not directly participating in the study but request to use the data for research purposes. This would allow carrying on studies that could not be conducted in single sites, in terms of enrolled subjects/patients and computational resources.

One of the first neuroimaging consortia is the International Consortium for Brain Mapping (ICBM) [16] which recognized the need to establish normative data on the brain from a wide range of human populations scanned in different parts of the world. The ICBM began with an effort to scan around 150 healthy subjects in Los Angeles, Montreal, and San Antonio and grew to include sites in Europe and Asia that broadened the age range and ethnic groups assessed. Later, the ICBM also extended the depth of the neuroimaging measures to include functional MRI and even postmortem histology as well as cytoarchitecture [17]. Further, the ICBM, given the high variability in brain anatomy, developed a range of "average" anatomical templates based on MRI scans of hundreds of healthy subjects. Softwares for brain imaging analysis contributed to disseminate these average brain templates and provided methods to relate new data to previously created atlases. This led to the wide adoption of statistical representations of imaging signals in standardized coordinate spaces, such as the MNI (Montreal Neurological Institute) [18], together with the development of automated registration algorithms [19].

2 Pros and Cons of Large-Scale Studies

Multicenter imaging studies have several advantages, with the most obvious one endorsing the statistical mantra that more subjects mean more power. As an example, in clinical treatment trials, multisite studies are usually the only way of achieving the large number of patients needed to statistically power the study. By taking advantage of a greater variety of patient types, etiologies, and range of symptoms, multisite studies can represent patient populations with better generality for population-level atlases of brain structure/function and, by their association with clinical variables, suggest treatment options with the widest possible efficacy [20]. Moreover, multicenter studies permit to obtain more imaging data per subject across time, thus improving the prediction of which normal or slightly impaired individuals will develop alterations in brain anatomy and over what time period. Last but not least, they can be considered as a tool for monitoring the prevalence or the progression of a brain disease across a geographically, culturally, and environmentally heterogeneous population. On the other hand, a number of caveats concerning multicenter collaborations must be acknowledged. First, these studies are economically challenging since expenses are not simply a linear sum of doing the same but reduced study at each site, the coordination, planning, and harmonization of methods and equipment (if foreseen) across sites [21], the infrastructure for sending, storing, and analyzing data in central locations, and the organizational load due to keeping everyone informed on any changes in the protocols form a necessary and costly overhead for these kind of studies.

The second issue, and probably the most problematic to face, is the intrinsic variation linked to the different MRI technologies. Indeed, using scanners that can vary in terms of magnetic field strength, vendor, model, gradients, coil, and software release (among others) can deeply influence how data are acquired and, consequently, interpreted. In order to fully appreciate this issue, it should be beared in mind that even the same make and model MRI system may demonstrate differing field inhomogeneity effects that have to be taken into account [22]. In fact, the signal measured during a conventional structural protocol (T1- or T2-weighted acquisition) is a function of several fundamental properties of the tissue including not only T1, T2, and proton density but also hardware-related effects such as radiofrequency coil sensitivity, electronic amplifier gains, and many others. Therefore, standardization is usually achieved through the use of common sequences with balanced acquisition parameters. Using this approach implies that acquisition parameters (e.g., echo time, repetition time, and flip angle) must be set according to the least state-of-the-art system, thus sacrificing the potential advantages, in terms of signal-to-noise ratio efficiency, offered by the high-quality technologies (improved gradient performance and dedicated multichannel radio frequency coils) available on newer systems [23].

Further, multisite studies involve patient confidentiality, thus implying that investigators must have access to private patient data and health information stored at other institutes while also being compliant with local and global privacy regulations on data anonymization [24]. At the same time, methods for ensuring smooth computational access to digital archives of neuroimaging and clinical data to be analyzed necessitate careful examination. Indeed, different factors come into play, including (1) technical requirements to ensure site-to-site network communication; (2) user authentication protocols; (3) resources for rapid design and deployment of analysis workflows, data ontological definition, description, and management, as well as large-scale data storage capabilities [20]; and (4) ethical issues dealing with sharing sensitive data such as clinical and sociodemographic patients information.

3 Categories of Multicenter Studies

Studies including a large number of subjects can be grouped into different categories, according to the level of control and planning that is used.

The most controlled are the *planned studies*, which are large-scale studies using a shared and coherent protocol for both subject recruitment (consistent inclusion/exclusion criteria and common diagnostic references) and data collection (deep calibration of scanners, harmonization of protocols). Such studies guarantee that samples are highly comparable across sites since sources of variance have been reduced as much as possible. However, their accomplishment requires, as stated above, a great effort both in terms of human/technological resources and of funding investments. One of the most representative examples of planned studies is the Alzheimer's Disease Neuroimaging Initiative (ADNI) [25], which was established to increase knowledge of the mechanisms of AD through the use of neuroimaging—thereby informing the development of treatment strategies aimed at slowing down or preventing neuronal death. The first phase of ADNI (ADNI I) began in 2004 under the leadership of Dr. Michael W. Weiner, funded as a private-public partnership with $27 million contributed by 20 companies and 2 foundations through the Foundation for the National Institutes of Health and $40 million from the National Institute on Aging (NIA). The cohort was composed of 200 controls, 400 patients with mild cognitive impairment (MCI), and 200 patients with AD. The second phase (ADNI-GO) has been funded by the American Recovery Act Funds and included additional 200 patients with early MCI. During the third phase

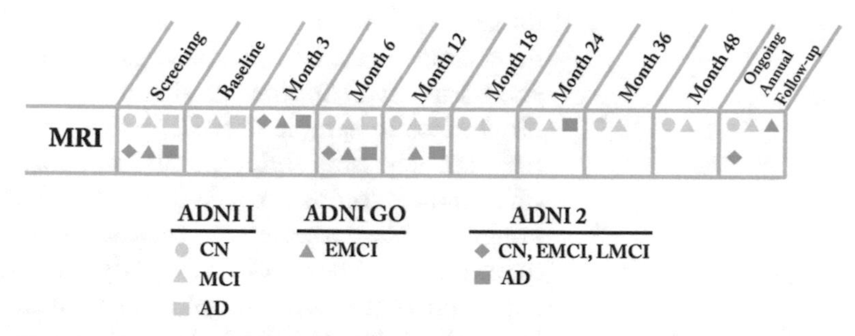

Fig. 1 Overview of the MRI data collected throughout the ADNI study. *CN* controls, *MCI* mild cognitive impairment, *AD* Alzheimer disease, *EMCI* early mild cognitive impairment, *LMCI* late mild cognitive impairment

(ADNI-2), funded by the NIS, industry, and foundations, further controls, early and late MCI, and AD patients (150, 100, 100, and 150, respectively) have been recruited. All patients have been repeatedly submitted to MRI scan in different follow-ups (see Fig. 1). ADNI has been instrumental in helping to identify clinical, neuroimaging, and biomarker outcome measures and longitudinal changes and the prediction of disease transitions. While more studies are obviously needed to determine the continued value of newer neuroimaging modalities for clinical trials with neuroprotective drugs, the ADNI project can be considered to have been a highly successful first step in large-scale neuroimaging and the sharing of that information with a larger community studying the efficacy of leading-edge treatment.

Less controlled is the second category of large-scale studies, i.e., the *aggregated mega-analyses*, in which existing datasets without previous coordination but with comparable imaging techniques and sample populations are merged and analyzed. They usually include a single imaging modality (e.g., T1-weighted-based structural imaging), without prior harmonization of MRI acquisition parameters. Furthermore, they are typically focused on a specific patient population with no common clinical criteria or available measures. Such studies allow collecting large samples with relatively little effort, if we exclude that made by the leader to coordinate the aggregation of all datasets and make the analyses. This process can be time-consuming, strenuous, and frustrating, but it is not comparable to the effort spent to recruit, test, and scan a large number of subjects. Aggregated mega-analyses have different limitations, with the most important being represented by the large variability in the images, due to the different scanning protocols adopted. As observed by Glover and co-workers [21], changes in acquisition parameters can produce specific image deformations as well as differences in terms of tissue contrast and thus affect estimates of any brain measure being under investigation. Variability among sites is usually controlled for in such studies by using site as nuisance dummy variable or factor in statistical modeling, although there is no clear evidence that this procedure completely accounts for inter-site variability. Another challenging issue is represented by the differences in sample characteristics. Indeed, single sites conduct their studies using different clinical and cognitive tools, which are often hardly comparable. As a consequence, the mismatch between available data decreases the magnitude of the sample and, consequently, the power of the study. A notable example of such approach is the recently released Consortium for Reliability and Reproducibility (CORR) dataset[1] which collated structural and functional imaging data from over 1600 subjects, available to the community.

The third kind of large-scale studies is *opportunistic studies*, which are often seen at single institutions (or combinations of institutions) that make their imaging data available for mining, without regard for similar sample populations or imaging protocols. The term *opportunistic* refers to the fact that the subjects are those scanned for other studies, using the imaging protocols properly set up for that study. These institutional-level data-sharing methods can require a great effort as well as high-level administrative involvement, support, and assurances to develop a managing system for all the imaging data collected at an institution. However, the repositories that result from it can be immense. As an example, the One Mind for Research project[2] is leveraging these sorts of efforts, with the goal of collating datasets from several thousand traumatic brain injury (TBI) subjects from participating trauma centers and emergency room locations, as well as developing a registry over time of 25,000 patients seen for a suspected TBI and their scans. Similarly, the Donders Institute for Cognitive Neuroscience have provided structural imaging data from their four MRI scanners for the Brain Imaging Genetics (BIG) study, from the pool of images from all college students being scanned for many other research projects [26]. Obviously, these approaches in many cases share the disadvantages of aggregated analyses such as varied and not harmonized imaging protocols in some cases, incomplete clinical pictures in others.

The last category of large-scale studies is *meta-analyses*, in which there is no control on the collection of imaging data or aggregation of data in one place. Meta-analyses can be either *post hoc* or *prospective*. Post hoc meta-analyses combine results across smaller studies (extracting published results and effect sizes from the literature), to derive conclusions about that body of research. A different approach is provided by *prospective* meta-analyses, in which data to be analyzed are not obtained from the published

[1] http://fcon_1000.projects.nitrc.org/indi/CoRR/html/index.html.

[2] One Mind for Research. In http://www.onemind.org/Our-Solutions/Gemini.

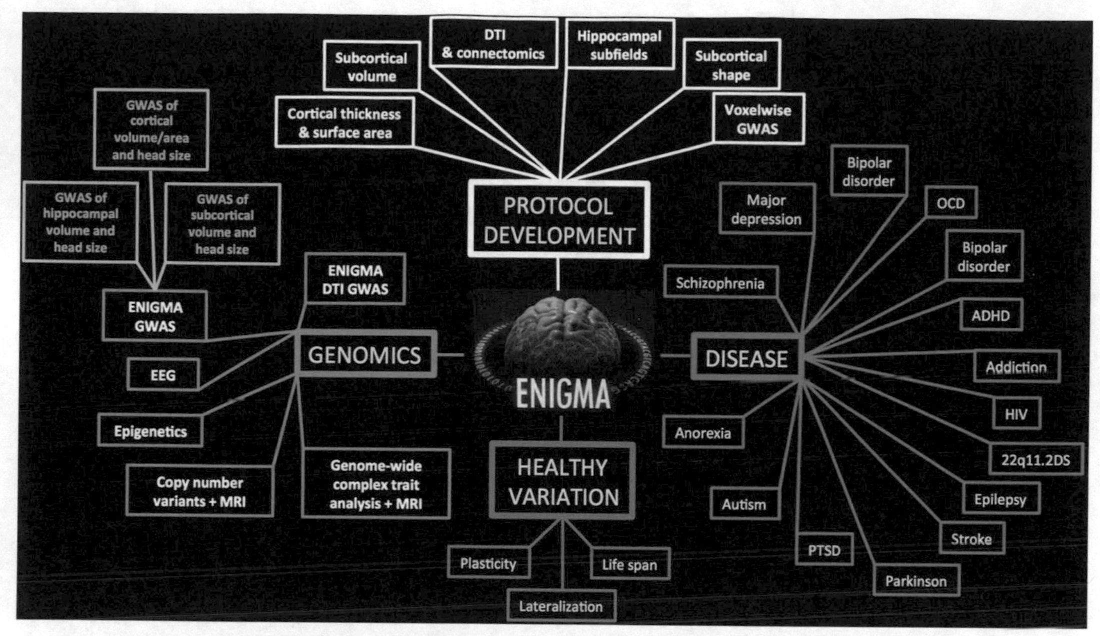

Fig. 2 ENIGMA working groups organized into four principal cores (protocol development, disease, healhty variation, genomics). *DTI* diffusion tensor imaging, *GWAS* genome-wide association study, *OCD* obsessive-compulsive disorder, *HIV* human immunodeficiency, *DS* deletion syndrome, *PTSD* post-traumatic stress disorder, *MRI* magnetic resonance imaging, *EEG* electroencephalography

literature. Rather, datasets that have been acquired independently are analyzed individually, using standardized processing pipelines and statistical models. The obtained individual results are then pooled together and analyzed through the same approach used for post hoc meta-analyses. The most known project adopting such approach is the ENIGMA ("Enhancing NeuroImaging Genetics through Meta-Analysis") consortium[3] which was born at the end of 2009 by a group of researchers with wide expertise in large-scale neuroimaging or large-scale genetic studies). ENIGMA is a network bringing together researchers in imaging genomics to understand brain structure, function, and diseases, based on brain imaging and genetic, but also clinical data and organized into four principal cores (protocol development, disease, healthy variation, genomics) depicted in Fig. 2.

Researchers involved in the consortium are typically asked to process their structural imaging data segmenting T1 weighted into various brain region volumes using a standardized pipeline (e.g., the FreeSurfer[4] software [27]), perform a common quality assessment protocol to remove poor quality data, and run ad hoc

[3] http://enigma.ini.usc.edu/.

[4] https://surfer.nmr.mgh.harvard.edu/.

developed scripts in R[5] to obtain standardized statistical outputs. The outputs (in terms of effect sizes) are then meta-analyzed by the project coordinator using the very same approach of classical meta-analyses. Although there is little control on data acquisition, all other steps are highly set up and double-checked, so that errors or inaccuracies are minimized. In particular, the quality control is strongly coded, and little is left to subjective feeling, thus leading to high homogeneity of data. The ENIGMA project now has a number of collaborative working groups varying in size, exploring these same issues in distinct neuropsychiatric disorders as well as focusing on specific imaging methodological issues. Individual sites are free to join the consortium at any time, according to the status of single subprojects. A number of studies have been published, highlighting gray matter morphometric alterations in a number of psychiatric pathologies, including attention deficit hyperactivity disorder [28], bipolar disorder [29, 30], obsessive-compulsive disorder [31], schizophrenia [32, 33], and major depressive disorder [34] with the inclusion of thousands of subjects in each study and the combined expertise of hundreds of professionals in these fields. Like other uncontrolled designs, the prospective meta-analysis approach can be affected by the variability in the collected data. For example, to date, there are no standard batteries to collect clinical, neuropsychological, and sociodemographic variables that can be applied to imaging studies of psychiatric diseases. One dataset may include broad cognitive measures, while another could be limited to the collection of very basic data such as the Mini-Mental State Examination [35]. Or, two datasets may comprise equally extensive cognitive measures acquired using different tests, thus leading to incomparability issues.

The cost of a crowd-sourced approach, such as the ENIGMA model, is in unpaid labor in many cases. ENIGMA and its subprojects are not planned multisite studies, with staff at every site funded to work on their part of the analyses. They exclusively rely upon volunteer army of researchers willing to participate because it is a unique opportunity and it is a data collection process that cannot be completed any other way.

Table 1 summarizes the three categories of large-scale studies highlighting their pros and cons.

All categories of large-scale studies share the same objective, that is, widen the sample to make it more representative and capture enough of the individual variability. Indeed, datasets collected in multiple research entities and different cities/countries are more likely to capture the variation of clinical populations than are smaller single-center investigations. Thus, large-scale studies allowed the collection of an incredible amount of data, immensely valuable

[5] https://cran.r-project.org/.

Table 1
Comparison of study category

Category	Homogeneity across datasets	Control over data acquisition	Control over data analysis	Ease of collecting a large sample
Planned study	High	Very high	Very high	Difficult
Aggregated mega-analysis	Moderate	High	Low	Moderately easy
Opportunistic studies	Moderate	Low	Moderate	Moderately difficult
Meta-analysis (post hoc)	Low	Low	Low	Easy
Meta-analysis (prospective)	Low	Low	High	Moderate

as ongoing resources for the research community. The availability of such a great load of information, however, not always goes at the same speed of the development of data analysis methods, and the original study designers of a given project may not have (or not yet) all computational resources or techniques available. Data integration, mediation, and mining are ongoing methodological issues within the field of neuroinformatics. As an example, the NIH-funded Human Connectome Project (HCP),[6] aimed at investigating region-to-region structural and functional connections of the human brain, has collected (at March 2017) 3 T MR imaging data from 1,206 healthy young adult participants. 3 T structural scans are available for 1,113 subjects. Forty-six subjects have 3 T HCP protocol retest data available, and 184 subjects have multimodal 7 T MR imaging data available. This large amount of data led to the release of ad hoc computing solutions such as OpenMOLE, a scientific workflow engine with a strong emphasis on workload distribution [36, 37]. Starting with the first quarterly (Q1) data release (March, 2013), HCP datasets, as well as complex pipelines to analyze the data, are being made freely available to the scientific community. Successful mapping of the human connectome in healthy adults will pave the way for future studies of brain circuitry during development and aging and in numerous brain disorders. In short, it will transform our understanding of the human brain in health and disease. Deciphering the amazingly complex wiring diagram of the human brain will reveal much about what makes us uniquely human and what makes every person different from all others.

[6] http://www.humanconnectome.org/data/.

4 Conclusions

Large-scale studies using neuroimaging data have been increasingly applied to study normal and pathological brain structure and function. The studies considerably vary in terms of data collection/analysis control, and each category of study has strengths and limitations. These important efforts will link the intellectual talent and resources of leading research centers toward a better understanding of brain development, aging, and disease. With more data being obtained with a view toward greater spatiotemporal precision, the informatics of neuroimaging will be an important consideration. Such efforts are imperative for guiding treatment recommendations for neuropsychiatric disorders as well as at the level of the individual patient. Multisite collaborations can be expected to strengthen the understanding of brain diseases that affect all walks of life, all ages, and all cultures. This helps to translate neuroimaging trial outcomes directly into clinical applications.

References

1. Giedd JN, Blumenthal J, Jeffries NO et al (1999) Brain development during childhood and adolescence: a longitudinal MRI study. Nat Neurosci 2:861–863. https://doi.org/10.1038/13158
2. Good CD, Johnsrude IS, Ashburner J et al (2001) A voxel-based morphometric study of ageing in 465 normal adult human brains. NeuroImage 14:21–36. https://doi.org/10.1006/nimg.2001.0786
3. Jernigan TL, Salmon DP, Butters N, Hesselink JR (1991) Cerebral structure on MRI, part II: specific changes in Alzheimer's and Huntington's diseases. Biol Psychiatry 29:68–81. https://doi.org/10.1016/0006-3223(91)90211-4
4. Gorell J, Ordidge R, Brown G et al (1995) Increased iron-related MRI contrast in the substantia nigra in Parkinson's disease. Neurology 45:1138–1143. https://doi.org/10.1212/WNL.45.6.1138
5. Helmick KM, Spells CA, Malik SZ et al (2015) Traumatic brain injury in the US military: epidemiology and key clinical and research programs. Brain Imaging Behav 9:358–366. https://doi.org/10.1007/s11682-015-9399-z
6. Stebbins GT, Nyenhuis DL, Wang C et al (2008) Gray matter atrophy in patients with ischemic stroke with cognitive impairment. Stroke 39:785–793. https://doi.org/10.1161/STROKEAHA.107.507392
7. Piven J, Arndt S, Bailey J et al (1995) An MRI study of brain size in autism. Am J Psychiatry 152:1145–1149. https://doi.org/10.1176/ajp.152.8.1145
8. Bremner JD, Narayan M, Anderson ER et al (2000) Hippocampal volume reduction in major depression. Am J Psychiatry 157:115–118. https://doi.org/10.1176/ajp.157.1.115
9. Kellner CH, Jolley RR, Holgate RC et al (1991) Brain MRI in obsessive-compulsive disorder. Psychiatry Res 36:45–49
10. Kuzniecky R, Murro A, King D et al (1993) Magnetic resonance imaging in childhood intractable partial epilepsies: pathologic correlations. Neurology 43:681–687
11. Andreasen NC, Ehrhardt JC, Swayze VW et al (1990) Magnetic resonance imaging of the brain in schizophrenia. The pathophysiologic significance of structural abnormalities. Arch Gen Psychiatry 47:35–44
12. Desmond JE, Glover GH (2002) Estimating sample size in functional MRI (fMRI) neuroimaging studies: statistical power analyses. J Neurosci Methods 118:115–128. https://doi.org/10.1016/S0165-0270(02)00121-8
13. Friston KJ, Holmes AP, Poline JB et al (1995) Analysis of fMRI time-series revisited. NeuroImage 2:45–53. https://doi.org/10.1006/nimg.1995.1007
14. Smith SM, Jenkinson M, Woolrich MW et al (2004) Advances in functional and structural MR image analysis and implementation as FSL. NeuroImage 23:S208–S219. https://doi.org/10.1016/j.neuroimage.2004.07.051

15. Fischl B, Sereno MI, Dale AM (1999) Cortical surface-based analysis. NeuroImage 9:195–207
16. Mazziotta JC, Toga AW, Evans A et al (1995) A probabilistic atlas of the human brain: theory and rationale for its development. The international consortium for brain mapping (ICBM). NeuroImage 2:89–101. https://doi.org/10.1006/nimg.1995.1012
17. Amunts K, Schleicher A, Bürgel U et al (1999) Broca's region revisited: cytoarchitecture and intersubject variability. J Comp Neurol 412:319–341. https://doi.org/10.1002/(SICI)1096-9861(19990920)412:2<319::AID-CNE10>3.0.CO;2-7
18. Evans AC, Collins DL, Milner B (1992) An MRI- based stereotactic Atlas from 250 young normal subjects. In: Soc. Neurosci. Abstr. p 408
19. Ashburner J, Andersson JL, Friston KJ (1999) High-dimensional image registration using symmetric priors. NeuroImage 9:619–628. https://doi.org/10.1006/nimg.1999.0437
20. Van Horn JD, Toga AW (2009) Multisite neuroimaging trials. Curr Opin Neurol 22:370–378. https://doi.org/10.1097/WCO.0b013e32832d92de
21. Glover GH, Mueller BA, Turner JA et al (2012) Function biomedical informatics research network recommendations for prospective multicenter functional MRI studies. J Magn Reson Imaging 36:39–54. https://doi.org/10.1002/jmri.23572
22. Friedman L, Glover GH, The FBIRN Consortium (2006) Reducing interscanner variability of activation in a multicenter fMRI study: controlling for signal-to-fluctuation-noise-ratio (SFNR) differences. NeuroImage 33:471–481 https://doi.org/10.1016/j.neuroimage.2006.07.012
23. Deoni SCL, Williams SCR, Jezzard P et al (2008) Standardized structural magnetic resonance imaging in multicentre studies using quantitative T1 and T2 imaging at 1.5 T. NeuroImage 40:662–671. https://doi.org/10.1016/j.neuroimage.2007.11.052
24. Zhou Z, Liu BJ (2005) HIPAA compliant auditing system for medical images. Comput Med Imaging Graph 29:235–241. https://doi.org/10.1016/j.compmedimag.2004.09.009
25. Mueller SG, Weiner MW, Thal LJ et al (2005) Ways toward an early diagnosis in Alzheimer's disease: the Alzheimer's Disease Neuroimaging Initiative (ADNI). Alzheimers Dement 1:55–66. https://doi.org/10.1016/j.jalz.2005.06.003
26. Chen J, Liu J, Calhoun VD et al (2014) Exploration of scanning effects in multi-site structural MRI studies. J Neurosci Methods 230:37–50. https://doi.org/10.1016/j.jneumeth.2014.04.023
27. Fischl B, Salat DH, Busa E et al (2002) Whole brain segmentation. Neuron 33:341–355. https://doi.org/10.1016/S0896-6273(02)00569-X
28. Hoogman M, Bralten J, Hibar DP et al (2017) Subcortical brain volume differences in participants with attention deficit hyperactivity disorder in children and adults: a cross-sectional mega-analysis. Lancet Psychiatry 4:310–319. https://doi.org/10.1016/S2215-0366(17)30049-4
29. Hibar DP, Westlye LT, van Erp TGM et al (2016) Subcortical volumetric abnormalities in bipolar disorder. Mol Psychiatry 21:1710–1716. https://doi.org/10.1038/mp.2015.227
30. Hibar DP, Westlye LT, Doan NT et al (2017) Cortical abnormalities in bipolar disorder: an MRI analysis of 6503 individuals from the ENIGMA Bipolar Disorder Working Group. Mol Psychiatry. https://doi.org/10.1038/mp.2017.73
31. Boedhoe PSW, Schmaal L, Abe Y et al (2017) Distinct subcortical volume alterations in pediatric and adult OCD: a worldwide meta- and mega-analysis. Am J Psychiatry 174:60–70. https://doi.org/10.1176/appi.ajp.2016.16020201
32. van Erp TGM, Hibar DP, Rasmussen JM et al (2016) Subcortical brain volume abnormalities in 2028 individuals with schizophrenia and 2540 healthy controls via the ENIGMA consortium. Mol Psychiatry 21:547–553. https://doi.org/10.1038/mp.2015.63
33. Walton E, Hibar DP, van Erp TGM et al (2017) Positive symptoms associate with cortical thinning in the superior temporal gyrus via the ENIGMA schizophrenia consortium. Acta Psychiatr Scand 135:439–447. https://doi.org/10.1111/acps.12718
34. Schmaal L, Veltman DJ, van Erp TGM et al (2015) Subcortical brain alterations in major depressive disorder: findings from the ENIGMA Major Depressive Disorder working group. Mol Psychiatry:1–7. https://doi.org/10.1038/mp.2015.69
35. Folstein MF, Folstein SE, McHugh PR (1975) "Mini-mental state". A practical method for

grading the cognitive state of patients for the clinician. J Psychiatr Res 12:189–198

36. Reuillon R, Leclaire M, Rey-Coyrehourcq S (2013) OpenMOLE, a workflow engine specifically tailored for the distributed exploration of simulation models. Futur Gener Comput Syst 29:1981–1990. https://doi.org/10.1016/j.future.2013.05.003
37. Passerat-Palmbach J, Reuillon R, Leclaire M et al (2017) Reproducible large-scale neuroimaging studies with the OpenMOLE Workflow Management System. Front Neuroinform. https://doi.org/10.3389/fninf.2017.00021

Part III

Brain Morphometry: Clinical Applications

Check for updates

Chapter 14

Brain Morphometry: Alzheimer's Disease

Matteo De Marco and Annalena Venneri

Abstract

Brains suffering from Alzheimer's disease show pronounced morphological modifications, with ample volumetric reduction of neural tissue. While particularly visible during the most severe disease stages, these changes are more subtle at the prodromal stage, which is the moment when a clinical diagnosis should be ideally reached. A large body of research has tried to disentangle the nature of such modifications, modeling the regional anatomical variability observed at various disease stages in samples and cohorts, implementing a number of different methodological avenues. The result is a complex picture in which brain morphology is not exclusively affected by disease processes, but is also under the influence of a large series of additional variables, which all contribute to the resulting phenotype via a tight network of multiple biological mechanisms. As a consequence, the study of morphological changes in AD highlights a sensible lack of clinical specificity. Despite these limitations, however, a large body of publications has highlighted the importance of brain morphology for the characterization of different phenotypic expressions of this disease and for the quantification of treatment effects.

Key words Gray matter, Cerebral cortex, Treatment effects, Cognitive functions, Clinical diagnosis, Mild cognitive impairment, T1-weighted imaging, Structural MRI, Disease trait

"(...) the monster got part of your wonderful brain. But what did you ever get from him?"

Inga, Young Frankenstein, 1974

1 The Use of MRI in the Clinical Study of Alzheimer's Disease

The etiological entity "Alzheimer's disease" (AD) is a label that refers to a complex and multifaceted diagnostic domain which alters the normal processes of aging experienced by the central nervous system, causing pathological neurodegeneration.

Throughout the years, this disease has assumed a profound importance in the clinical setting and has been generating a thriving interest worldwide in the field of research. Despite having been thoroughly explored by various disciplines through multiple approaches, it is possible to identify two major socioeconomical aspects which represent the main motivational drives behind such

Gianfranco Spalletta et al. (eds.), *Brain Morphometry*, Neuromethods, vol. 136,
https://doi.org/10.1007/978-1-4939-7647-8_14,

persistent attention for conceptualizing, describing, and characterizing the multidimensionality of AD: (1) the widely recognized symptomatology which cripples the well-being of patients and their families [1] and (2) the epidemiological perspectives for the future, numbers which are predicted to rise steeply and translate into large cost increases to sustain the demands related to this disease on healthcare systems worldwide [2]. As a consequence, clinical and research settings have joined forces to progress in the identification of adequate diagnostic/prognostic tools and potential efficient avenues of treatment. In order to do so, it has been necessary to operationalize AD within convenient interpretational frameworks. At present, for instance, a major clinically approved treatment prescribed to patients who receive a diagnosis of AD is represented by pharmacological enhancement of cholinergic neurotransmission [3]. Based on this specific rationale, AD is operationalized as a condition in which cholinergic synapses are dysfunctional and can thus be regulated by an appropriate intervention which aims at restoring this specific synaptic pathway. With this regard, the extent to which the neurotransmission is compromised is seen as a "marker" of the disease. A marker (or, more specifically in this case, a "biological marker" or "biomarker") can be defined as "objective indications of medical state observed from outside the patient, which can be measured accurately and reproducibly" [4]. Along this line, a number of markers for AD have been identified, each of which tends to serve a specific clinical purpose. In this regard, a cholinergic marker suits a treatment framework. On the other hand, quantifying the deposition of the typical peptidic specimens is crucial when it comes to ascertaining the presence of AD pathophysiology. Unfortunately, an definite diagnosis of AD can only be reached (or ruled out) after a postmortem histological examination of brain tissue. To exploit this same theoretical rationale, however, a large amount of research has been carried out to help refine clinical routines based on the measurement of these neurotoxic proteic accumulations in vivo, either in the form of concentration levels in the cerebrospinal fluid [5] or as the pattern of deposition captured by positron emission tomography neuromolecular imaging [6, 7]. In this respect, while cholinergic transmission is a marker that is associated with viable therapeutics, cerebrospinal or PET levels of AD-specific proteins are markers associated with the unique pathophysiological characteristics that allow a definitive diagnosis, although the reliability of these markers in the absence of clinical symptoms has been repeatedly questioned. Along this same line, there are other types of markers that have been extensively investigated and incorporated into the clinically established diagnostic formulas. One of these is certainly structural magnetic resonance imaging (sMRI).

The use of sMRI in support of a diagnosis of Alzheimer's disease was originally theorized more than 30 years ago [8], and, since then, more detailed description of the sMRI-based structural features of AD have been published in subsequent years [9]. The theoretical advances offered by sMRI, alongside those of X-ray-based computerized tomography, identified soon the hippocampal complex as a region subjected to substantial volumetric decrease in AD [10]. This radiological finding was consistent with the evidence reported by histological studies, which found that this region encountered a progressive functional disconnection from other subcortical hubs and from the associative cortex and, that this, in turn, could be interpreted as the organic reason behind the evident problems seen in amnestic functioning [11]. In addition to the evident involvement of the hippocampus, other early studies based on sMRI found that the volumetric loss seen in AD extended to multiple cortical and subcortical regions (e.g., [12]).

The initial investigations exploring morphometric changes triggered by AD give some indication on how the use of sMRI started following a dual track: (1) a clinical drive with the objective of exploiting brain morphometry for the description of the structural features of the disease, their trajectories along the axis of time, the identification of reliable markers of the disease in preclinical stages, the potential influence exerted by treatment avenues, and the extent to which these features can account for the multiple clinical phenotypes that characterize AD and (2) a concurrent methodological drive, aiming at improving the quantity and quality of the information made available from a sMRI acquisition for subsequent extraction and processing. Despite being substantially different, these two approaches have complemented each other in research and clinical work on AD, allowing clinicians to take advantage of technological and methodological advances and, at the same time, enabling researchers to tailor their expertise based on the specific needs of the AD clinical population. Since the literature of reference for both these approaches is immense, the following two sections will try to cover, in a concise way, the multidimensional nature of this research field following each of the aforementioned approaches.

2 Morphometric Features of the Brain and AD: Focus on Methodology

Under normal circumstances, the images of the brain of an individual referred to sMRI for clinical reasons are inspected by a neuroradiologist, who is expected to review them. Given the nature of their job and the variability of the phenomena they are called to address (e.g., hemorrhagic and ischemic events, neurodegenerative processes, traumatic lesions, brain tumors, abnormal conditions such as hydrocephalus, etc.), the clinical interpretation of brain morphology is prevalently based on pattern recognition. This "pattern" would be extrapolated via an accurate inspection of the images that offer the best view of the structures affected by the disease and by the concurrent analysis of the sequences sensitive to other features, necessary to comply with differential diagnostic procedures, when applicable. In the case of AD, the image modalities which offer the best results are a high-resolution T1-weighted (T1W) and a T2-weighted, fluid attenuated inversion recovery or proton density-weighted image, respectively. One or more of these non-T1W images are helpful for the detection of cerebrovascular disease as primary etiological entity, as vascular dementia is the second most common type of dementia after AD [13]. While the diagnostic procedures to suspect AD from a T1W scan, nowadays, are still based on visual examination [14], the methods of formal quantification of cortical and hippocampal atrophy have instead evolved. Visual-rating systems of ventricular dilation and sulcal/ hippocampal atrophy based on simple numerical scales [15] have been partially replaced by more sophisticated in-house routines such as, for instance, the manual tracing of structural boundaries for volume calculation [16]. The progressive technologizing of the procedures devised to extract numerical information from sMRI pursues the fact that, ultimately, even in a clinical setting, a brain scan is not just a set of pictures but an invaluable source of data [17].

One of the most fruitful uses of T1W images is the focus on the specific maps of gray matter, white matter, and cerebrospinal fluid. This feature has been exploited in the study of AD to separate the three tissue classes, using manual, pixel-by-pixel automatized, or template-based automatized segmentation pipelines and helps focus on the specific map of interest [18]. The map of gray matter shows the largest morphometric changes along the progression of the disease [19], while white matter morphometry is less affected, although its microstructural properties show some relevant modifications [20]. Based on this, a large number of studies have focused on gray matter maps, using a voxel-by-voxel approach and focusing on the volumetric amount of voxels characterizing each region [21]. Alternatively, cortical maps have also been manipulated as a function of their layer thickness, using voxel-based or surface-based, vertex-by-vertex methods [22].

The variability found in patients at various stages of the disease has been the object of study for a large proportion of:

- Cross-sectional group comparisons between two different diagnostic/clinical pictures, to shed light on the structures that play a role in defining the clinical peculiarity that differentiates the two groups (e.g., patients vs. controls, patients with vs. patients without a certain psychiatric symptom, or AD patients carrying a genetic risk factor vs. AD patients genetically protected)
- Longitudinal models to monitor the changes of a specific morphometric feature over the course of the disease and assess the

effect of a particular experimental manipulation (e.g., randomized trials to evaluate the effect of an intervention)

- Statistical associations between structural properties of the brain and specific clinical variables, to theorize the role of the regions in the development of a specific symptom or in the retention of a specific ability (e.g., the association between gray matter variability and attentional processes)

The use of volumetric and thickness measures in combination with similar inferential models represents a very large proportion of clinical research on AD. Although these statistical models are excellent choices to explore characteristics of the disease in a general sample, they are of little utility in the clinical setting, where a diagnostic decision has to be made for each single subject. As addressed more in detail in the next section, the initial stages of AD may be particularly challenging, and the distinction from the subclinical changes normally observed in non-pathological aging extremely difficult when relying only on visual inspection or basic quantitation of structural imaging. For this reason, other methodological techniques have been recently implemented for the individual classification of potential AD patients, carrying out feature extraction from sMRI (in combination with demographic information and/or other clinical and neuroimaging features) and using multivariate techniques and machine learning algorithms to assign each individual to one of two or more diagnostic labels, usually "healthy adults" and "patients," of diverse severity [23]. These methods based on subject classification are a clear example of procedures for which, potentially, no theoretical assumption would be needed to short-list a set of regions to focus on and no risk of selection bias would be run. However, there are several other studies investigating brain morphometry in AD which stem from a completely different approach, with previous findings suggesting a specific experimental hypothesis. For instance, Kim and colleagues [24] suggested a morphometric marker which focuses on cortical thinning in the sole areas that are concurrently affected by the benchmark peptidic specimen of the disease (neurotoxic plaques of beta-amyloid protein). Or, as a second example, Montembeault et al. [25] investigated patterns of structural covariance in AD patients based on the spatial maps of long-range functional-connectivity brain networks normally seen with resting-state functional MRI.

In summary, from a merely methodological, nonclinical perspective, the use and modeling of sMRI has had a large range of application in the study of AD. The study of brain morphometry, prevalently based on the exploration of gray matter volumes and thickness (the use of deformation-based techniques such as tensor-based morphometry has been minimal) either limited to single brain regions of relevance or extended to the entire cortex, has been exploited to study various aspects of the disease: group differences, longitudinal changes, and the role of specific clinical variables. Additionally, over recent years, classification methods have been implemented in research to try and identify the morphometric signature of AD for patient-by-patient classification. Methodologies have been designed either in an exploratory way, via whole-brain analyses or through the inclusion of all available data entries, or in a more hypothesis-driven fashion, where the input of clinicians has driven the choice of the morphometric features to explore.

In addition to this brief methodology-based overview of AD and brain morphometry, there are numerous studies which have either contributed to improve and perfect specific details of procedural routines (e.g., the introduction of multiple templates rather than a single one for image registration [26]) or have come up with novel and increasingly more sophisticated methodologies such as enhancement of descriptors [27, 28] and feature computation based on latent variable models (e.g., [29]). Concurrently, the spatial resolution of the structural images has been enhanced with the introduction of high-strength magnetic fields. These have allowed researchers to reduce the size of the acquisition voxel considerably and to achieve more detailed information on brain morphometry of AD patients [30].

In conclusion, the use of morphometric information in research and in the clinical management of AD has been achieved as a result of numerous technical and methodological improvements. Although all these advances could make a significant contribution to daily hospital routines, their practical application, however, has been limited, mainly because the implementation of novel procedures requires significant expertise which clinicians do not always have.

3 Morphometric Features of the Brain and AD: Theoretical Advances

3.1 Irregular Relationship Between Morphometry and Disease Processes

The theoretical universe around AD is studded with a myriad of biological mechanisms and hypotheses. A large number of variables play a causative role in the modification of key aspects of the disease, while others simply act as risk/protection factors, modifying the likelihood of developing certain features. In a small percentage of cases, AD is caused by genetic mutations, whereas most of the diagnoses are sporadic. These two types of etiology show a differential clinical presentation, with familial cases tending toward an earlier onset of symptoms and a longer disease duration [31]. In addition, and independently of genetics, early- and late-onset forms of AD show considerable differences in their clinical presentations [32]. Furthermore, there is a considerable variability in the type of cognitive domains affected by AD, with a non-negligible proportion of patients who present an “atypical” cognitive phenotype, characterized by initial non-amnestic deficits [33]. To complete the picture, there is also substantial variability in the pathophysiological processes of the disease. In fact, the typical neurofibrillary pathology due to abnormal accumulation of the hyperphosphorylated TAU (pTAU) protein seen in the hippocampal neurons of AD brains is actually not detected in one out of four patients [34]. All these pieces of evidence converge in describing AD as a heterogeneous pathology, rather than a diagnosis having constant inter-patient profile and predictable evolution. This clinical and pathophysiological variability, inevitably, influences the reliability of the different markers of the disease and complicates the exhaustivity of diagnostic (and, as a consequence, prognostic) formulas. Specific changes in brain structure are one of the most "classic" biomarkers of AD, with a volumetric decrease in the left hippocampal complex being the most frequent signature of progression from the prodromal stage of mild cognitive impairment (MCI) to a clinically established phase of dementia (Fig. 1) [19]. Among the sets of detrimental changes that occur in the brain due to AD, atrophy is a primary feature, with a decrease affecting the hippocampus, as this region is neither subjected to significant deposition of beta-amyloid peptide nor to a comparable reduction of glucose metabolism [35]. Regional BOLD activation of the hippocampal complex, on the other hand, seems to follow a quadratic trend, increasing in the MCI phase [36] and decreasing in early AD dementia [37]. Concurrently, there appears to be a trade-off between hippocampal metabolism and extra hippocampal connectivity, with MCI and

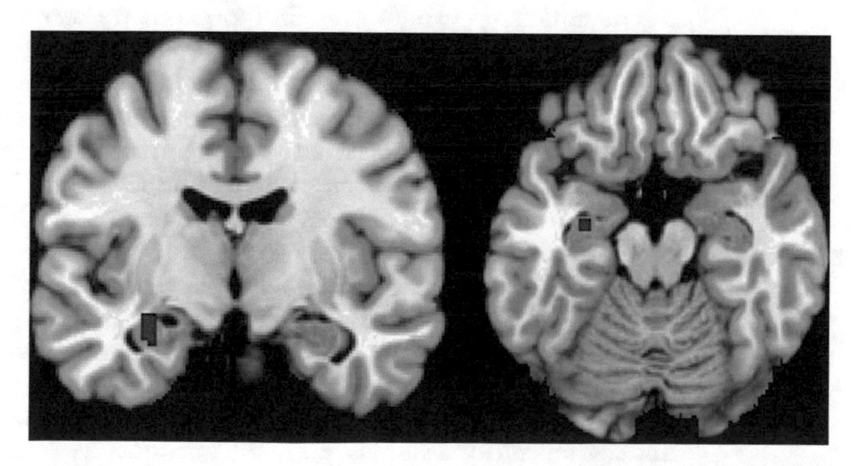

Fig. 1 The results of a meta-analysis comparing gray matter maps of patients with amnestic mild cognitive impairment who progressed to a condition of dementia of the Alzheimer's type and non-progressing patients. Six studies were included in the analytical procedures, for a total of 429 patients (142 converters). The findings indicate that the left hippocampus and parahippocampal gyrus are crucial predictors of future decline and development of clinically established dementia of the Alzheimer's type. Reprinted from Neurobiology of Aging, Vol 32, Ferreira, Diniz, Forlenza, Busatto and Zanetti, "Neurostructural predictors of Alzheimer's disease: a meta-analysis of VBM studies", 1733–1741, 2011, with permission from Elsevier

AD patients showing a negative association between hippocampal-parietal connectivity and hippocampal metabolism [38]. This evidence is consistent with the cellular pathology described at a micro-structural level, which describes the hippocampus as a structure subjected to disconnection and, as a consequence, computational isolation [11]. This multidimensional description of the events occurring at the level of the hippocampus in the brain of MCI and AD patients exemplifies the unilateral nature of brain morphometry when it comes to describing disease processes. In fact, physical changes in brain structure underlie several processes which do not always follow the same direction as with the evidence seen in brain function.

The MCI stage of AD triggers volumetric reductions in other areas aside from the mediotemporal lobe and affects the inferior/middle portion of the temporal cortex, the inferior parietal lobe, precuneus, and posterior cingulate regions [39]. A parallel thinning of the cortex is also visible in these areas [40]. This set of regions normally shows a pattern of high intrinsic functional connectivity, and its hemodynamics are known as the posterior portion of the default mode network, a circuit that deactivates during task performance and which shows disease-specific downregulation during the course of AD [41]. Interestingly, the core cortical regions of this circuit (specifically, the inferior parietal lobe and the posterior cingulate) are also two areas that show a major pathological deposition of beta-amyloid neurotoxic plaques [42]. As a consequence, in these two areas, changes in brain morphometry accompany changes in functional connectivity and peptidic pathology.

In summary, changes in brain morphometry seen in the prodromal phase of AD affect both the mediotemporal complex as well as other cortical areas. While the reduction seen in the temporal, parietal, and postero-limbic regions is paralleled by a concurrent reduction of network connectivity and a concurrent increase in local beta-amyloid deposition, the reduction seen in the hippocampus is mirrored by increases in local metabolism and little or no beta-amyloid deposition. These two opposite but concurrent pieces of evidence indicate that regional gray matter changes triggered by AD have to be contextualized together with other types of evidence, in order to capture fully the nature of changes which occur in the brain of patients. As a consequence, the association between peptidic pathology and brain atrophy is, under general terms, undetermined (for a summary, see the introductory part of [43]).

The typical morphometric modifications only capture the general trend visible in the population of patients who are in the earliest clinical stage of typical AD, but are of little or no utility in diagnostic settings. The current criteria for AD, in fact, do not assign any core role to structural biomarkers [44, 45], while the latest criteria for the diagnosis of prodromal AD when this is in the phase of MCI indicate that changes in brain volumetry might be indicative of AD neuronal injury, but are not disease specific [46]. There are two major conceptual reasons underlying this lack of specificity. First, the biological mechanisms which induce changes in brain morphometry have not been completely clarified. Neuronal death is not the sole cause behind brain shrinkage. In fact, other mechanisms such as reduction of cytoplasmic volume or thinning out of the synaptic architecture have been documented [47], and these processes do generate gray matter decrements physically similar to those caused by apoptosis. In addition, there is not just an inhomogeneous impact of AD on the different cerebral lobes, but there are also innate regional differences in the biology defining the morphometric properties of the various lobes. In fact, perikaryon volumes are significantly larger in the frontal lobe compared to the occipital, temporal, and parietal areas, and the occipital lobe, vice versa, features significantly lower perikaryon volume [48]. Similarly, the complexity of the dendritic architecture and the concurrent synaptic scaffold differs from area to area [49, 50], and this contributes to the definition of regional morphometry. Unfortunately, the physical structure of the brain as measured with MRI is completely devoid of any information concerning region-to-region differences in the biological substrate. As a consequence, the same physical changes in brain morphometry might be triggered by other biological mechanisms, determining a serious lack of specificity. In fact, reduction of mediotemporal volumes is also seen in other conditions, such as post-traumatic stress disorder [51], hippocampal sclerosis due to epilepsy [52], late-life depression [53], and type II diabetes [54], and is also part of the normal processes of physiological aging [55]. As with physiological aging, it is important to highlight that the entire cortex is subjected to a physiological shrinkage, and recent evidence suggests that the morphometric reduction seen in AD is at least in part the result of the underlying processes of aging [56, 57].

The nature of the biological mechanisms causing brain structure downregulation in AD is not clear. Recent evidence obtained from patients in preclinical stages indicates that cortical thinning is associated with cerebrospinal levels of pTAU, while, on the other hand, hippocampal shrinkage is associated with decreased cerebrospinal levels of beta-amyloid [58]. This latter association, however, has not been always replicated (see Gispert et al. [43] for a concise review on the topic). Results from studies of the ADNI cohort show that MCI patients with amyloid pathology (as measured via neuromolecular imaging) show more longitudinal atrophy in a large set of regions, including the hippocampus, than MCI patients free from amyloid burden. Vice versa, the presence of amyloid in healthy adults was associated with increased longitudinal atrophy in the sole posterior cingulate cortex, with no specific involvement of the mediotemporal formation [59]. Moreover, other evidence suggests that there is no signature of amyloid deposition in healthy adults [60]. Based on all these findings it is hard to draw a complete and satisfactory picture. Evidence suggests that the brain morphometry alterations seen in patients are not just the result of the biology of AD. Other, not necessarily pathological, phenomena give their contribution to the final picture, and it would be the interplay of all these factors which shapes the resulting phenotypic characterization.

These variables which do not play any causative role have been studied in populations of adults free from symptomatic neurodegenerative conditions, in order to identify their effects, net of the clinical impact of pathology. Two large categories can be identified.

- Genetic factors

  The pattern of brain structure among healthy individuals is significantly associated with a large number of variables completely determined by DNA expression, which, as a consequence, are unmodifiable. Some of these are to be considered as variability generators remaining completely within the domain of normality, like gender [61] or blood type (Fig. 2) [62]. Others

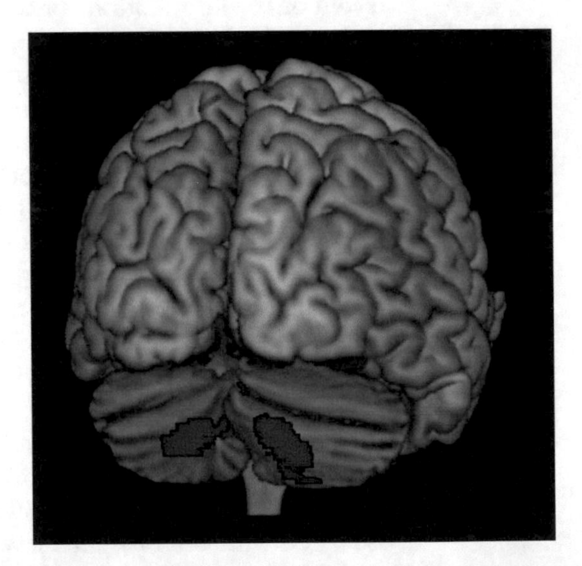

Fig. 2 The pattern of statistically significant differences between a sample of healthy individuals with an "A" blood group and a sample of healthy individuals with an "0" blood group. Differences were found in morphometric properties of the posterior cerebellar cortex, where adults with an "0" blood group were found to have more gray matter. Additional analyses run to compare individuals with an "0" blood group and individuals with a more generic "non-0" blood group (thus including individuals with an "A," "B," or "AB" blood group) revealed that, in addition to the cerebellum, mediotemporal regions (including the hippocampus) were associated with blood group-related volumetric variability. Reprinted from Brain Research Bulletin, Vol 116, De Marco and Venneri, "'0' blood type is associated with larger grey-matter volumes in the cerebellum", 1–6, 2015, with permission from Elsevier

are genetic variables which are neither necessary nor sufficient to induce the onset of AD but represent significant risk factors, such as the presence of the Apolipoprotein E ε_4 allele [63]. A third and extremely important category is represented by genetic variables that *cause* the disease in a very small percentage of cases and induce morphometric differences even in the pre-symptomatic stages, thus years before their causality is expressed [64].

– Lifestyle factors

It has been demonstrated that brain morphology is associated with an immense amount of variables that are the direct or indirect result of lifestyle choices. These range from medical conditions, such as chronic hypertension [65] or type II diabetes [66], to aspects of neurological-cognitive nature. A number of theoretical frameworks have been put forward to account for different aspects of structural variability. The two most important concepts are those of *reserve* and *plasticity*. Whereas neural reserve simply refers to the amount of "neural supply" which can be recruited for computations following neural damage (i.e., "the more brain, the better"), cognitive reserve conceptualizes the interindividual differences in processing tasks to cope with the amount of damage through the exploitation of efficiency/ flexibility and compensation mechanisms [67]. These "extra skills" of which the brain can be more or less capable are conveyed and boosted by a series of lifestyle factors which also are themselves associated with brain structure, including educational attainment [68], or lifelong practice in specific skills. It has been, in fact, demonstrated that differences in brain morphology exist between experts in a field of activities such as taxi drivers or musicians and naïve individuals [69, 70]. As a consequence, both innate and environmental factors concur and plausibly interact in being the source of morphometric variability within the general population and within the population of patients, conferring both different levels of susceptibility and different levels of resistance to pathology.

If reserve works as a "brake pedal," plasticity represents instead a sort of mild "counter-accelerator." This idea refers to a set of processes by which the nervous system succeeds at remolding itself, to remain as fully functional as possible after the changes triggered by external inputs (e.g., the detrimental alterations induced by AD). Neuroplastic modifications are intrinsically structural in nature and are the result of microstructural mechanisms such as neurogenesis, gliogenesis, angiogenesis, synaptogenesis, myelination, or dendritic rewiring [71]). Neuroplastic processes are believed to be the main causative factor behind the morphometric changes observed after cognitive training (e.g., [72]) and are, in all likelihood, also responsible for the neuroanatomical uniqueness found in the cortex of taxi drivers and musicians, as found by research mentioned above. Within this context, reserve and plasticity contribute to shield brain structure from modifications triggered by AD. Two fictional patients affected by an equally severe amount of AD pathology might show significant discrepancies in brain anatomy, if substantial differences existed in premorbid levels of cognitive reserve. Similarly, and net of reserve, morphometric properties of the brain might be different in one of these two fictional patients, if the progress of AD were slowed down by intense levels of activities aimed at triggering retained neuroplasticity. One of the most fruitful operative conceptualizations of AD describes the disease as a condition which progressively disrupts the neuroplastic capacities of the brain [73]. Based on Mesulam's postulate, disease monitoring would be carried out by estimating the extent to which morphometric changes due to retained neuroplasticity are inducible. Although this possibility has not yet been explicitly tested, there are a number of studies which have reported structural changes triggered by cholinergic treatment (*see* Sect. 3.3), in support of the idea that modifications of brain morphometry in regions susceptible to disease mechanisms can also be an index of how prone the system is to row against the pathology.

In summary, the simple detailed description of brain morphometry in AD is itself not sufficient neither to define a clear association with the cellular mechanisms of the disease nor to predict a univocal clinical picture. The reason behind this fogginess lies in the large variability accounted for by a set of genetic and environmental factors, which are capable of altering brain structure significantly, along the axis of disease progression. Moreover, the picture is complicated even more by the fact that AD is increasingly revealing itself as a heterogeneous disease. Neuropathological evidence indicates that 25% of all cases have sparing of the hippocampal complex [34] (proportion set at 22% by volumetric analysis of the ADNI cohort, see [74]), and it has also been suggested that the age of onset (early vs. late) might play a crucial role in separating "typical" and "atypical" cases [75]. When it comes to brain morphometry, the patients' age is not a trivial variable. In fact, the global volumetric properties of the healthy human cortex follow an age-dependent linear reduction [76], and the development of specific regions is often curvilinear. For instance, the thickness of the entorhinal cortex follows a quadratic trend [77]. Arguably, the onset of Braak stage I pathology in a thicker/thinner entorhinal cortex might lead to different disease trajectories, with a more pronounced visible loss of function in those patients whose cortex tends to be thinner. As a result, any clinical judgment on the severity of an atrophic pattern should be carefully evaluated based on the precise awareness of a range of linear/curvilinear reference values. On this note, age- and gender-adjusted normative values might be of clinical help for the identification of AD from sMRI [78]. However, as motivated above, these morphometric norms should be further tailored on the specific status of the patients, by taking into consideration possible innate (e.g., Apolipoprotein E genotype) and experience-based factors (e.g., indices of reserve and neuroplastic capacities) which could modify regional anatomy as measured by sMRI. This large variability is reflected by the uncertain findings of those studies which, using a longitudinal design, have attempted to identify patterns of a potential signature of conversion from MCI to AD dementia. While some studies found differences in hippocampal volume between MCI converters and non-converters (e.g., [79]), suggesting that atrophy of this structure could be a good indicator of underlying AD pathology, there are other studies which did not replicate this finding, highlighting instead the role of other extra-hippocampal mediotemporal structures and cortical regions [80, 81]. As indicated by the rich literature on this broad and complicated research field, the complex intertwining of multiple clinical and individual-related variables makes it extremely difficult to identify a unique and irrefutable morphometric index which allows a clear identification of Alzheimer-type pathological processes in all patients, indiscriminately. This is one of the major reasons the most recent studies seeking classificatory algorithms to separate patients from controls are not exclusively based on sMRI features, but include other types of indices such as concentration levels of disease-specific proteins, brain function, and cognitive variables [82–84].

3.2 Brain Morphometry as Vehicle to Explore Clinical Traits and Aspects of Susceptibility

Even though its potential clinical applications are limited, the study of brain structure in AD has been pursued more successfully for a parallel purpose of more explorative nature. The large variability described among patients diagnosed with AD dementia (as mentioned above) and among patients diagnosed with MCI of the amnestic or non-amnestic type prodromal to AD dementia is associated with clinical (or subclinical) morphometric variability, which is, in turn, associated with the variability seen in behavioral domains. This biology-to-morphometry-to-behavior connection is exemplified by the Imaging Genetics framework [85]. The main goal of this model is attempting to account for the effects diverse allelic expression exerts on neuroimaging variables and, as a consequence, behavior (e.g., cognitive functioning). By working out in reverse the terminal part of this same rationale, part of the behavioral variability can be operationalized as the effect of variability seen in regional aspects of brain structure. This can be studied, for instance, adopting a voxel-based correlation methodology [86] or confronting via statistical models two otherwise comparable groups of patients who differ in a sole specific feature. Thanks to this type of approach aimed at exploring the role of brain regions, it has been possible to clarify what structures are or may be responsible for the presence of specific behavioral traits in AD, and it has been possible to speculate (proceeding one further step backward along the frameworks) what biological mechanisms might sustain specific phenotypes. There are, for instance, cognitive symptoms which are not strictly specific to AD, and the study of a cross-diagnosis brain structure involvement in naming difficulties reveals different regional associations between naming deficits and regional volumetric loss. Although patients with frontotemporal lobar degeneration or cortico-basal syndrome also show impaired performance in naming tasks, regional patterns of cortical volumes associated with this common deficit appear to differ across the three diagnostic categories [87]. In a similar fashion, Pennington and colleagues [88] studied a sample of patients with AD and a second sample of patients with frontotemporal dementia (behavioral variant), and, although there was no significant group difference in any aspect of verbal/visuospatial recall/recognition, they described a diagnosis-dependent pattern of association between memory retrieval and regional patterns of atrophy. Specifically, memory scores correlated with prefrontal atrophy in the subgroup of patients with frontotemporal dementia and with both hippocampal and prefrontal atrophy in the subgroup of patients with AD. While the investigation of morphometric correlates across diagnoses allows the formulation of hypotheses on the disease-specific mechanisms which underlie specific symptoms, the study of a symptom along the axis of AD progression, comparing groups of patients of different severity, may shed additional light on the relationship between AD pathology and clinical presentation. Cross-sectional results show that the pattern of association between retained episodic memory/executive functions and gray matter is subjected to qualitative modifications, as patients worsen and convert to dementia [89]. Similar differences have been found also in semantic processing and throughout an ampler disease spectrum [90]. These progressive changes reflect both pathological processes affecting areas which are crucial in sustaining these tasks and also, at least in part, neuroplastic alterations. The brain of AD patients who are at the MCI stage still retains capacity for neuroplasticity [91], but it is challenging to understand to what extent what emerges as a structural remapping is actually beneficial or maladaptive. Within this framework, the study of brain structure in AD has been fruitful to clarify the role of specific structures supporting performance in specific aspects of cognition like verbal fluency [92], in specific tasks for which multiple cognitive functions need to cooperate, like the clock-drawing test [93], or in case of the presence of other, non-cognitive salient disease traits. For instance, albeit AD does not primarily affect the neuropsychiatric balance of patients, there are patients who develop atypical behavioral symptoms such

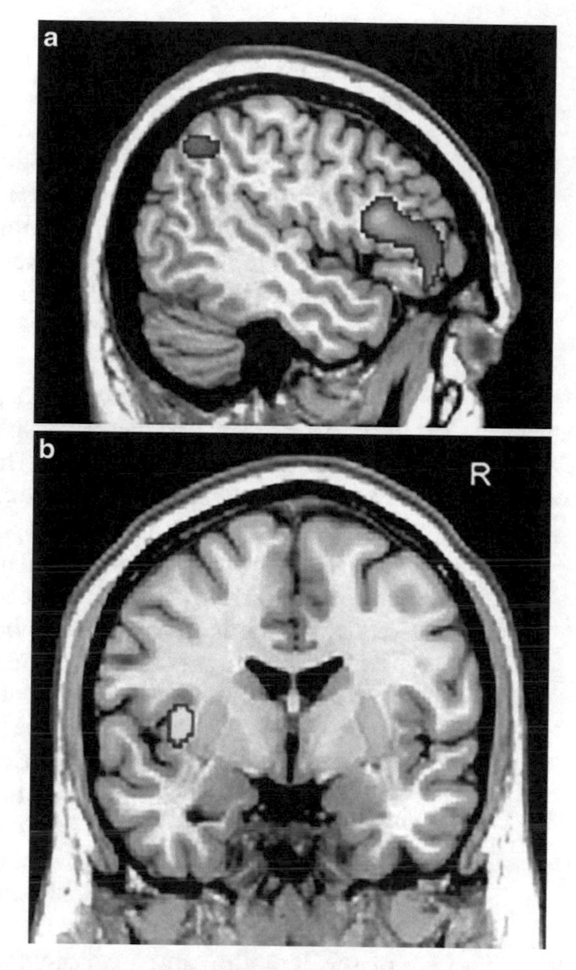

Fig. 3 A sagittal and coronal slice showing the positive statistical association between gray matter volume and delusional behaviors in patients with a clinical diagnosis of mild Alzheimer's disease (average group-level mini-mental state examination score, 23.30). In this linear model, the regional variability of gray matter volumes was modeled as a function of the presence of a selective disruption of psychiatric well-being, as quantified by caregivers with a specific questionnaire. The presence of high levels of delusions was found in concomitance with lower gray matter densities in inferior-frontal and parietal regions (**a**) and in the left claustrum (**b**). Reprinted from Brain, Vol 131, Bruen, McGeown, Shanks and Venneri, "Neuroanatomical correlates of neuropsychiatric symptoms in Alzheimer's disease", 2455–2463, 2008, with permission from Oxford University Press

as delusions and agitations early in the course of the disease. The presence of these features has been shown to be associated with reduction in brain volumes of regions which are part of specific high-order neural networks (Fig. 3) [94].

Apart from understanding the structural nature of specific behavioral traits, the study of brain morphometry in AD is relevant for the comprehension of specific aspects of disease mechanisms which may either be causally related to AD or, rather, simple "by-product processes." We mentioned in Sect. 3.1 a list of genetic and environmental factors which concur to define regional morphometry in healthy and abnormal aging. When these variables show a clear-cut association with disease mechanisms, it is informative to explore this pattern further and understand whether specific clinical traits are actually due to these variables. In such case, the study of brain morphometry would allow the identification of structural endophenotypes that are helpful in the characterization of patients and, possibly, in the definition of a prognosis. In agreement with the Imaging Genetics model [85], the study of brain anatomy has also been useful to describe specific trajectories which may be induced by genetic variability. The study of individuals carrying a mutation in the amyloid precursor protein or presenilin 1 genetic loci, for instance, offers the opportunity to differentiate the processes of atrophy seen in this small proportion of patients from the cohort who have developed the sporadic form [95]. Similarly, the study of genetic risk factors has been informative in understanding specific disease processes that might not be present in all patients but only in those subjected to additional genetically determined susceptibility. On this note, the Apolipoprotein E ε_4 allele (the most relevant genetic risk factor for developing the sporadic, late-onset AD-type form of the disease) has been shown to be profoundly implicated in the main pathophysiological processes of AD [96]. Nonetheless, there are AD patients who do not carry this isoform, and a number of studies have described an allele-dependent disease signature consisting of a more profound atrophy of the temporal and mediotemporal regions [97, 98]. Since this gene is often included as part of the diagnostic characterization of patients, it might be relevant to tailor clinical judgment based on the presence/absence of a genetic factor which, year by year, influences the trajectory of disease mechanisms in a cumulative way. Using similar methodologies, additional studies have also focused on other genes, like CR1 and PICALM [99], which had emerged as statistically relevant loci in genome-wise association studies (for a meta-analysis, see [100]). On this note, the study of the effect triggered by genetic factors on morphometric variability can also be addressed in a hypothesis-free fashion. Hibar and colleagues [101] modeled the effect of over 18,000 genes to reveal the impact of specific loci on regional structure of a large mixed sample of healthy adults and patients diagnosed with MCI or AD dementia, using multivariate techniques and tensor-based morphometry. The findings showed the presence of statistical associations between localized clusters and a series of genes which had been previously unexplored in the study of neurodegeneration. In a subsequent study, the same team took the association between genetics and morphometry to the next level, modeling the pattern of gene-to-gene interactions which might supersede the main effects of a single gene [102].

If on one hand genetic aspects associated in some way with AD are relatively simple to explore in these conceptual terms, on the other hand, the study of endophenotypes associated with those variables mentioned in Sect. 3.1 as environmental factors may prove more difficult. Type II diabetes and obesity, for instance, are conditions which cause susceptibility of the cerebral blood vessel architecture. The vascular processes in Alzheimer's disease are slowly being clarified at a microstructural level [103], and, in parallel, there has been an attempt to disentangle the causal processes behind the associations found between brain structure and vascular risk factors (e.g., [104]). It is likely that the accurate characterization of the morphometric endophenotypes associated with these relatively common medical conditions will, in the future, provide clinicians with additional information of diagnostic and prognostic importance. At the same time, on the other hand, morphometric endophenotypes of concepts such as reserve or plasticity are more difficult to identify. Reserve, for instance, is sustained by some form of neural implementation [105], but it is unknown whether higher cognitive reserve (intended in its overall conceptualization and not as a single aspect, for instance, bilingualism) is associated with a specific endophenotype which could bear relevant information to be used in clinical settings.

In conclusion, the use of sMRI has been helpful in the definition of the mechanisms underlying specific disease-specific traits and in the clinical characterization of variables that are associated with AD and may complicate the clinical picture of patients.

3.3 Brain Morphometry as an Index of Treatment Effect

Although the study of morphometry does not univocally reflect the mechanisms of disease, observing a meaningful modification induced by an intervention is evidence of treatment effectiveness. As a consequence, the use of sMRI may prove useful in supporting efficacy of a multitude of pharmacological and non-pharmacological therapeutics (even though absence of change does not imply treatment inefficacy).

Nowadays, the research addressing pharmacological agents for AD is fervent. However, very few compounds make it to human-participant testing. As a consequence, only a small number of studies have investigated the changes of morphometric properties of the brain triggered by these drugs. The major approved pharmacological treatment protocol for patients with AD is represented by second-generation drugs which aim at upregulating the cholinergic system. Deficit of cholinergic neurotransmission is associated with the presence of cognitive and behavioral symptoms. As a consequence, improving the cholinergic balance might induce structural alterations in the areas which sustain a more efficient neuronal communication. A look at the literature shows, however, that the pattern of results is not homogeneous. In fact, a study carried out on patients with mild AD yielded nonsignificant changes triggered by donepezil (the most used molecule) at reducing hippocampal atrophy [106]. Vice versa, other two studies testing comparable samples found positive results [107, 108]. Similar findings were also reported in cohorts of patients with amnestic MCI. A recent study found protective effects of donepezil at reducing the volumetric decline of the hippocampus of MCI patients [109], but other two studies did not replicate these findings [110, 111]. In the study by Schuff and colleagues [111], however, donepezil was found to mitigate the rate of longitudinal atrophy of the global cortex.

The existence of such large heterogeneity may be due to the mixed nature of these groups of patients, which normally include patients who do and patients who do not respond to cholinergic upregulations. On this note, morphometric measures have also been used to differentiate these two subgroups and understand potential differences which may translate into high-vs.-low suitability of patients toward this specific type of treatment. Curiously, two studies found that responders had larger baseline levels of atrophy affecting the substantia innominata [112, 113], while in another study it was non-responders who were found to have more pronounced atrophy of the nucleus basalis [114]. Another study related the therapeutic success of donepezil to baseline hippocampal volumes [115].

Following these same investigational schemes, measures of brain structure have been implemented experimentally to study other cholinergic drugs such as galantamine [116] and rivastigmine [117] or the glutamatergic drug memantine administered to patients already under stable cholinergic treatment [118] and, additionally, to study the effect on progression of atrophy of the most novel experimental treatment protocols such as immunization for beta-amyloid [119], intravenous administration of immunoglobulin [120], or nerve growth factor [121]. Overall, the available evidence suggests that sMRI measurements to detect treatment effects may be not the most useful imaging outcome measure, and whether or not sMRI features are good surrogate outcome measures of treatment efficacy in AD is questionable; especially the available findings suggest that features based on hippocampal volumetry are most likely unsuitable to detect subtle treatment effects in patients with prodromal or clinically established AD.

A further field of applicability, finally, is represented by the assessment of non-pharmacological protocols of intervention. On this note, cognitive stimulation has developed as a major avenue of research, with a large number of studies designed to test the efficiency of pen-and-paper or computerized packages of training exercises. Although a large part of this research field is conceived with the intent to address a clinical problem (hence outcome measures being represented by clinical, rather than neuroimaging variables), a proportion of these studies focuses on the specific neural mechanisms by which these types of intervention would be efficient. A number of experimental studies have been published to test the effect various paradigms of cognitive stimulation have on the anatomy of the brain [122]. Along these lines, even physical exercise has been considered as a potential therapeutic option in aging and neurodegeneration, and experimental evidence has shown how this type of activity does trigger structural modifications of the brain [123].

4 Conclusion

In summary, while the clinical applicability of sMRI and the study of cerebral morphometry have been limited for the diagnosis of AD, it has proved to aid efficiently the study and exploration of specific aspects associated with the disease, aspects which can then be further investigated using approaches which manage to capture the biology of the brain more accurately. As part of this latter category, functional connectivity is an example of a technique which relies on the framework of system neuroscience, a view of brain functioning which is hierarchically more dynamic than the past approach based on a function-brain area localization. Although recent approaches aimed at incorporating sMRI to a system-based framework (e.g., the study of structural covariance [124]), most aspects of brain morphometry describe physical features which do not always reflect the biological changes triggered by this disease.

References

1. Opara JA (2012) Activities of daily living and quality of life in Alzheimer disease. J Med Life 5:162–167
2. Brookmeyer R, Johnson E, Ziegler-Graham-K, Arrighi HM (2007) Forecasting the global burden of Alzheimer's disease. Alzheimers Dement 3:186–191
3. Ehret MJ, Chamberlin KW (2015) Current practices in the treatment of Alzheimer disease: where is the evidence after the phase III trials? Clin Ther 37:1604–1616
4. Strimbu K, Tavel JA (2010) What are biomarkers? Curr Opin HIV AIDS 5:463–466
5. Blennow K, Zetterberg H (2015) The past and the future of Alzheimer's disease CSF biomarkers-a journey toward validated biochemical tests covering the whole spectrum of molecular events. Front Neurosci 9:345
6. Adlard PA, Tran BA, Finkelstein DI et al (2014) A review of β-amyloid neuroimaging in Alzheimer's disease. Front Neurosci 8:327
7. Okamura N, Harada R, Furumoto S, Arai H, Yanai K, Kudo Y (2014) Tau PET imaging in Alzheimer's disease. Curr Neurol Neurosci Rep 14:500
8. Besson JAO, Corrigan FM, Foreman EI, Eastwood LM, Smith FW, Ashcroft GW (1984) Proton NMR observations in dementia. Magn Reson Med 1:106–107
9. Brun A, Englund E (1986) Brain changes in dementia of Alzheimer's type relevant to new imaging diagnostic methods. Prog Neuro-Psychopharmacol Biol Psychiatry 10:297–308
10. George AE, de Leon MJ, Stylopoulos LA et al (1990) CT diagnostic features of Alzheimer

disease: importance of the choroidal/hippocampal fissure complex. Am J Neuroradiol 11:101–107
11. Hyman BT, Van Hoesen GW, Damasio AR, Barnes CL (1984) Alzheimer's disease: cell-specific pathology isolates the hippocampal formation. Science 225:1168–1170
12. Jernigan TL, Salmon DP, Butters N, Hesselink JR (1991) Cerebral structure on MRI, part II: specific changes in Alzheimer's and Huntington's diseases. Biol Psychiatry 29:68–81
13. Alzheimer's Society UK. Dementia Infographic (2013). https://www.alzheimers.org.uk/site/scripts/download_info.php?fileID=1409
14. Tuokkola T, Koikkalainen J, Parkkola R, Karrasch M, Lötjönen J, Rinne JO (2016) Visual rating method and tensor-based morphometry in the diagnosis of mild cognitive impairment and Alzheimer's disease: a comparative magnetic resonance imaging study. Acta Radiol 57:348–355
15. Pasquier F, Leys D, Weerts JG, Mounier-Vehier F, Barkhof F, Scheltens P (1996) Inter- and intraobserver reproducibility of cerebral atrophy assessment on MRI scans with hemispheric infarcts. Eur Neurol 36:268–272
16. Juottonen K, Laakso MP, Partanen K, Soininen H (1999) Comparative MR analysis of the entorhinal cortex and hippocampus in diagnosing Alzheimer disease. Am J Neuroradiol 20:139–144
17. Gillies RJ, Kinahan PE, Hricak H (2016) Radiomics: images are more than pictures, they are data. Radiology 278:563–577
18. Tanabe JL, Amend D, Schuff N et al (1997) Tissue segmentation of the brain in Alzheimer disease. Am J Neuroradiol 18:115–123
19. Ferreira LK, Diniz BS, Forlenza OV, Busatto GF, Zanetti MV (2011) Neurostructural predictors of Alzheimer's disease: a meta-analysis of VBM studies. Neurobiol Aging 32:1733–1741
20. Yoon B, Shim YS, Hong YJ et al (2011) Comparison of diffusion tensor imaging and voxel-based morphometry to detect white matter damage in Alzheimer's disease. J Neurol Sci 302:89–95
21. Busatto GF, Diniz BS, Zanetti MV (2008) Voxel-based morphometry in Alzheimer's disease. Expert Rev Neurother 8:1691–1702
22. Li C, Wang J, Gui L, Zheng J, Liu C, Du H (2011) Alterations of whole-brain cortical area and thickness in mild cognitive impairment and Alzheimer's disease. J Alzheimers Dis 27:281–290
23. Arbabshirani MR, Plis S, Sui J, Calhoun VD (2017) Single subject prediction of brain disorders in neuroimaging: promises and pitfalls. NeuroImage 145:137–165
24. Kim CM, Hwang J, Lee JM et al (2015) Amyloid beta-weighted cortical thickness: a new imaging biomarker in Alzheimer's disease. Curr Alzheimer Res 12:563–571
25. Montembeault M, Rouleau I, Provost JS, Brambati SM, Alzheimer's Disease Neuroimaging Initiative (2016) Altered gray matter structural covariance networks in early stages of Alzheimer's disease. Cereb Cortex 26:2650–2662
26. Cardoso MJ, Leung K, Modat M et al (2013) STEPS: similarity and truth estimation for propagated segmentations and its application to hippocampal segmentation and brain parcelation. Med Image Anal 17:671–684
27. Ben Ahmed O, Mizotin M, Benois-Pineau J et al (2015) Alzheimer's disease diagnosis on structural MR images using circular harmonic functions descriptors on hippocampus and posterior cingulate cortex. Comput Med Imaging Graph 44:13–25
28. Miller MI, Younes L, Ratnanather JT et al (2015) Amygdalar atrophy in symptomatic Alzheimer's disease based on diffeomorphometry: the BIOCARD cohort. Neurobiol Aging 36:S3–S10
29. Zhang Y, Dong Z, Phillips P et al (2015) Detection of subjects and brain regions related to Alzheimer's disease using 3D MRI scans based on eigenbrain and machine learning. Front Comput Neurosci 9:66
30. Wisse LE, Biessels GJ, Heringa SM et al (2014) Hippocampal subfield volumes at 7T in early Alzheimer's disease and normal aging. Neurobiol Aging 35:2039–2045
31. Joshi A, Ringman JM, Lee AS, Juarez KO, Mendez MF (2012) Comparison of clinical characteristics between familial and non-familial early onset Alzheimer's disease. J Neurol 259:2182–2188
32. Tellechea P, Pujol N, Esteve-Belloch P et al (2015) Early- and late-onset Alzheimer disease: are they the same entity? [Article in English, Spanish] Neurologia. pii: S0213-4853(15)00210-8. doi: https://doi.org/10.1016/j.nrl.2015.08.002
33. Karantzoulis S, Galvin JE (2011) Distinguishing Alzheimer's disease from other major forms of dementia. Expert Rev Neurother 11:1579–1591

34. Murray ME, Graff-Radford NR, Ross OA, Petersen RC, Duara R, Dickson DW (2011) Neuropathologically defined subtypes of Alzheimer's disease with distinct clinical characteristics: a retrospective study. Lancet Neurol 10:785–796
35. La Joie R, Perrotin A, Barré L et al (2012) Region-specific hierarchy between atrophy, hypometabolism, and β-amyloid (Aβ) load in Alzheimer's disease dementia. J Neurosci 32:16265–16273
36. Putcha D, Brickhouse M, O'Keefe K et al (2011) Hippocampal hyperactivation associated with cortical thinning in Alzheimer's disease signature regions in non-demented elderly adults. J Neurosci 31:17680–17688
37. Sperling RA, Dickerson BC, Pihlajamaki M et al (2010) Functional alterations in memory networks in early Alzheimer's disease. NeuroMolecular Med 12:27–43
38. Tahmasian M, Pasquini L, Scherr M et al (2015) The lower hippocampus global connectivity, the higher its local metabolism in Alzheimer disease. Neurology 84:1956–1963
39. McDonald CR, McEvoy LK, Gharapetian L et al (2009) Regional rates of neocortical atrophy from normal aging to early Alzheimer disease. Neurology 73:457–465
40. Im K, Lee JM, Seo SW et al (2008) Variations in cortical thickness with dementia severity in Alzheimer's disease. Neurosci Lett 436:227–231
41. Greicius MD, Srivastava G, Reiss AL, Menon V (2004) Default-mode network activity distinguishes Alzheimer's disease from healthy aging: evidence from functional MRI. Proc Natl Acad Sci U S A 101:4637–4642
42. Sperling RA, Laviolette PS, O'Keefe K et al (2009) Amyloid deposition is associated with impaired default network function in older persons without dementia. Neuron 63:178–188
43. Gispert JD, Rami L, Sánchez-Benavides G et al (2015) Nonlinear cerebral atrophy patterns across the Alzheimer's disease continuum: impact of APOE4 genotype. Neurobiol Aging 36:2687–2701
44. Dubois B, Feldman HH, Jacova C et al (2014) Advancing research diagnostic criteria for Alzheimer's disease: the IWG-2 criteria. Lancet Neurol 13(6):614–629
45. McKhann GM, Knopman DS, Chertkow H et al (2011) The diagnosis of dementia due to Alzheimer's disease: recommendations from the National Institute on Aging-Alzheimer's Association workgroups on diagnostic guidelines for Alzheimer's disease. Alzheimers Dement 7:263–269
46. Albert MS, DeKosky ST, Dickson D et al (2011) The diagnosis of mild cognitive impairment due to Alzheimer's disease: recommendations from the National Institute on Aging-Alzheimer's Association workgroups on diagnostic guidelines for Alzheimer's disease. Alzheimers Dement 7:270–279
47. Peters R (2006) Ageing and the brain. Postgrad Med J 82:84–88
48. Stark AK, Toft MH, Pakkenberg H et al (2007) The effect of age and gender on the volume and size distribution of neocortical neurons. Neuroscience 150:121–130
49. Jacobs B, Schall M, Prather M et al (2001) Regional dendritic and spine variation in human cerebral cortex: a quantitative golgi study. Cereb Cortex 11:558–571
50. Morrison JH, Baxter MG (2012) The ageing cortical synapse: hallmarks and implications for cognitive decline. Nat Rev Neurosci 13:240–250
51. Wignall EL, Dickson JM, Vaughan P et al (2004) Smaller hippocampal volume in patients with recent-onset posttraumatic stress disorder. Biol Psychiatry 56:832–836
52. Labate A, Cerasa A, Gambardella A, Aguglia U, Quattrone A (2008) Hippocampal and thalamic atrophy in mild temporal lobe epilepsy: a VBM study. Neurology 71:1094–1101
53. Bell-McGinty S, Butters MA, Meltzer CC, Greer PJ, Reynolds CF 3rd, Becker JT (2002) Brain morphometric abnormalities in geriatric depression: long-term neurobiological effects of illness duration. Am J Psychiatry 159:1424–1427
54. Moulton CD, Costafreda SG, Horton P, Ismail K, CH F (2015) Meta-analyses of structural regional cerebral effects in type 1 and type 2 diabetes. Brain Imaging Behav 9:651–662
55. Tarroun A, Bonnefoy M, Bouffard-Vercelli J, Gedeon C, Vallee B, Cotton F (2007) Could linear MRI measurements of hippocampus differentiate normal brain aging in elderly persons from Alzheimer disease? Surg Radiol Anat 29:77–81
56. Walhovd KB, Westlye LT, Amlien I et al (2011) Consistent neuroanatomical age-related volume differences across multiple samples. Neurobiol Aging 32:916–932
57. Fjell AM, Westlye LT, Grydeland H et al (2014) Accelerating cortical thinning: unique to dementia or universal in aging? Cereb Cortex 24:919–934

58. Wang L, Benzinger TL, Hassenstab J et al (2015) Spatially distinct atrophy is linked to β-amyloid and tau in preclinical Alzheimer disease. Neurology 84:1254–1260
59. Araque Caballero MÁ, Brendel M, Delker A et al (2015) Alzheimer's disease neuroimaging initative (ADNI). mapping 3-year changes in gray matter and metabolism in Aβ-positive nondemented subjects. Neurobiol Aging 36:2913–2924
60. Whitwell JL, Tosakulwong N, Weigand SD et al (2013) Does amyloid deposition produce a specific atrophic signature in cognitively normal subjects? Neuroimage Clin 2:249–257
61. Rabinowicz T, Dean DE, Petetot JM, de Courten-Myers GM (1999) Gender differences in the human cerebral cortex: more neurons in males; more processes in females. J Child Neurol 14:98–107
62. De Marco M, Venneri A (2015) 'O' blood type is associated with larger grey-matter volumes in the cerebellum. Brain Res Bull 116:1–6
63. O'Dwyer L, Lamberton F, Matura S et al (2012) Reduced hippocampal volume in healthy young ApoE4 carriers: an MRI study. PLoS One 7:e48895
64. Sala-Llonch R, Lladó A, Fortea J et al (2015) Evolving brain structural changes in PSEN1 mutation carriers. Neurobiol Aging 36:1261–1270
65. Allan CL, Zsoldos E, Filippini N et al (2015) Lifetime hypertension as a predictor of brain structure in older adults: cohort study with a 28-year follow-up. Br J Psychiatry 206:308–315
66. Whitlow CT, Sink KM, Divers J et al (2015) Effects of type 2 diabetes on brain structure and cognitive function: African American-Diabetes Heart Study MIND. Am J Neuroradiol 36:1648–1653
67. Stern Y (2009) Cognitive reserve. Neuropsychologia 47:2015–2028
68. Cho H, Jeon S, Kim C et al (2015) Higher education affects accelerated cortical thinning in Alzheimer's disease: a 5-year preliminary longitudinal study. Int Psychogeriatr 27:111–120
69. Gaser C, Schlaug G (2003) Brain structures differ between musicians and non-musicians. J Neurosci 23:9240–9245
70. Maguire EA, Spiers HJ, Good CD, Hartley T, Frackowiak RS, Burgess N (2003) Navigation expertise and the human hippocampus: a structural brain imaging analysis. Hippocampus 13:250–259
71. Lövdén M, Bäckman L, Lindenberger U, Schaefer S, Schmiedek F (2010) A theoretical framework for the study of adult cognitive plasticity. Psychol Bull 136:659–676
72. Takeuchi H, Taki Y, Hashizume H et al (2011) Effects of training of processing speed on neural systems. J Neurosci 31:12139–12148
73. Mesulam MM (1999) Neuroplasticity failure in Alzheimer's disease: bridging the gap between plaques and tangles. Neuron 24:521–529
74. Byun MS, Kim SE, Park J et al (2015) Heterogeneity of regional brain atrophy patterns associated with distinct progression rates in Alzheimer's disease. PLoS One 10:e0142756
75. Masliah E, Hansen LA (2011) Alzheimer disease: AD pathology—emerging subtypes or age-of-onset spectrum? Nat Rev Neurol 8:11–12
76. Ge YL, Grossman RI, Babb JS, Rabin ML, Mannon LJ, Kolson DL (2002) Age-related total gray matter changes in normal adult brain. Part I: volumetric MR imaging analysis. Am J Neuroradiol 23:1327–1333
77. Hasan KM, Mwangi B, Cao B et al (2016) Entorhinal cortex thickness across the human lifespan. J Neuroimaging 26:95–102
78. Jack CR Jr, Petersen RC, Xu YC et al (1997) Medial temporal atrophy on MRI in normal aging and very mild Alzheimer's disease. Neurology 49:786–794
79. Risacher SL, Saykin AJ, West JD et al (2009) Baseline MRI predictors of conversion from MCI to probable AD in the ADNI cohort. Curr Alzheimer Res 6:347–361
80. Spulber G, Niskanen E, Macdonald S et al (2012) Evolution of global and local grey matter atrophy on serial MRI scans during the progression from MCI to AD. Curr Alzheimer Res 9:516–524
81. Venneri A, Gorgoglione G, Toraci C, Nocetti L, Panzetti P, Nichelli P (2011) Combining neuropsychological and structural neuroimaging indicators of conversion to Alzheimer's disease in amnestic mild cognitive impairment. Curr Alzheimer Res 8:789–797
82. Beltrachini L, De Marco M, Taylor ZA, Lotjonen J, Frangi AF, Venneri A (2015) Integration of cognitive tests and resting state fMRI for the individual identification of mild cognitive impairment. Curr Alzheimer Res 12:592–603
83. Westman E, Muehlboeck JS, Simmons A (2012) Combining MRI and CSF measures for classification of Alzheimer's disease and

prediction of mild cognitive impairment conversion. NeuroImage 62:229–238

84. Yun HJ, Kwak K, Lee JM, Alzheimer's Disease Neuroimaging Initiative (2015) Multimodal discrimination of Alzheimer's disease based on regional cortical atrophy and hypometabolism. PLoS One 10:e0129250
85. Mattay VS, Goldberg TE, Sambataro F, Weinberger DR (2008) Neurobiology of cognitive aging: insights from imaging genetics. Biol Psychol 79:9–22
86. Tyler LK, Marslen-Wilson W, Stamatakis EA (2005) Dissociating neuro-cognitive component processes: voxel-based correlational methodology. Neuropsychologia 43:771–778
87. Grossman M, McMillan C, Moore P et al (2004) What's in a name: voxel-based morphometric analyses of MRI and naming difficulty in Alzheimer's disease, frontotemporal dementia and corticobasal degeneration. Brain 127:628–649
88. Pennington C, Hodges JR, Hornberger M (2011) Neural correlates of episodic memory in behavioral variant frontotemporal dementia. J Alzheimers Dis 24:261–268
89. Nho K, Risacher SL, Crane PK et al (2012) Voxel and surface-based topography of memory and executive deficits in mild cognitive impairment and Alzheimer's disease. Brain Imaging Behav 6:551–567
90. Rodríguez-Ferreiro J, Cuetos F, Monsalve A, Martínez C, Perez AJ, Venneri A (2012) Establishing the relationship between cortical atrophy and semantic deficits in Alzheimer's disease and mild cognitive impairment patients through voxel-based morphometry. J Neuroling 25:139–149
91. Belleville S, Clément F, Mellah S, Gilbert B, Fontaine F, Gauthier S (2011) Training-related brain plasticity in subjects at risk of developing Alzheimer's disease. Brain 134:1623–1634
92. Venneri A, McGeown WJ, Hietanen HM, Guerrini C, Ellis AW, Shanks MF (2008) The anatomical bases of semantic retrieval deficits in early Alzheimer's disease. Neuropsychologia 46:497–510
93. Thomann PA, Toro P, Dos Santos V, Essig M, Schröder J (2008) Clock drawing performance and brain morphology in mild cognitive impairment and Alzheimer's disease. Brain Cogn 67:88–93
94. Bruen PD, McGeown WJ, Shanks MF, Venneri A (2008) Neuroanatomical correlates of neuropsychiatric symptoms in Alzheimer's disease. Brain 131:2455–2463
95. Scahill RI, Ridgway GR, Bartlett JW et al (2013) Genetic influences on atrophy patterns in familial Alzheimer's disease: a comparison of APP and PSEN1 mutations. J Alzheimers Dis 35:199–212
96. Mahley RW, Weisgraber KH, Huang Y (2006) Apolipoprotein E4: a causative factor and therapeutic target in neuropathology, including Alzheimer's disease. Proc Natl Acad Sci U S A 103:5644–5651
97. Filippini N, Rao A, Wetten S et al (2009) Anatomically-distinct genetic associations of APOE epsilon4 allele load with regional cortical atrophy in Alzheimer's disease. NeuroImage 44:724–728
98. Pievani M, Galluzzi S, Thompson PM, Rasser PE, Bonetti M, Frisoni GB (2011) APOE4 is associated with greater atrophy of the hippocampal formation in Alzheimer's disease. NeuroImage 55:909–919
99. Biffi A, Anderson CD, Desikan RS et al (2010) Genetic variation and neuroimaging measures in Alzheimer disease. Arch Neurol 67:677–685
100. Lambert JC, Ibrahim-Verbaas CA et al (2013) Meta-analysis of 74,046 individuals identifies 11 new susceptibility loci for Alzheimer's disease. Nat Genet 45:1452–1458
101. Hibar DP, Stein JL, Kohannim O et al (2011) Voxelwise gene-wide association study (vGeneWAS): multivariate gene-based association testing in 731 elderly subjects. NeuroImage 56:1875–1891
102. Hibar DP, Stein JL, Jahanshad N et al (2015) Genome-wide interaction analysis reveals replicated epistatic effects on brain structure. Neurobiol Aging 36:S151–S158
103. Di Marco LY, Venneri A, Farkas E, Evans PC, Marzo A, Frangi AF (2015) Vascular dysfunction in the pathogenesis of Alzheimer's disease—a review of endothelium-mediated mechanisms and ensuing vicious circles. Neurobiol Dis 82:593–606
104. Wang R, Fratiglioni L, Laveskog A et al (2014) Do cardiovascular risk factors explain the link between white matter hyperintensities and brain volumes in old age? A population-based study. Eur J Neurol 21:1076–1082
105. Steffener J, Stern Y (2012) Exploring the neural basis of cognitive reserve in aging. Biochim Biophys Acta 1822:467–473
106. Wang L, Harms MP, Staggs JM et al (2010) Donepezil treatment and changes in hippocampal structure in very mild Alzheimer disease. Arch Neurol 67:99–106

107. Krishnan KR, Charles HC, Doraiswamy PM et al (2003) Randomized, placebo-controlled trial of the effects of donepezil on neuronal markers and hippocampal volumes in Alzheimer's disease. Am J Psychiatry 160:2003–2011
108. Hashimoto M, Kazui H, Matsumoto K, Nakano Y, Yasuda M, Mori E (2005) Does donepezil treatment slow the progression of hippocampal atrophy in patients with Alzheimer's disease? Am J Psychiatry 162:676–682
109. Dubois B, Chupin M, Hampel H et al (2015) Donepezil decreases annual rate of hippocampal atrophy in suspected prodromal Alzheimer's disease. Alzheimers Dement 11:1041–1049
110. Jack CR Jr, Petersen RC, Grundman M et al (2008) Longitudinal MRI findings from the vitamin E and donepezil treatment study for MCI. Neurobiol Aging 29:1285–1295
111. Schuff N, Suhy J, Goldman R et al (2011) An MRI substudy of a donepezil clinical trial in mild cognitive impairment. Neurobiol Aging 32:2318.e31–2318.e41
112. Tanaka Y, Hanyu H, Sakurai H, Takasaki M, Abe K (2003) Atrophy of the substantia innominata on magnetic resonance imaging predicts response to donepezil treatment in Alzheimer's disease patients. Dement Geriatr Cogn Disord 16:119–125
113. Kanetaka H, Hanyu H, Hirao K et al (2008) Prediction of response to donepezil in Alzheimer's disease: combined MRI analysis of the substantia innominata and SPECT measurement of cerebral perfusion. Nucl Med Commun 29:568–573
114. Bottini G, Berlingeri M, Basilico S et al (2012) GOOD or BAD responder? Behavioural and neuroanatomical markers of clinical response to donepezil in dementia. Behav Neurol 25:61–72
115. Csernansky JG, Wang L, Miller JP, Galvin JE, Morris JC (2005) Neuroanatomical predictors of response to donepezil therapy in patients with dementia. Arch Neurol 62:1718–1722
116. Prins ND, van der Flier WA, Knol DL et al (2014) The effect of galantamine on brain atrophy rate in subjects with mild cognitive impairment is modified by apolipoprotein E genotype: post-hoc analysis of data from a randomized controlled trial. Alzheimers Res Ther 6:47
117. Venneri A, McGeown WJ, Shanks MF (2005) Empirical evidence of neuroprotection by dual cholinesterase inhibition in Alzheimer's disease. Neuroreport 16:107–110
118. Weiner MW, Sadowsky C, Saxton J et al (2011) Magnetic resonance imaging and neuropsychological results from a trial of memantine in Alzheimer's disease. Alzheimers Dement 7:425–435
119. Fox NC, Black RS, Gilman S et al (2005) AN1792(QS-21)-201 Study. Effects of Abeta immunization (AN1792) on MRI measures of cerebral volume in Alzheimer disease. Neurology 64:1563–1572
120. Kile S, Au W, Parise C et al (2017) IVIG treatment of mild cognitive impairment due to Alzheimer's disease: a randomised double-blinded exploratory study of the effect on brain atrophy, cognition and conversion to dementia. J Neurol Neurosurg Psychiatry 88 (2):106–112. [Epub ahead of print]
121. Ferreira D, Westman E, Eyjolfsdottir H et al (2015) Brain changes in Alzheimer's disease patients with implanted encapsulated cells releasing nerve growth factor. J Alzheimers Dis 43:1059–1072
122. De Marco M, Shanks MF, Venneri A (2014) Cognitive stimulation: the evidence base for its application in neurodegenerative disease. Curr Alzheimer Res 11:469–483
123. Erickson KI, Weinstein AM, Lopez OL (2012) Physical activity, brain plasticity, and Alzheimer's disease. Arch Med Res 43:615–621
124. Hafkemeijer A, Möller C, Dopper EG et al (2016) Differences in structural covariance brain networks between behavioral variant frontotemporal dementia and Alzheimer's disease. Hum Brain Mapp 37:978–988

Chapter 15

Structural MRI in Neurodegenerative Non-Alzheimer's Dementia

Margherita Di Paola and Ali K. Bourisly

Abstract

Neurodegenerative non-Alzheimer's dementias (NADs) are an increasingly relevant problem on the global scale, due to the progressive aging of the world's population. Diagnosis of NADs remains a challenge due to the overlap of clinical and neuroimaging features of different NADs. In this chapter, after showing the conventional magnetic resonance imaging (MRI) finding, the pathologies of different NADs along with studies that have applied the advanced MRI neuroimaging techniques to study different NADs will be presented. The NADs discussed here, include Huntington's Disease, Progressive Supranuclear Palsy, Multiple System Atrophy, CorticoBasal Dementia, Parkinson's Disease with Dementia, and Dementia with Lewy bodies.

Key words Huntington's Disease (HD), Progressive Supranuclear Palsy (PSP), Multiple System Atrophy (MSA), CorticoBasal Dementia (CBD), Parkinson's Disease with Dementia (PDD), Dementia with Lewy bodies (DLBs), Structural MRI, Non-Alzheimer's dementia

Abbreviations

ADC	Apparent Diffusion Coefficient
CBD	CorticoBasal Degeneration
DTI	Diffusion Tensor Imaging
FA	Fractional Anisotropy
FAt	corrected Fractional Anisotropy value
HD	Huntington's Disease
MD	Mean Diffusivity
MRI	Magnetic Resonance Imaging
MRPI	Magnetic Resonance Parkinsonism Index
MSA	Multiple System Atrophy
NADs	Non-Alzheimer's Dementias
PDD	Parkinson's Disease with Dementia
PD	Parkinson's Disease
pre-HD	presymptomatic HD
PSP	Progressive Supranuclear Palsy
RD	Radial Diffusivity

Gianfranco Spalletta et al. (eds.), *Brain Morphometry*, Neuromethods, vol. 136,
https://doi.org/10.1007/978-1-4939-7647-8_15, © Springer Science+Business Media, LLC 2018

ROI	Region Of Interest
rADC	regional Apparent Diffusion Coefficient
TBSS	Tract-Based Spatial Statistic
VBM	Voxel-Based Morphometry

1 Introduction

Neurodegenerative non-Alzheimer's dementias (NADs) include different disease types, such as Huntington's Disease, Progressive Supranuclear Palsy, Multiple System Atrophy, CorticoBasal Dementia, Parkinson's Disease with Dementia, and Dementia with Lewy Bodies. From years of research studies and clinical practice, we currently know that such NADs can be clinically indistinguishable, particularly in the very early stages, and that brain pathologic changes are present long before the development of functional impairment.

Due to the nature of the aforementioned NADs diseases, neuroimaging plays an increasing role in their diagnosis and prognosis, suggesting possible differential diagnosis as well as tracking their evolution. However, this role remains a challenge due to the considerable clinical overlap among these diseases in the early and the presymptomatic stages. Therefore, neuroimaging has yet to offer definitive diagnosis for the NADs.

Vast and different research efforts have been allocated toward applying different multimodal neuroimaging approaches that combine conventional neuroimaging with, relatively, newer neuroimaging biomarker techniques, in an attempt to individuate the unique brain signature of each NADs. Currently, there are a number of techniques that are available for studying the changes associated with neurodegenerative diseases. These techniques include gray and white matter metrics, neurotransmitter function, task-related synaptic activity, and chemical signature. However, most of these techniques are not clinically implemented yet.

In this chapter, the focus will be on advanced magnetic resonance imaging (MRI) techniques that have been utilized to study Huntington's Disease, Progressive Supranuclear Palsy, Multiple System Atrophy, CorticoBasal Dementia, Parkinson's Disease with Dementia, and Dementia with Lewy Bodies. The focus of this chapter will also address how such imaging techniques relate to the pathological core of the respective NADs disease.

2 Huntington's Disease

Huntington's Disease (HD), also known as Huntington's chorea, is a progressive neurodegenerative disorder caused by a trinucleotide (CAG) repeat expansion in the huntingtin gene, with a strong inheritance correlations [1]. HD is characterized by abnormal choreiform movements (involuntary movements consisting of non-repetitive and non-periodic jerking of the face, trunk, or limbs).

The mean age range of onset of HD is 35–44 years of age with a median survival time of 15–18 years after onset [2]. Psychological and psychiatric symptoms of the disease include early onset of dementia, affective psychosis, and development of significant changes in personality. On the behavioral level, HD can result in various disturbances that can include apathy, aggression, sexual dysfunction, and increased appetite [2].

Progressive and global decline in cognitive abilities can also occur in individuals diagnosed with HD. Such a decline can include impairment in visuospatial abilities, forgetfulness, slowing of thought processes, and decline in the ability to acquire and manipulate knowledge. The overall cognitive and behavioral syndrome in individuals with HD is more similar to frontotemporal dementia than to Alzheimer's Disease [3].

2.1 Pathology Core

The most vulnerable brain structure in HD is the striatum (caudate nucleus and putamen). The pathological changes in HD include inflammation, neuronal loss, and nuclear inclusions of the protein huntingtin, as well as aberrant iron accumulation [4].

2.2 Conventional MRI

Using conventional MRI techniques (e.g., T1 weighted), caudate head atrophy can be found. This atrophy results in the enlargement of the frontal horns giving them a box-like configuration. Quantification of such a method is also possible by utilizing a number of measurement which include frontal horn width to the intercaudate distance ratio (FH/CC) and the intercaudate distance to inner table width ratio (CC/TT) [5] (Fig. 1a).

2.3 Advanced MRI

The most evident neuropathological changes in HD take place in the basal ganglia [6], and thus a plethora of MRI studies have focused on this brain region. By applying both automatic and manual region of interest (ROI) delineation, progressive reduction in the striatal volume can be found in both presymptomatic HD (pre-HD) and symptomatic HD patients and even up to 20 years before any clinical HD symptoms appear [7–9]. At later stages of the disease, atrophy has been shown to expand to the nucleus accumbens and globus pallidus, while greater atrophy is associated with the putamen compared to the caudate [7, 8, 10]. Such striatal atrophy showed significant correlation with impaired cognitive and motor functions [8, 10]. Moreover striatal baseline volume [11] and atrophy rate [9] can predict conversion of disease state, and there exists a general volume reductions in the basal ganglia, though such reductions are not notable as early as reductions in the caudate and putamen [10, 12–15].

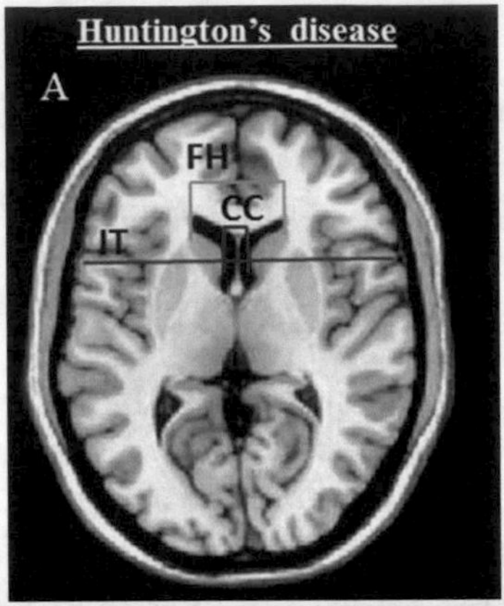

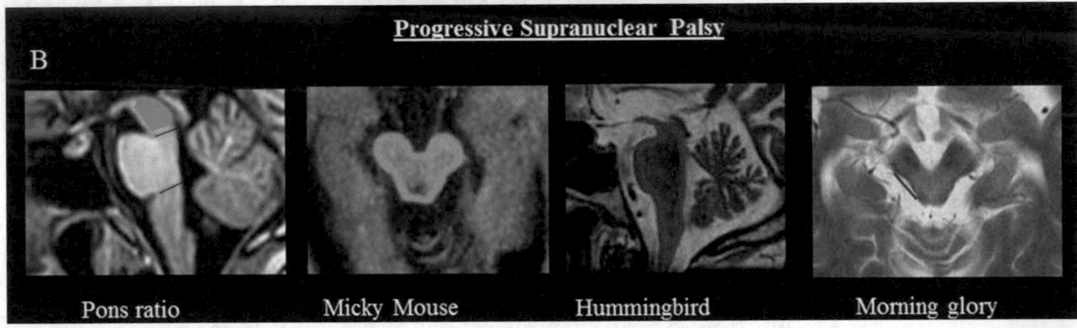

Fig. 1 Conventional MRI sign. The figure shows the main pathological MRI signals in Huntington's disease and in progressive supranuclear palsy. Panel (**a**) shows Huntington's disease, frontal horn width (FH), intercaudate distance (CC) ratio, and inner table width ratio (IT); panel (**b**) shows progressive supranuclear palsy pons ratio measures, mickey mouse, hummingbird, and morning glory signs

More recent researches suggest that there is also a widespread cortical involvement in the degenerative process of HD. Gray matter loss [9, 16], whole-brain atrophy [8, 9, 17–19], and expansion of the cerebrospinal fluid [8, 9, 17, 20] have proved to be sensitive to disease stage and disease-related changes in early HD. However, the sensitivity of such quantitative measures is not very well established for pre-HD stages. Even with that, there is now some considerable amount of evidence that shows that white matter atrophy is evident in both pre-HD [8, 17, 21, 22] and HD [23, 19, 24]. Atrophy, also, has been shown to expand to the corpus callosum, white matter close to the striatum, and frontal lobe white matter [7, 25–28]. In agreement with white matter atrophy, diffusion tensor imaging (DTI) studies have shown reduced white matter integrity. Areas involved are in the frontal lobe, precentral gyrus, postcentral gyrus, corpus callosum, anterior and posterior limbs of the internal capsule, putamen, and globus pallidus in patients with pre-HD and manifest HD [27–29].

Recently Phillips et al. found, in pre-HD compared to controls, a general increase in axial diffusivity and radial diffusivity (RD) of the superficial white matter which is composed by late-myelinating intracortical and short-range association fibers across occipital and parietal lobes as well as the superior temporal lobes. Such an increased diffusivity in the superficial white matter covered the whole brain in HD and correlated with disease burden and genotype (CAG repeat number) [30].

As for cortical thickness, cortical thinning was notable for both HD [23, 31] and pre-HD [23, 32, 33].

The cortical thickness is affected early during the course of the disease and does seem to show topographical selectivity, proceeding from posterior to anterior cortical regions during progression of the disease [31, 34].

2.4 Progressive Supranuclear Palsy

Progressive Supranuclear Palsy (PSP), also known as the Steele-Richardson-Olszewski syndrome [35] is one of the most common forms of atypical Parkinsonism disorders and is one of the so-called tauopathy subsets.[1] PSP is characterized by sporadic and progressive neurodegeneration with an average onset age of 40. It presents with supranuclear vertical gaze palsy, levodopa-unresponsive parkinsonism, prominent postural instability with falls, and cognitive disturbances. More recently, several other phenotypic variants of PSP have been described. These variants vary in pattern and severity of atrophy and clinical features. Examples of PSP phenotypic variants include PSP-PAGF (pure akinesia with gait freezing), PSP-PNFA (PSP with progressive nonfluent aphasia), PSP-C (PSP-cerebellar), and PSP-CBS (PSP-corticobasal syndrome) [35–38]. Common cognitive manifestations are early executive dysfunction, speech disturbances, decreased cognition, emotional lability, apathy, depression, apathy, disinhibition, dysphoria, and anxiety [37, 38].

2.5 Pathology Core

Pathologically PSP is characterized by intracellular somatodendritic aggregates of the microtubule associated tau protein (in the form of neurofibrillary tangles in neurons), neuropil threads, coiled bodies in oligodendrocytes, and tufted astrocytes in the midbrain, basal ganglia, superior cerebellar peduncle, and motor cortex [39, 40].

2.6 Conventional MRI

Midbrain-to-pons area ratio is a measure that can help to predict the presence of PSP, by conventional MRI techniques. Its use (on midline sagittal images) was found to be useful in confirming a small midbrain in a PSP setting [41]; such a reduction in midbrain size results in what is called the hummingbird sign. Moreover, Oba et al. described a method in which the area of both the midbrain and pons are calculated in order to obtain the midbrain-to-pons

[1] Tauopathies are a heterogeneous group of neurodegenerative diseases characterized by abnormal metabolism of tau proteins leading to intracellular accumulation and formation of neurofibrillary tangles (NFT). These neurofibrillary tangles are deposited in the cytosol of neurons and glial cells.

ratio (a normal value for the ratio is ~0.24 and a reduced value is ~0.12) [41] (Fig. 1b).

Another MRI parameter is the magnetic resonance parkinsonism index (MRPI) in patients with clinically unclassifiable parkinsonism. MRPI is calculated using [(area of pons in midsagittal plane)/(area of midbrain in midsagittal plane)] × [(width of the middle cerebellar peduncle)/(width of the superior cerebellar peduncle)] [42, 43]. If this formula results in a value above ~13.6, then the results are abnormal [43].

Other conventional MRI signs of PSP include those of midbrain atrophy: the mickey mouse appearance results from the reduced anteroposterior midbrain diameter (~<12 mm) [38, 44], the hummingbird sign results from the flattening of the superior aspect of the midbrain [45], and the morning glory (a flower name) sign results from the loss of the lateral convex margin of the tegmentum of the midbrain [46] (Fig. 1b).

2.7 Advanced MRI

In patients with PSP, significant volumetric reductions were measured in the frontal lobes, striatum, midbrain, and thalamus using atrophy measurements of predefined ROIs and MRI volumetry [47–51]. Along with the meta-analysis studies [52–55], other studies [56–61] identified significant regional gray matter reductions in the following regions: thalamus, basal ganglia, midbrain, insular cortex, frontal cortex, and pons.

Canu et al. reported gray matter atrophy of the basal ganglia as well as of the midbrain and pons, and several motor and extramotor regions, in five PSP patients [62]. Also, in a voxel-based morphometry (VBM) study, Piattella et al. found gray matter volume reduction that involved the thalamus, putamen, pallidum, nucleus accumbens, frontal cortex, cerebellum, and brainstem [63].

Padovani et al., using a DTI ROI-based approach, described reduced fractional anisotropy (FA) in the left arcuate fasciculus, the anterior corpus callosum, the posterior thalamic radiation, the internal capsule, and the superior longitudinal fascicle, in 14 patients with mild PSP [59]. Using DTI, Wang et al. reported an increase in the mean, axial, and radial diffusivity in the putamen midbrain in 17 PSP patients [64]. Furthermore, Tsukamoto et al. found increased regional apparent diffusion coefficient[2] (rADC) values in the midbrain, caudate nucleus, globus pallidus, and superior cerebellar peduncle in 20 PSP patients [66].

[2] Apparent diffusion coefficient (ADC) is a measure of the magnitude of diffusion (of water molecules) within tissue and is commonly clinically calculated using MRI with diffusion weighted imaging (DWI) [65].

In a recent study [67], an ROI-based bi-tensor model[3] was used to examine free water and FA in the following brain areas: midbrain, basal ganglia, thalamus, corpus callosum, and cerebellum in patients with different neurodegenerative diseases, including PSP. Compared with controls, PSP had increased FA in the caudate nucleus, putamen, thalamus, vermis, and reduced corpus callosum and superior cerebellar peduncle. On the same note, the study found that PSP showed striking reduction in the superior peduncle, and this finding may have potential to differentiate PSP from other neurodegenerative diseases [67].

In a study [69] that included 18 PSP patients, tract degeneration was found in both the superior longitudinal fasciculus and the dentatorubrothalamic tract. Another study [70] found changes in FA and mean diffusivity (MD) DTI indices in PSP patients compared to controls. A further study on 18 patients with PSP reported white matter changes occurring at the level of the corpus callosum, midbrain, fornix, inferior fronto-occipital fasciculus, anterior thalamic radiation, superior cerebellar peduncle, uncinate fasciculus, superior longitudinal fasciculus, cingulate gyrus, and corticospinal tract [71].

As for tract-based spatial statistics (TBSS) DTI, authors [72] found that PSP patients had altered DTI measures at the level of the superior cerebellar peduncle, cerebellar white matter, vermis of the cerebellum, fornix, body of the corpus callosum, and olfactory regions. Selective evidence of TBSS showed abnormal diffusivity at the level of superior cerebellar peduncle in three PSP patients [73]. In a TBSS-DTI study [63], white matter abnormalities in 16 PSP patients were reported. This study showed widespread changes in white matter bundles, which primarily affected the thalamic radiations, cerebellar peduncles, corticospinal tracts, corpus callosum, and longitudinal fasciculi. Furthermore, Surova et al. [51] investigated disease specific changes in eight PSP patients using TBSS; the authors observed an increase in MD, in PSP compared to controls, in the following brain regions: midbrain, superior cerebellar, and thalamus. Similar findings were found in other studies [69, 72, 74–76].

Comparing PSP with other neurodegenerative diseases as Multiple System Atrophy (MSA)—Parkinson type (MSA-P) a study [77] found that compared to MSA-P, PSP showed higher iron concentrations in the thalamus, red nucleus, substantia nigra, and globus pallidus. Meanwhile both the thalamus and the globus pallidus were found to be the most valuable nuclei in differentiating PSP from MSA-P. A voxel-based analysis of iron content in the

[3] Bi-tensor model separates the diffusion properties of water in brain tissue from those of water in extracellular space [68]. Thus, the major contribution of bi-tensor model compared with the well-known and largely used single tensor model is the former allows to control for partial volume effects with extracellular free water when quantifying the fractional anisotropy metric and thus provide a corrected fractional anisotropy value (FAt).

same participants showed hypodense signals in the anterior and medial aspects of the globus pallidus and thalamus of PSP patients [77].

As for cortical thickness studies on PSP, only one study [78] was found, and it showed widespread cortical thinning and volumetric loss within the frontal lobe, particularly the superior frontal gyrus, increased surface area in the pericalcarine.

3 Corticobasal Degeneration

The first description of CorticoBasal Degeneration (CBD) has been attributed to a paper titled *Corticodentatonigral Degeneration with Neuronal Achromasia*[4] by Rebeiz et al. in 1968. Since that description, many CBD cases have been found, but different terms were used to describe these findings (i.e., corticonigral degeneration with nuclear achromasia, cortical basal ganglionic degeneration, and others). In 1989 all such terms have been replaced by CBD [79]. Asymmetric motor deficit characterized by rigidity, akinesia, involuntary movements, and apraxia (alien limb phenomenon) is usually characteristic to CBD [80]. Dementia and aphasia are also frequently observed in CBD [81, 82]. CBD is regarded as a rare neurodegenerative disorder, with unknown true prevalence and usually presents in the age range of 60–80 [83].

3.1 Pathology Core

The characteristic histopathological findings of CBD are neuronal loss and numerous swollen achromatic neurons [84]. Although these features are seen throughout the brain, certain brain regions (including the frontoparietal cortex) and subcortical structures (striatum and substantia nigra) are more severely affected than others. Atrophy may be asymmetric, and usually the occipital and temporal lobes are spared [85].

3.2 Conventional MRI

One of the most consistent features of CBD at T1-weighted MR image is asymmetric cortical atrophy of the superior parietal lobule and superior frontal gyri together with atrophy of basal ganglia [84]. Using a database from Mayo Clinic Hospital, Boeve et al. performed a study on 13 cases of CBD. Based on the results, the authors gathered that clinical features of CBD are associated with heterogeneous pathologies. The authors also concluded that CBD can occur despite the absence of degeneration in the basal ganglia and substantia nigra, while asymmetric parieto-frontal degeneration being a common feature of CBD [86].

A T2-weighted MR image of a CBD patient has been found to show hyperintensities of white matter [87]. In the case report by

[4] Achromasia of cells or tissues is the loss of the usual reaction to stains.

Doi T et al., the T2-weighted image revealed asymmetric hyperintensities of the frontal white matter in the fourth year of the disease and focal atrophy of the bilateral precentral gyrus in the fifth year of the disease [87]. In another study by Huang et al. on CBD patients, it was found that conventional MRI, T1- and T2-weighted scans, showed mild to moderate cortical atrophy in the right hemisphere [88] (contralateral to the more severely affected limb).

3.3 Advanced MRI

Boxer et al., using VBM, found that CBD shows distinct brain atrophy patterns. They showed that CBD patients had asymmetric patterns of brain atrophy that included the striatum, premotor cortex, and superior parietal lobules [57]. Further MRI studies [57, 89, 90] confirmed frontoparietal gray matter atrophy to occur in CBD. Moreover, frontal and parietal region dysfunction has been presumed to account for apraxia, which is a typical feature of the disease [91]. Also, other studies [57, 92] have shown putamen and pallidum atrophy in CBD. A more recent VBM meta-analysis [52] on 165 patients identified a characteristic pattern of atrophy in the superior parietal lobule. A study by Erbetta et al. found similar asymmetry using DTI [93].

4 Multiple System Atrophy

Multiple System Atrophy (MSA) is a sporadic, progressive neurodegenerative disease. Although the etiology remains unclear, this disease of the nervous system appears to result from α-synuclein. In fact, along with Parkinson's disease (PD) and Dementia with Lewy bodies (DLBs), MSA is an α-synucleinopathy[5] [95]. MSA is characterized by cerebellar ataxia and autonomic (particularly urogenital) dysfunction. Indeed among the usual first symptoms of the disease are bladder dysfunction (in both males and females) followed by erectile dysfunction (in males) [96]. Onset of the disease is usually >30 years. Cognitive manifestations of MSA include constructional function, visuospatial, executive function, and verbal fluency deficits [97]. Dementia is uncommon in MSA, and if present, it mainly occurs in MSA-P patients (see below) [98, 99].

There are two types of MSA according to predominating symptoms: MSA-C has predominance of cerebellar symptoms (i.e., olivopontocerebellar atrophy), and MSA-P has predominance of parkinsonian signs (i.e., striatonigral degeneration) [96]. The MSA-C and MSA-P designations refer to the predominant feature at the time of the evaluation of the patient, which can change with time [96].

[5] Synucleinopathies (also called α-synucleinopathies) are neurodegenerative diseases characterized by the abnormal accumulation of aggregates of alpha-synuclein protein in neurons, nerve fibers, or glial cells [94].

4.1 Pathology Core

Neuropathology of MSA consists of gliosis, demyelination, and neuronal loss of primarily the striatonigral and/or olivopontocerebellar structures [100].

4.2 Conventional MRI

The most common MRI signal is visible in T2-weighted MRI scan. It results in images with T2 hyperintensities in the form of a cross on the axial images through the pons (hot cross bun sign) (Fig. 2a). This sign represents selective degeneration of the pontocerebellar tracts. This hot cross bun sign is very much common in MSA-C. Other MRI signals that also represent selective degeneration of the pontocerebellar tracts are T2 hyperintensities at the level of the cerebellar peduncles and cerebellum.

As from MSA-P, the most common MRI signal is that of volumetric reduction of the putamen. Using a gradient recalled echo (GRE) MRI pulse sequence, the T2 signal shows an

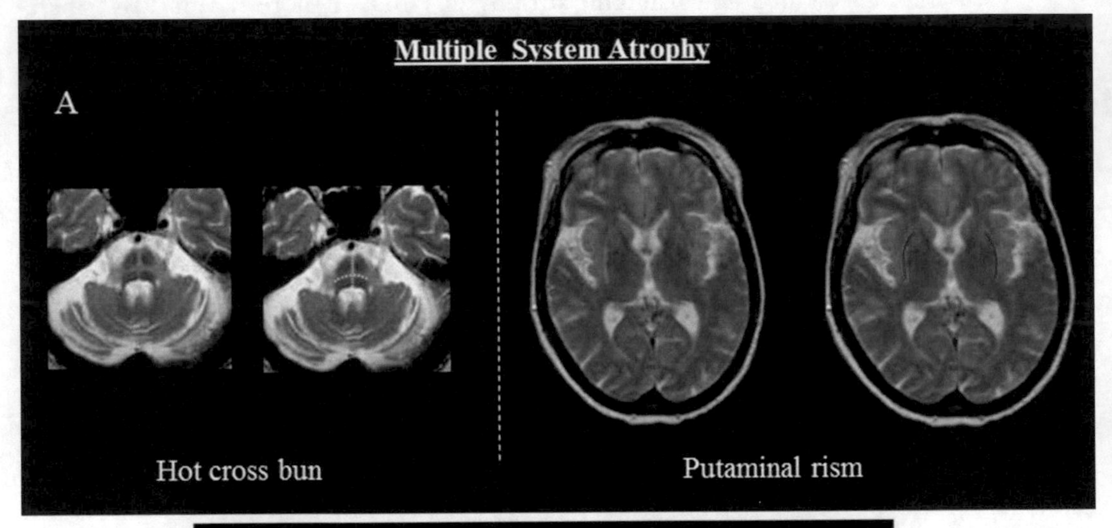

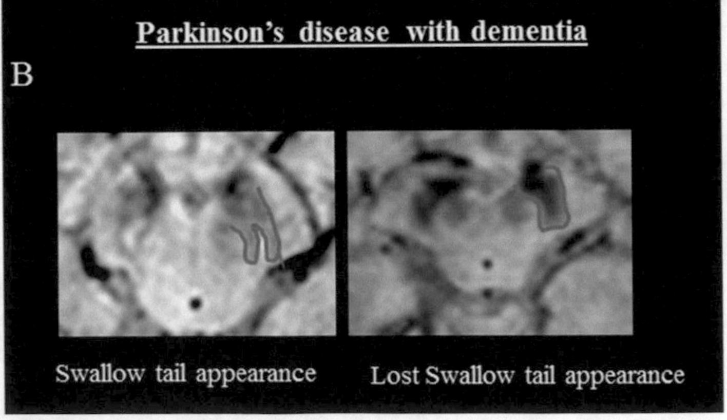

Fig. 2 Conventional MRI sign. The figure shows the main pathological MRI signals in multiple system atrophy and Parkinson's disease with dementia. Panel (**a**) shows multiple system atrophy, hot cross bun, and putaminal rim signs; panel (**b**) shows Parkinson's disease with dementia swallow tail appearance sign

abnormally hyperintense rim surrounding the lateral margin of the putamen (putaminal rim sign) [101, 102] (Fig. 2a).

A recent study [103] that used high-field MRI (7 T) have shown promising results in evaluating the hyperintensities of the substantia nigra in PD, MSA with predominant parkinsonism, and PSP. Compared to controls, the nigral hyperintensity was not visible in all the respective group patients. This finding suggests that imaging of parkinsonism may benefit from using high-field MRI (i.e., 7 T).

4.3 Advanced MRI

In a study [95] using semiautomatic ROI segmentation, volumetric reductions in the thalamus, putamen, pallidum, hippocampus, brainstem, and cerebellum in MSA-P compared to PD patients and controls was found. Moreover, only few VBM studies investigated gray matter loss. A study on MSA-P revealed reduced volume in the middle and inferior cerebellar peduncles, the basal ganglia, the pons, and throughout the cerebrum compared to controls and PD [104]. MSA-P also showed decreased gray matter in the caudate nuclei, putamen, thalami, anterior cerebellar lobes, and the cerebral cortex in general, along with white matter atrophy in the midbrain, pons, and peduncles [105]. In a VBM meta-analysis, Yu et al. identified distinctive regions of atrophy for MSA-P that included both the putamen and claustrum [52].

In a DTI ROI-based approach retrospective study [66], rADC values for MSA patients were significantly higher in the middle cerebellar peduncle, pons, cerebellar white matter, and cerebellar dentate nucleus compared to PSP, PD, and controls. In addition, rADC values in the posterior putamen were significantly higher in MSA than in PSP and controls. Furthermore, a study [106] that combined conventional imaging and DTI ROI-based approach to differentiate 26 PD and 13 MSA patients found the mean middle cerebellar peduncles width, anteroposterior diameter of the pons, and the FA of the middle cerebellar peduncles to be the most useful parameters that are able to distinguish between MSA and PD. In another study [107] that also combined conventional MRI and DTI for studying patients with MSA-C and controls, DTI identified microstructural abnormalities in pontine longitudinal and transverse fibers in patients that showed no abnormalities on conventional MRI images.

Researchers [67] also used a bi-tensor model[6] to examine free water and FAt in MSA patients and found that, compared to controls, MSA had increased FAt in putamen and caudate nucleus. As for single tensor models, generally speaking, MSA was found to have lower FA values and higher axial diffusivity, higher RD, higher MD, higher apparent diffusion coefficient in the middle cerebellar

[6] See note number 3.

peduncle, lower FA, higher apparent diffusion coefficient, and higher trace (D) values in the putamen [66, 108–116].

Higher ADC values in MSA-P compared to PD were found in the superior parts of the corona radiata, lateral periputaminal white matter, and the anterior limb of the inner capsule. The localization of anatomical changes seems to suggest an involvement of pyramidal motor pathways in MSA-P [110]. In a TBSS study of 11 probable MSA-P and five MSA-C [70], the authors found that MSA patients compared to controls showed increased MD in the corticospinal tract, medial lemniscus, and middle and inferior cerebellar peduncles. Similar abnormal diffusivity patterns were also found in the bilateral corticospinal tract and right anterior thalamic radiation in patients with MSA-C and MSA-P compared to controls. Direct comparison of the two MSA variants showed higher axial diffusivity values of superior longitudinal fasciculus in MSA-P than in MSA-C. These results may explain why functional deterioration in MSA-P is faster compared to MSA-C [117].

In a DTI study [118] of 41 MSA-C patients and 15 cortical cerebellar atrophy (CCA) patients, FA values of both cerebellar afferent tracts (olivocerebellar tract and pontocerebellar tracts) showed significant reduction in MSA-C compared to CCA. No differences in cerebello-rubro tract (efferent) were found. In patients with moderate to advanced MSA, a study [73] found middle cerebellar peduncle degeneration along with pontocerebellar tract degeneration and decreased FA and increased ADC in these patients. Another study [119] found that FA of pontine transverse fibers in MSA patients decreased with the development of the cross sign and that the pontine transverse fibers degenerate as the cross sign develops, and the degeneration of the pontine longitudinal fibers begin, or even accelerate, when the cross sign is apparent.

As for cortical thickness, a study [120] has identified different topographic distributions of cortical thinning patterns in non-demented MSA subtypes. In 53 probable MSA patients (29 MSA-C, 24 MSA-P) and 35 controls, the authors of this study found clusters (i.e., lingual gyrus, left ventromedial prefrontal, prefrontal cortex, and right parahippocampal) exhibiting significant cortical thinning in MSA-C and other clusters (right primary sensory motor and left ventromedial prefrontal cortex) exhibiting a thinning tendency in MSA-P compared with the control group. In another study [121], the authors tried to distinguish cortical thickness changes in MSA with and without dementia. They found cortical thickness decreased in the precuneus/cuneus, uncus, and posterior cingulate in MSA with dementia compared to the controls and in parahippocampal and lingual cortices compared to MSA without dementia. Conversely, a study on cortical difference on 18 probable MSA patients (11 categorized as MSA-P and 7 as

MSA-C) concluded that MSA did not display significant changes in cortical morphology [78].

As for iron deposits, a study [77] found that, compared to controls, MSA-P showed significantly higher levels of iron deposition in all the subregions investigated: the substantia nigra, red nucleus, head of the caudate nucleus, putamen, globus pallidus, and thalamus. Compared with PSP, MSA-P showed higher iron levels in the putamen. The area under curve (AUC) indicated that the putamen was the most valuable nucleus in differentiating between MSA-P and PSP.

5 Parkinson's Disease with Dementia

Parkinson's Disease (PD), also known as idiopathic parkinsonism, is one of the most common neurodegenerative diseases (a synucleinopathy[7]). PD presents with motor symptom movement disorder characterized by resting tremor, rigidity, bradykinesia, and postural instability. Dementia can be a late feature. When present it is known as Parkinson disease with later developing dementia. Moreover, the prevalence of Parkinson's Disease with Dementia (PDD) has been reported to be at least 75% among PD patients [122].

5.1 Pathology Core

The central pathological implication of PD has been shown to be intracellular accumulation of α-synuclein [123].Whereby these α-synuclein accumulations result in the formation of characteristic inclusions called Lewy bodies (LBs) [123]. It is worth noting that although PDD and Dementia with Lewy bodies exhibit very similar brain pathology, more recent studies have shown extensive deposition of both Aβ and α-synuclein in the hippocampus and striatum for Dementia with Lewy bodies compared to only α-synuclein in PDD [124, 125]. There has been a more recent consortium criteria for classifying Lewy body diseases (LBD). Moreover, a relatively new consortium criteria for classifying Lewy body diseases suggest that the clinical presentations in Dementia with Lewy bodies patients is of dementia followed by parkinsonism, while in PDD patients, it is of parkinsonism followed by dementia [60]. The main brain region affected by PDD is the substantia nigra [61]. However, a number of other brain regions are also involved (i.e., the brainstem, basal ganglia, autonomic nervous system, and cerebral cortex) [124].

5.2 Conventional MRI

Up till now the most common brain area that exhibits change in PD, using conventional MRI, seems to be the substantia nigra, [126] resulting in an alteration of susceptibility signal pattern due

[7] See note number 5.

to loss of neuromelanin[8] and iron accumulation. In PD patients, the "swallow tail" appearance of the dorsolateral substantia nigra is no more present [128] (Fig. 2b).

5.3 Advanced MRI

While evidence pointing to gray matter atrophy and, more specifically, hippocampal atrophy have been consistently reported in PDD [129, 130], the neuropathological basis of PDD remains controversial [131]. PDD has been associated with atrophy in the amygdala [14] and degeneration in subcortical structures; also the presence of Alzheimer-type neuropathology has been pointed as a likely contributor to dementia [131].

Analysis of cortical regions with VBM showed marked volume decrease in the anterior cingulate in patients with PDD compared with controls [131]. It has been shown that atrophy in PDD is not only different from PD without dementia but takes a similar form as Alzheimer's disease, affecting the medial temporal lobe [132, 133]. Also, in patients with PDD, reduced gray matter volume in temporal, occipital, right frontal, and left parietal lobes were found [132]. In another study, Beyer et al. found general and widespread reduced density of the cortical gray matter in PDD compared with controls and PD without dementia patients [134]. Furthermore Apaydin et al. concluded that Lewy body disease is the primary pathologic substrate for dementia developing later in PD [135].

DTI studies showed that white matter fibers can be involved in PDD including the cingulate bundles [136, 137], the inferior and superior longitudinal fasciculus, the uncinate fasciculus, and the inferior fronto-occipital fasciculus [138].

It was also found widespread cortical thinning in the frontal, premotor, left temporal, parietal, and right lateral occipital regions and that cortical thickness was associated with disease stage in PD [139]. With regard to cortical thinning in PDD, Zarei et al. investigated the cortical thinning in PD across disease stages. They have showed that the mean cortical thickness along with hippocampal volume can be used to help identify PDD patients [140].

6 Dementia with Lewy Bodies

Dementia with Lewy bodies (DLBs) is a neurodegenerative disease and is also an alpha-synucleinopathy. DLBs is sporadic and presents in older patients (typical onset: 50–70 years of age) [141]. It is

[8] Neuromelanin is a dark pigment found in the brain, which is structurally related to melanin. Neuromelanin is expressed in large quantities in catecholaminergic cells of the substantia nigra pars compacta and locus coeruleus, giving dark color to the structures [127]. Neuromelanin containing neurons in the substantia nigra undergoes neurodegeneration during Parkinson's disease. Neuromelanin concentration increases with age, suggesting a role in neuroprotection or senescence.

reported as the second most common form of neurodegenerative dementia following Alzheimer's Disease. Autopsy studies revealed that 15–25% of demented patients had Lewy bodies [142].

6.1 Pathology Core

With regard to neuropathology, it seems that there are no apparent differences between PDD (discussed earlier) and DLBs, with main alteration in the brainstem and cortex. Therefore, the most productive approach to distinguish between DLBs and PDD is through a descriptive clinicopathological classification approach. The temporal order of symptoms is widely used to distinguish between DLBs and PDD [143]. Patients with Parkinson disease develop dementia years after onset of parkinsonian symptoms, while inDLBs patients, dementia precedes or accompanies parkinsonism (or at least becomes clinically evident within 12 months of presentation) [144].

6.2 Conventional MRI

T2-weighted and proton-density images in a study [145] that included DLBs patients, using a 1.0 tesla MRI system, found that DLBs patients showed hyperintensities that are similar to those observed in Alzheimer's Disease and vascular dementia (i.e., periventricular and white matter hyperintensities). In addition, more recently Burton et al. investigated white matter hyperintensity changes in patients diagnosed with dementia. This study included fluid-attenuated inversion recovery images of DLBs patients. It was found that significant white matter hyperintensity progression occurs in degenerative dementias of all types [146].

6.3 Advanced MRI

Using VBM, Burton et al. found the patients with DLB showed significant decrease in volume in both frontal and temporal lobes as well as the insular cortex. The same study also showed significant distinction between DLBs and Alzheimer's Disease, in that Alzheimer's Disease patients were found to have significant decrease in volume of the medial temporal lobe, amygdala, thalamus, and hippocampus [147]. Judging from various studies, Tateno et al. concluded that the relative volumetric preservation of the medial temporal lobe could be used to aid in the diagnosis of DLBs and especially for discriminative between Alzheimer's Disease and DLBs [148]. Furthermore, comparing DLBs and Alzheimer's Disease in ROI studies, DLBs showed relative preservation of both temporal lobe structures and the hippocampus [147].

A VBM meta-analysis study [149] of seven studies comprising of 218 DLBs patients found that DLBs patients showed consistent gray matter reduction in the right lateral temporal/insular cortex and the left lenticular nucleus/insular cortex. Furthermore, another structural VBM study [150] found that DLBs was characterized by occipital and parietal atrophy.

In order to point out and spot changes in diffusivity patterns in patients with DLBs and Alzheimer's Disease Kantarci et al. performed DTI study. Based upon the results, and with respect to

DLBs, the authors showed increased diffusivity of the amygdala, while tissue loss is absent in DLBs and that diffusivity measures are complementary to conventional structural MRI [151]. Moreover, in a more recent study [152], TBSS-DTI was used to investigate patterns of FA and MD in DLBs patients. This study found that DLBs patients had reduced FA primarily in parieto-occipital white matter tracts, and in comparison to Alzheimer's Disease, DLBs showed reduced FA in the left thalamus and pons. This study, also, found that DTI patterns of DLBs were unique compared to Alzheimer's Disease and suggested the potential importance of white matter tract change with regard to DLBs pathogenesis and that DTI may be a very useful modality to identify early changes in DLBs [152].

On cortical thickness, Alzheimer's Disease was shown to have larger atrophy over a 1 year time period compared to DLBs and healthy controls, while there was no significant difference in percent whole brain volume change between DLBs and healthy controls [153]. It was found that Ballmaier et al. used MRI and cortical pattern matching to identify cortical gray matter differences between Alzheimer's Disease and DLBs patients. The results showed larger gray matter loss in the anterior cingulate for DLBs patients and that the orbitofrontal cortices and temporal lobe relative preservation may contribute toward distinguishing between Alzheimer's Disease and DLBs. On a final note, the authors suggested further investigations into the role of gender for both DLBs and Alzheimer's Disease [154].

7 Conclusion

Neurodegenerative non-Alzheimer's dementias (NADs) are comprised of different neuropathologies with temporal variance (from rapid onset to gradual progressive disease spectrum). Often there is considerable overlap of clinical features among NADs; as for current neuroimaging features, they tend to be non-specific during the initial stage of the disease. Thus far conventional neuroimaging techniques suggest possible differential diagnoses and can track evolving changes from follow-up studies [98], while they have yet to offer definitive diagnoses.

Thanks to research on the respective topics, newer imaging techniques have brought about a dramatic improvement in the understanding of neurodegenerative diseases in general. However, there do remain a number of significant limitations. These limitations include sample and neuroimaging features. The characteristics of the reported samples vary widely across the literature, with differences in the clinical criteria used, disease duration, mixing of disease subtypes, and limited sample size. The other limitation has to do with MRI scanner strength (i.e., 1.5 T vs. 3 T), scanner manufacturer (i.e., Siemens, Phillips, GE), image quality, image acquisition, variability in preprocessing and post-processing methods, ROI landmarks, statistical analysis, quality control procedures, and data presentation.

These limitations highlight the need for standardization. Thus a next step can be to establish a "best practice" guideline and/or methodological framework to standardize clinical and research procedure(s) used in MRI neuroimaging studies for neurodegenerative non-Alzheimer's dementia (i.e., defining acquisition sequences and parameters and developing data quality assurance protocols and automated quality assurance analytical tools, quality control, image analysis pipelines, statistical analysis methods that may include machine learning technologies for biomarker research, data management services, archiving, customized data sharing).

Standardization of protocol(s) and algorithm(s) for MR-based neuroimaging studies in NADs does constitute an open issue in the research community. Such standardization can provide more reliability of multicenter collaboration/data that combine newer imaging techniques with conventional neuroimaging techniques to find disease biomarkers, establish in vivo early-stage pathologic diagnoses, assess disease course, and assess different applications of novel disease modifying drugs for treatment and response evaluation. Collaboration between researchers and clinicians is key toward making such standardization possible and allows for optimum results.

Acknowledgments

The present chapter was supported by the Italian Ministry of Health. Contract grant number: 204/GR-2009-1606835

References

1. Walker FO (2007) Huntington's disease. Semin Neurol 27:143–150
2. Reilmann R, Leavitt BR, Ross CA (2014) Diagnostic criteria for Huntington's disease based on natural history. Mov Disord 29:1335–1341
3. Warby SC, Graham RK, Hayden MR (1993) Huntington disease. In: Pagon RA, Adam MP, Ardinger HH et al (eds) GeneReviews®. University of Washington, Seattle, WA
4. Vonsattel JP (2008) Huntington disease models and human neuropathology: similarities and differences. Acta Neuropathol 115:55–69
5. Ho VB, Chuang HS, Rovira MJ, Koo B (1995) Juvenile Huntington disease: CT and MR features. Am J Neuroradiol 16:1405–1412
6. Halliday GM, McRitchie DA, Macdonald V, Double KL, Trent RJ, McCusker E (1998) Regional specificity of brain atrophy in Huntington's disease. Exp Neurol 154:663–672
7. Georgiou-Karistianis N, Scahill R, Tabrizi SJ, Squitieri F, Aylward E (2013) Structural MRI in Huntington's disease and recommendations for its potential use in clinical trials. Neurosci Biobehav Rev 37:480–490
8. Paulsen JS, Nopoulos PC, Aylward E, Ross CA, Johnson H, Magnotta VA, Juhl A, Pierson RK, Mills J, Langbehn D, Nance M, Investigators P-H, Coordinators of the Huntington's Study G (2010) Striatal and white

matter predictors of estimated diagnosis for Huntington disease. Brain Res Bull 82:201–207

9. Tabrizi SJ, Reilmann R, Roos RA, Durr A, Leavitt B, Owen G, Jones R, Johnson H, Craufurd D, Hicks SL, Kennard C, Landwehrmeyer B, Stout JC, Borowsky B, Scahill RI, Frost C, Langbehn DR, investigators T-H (2012) Potential endpoints for clinical trials in premanifest and early Huntington's disease in the TRACK-HD study: analysis of 24 month observational data. Lancet Neurol 11:42–53
10. Majid DS, Aron AR, Thompson W, Sheldon S, Hamza S, Stoffers D, Holland D, Goldstein J, Corey-Bloom J, Dale AM (2011) Basal ganglia atrophy in prodromal Huntington's disease is detectable over one year using automated segmentation. Mov Disord 26:2544–2551
11. Aylward EH, Liu D, Nopoulos PC, Ross CA, Pierson RK, Mills JA, Long JD, Paulsen JS, Investigators P-H, Coordinators of the Huntington Study G (2012) Striatal volume contributes to the prediction of onset of Huntington disease in incident cases. Biol Psychiatry 71:822–828
12. Ginestroni A, Battaglini M, Diciotti S, Della Nave R, Mazzoni LN, Tessa C, Giannelli M, Piacentini S, De Stefano N, Mascalchi M (2010) Magnetization transfer MR imaging demonstrates degeneration of the subcortical and cortical gray matter in Huntington disease. Am J Neuroradiol 31:1807–1812
13. van den Bogaard SJ, Dumas EM, Acharya TP, Johnson H, Langbehn DR, Scahill RI, Tabrizi SJ, van Buchem MA, van der Grond J, Roos RA, Group T-HI (2011) Early atrophy of pallidum and accumbens nucleus in Huntington's disease. J Neurol 258:412–420
14. Jurgens CK, van de Wiel L, van Es AC, Grimbergen YM, Witjes-Ane MN, van der Grond J, Middelkoop HA, Roos RA (2008) Basal ganglia volume and clinical correlates in 'preclinical' Huntington's disease. J Neurol 255:1785–1791
15. Sanchez-Castaneda C, Cherubini A, Elifani F, Peran P, Orobello S, Capelli G, Sabatini U, Squitieri F (2013) Seeking Huntington disease biomarkers by multimodal, cross-sectional basal ganglia imaging. Hum Brain Mapp 34:1625–1635
16. Ruocco HH, Bonilha L, Li LM, Lopes-Cendes I, Cendes F (2008) Longitudinal analysis of regional grey matter loss in Huntington disease: effects of the length of the expanded CAG repeat. J Neurol Neurosurg Psychiatry 79:130–135
17. Aylward EH, Nopoulos PC, Ross CA, Langbehn DR, Pierson RK, Mills JA, Johnson HJ, Magnotta VA, Juhl AR, Paulsen JS, Investigators P-H, Coordinators of Huntington Study G (2011) Longitudinal change in regional brain volumes in prodromal Huntington disease. J Neurol Neurosurg Psychiatry 82:405–410
18. Henley SM, Wild EJ, Hobbs NZ, Scahill RI, Ridgway GR, Macmanus DG, Barker RA, Fox NC, Tabrizi SJ (2009) Relationship between CAG repeat length and brain volume in premanifest and early Huntington's disease. J Neurol 256:203–212
19. Rosas HD, Koroshetz WJ, Chen YI, Skeuse C, Vangel M, Cudkowicz ME, Caplan K, Marek K, Seidman LJ, Makris N, Jenkins BG, Goldstein JM (2003) Evidence for more widespread cerebral pathology in early HD: an MRI-based morphometric analysis. Neurology 60:1615–1620
20. Hobbs NZ, Barnes J, Frost C, Henley SM, Wild EJ, Macdonald K, Barker RA, Scahill RI, Fox NC, Tabrizi SJ (2010) Onset and progression of pathologic atrophy in Huntington disease: a longitudinal MR imaging study. Am J Neuroradiol 31:1036–1041
21. Ciarmiello A, Cannella M, Lastoria S, Simonelli M, Frati L, Rubinsztein DC, Squitieri F (2006) Brain white-matter volume loss and glucose hypometabolism precede the clinical symptoms of Huntington's disease. J Nucl Med 47:215–222
22. Thieben MJ, Duggins AJ, Good CD, Gomes L, Mahant N, Richards F, McCusker E, Frackowiak RS (2002) The distribution of structural neuropathology in pre-clinical Huntington's disease. Brain 125:1815–1828
23. Tabrizi SJ, Langbehn DR, Leavitt BR, Roos RA, Durr A, Craufurd D, Kennard C, Hicks SL, Fox NC, Scahill RI, Borowsky B, Tobin AJ, Rosas HD, Johnson H, Reilmann R, Landwehrmeyer B, Stout JC, investigators T-H (2009) Biological and clinical manifestations of Huntington's disease in the longitudinal TRACK-HD study: cross-sectional analysis of baseline data. Lancet Neurol 8:791–801
24. Beglinger LJ, Nopoulos PC, Jorge RE, Langbehn DR, Mikos AE, Moser DJ, Duff K, Robinson RG, Paulsen JS (2005) White matter volume and cognitive dysfunction in early Huntington's disease. Cogn Behav Neurol 18:102–107
25. Tabrizi SJ, Scahill RI, Durr A, Roos RA, Leavitt BR, Jones R, Landwehrmeyer GB, Fox NC, Johnson H, Hicks SL, Kennard C,

Craufurd D, Frost C, Langbehn DR, Reilmann R, Stout JC, Investigators T-H (2011) Biological and clinical changes in premanifest and early stage Huntington's disease in the TRACK-HD study: the 12-month longitudinal analysis. Lancet Neurol 10:31–42

26. Hobbs NZ, Pedrick AV, Say MJ, Frost C, Dar Santos R, Coleman A, Sturrock A, Craufurd D, Stout JC, Leavitt BR, Barnes J, Tabrizi SJ, Scahill RI (2011) The structural involvement of the cingulate cortex in premanifest and early Huntington's disease. Mov Disord 26:1684–1690
27. Di Paola M, Phillips OR, Sanchez-Castaneda-C, Di Pardo A, Maglione V, Caltagirone C, Sabatini U, Squitieri F (2014) MRI measures of corpus callosum iron and myelin in early Huntington's disease. Hum Brain Mapp 35:3143–3151
28. Di Paola M, Luders E, Cherubini A, Sanchez-Castaneda C, Thompson PM, Toga AW, Caltagirone C, Orobello S, Elifani F, Squitieri F, Sabatini U (2012) Multimodal MRI analysis of the corpus callosum reveals white matter differences in presymptomatic and early Huntington's disease. Cereb Cortex 22:2858–2866
29. Magnotta VA, Kim J, Koscik T, Beglinger LJ, Espinso D, Langbehn D, Nopoulos P, Paulsen JS (2009) Diffusion tensor imaging in preclinical Huntington's disease. Brain Imaging Behav 3:77–84
30. Phillips O, Joshi SH, Squitieri F, Sanchez-Castaneda C, Narr K, Shattuck D, Caltagirone C, Sabatini U, Di Paola M (2016) Major superficial white matter abnormalities in Huntington disease. Front Neurosci 10:197
31. Rosas HD, Liu AK, Hersch S, Glessner M, Ferrante RJ, Salat DH, van der Kouwe A, Jenkins BG, Dale AM, Fischl B (2002) Regional and progressive thinning of the cortical ribbon in Huntington's disease. Neurology 58:695–701
32. Nopoulos P, Magnotta VA, Mikos A, Paulson H, Andreasen NC, Paulsen JS (2007) Morphology of the cerebral cortex in preclinical Huntington's disease. Am J Psychiatry 164:1428–1434
33. Rosas HD, Hevelone ND, Zaleta AK, Greve DN, Salat DH, Fischl B (2005) Regional cortical thinning in preclinical Huntington disease and its relationship to cognition. Neurology 65:745–747
34. Rosas HD, Salat DH, Lee SY, Zaleta AK, Pappu V, Fischl B, Greve D, Hevelone N, Hersch SM (2008) Cerebral cortex and the clinical expression of Huntington's disease: complexity and heterogeneity. Brain 131:1057–1068
35. Litvan I, Agid Y, Calne D, Campbell G, Dubois B, Duvoisin RC, Goetz CG, Golbe LI, Grafman J, Growdon JH, Hallett M, Jankovic J, Quinn NP, Tolosa E, Zee DS (1996) Clinical research criteria for the diagnosis of progressive supranuclear palsy (Steele-Richardson-Olszewski syndrome): report of the NINDS-SPSP international workshop. Neurology 47:1–9
36. Dabrowska M, Schinwelski M, Sitek EJ, Muraszko-Klaudel A, Brockhuis B, Jamrozik Z, Slawek J (2015) The role of neuroimaging in the diagnosis of the atypical parkinsonian syndromes in clinical practice. Neurol Neurochir Pol 49:421–431
37. McGinnis SM (2012) Neuroimaging in neurodegenerative dementias. Semin Neurol 32:347–360
38. Caramia M, Mosti S, Cologno D, Di Paola M, Caltagirone C (2003) Disturbi comportamentali e psichici in un caso di paralisi sopra nucleare progressiva. Nuova Riv Neurol 13:25–33
39. Dickson DW (1999) Neuropathologic differentiation of progressive supranuclear palsy and corticobasal degeneration. J Neurol 246 (Suppl 2):II6–I15
40. Tsuboi Y, Slowinski J, Josephs KA, Honer WG, Wszolek ZK, Dickson DW (2003) Atrophy of superior cerebellar peduncle in progressive supranuclear palsy. Neurology 60:1766–1769
41. Oba H, Yagishita A, Terada H, Barkovich AJ, Kutomi K, Yamauchi T, Furui S, Shimizu T, Uchigata M, Matsumura K, Sonoo M, Sakai M, Takada K, Harasawa A, Takeshita K, Kohtake H, Tanaka H, Suzuki S (2005) New and reliable MRI diagnosis for progressive supranuclear palsy. Neurology 64:2050–2055
42. Quattrone A, Nicoletti G, Messina D, Fera F, Condino F, Pugliese P, Lanza P, Barone P, Morgante L, Zappia M, Aguglia U, Gallo O (2008) MR imaging index for differentiation of progressive supranuclear palsy from Parkinson disease and the Parkinson variant of multiple system atrophy. Radiology 246:214–221
43. Morelli M, Arabia G, Novellino F, Salsone M, Giofre L, Condino F, Messina D, Quattrone A (2011) MRI measurements predict PSP in unclassifiable parkinsonisms: a cohort study. Neurology 77:1042–1047
44. Righini A, Antonini A, De Notaris R, Bianchini E, Meucci N, Sacilotto G, Canesi M, De Gaspari D, Triulzi F, Pezzoli G (2004) MR imaging of the superior profile

of the midbrain: differential diagnosis between progressive supranuclear palsy and Parkinson disease. Am J Neuroradiol 25:927–932

45. Groschel K, Kastrup A, Litvan I, Schulz JB (2006) Penguins and hummingbirds: midbrain atrophy in progressive supranuclear palsy. Neurology 66:949–950
46. Adachi M, Kawanami T, Ohshima H, Sugai Y, Hosoya T (2004) Morning glory sign: a particular MR finding in progressive supranuclear palsy. Magn Reson Med Sci 3:125–132
47. Paviour DC, Price SL, Jahanshahi M, Lees AJ, Fox NC (2006) Regional brain volumes distinguish PSP, MSA-P, and PD: MRI-based clinico-radiological correlations. Mov Disord 21:989–996
48. Cordato NJ, Duggins AJ, Halliday GM, Morris JG, Pantelis C (2005) Clinical deficits correlate with regional cerebral atrophy in progressive supranuclear palsy. Brain 128:1259–1266
49. Groschel K, Hauser TK, Luft A, Patronas N, Dichgans J, Litvan I, Schulz JB (2004) Magnetic resonance imaging-based volumetry differentiates progressive supranuclear palsy from corticobasal degeneration. NeuroImage 21:714–724
50. Looi JC, Macfarlane MD, Walterfang M, Styner M, Velakoulis D, Latt J, van Westen D, Nilsson C (2011) Morphometric analysis of subcortical structures in progressive supranuclear palsy: in vivo evidence of neostriatal and mesencephalic atrophy. Psychiatry Res 194:163–175
51. Surova Y, Nilsson M, Latt J, Lampinen B, Lindberg O, Hall S, Widner H, Nilsson C, van Westen D, Hansson O (2015) Disease-specific structural changes in thalamus and dentatorubrothalamic tract in progressive supranuclear palsy. Neuroradiology 57:1079–1091
52. Yu F, Barron DS, Tantiwongkosi B, Fox P (2015) Patterns of gray matter atrophy in atypical parkinsonism syndromes: a VBM meta-analysis. Brain Behav 5:e00329
53. Shi HC, Zhong JG, Pan PL, Xiao PR, Shen Y, LJ W, Li HL, Song YY, He GX, Li HY (2013) Gray matter atrophy in progressive supranuclear palsy: meta-analysis of voxel-based morphometry studies. Neurol Sci 34:1049–1055
54. Shao N, Yang J, Li J, Shang HF (2014) Voxelwise meta-analysis of gray matter anomalies in progressive supranuclear palsy and Parkinson's disease using anatomic likelihood estimation. Front Hum Neurosci 8:63
55. Yang J, Shao N, Li J, Shang H (2014) Voxelwise meta-analysis of white matter abnormalities in progressive supranuclear palsy. Neurol Sci 35:7–14
56. Agosta F, Kostic VS, Galantucci S, Mesaros S, Svetel M, Pagani E, Stefanova E, Filippi M (2010) The in vivo distribution of brain tissue loss in Richardson's syndrome and PSP-parkinsonism: a VBM-DARTEL study. Eur J Neurosci 32:640–647
57. Boxer AL, Geschwind MD, Belfor N, Gorno-Tempini ML, Schauer GF, Miller BL, Weiner MW, Rosen HJ (2006) Patterns of brain atrophy that differentiate corticobasal degeneration syndrome from progressive supranuclear palsy. Arch Neurol 63:81–86
58. Brenneis C, Seppi K, Schocke M, Benke T, Wenning GK, Poewe W (2004) Voxel based morphometry reveals a distinct pattern of frontal atrophy in progressive supranuclear palsy. J Neurol Neurosurg Psychiatry 75:246–249
59. Padovani A, Borroni B, Brambati SM, Agosti C, Broli M, Alonso R, Scifo P, Bellelli G, Alberici A, Gasparotti R, Perani D (2006) Diffusion tensor imaging and voxel based morphometry study in early progressive supranuclear palsy. J Neurol Neurosurg Psychiatry 77:457–463
60. Lehericy S, Hartmann A, Lannuzel A, Galanaud D, Delmaire C, Bienaimee MJ, Jodoin N, Roze E, Gaymard B, Vidailhet M (2010) Magnetic resonance imaging lesion pattern in Guadeloupean parkinsonism is distinct from progressive supranuclear palsy. Brain 133:2410–2425
61. Price S, Paviour D, Scahill R, Stevens J, Rossor M, Lees A, Fox N (2004) Voxel-based morphometry detects patterns of atrophy that help differentiate progressive supranuclear palsy and Parkinson's disease. NeuroImage 23:663–669
62. Canu E, Agosta F, Baglio F, Galantucci S, Nemni R, Filippi M (2011) Diffusion tensor magnetic resonance imaging tractography in progressive supranuclear palsy. Mov Disord 26:1752–1755
63. Piattella MC, Upadhyay N, Bologna M, Sbardella E, Tona F, Formica A, Petsas N, Berardelli A, Pantano P (2015) Neuroimaging evidence of gray and white matter damage and clinical correlates in progressive supranuclear palsy. J Neurol 262:1850–1858
64. Wang J, Wai Y, Lin WY, Ng S, Wang CH, Hsieh R, Hsieh C, Chen RS, CS L (2010) Microstructural changes in patients with progressive supranuclear palsy: a diffusion tensor

imaging study. J Magn Reson Imaging 32:69–75

65. Sener RN (2001) Diffusion MRI: apparent diffusion coefficient (ADC) values in the normal brain and a classification of brain disorders based on ADC values. Comput Med Imaging Graph 25:299–326
66. Tsukamoto K, Matsusue E, Kanasaki Y, Kakite S, Fujii S, Kaminou T, Ogawa T (2012) Significance of apparent diffusion coefficient measurement for the differential diagnosis of multiple system atrophy, progressive supranuclear palsy, and Parkinson's disease: evaluation by 3.0-T MR imaging. Neuroradiology 54:947–955
67. Planetta PJ, Ofori E, Pasternak O, Burciu RG, Shukla P, DeSimone JC, Okun MS, McFarland NR, Vaillancourt DE (2016) Free-water imaging in Parkinson's disease and atypical parkinsonism. Brain 139:495–508
68. Pasternak O, Sochen N, Gur Y, Intrator N, Assaf Y (2009) Free water elimination and mapping from diffusion MRI. Magn Reson Med 62:717–730
69. Whitwell JL, Avula R, Master A, Vemuri P, Senjem ML, Jones DT, Jack CR Jr, Josephs KA (2011) Disrupted thalamocortical connectivity in PSP: a resting-state fMRI, DTI, and VBM study. Parkinsonism Relat Disord 17:599–605
70. Worker A, Blain C, Jarosz J, Chaudhuri KR, Barker GJ, Williams SC, Brown RG, Leigh PN, Dell'Acqua F, Simmons A (2014) Diffusion tensor imaging of Parkinson's disease, multiple system atrophy and progressive supranuclear palsy: a tract-based spatial statistics study. PLoS One 9:e112638
71. Tessitore A, Giordano A, Caiazzo G, Corbo D, De Micco R, Russo A, Liguori S, Cirillo M, Esposito F, Tedeschi G (2014) Clinical correlations of microstructural changes in progressive supranuclear palsy. Neurobiol Aging 35:2404–2410
72. Knake S, Belke M, Menzler K, Pilatus U, Eggert KM, Oertel WH, Stamelou M, Hoglinger GU (2010) In vivo demonstration of microstructural brain pathology in progressive supranuclear palsy: a DTI study using TBSS. Mov Disord 25:1232–1238
73. Nilsson C, Markenroth Bloch K, Brockstedt S, Latt J, Widner H, Larsson EM (2007) Tracking the neurodegeneration of parkinsonian disorders—a pilot study. Neuroradiology 49:111–119
74. Blain CR, Barker GJ, Jarosz JM, Coyle NA, Landau S, Brown RG, Chaudhuri KR, Simmons A, Jones DK, Williams SC, Leigh PN (2006) Measuring brain stem and cerebellar damage in parkinsonian syndromes using diffusion tensor MRI. Neurology 67:2199–2205
75. Agosta F, Galantucci S, Svetel M, Lukic MJ, Copetti M, Davidovic K, Tomic A, Spinelli EG, Kostic VS, Filippi M (2014) Clinical, cognitive, and behavioural correlates of white matter damage in progressive supranuclear palsy. J Neurol 261:913–924
76. Tha KK, Terae S, Yabe I, Miyamoto T, Soma H, Zaitsu Y, Fujima N, Kudo K, Sasaki H, Shirato H (2010) Microstructural white matter abnormalities of multiple system atrophy: in vivo topographic illustration by using diffusion-tensor MR imaging. Radiology 255:563–569
77. Han YH, Lee JH, Kang BM, Mun CW, Baik SK, Shin YI, Park KH (2013) Topographical differences of brain iron deposition between progressive supranuclear palsy and parkinsonian variant multiple system atrophy. J Neurol Sci 325:29–35
78. Worker A, Blain C, Jarosz J, Chaudhuri KR, Barker GJ, Williams SC, Brown R, Leigh PN, Simmons A (2014) Cortical thickness, surface area and volume measures in Parkinson's disease, multiple system atrophy and progressive supranuclear palsy. PLoS One 9:e114167
79. Wadia PM, Lang AE (2007) The many faces of corticobasal degeneration. Parkinsonism Relat Disord 13(Suppl 3):S336–S340
80. Grijalvo-Perez AM, Litvan I (2014) Corticobasal degeneration. Semin Neurol 34:160–173
81. Bergeron C, Davis A, Lang AE (1998) Corticobasal ganglionic degeneration and progressive supranuclear palsy presenting with cognitive decline. Brain Pathol 8:355–365
82. Grimes DA, Lang AE, Bergeron CB (1999) Dementia as the most common presentation of cortical-basal ganglionic degeneration. Neurology 53:1969–1974
83. Mahapatra RK, Edwards MJ, Schott JM, Bhatia KP (2004) Corticobasal degeneration. Lancet Neurol 3:736–743
84. Tokumaru AM, O'Uchi T, Kuru Y, Maki T, Murayama S, Horichi Y (1996) Corticobasal degeneration: MR with histopathologic comparison. Am J Neuroradiol 17:1849–1852
85. Bergeron C, Pollanen MS, Weyer L, Black SE, Lang AE (1996) Unusual clinical presentations of cortical-basal ganglionic degeneration. Ann Neurol 40:893–900
86. Boeve BF, Maraganore DM, Parisi JE, Ahlskog JE, Graff-Radford N, Caselli RJ, Dickson DW, Kokmen E, Petersen RC (1999)

Pathologic heterogeneity in clinically diagnosed corticobasal degeneration. Neurology 53:795–800

87. Doi T, Iwasa K, Makifuchi T, Takamori M (1999) White matter hyperintensities on MRI in a patient with corticobasal degeneration. Acta Neurol Scand 99:199–201
88. Huang KJ, MK L, Kao A, Tsai CH (2007) Clinical, imaging and electrophysiological studies of corticobasal degeneration. Acta Neurol Taiwanica 16:13–21
89. Josephs KA, Whitwell JL, Dickson DW, Boeve BF, Knopman DS, Petersen RC, Parisi JE, Jack CR Jr (2008) Voxel-based morphometry in autopsy proven PSP and CBD. Neurobiol Aging 29:280–289
90. Yekhlef F, Ballan G, Macia F, Delmer O, Sourgen C, Tison F (2003) Routine MRI for the differential diagnosis of Parkinson's disease, MSA, PSP, and CBD. J Neural Transm (Vienna) 110:151–169
91. Peigneux P, Salmon E, Garraux G, Laureys S, Willems S, Dujardin K, Degueldre C, Lemaire C, Luxen A, Moonen G, Franck G, Destee A, Van der Linden M (2001) Neural and cognitive bases of upper limb apraxia in corticobasal degeneration. Neurology 57:1259–1268
92. Hauser RA, Murtaugh FR, Akhter K, Gold M, Olanow CW (1996) Magnetic resonance imaging of corticobasal degeneration. J Neuroimaging 6:222–226
93. Erbetta A, Mandelli ML, Savoiardo M, Grisoli M, Bizzi A, Soliveri P, Chiapparini L, Prioni S, Bruzzone MG, Girotti F (2009) Diffusion tensor imaging shows different topographic involvement of the thalamus in progressive supranuclear palsy and corticobasal degeneration. Am J Neuroradiol 30:1482–1487
94. McCann H, Stevens CH, Cartwright H, Halliday GM (2014) ς-synucleinopathy phenotypes. Parkinsonism Relat Disord 20:S62–S67
95. Messina D, Cerasa A, Condino F, Arabia G, Novellino F, Nicoletti G, Salsone M, Morelli M, Lanza PL, Quattrone A (2011) Patterns of brain atrophy in Parkinson's disease, progressive supranuclear palsy and multiple system atrophy. Parkinsonism Relat Disord 17:172–176
96. Gilman S, Wenning GK, Low PA, Brooks DJ, Mathias CJ, Trojanowski JQ, Wood NW, Colosimo C, Durr A, Fowler CJ, Kaufmann H, Klockgether T, Lees A, Poewe W, Quinn N, Revesz T, Robertson D, Sandroni P, Seppi K, Vidailhet M (2008) Second consensus statement on the diagnosis of multiple system atrophy. Neurology 71:670–676
97. Kawai Y, Suenaga M, Takeda A, Ito M, Watanabe H, Tanaka F, Kato K, Fukatsu H, Naganawa S, Kato T, Ito K, Sobue G (2008) Cognitive impairments in multiple system atrophy: MSA-C MSA-P. Neurology 70:1390–1396
98. Patro SN, Glikstein R, Hanagandi P, Chakraborty S (2015) Role of neuroimaging in multidisciplinary approach towards non-Alzheimer's dementia. Insights Imaging 6:531–544
99. Litvan I, Mega MS, Cummings JL, Fairbanks L (1996) Neuropsychiatric aspects of progressive supranuclear palsy. Neurology 47:1184–1189
100. Trojanowski JQ, Revesz T, Neuropathology Working Group on MSA (2007) Proposed neuropathological criteria for the post mortem diagnosis of multiple system atrophy. Neuropathol Appl Neurobiol 33:615–620
101. Lee JY, Yun JY, Shin CW, Kim HJ, Jeon BS (2010) Putaminal abnormality on 3-T magnetic resonance imaging in early parkinsonism-predominant multiple system atrophy. J Neurol 257:2065–2070
102. Lee WH, Lee CC, Shyu WC, Chong PN, Lin SZ (2005) Hyperintense putaminal rim sign is not a hallmark of multiple system atrophy at 3T. AJNR Am J Neuroradiol 26:2238–2242
103. Kim JM, Jeong HJ, Bae YJ, Park SY, Kim E, Kang SY, Oh ES, Kim KJ, Jeon B, Kim SE, Cho ZH, Kim YB (2016) Loss of substantia nigra hyperintensity on 7 Tesla MRI of Parkinson's disease, multiple system atrophy, and progressive supranuclear palsy. Parkinsonism Relat Disord. https://doi.org/10.1016/j.parkreldis.2016.01.023
104. Planetta PJ, Kurani AS, Shukla P, Prodoehl J, Corcos DM, Comella CL, McFarland NR, Okun MS, Vaillancourt DE (2015) Distinct functional and macrostructural brain changes in Parkinson's disease and multiple system atrophy. Hum Brain Mapp 36:1165–1179
105. Tzarouchi LC, Astrakas LG, Konitsiotis S, Tsouli S, Margariti P, Zikou A, Argyropoulou MI (2010) Voxel-based morphometry and voxel-based relaxometry in parkinsonian variant of multiple system atrophy. J Neuroimaging 20:260–266
106. Nair SR, Tan LK, Mohd Ramli N, Lim SY, Rahmat K, Mohd Nor H (2013) A decision tree for differentiating multiple system atrophy from Parkinson's disease using 3-T MR imaging. Eur Radiol 23:1459–1466

107. Yang H, Wang X, Liao W, Zhou G, Li L, Ouyang L (2015) Application of diffusion tensor imaging in multiple system atrophy: the involvement of pontine transverse and longitudinal fibers. Int J Neurosci 125:18–24
108. Nicoletti G, Lodi R, Condino F, Tonon C, Fera F, Malucelli E, Manners D, Zappia M, Morgante L, Barone P, Barbiroli B, Quattrone A (2006) Apparent diffusion coefficient measurements of the middle cerebellar peduncle differentiate the Parkinson variant of MSA from Parkinson's disease and progressive supranuclear palsy. Brain 129:2679–2687
109. Chung EJ, Kim EG, Bae JS, Eun CK, Lee KS, Oh M, Kim SJ (2009) Usefulness of diffusion-weighted MRI for differentiation between Parkinson's disease and Parkinson variant of multiple system atrophy. J Mov Disord 2:64–68
110. Cnyrim CD, Kupsch A, Ebersbach G, Hoffmann KT (2014) Diffusion tensor imaging in idiopathic Parkinson's disease and multisystem atrophy (Parkinsonian type). Neurodegener Dis 13:1–8
111. Seppi K, Schocke MF, Esterhammer R, Kremser C, Brenneis C, Mueller J, Boesch S, Jaschke W, Poewe W, Wenning GK (2003) Diffusion-weighted imaging discriminates progressive supranuclear palsy from PD, but not from the Parkinson variant of multiple system atrophy. Neurology 60:922–927
112. Kanazawa M, Shimohata T, Terajima K, Onodera O, Tanaka K, Tsuji S, Okamoto K, Nishizawa M (2004) Quantitative evaluation of brainstem involvement in multiple system atrophy by diffusion-weighted MR imaging. J Neurol 251:1121–1124
113. Schocke MF, Seppi K, Esterhammer R, Kremser C, Mair KJ, Czermak BV, Jaschke W, Poewe W, Wenning GK (2004) Trace of diffusion tensor differentiates the Parkinson variant of multiple system atrophy and Parkinson's disease. NeuroImage 21:1443–1451
114. Ito S, Shirai W, Hattori T (2007) Evaluating posterolateral linearization of the putaminal margin with magnetic resonance imaging to diagnose the Parkinson variant of multiple system atrophy. Mov Disord 22:578–581
115. Pellecchia MT, Barone P, Mollica C, Salvatore E, Ianniciello M, Longo K, Varrone A, Vicidomini C, Picillo M, De Michele G, Filla A, Salvatore M, Pappata S (2009) Diffusion-weighted imaging in multiple system atrophy: a comparison between clinical subtypes. Mov Disord 24:689–696
116. Baudrexel S, Seifried C, Penndorf B, Klein JC, Middendorp M, Steinmetz H, Grunwald F, Hilker R (2014) The value of putaminal diffusion imaging versus 18-fluorodeoxyglucose positron emission tomography for the differential diagnosis of the Parkinson variant of multiple system atrophy. Mov Disord 29:380–387
117. Ji L, Zhu D, Xiao C, Shi J (2014) Tract based spatial statistics in multiple system atrophy: a comparison between clinical subtypes. Parkinsonism Relat Disord 20:1050–1055
118. Fukui Y, Hishikawa N, Sato K, Nakano Y, Morihara R, Ohta Y, Yamashita T, Abe K (2015) Characteristic diffusion tensor tractography in multiple system atrophy with predominant cerebellar ataxia and cortical cerebellar atrophy. J Neurol. https://doi.org/10.1007/s00415-015-7934-x
119. Makino T, Ito S, Kuwabara S (2011) Involvement of pontine transverse and longitudinal fibers in multiple system atrophy: a tractography-based study. J Neurol Sci 303:61–66
120. Kim JS, Youn J, Yang JJ, Lee DK, Lee JM, Kim ST, Kim HT, Cho JW (2013) Topographic distribution of cortical thinning in subtypes of multiple system atrophy. Parkinsonism Relat Disord 19:970–974
121. Kim HJ, Jeon BS, Kim YE, Kim JY, Kim YK, Sohn CH, Yun JY, Jeon S, Lee JM, Lee JY (2013) Clinical and imaging characteristics of dementia in multiple system atrophy. Parkinsonism Relat Disord 19:617–621
122. Melzer TR, Watts R, MacAskill MR, Pitcher TL, Livingston L, Keenan RJ, Dalrymple-Alford JC, Anderson TJ (2012) Grey matter atrophy in cognitively impaired Parkinson's disease. J Neurol Neurosurg Psychiatry 83:188–194
123. Tsuboi Y, Uchikado H, Dickson DW (2007) Neuropathology of Parkinson's disease dementia and dementia with Lewy bodies with reference to striatal pathology. Parkinsonism Relat Disord 13(Suppl 3):S221–S224
124. Ghosh BC, Calder AJ, Peers PV, Lawrence AD, Acosta-Cabronero J, Pereira JM, Hodges JR, Rowe JB (2012) Social cognitive deficits and their neural correlates in progressive supranuclear palsy. Brain 135:2089–2102
125. Takahashi R, Ishii K, Kakigi T, Yokoyama K, Mori E, Murakami T (2011) Brain alterations and mini-mental state examination in patients with progressive supranuclear palsy: voxel-based investigations using f-fluorodeoxyglucose positron emission tomography and magnetic resonance imaging. Dement Geriatr Cogn Dis Extra 1:381–392

126. Davie CA (2008) A review of Parkinson's disease. Br Med Bull 86:109–127
127. Fedorow H, Tribl F, Halliday G, Gerlach M, Riederer P, Double KL (2005) Neuromelanin in human dopamine neurons: comparison with peripheral melanins and relevance to Parkinson's disease. Prog Neurobiol 75:109–124
128. Schwarz ST, Afzal M, Morgan PS, Bajaj N, Gowland PA, Auer DP (2014) The 'swallow tail' appearance of the healthy nigrosome - a new accurate test of Parkinson's disease: a case-control and retrospective cross-sectional MRI study at 3T. PLoS One 9:e93814
129. Bouchard TP, Malykhin N, Martin WR, Hanstock CC, Emery DJ, Fisher NJ, Camicioli RM (2008) Age and dementia-associated atrophy predominates in the hippocampal head and amygdala in Parkinson's disease. Neurobiol Aging 29:1027–1039
130. Brenner SR (2008) Gray matter atrophy in Parkinson disease with dementia and dementia with Lewy bodies. Neurology 70:2265–2265
131. Summerfield C, Junque C, Tolosa E, Salgado-Pineda P, Gomez-Anson B, Marti MJ, Pastor P, Ramirez-Ruiz B, Mercader J (2005) Structural brain changes in Parkinson disease with dementia: a voxel-based morphometry study. Arch Neurol 62:281–285
132. Burton EJ, McKeith IG, Burn DJ, Williams ED, O'Brien JT (2004) Cerebral atrophy in Parkinson's disease with and without dementia: a comparison with Alzheimer's disease, dementia with Lewy bodies and controls. Brain 127:791–800
133. Laakso MP, Partanen K, Riekkinen P, Lehtovirta M, Helkala EL, Hallikainen M, Hanninen T, Vainio P, Soininen H (1996) Hippocampal volumes in Alzheimer's disease, Parkinson's disease with and without dementia, and in vascular dementia: an MRI study. Neurology 46:678–681
134. Beyer MK, Aarsland D (2008) Grey matter atrophy in early versus late dementia in Parkinson's disease. Parkinsonism Relat Disord 14:620–625
135. Apaydin H, Ahlskog JE, Parisi JE, Boeve BF, Dickson DW (2002) Parkinson disease neuropathology: later-developing dementia and loss of the levodopa response. Arch Neurol 59:102–112
136. Matsui H, Nishinaka K, Oda M, Niikawa H, Kubori T, Udaka F (2007) Dementia in Parkinson's disease: diffusion tensor imaging. Acta Neurol Scand 116:177–181
137. Meijer FJ, Bloem BR, Mahlknecht P, Seppi K, Goraj B (2013) Update on diffusion MRI in Parkinson's disease and atypical parkinsonism. J Neurol Sci 332:21–29
138. Hattori T, Orimo S, Aoki S, Ito K, Abe O, Amano A, Sato R, Sakai K, Mizusawa H (2012) Cognitive status correlates with white matter alteration in Parkinson's disease. Hum Brain Mapp 33:727–739
139. Pereira JB, Ibarretxe-Bilbao N, Marti MJ, Compta Y, Junque C, Bargallo N, Tolosa E (2012) Assessment of cortical degeneration in patients with Parkinson's disease by voxel-based morphometry, cortical folding, and cortical thickness. Hum Brain Mapp 33:2521–2534
140. Zarei M, Ibarretxe-Bilbao N, Compta Y, Hough M, Junque C, Bargallo N, Tolosa E, Marti MJ (2013) Cortical thinning is associated with disease stages and dementia in Parkinson's disease. J Neurol Neurosurg Psychiatry 84:875–881
141. Taylor JP, O'Brien J (2012) Neuroimaging of dementia with Lewy bodies. Neuroimaging Clin N Am 22:67–81. viii
142. McKeith IG (2006) Consensus guidelines for the clinical and pathologic diagnosis of dementia with Lewy bodies (DLB): report of the consortium on DLB international workshop. J Alzheimers Dis 9:417–423
143. McKeith IG, Burn DJ, Ballard CG, Collerton D, Jaros E, Morris CM, McLaren A, Perry EK, Perry R, Piggott MA, O'Brien JT (2003) Dementia with Lewy bodies. Semin Clin Neuropsychiatry 8:46–57
144. Harding AJ, Broe GA, Halliday GM (2002) Visual hallucinations in Lewy body disease relate to Lewy bodies in the temporal lobe. Brain 125:391–403
145. Barber R, Scheltens P, Gholkar A, Ballard C, McKeith I, Ince P, Perry R, O'Brien J (1999) White matter lesions on magnetic resonance imaging in dementia with Lewy bodies, Alzheimer's disease, vascular dementia, and normal aging. J Neurol Neurosurg Psychiatry 67:66–72
146. Burton EJ, McKeith IG, Burn DJ, Firbank MJ, O'Brien JT (2006) Progression of white matter hyperintensities in Alzheimer disease, dementia with Lewy bodies, and Parkinson disease dementia: a comparison with normal aging. Am J Geriatr Psychiatry 14:842–849
147. Burton EJ, Karas G, Paling SM, Barber R, Williams ED, Ballard CG, McKeith IG, Scheltens P, Barkhof F, O'Brien JT (2002) Patterns of cerebral atrophy in dementia with Lewy bodies using voxel-based morphometry. NeuroImage 17:618–630

148. Tateno M, Kobayashi S, Saito T (2009) Imaging improves diagnosis of dementia with lewy bodies. Psychiatry Investig 6:233–240
149. Zhong J, Pan P, Dai Z, Shi H (2014) Voxelwise meta-analysis of gray matter abnormalities in dementia with Lewy bodies. Eur J Radiol 83:1870–1874
150. Borroni B, Premi E, Formenti A, Turrone R, Alberici A, Cottini E, Rizzetti C, Gasparotti R, Padovani A (2015) Structural and functional imaging study in dementia with Lewy bodies and Parkinson's disease dementia. Parkinsonism Relat Disord 21:1049–1055
151. Kantarci K, Avula R, Senjem ML, Samikoglu AR, Zhang B, Weigand SD, Przybelski SA, Edmonson HA, Vemuri P, Knopman DS, Ferman TJ, Boeve BF, Petersen RC, Jack CR Jr (2010) Dementia with Lewy bodies and Alzheimer disease: neurodegenerative patterns characterized by DTI. Neurology 74:1814–1821
152. Watson R, Blamire AM, Colloby SJ, Wood JS, Barber R, He J, O'Brien JT (2012) Characterizing dementia with Lewy bodies by means of diffusion tensor imaging. Neurology 79:906–914
153. Mak E, Su L, Williams GB, Watson R, Firbank M, Blamire AM, O'Brien JT (2015) Longitudinal assessment of global and regional atrophy rates in Alzheimer's disease and dementia with Lewy bodies. Neuroimage Clin 7:456–462
154. Ballmaier M, O'Brien JT, Burton EJ, Thompson PM, Rex DE, Narr KL, McKeith IG, DeLuca H, Toga AW (2004) Comparing gray matter loss profiles between dementia with Lewy bodies and Alzheimer's disease using cortical pattern matching: diagnosis and gender effects. NeuroImage 23:325–335

Check for updates

Chapter 16

Brain Morphometry: Parkinson's Disease

Patrice Péran, Federico Nemmi, and Gaetano Barbagallo

Abstract

Concerning Parkinson's disease (PD), the utility of brain morphometry is open to debate. Indeed, standard neuroimaging techniques have had a marginal role in the diagnosis and follow-up of PD patients. The diagnosis of PD is essentially based on clinical data (neurological examination and evaluation of therapeutic response). To explore the contribution of brain morphometry in PD, we discuss changes induced by PD as a three-level system, (1) brainstem (neuronal degeneration), (2) striatum (dopaminergic deafferentation), and (3) cortical (functional deafferentation and neuronal degeneration). High-field MRI associated to advanced MRI techniques opened new perspectives to investigate neurodegenerative lesions associated to PD.

Key words Shape analysis, Substantia nigra, Putamen, Caudate, Nucleus, Brainstem

1 Introduction

Concerning Parkinson's disease (PD), the utility of brain morphometry is open to debate. Indeed, standard neuroimaging techniques have had a marginal role in the diagnosis and follow-up of PD patients. The diagnosis of PD is essentially based on clinical data (neurological examination and evaluation of therapeutic response) (UK Parkinson's Disease Brain Bank criteria) [1]. In the early phase the MR examination may be negative. Standard MR has a secondary role that is in fact limited to gross differential diagnosis with other neurological disorders. High-field MRI associated to advanced MRI techniques opened new perspectives to investigate neurodegenerative lesions associated to PD.

This promising expectation must not make us forget the PD pathophysiology. Sporadic PD is a degenerative disorder characterized mainly by progressive and focal loss of the dopaminergic neurons of substantia nigra (SN) pars compacta. Their depletion induces functional changes in the circuit of the basal ganglia (*dopaminergic deafferentation*), whose activity is modulated by SN, and eventually functional deafferentation of fronto-striatal circuit

Gianfranco Spalletta et al. (eds.), *Brain Morphometry*, Neuromethods, vol. 136,
https://doi.org/10.1007/978-1-4939-7647-8_16, © Springer Science+Business Media, LLC 2018

[2]. However, the whole pathological process involves also other structures in the brainstem at the first stage or the cerebral cortex at the last stage [3].

To explore the contribution of brain morphometry in PD, we discuss changes induced by PD as a three-level system, (1) brainstem (*neuronal degeneration*), (2) striatum (*dopaminergic deafferentation*), and (3) cortical (*functional deafferentation and neuronal degeneration*).

2 Brainstem Morphometry and PD

In 2003, Braak et al. showed a chronological progression of neurodegenerative lesion, mainly Lewy bodies (LB), beginning in the caudal medulla oblongata and progressing in an ascending course [3]. Recent clinicopathological studies show that at least 80% of cases follow the proposed scheme of Braak et al. and that LBs are detected in most of the vulnerable brainstem nuclei in more than 90% of the PD cases during the course of the disease [4]. These later lesions could be, at least in part, at the origin of non-motor symptoms in PD [5]. Medulla oblongata nuclei morphometry by means of MRI remains methodologically tricky because of the small size and the diffuse aspect of these nuclei. For the first time by means of MRI, the Onchi's group showed in vivo evidence that brainstem damage may be the first identifiable stage of PD neuropathology using voxel-based morphometry protocol [6].

Another part of brainstem is involved in PD: the mesencephalon. Indeed, the loss of dopaminergic neurons in the substantia nigra (SN) pars compacta is still considered as the main damage caused by PD.

SN morphometry is particularly difficult due to its intrinsic characteristics and its reduced size. Indeed, T1-3D-weighted images generally used in morphometry cannot permit the SN identification. The neuromelanin present in dopaminergic neurons helps to enhance the contrast between the structures and surrounding tissue in T2-weighted image or T2*-weighted image (Fig. 1). Based on these latter images, the substantia nigra could be manually identified as the band of signal hypointensity in the midbrain on continuous slices. The morphometric indexes differ according to the study published. For example, Adachi and collaborators used a measurement of the maximal thickness of the substantia nigra, but they failed to find a significant reduction in PD patients [7]. Several studies calculated the SN volume (Fig. 1) after manual segmentation (e.g., [8–11]). Unfortunately, the results comparing SN volume in PD patients and healthy controls were contradictory. Some previous studies failed to find a significant difference [10, 11] between groups in the whole SN, while a recent study found

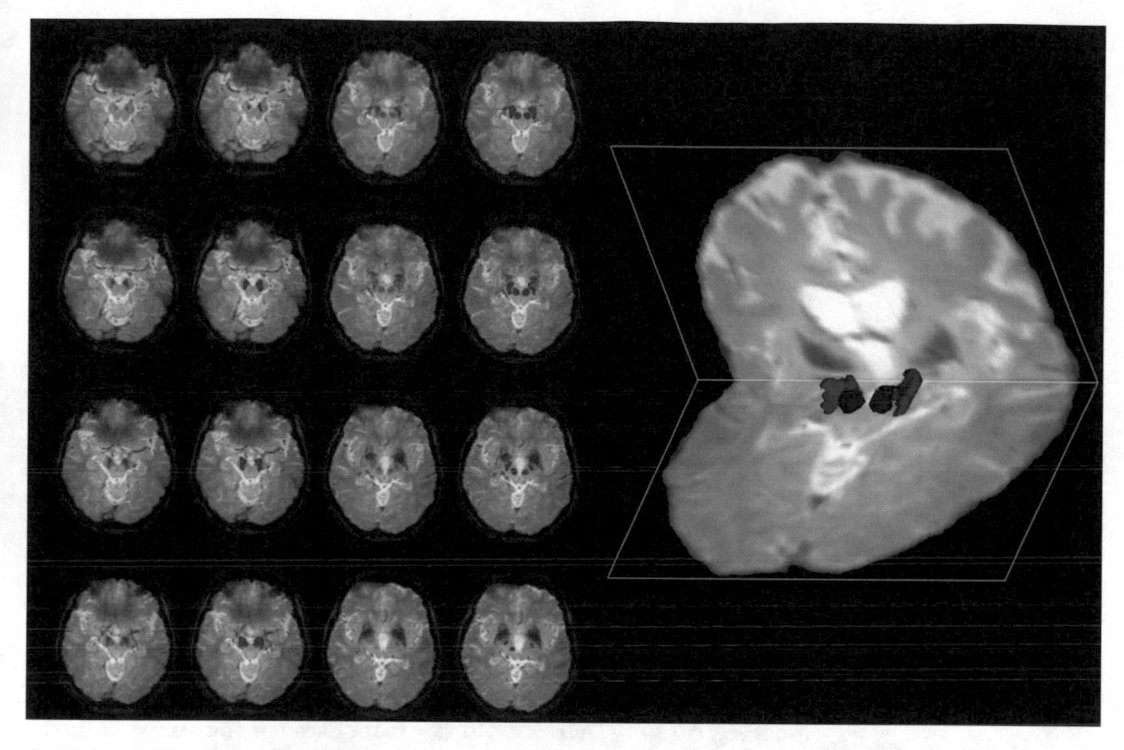

Fig. 1 Manual segmentation based on T2*-weighted images of red nucleus (red) and substantia nigra (blue). At the right side, 3D reconstruction of both structures

volumetric reduction of the SN pars compacta in the earliest stages of PD [9].

The MRI investigation of the brainstem, whether the medulla oblongata or the midbrain, is crucial to assist the diagnosis at the onset of parkinsonian premotor signs. The development of new methods of acquisition and data processing for extracting morphometric index is a current axis research. Moreover, the advent of ultrahigh-field MRI (more than 3 T) could also provide higher spatial resolution and consequently greater and more accurate anatomical definition of brainstem structures (e.g., [12]).

3 Striatum Morphometry and PD

In the context of PD, the focus of imaging research has long been on structural sequence in order to investigate the morphometry of the basal ganglia, with a particular attention to the caudate nucleus [13, 14]. Consistent volumetric alteration has proven hard to find in PD patients. While several studies have inquired volumetric difference between PD patients and healthy controls, the results have been conflicting. Some studies reported volumetric difference in the striatum and in the thalamus [15–17] with other studies failing to find any difference [10, 18].

One possibility for this lack of converging evidence about volumetric abnormalities is that the degeneration in the subcortical nuclei could be confined to specific subregions, rather than being widespread. If this would be the case, more fine-grained morphometric approach able to perform voxel-wise comparisons could be better suited for assessing differences between PD patients and healthy controls. One of the techniques that has been most commonly used to perform voxel-wise comparisons of morphometry between different groups is the voxel-based morphometry (VBM), which allows to analyze the gray matter volume in each and single voxel [19]. However, even VBM has yield conflicting evidence as for volumetric reduction in PD patients relative to healthy controls. While some studies have found volumetric reduction within the hippocampus [20] and the caudate nucleus [21], other groups were not able to replicate these findings [22, 23].

In the recent years, a new type of morphometric analysis based on T1-structural imaging has been developed: shape analysis. As in VBM, this analysis retains the ability to compare different population point by point (i.e., retaining special specificity), but instead of focusing on gray matter volume, it focuses on the shape of brain structures. This method has two appealing features: it can detect differences in shape even if the overall volume of a brain structure is similar between two groups, allowing for a true morphometric approach; it models the shape of the structure being studied as smooth and continuous, allowing for a (limited) subvoxel resolution [24]. This method has received attention in recent years, especially for PD patients. In particular, four studies have used shape analysis in order to inquire the shape differences in the subcortical nuclei of PD patients compared to healthy controls, as well as to study the association between shape variability in the same nuclei and cognitive measure or clinical scale [25–28]. Apostolova and colleagues used radial distance as proxy for shape and focused on the hippocampus, caudate nucleus, and ventricles. They found a significant difference in shape between PD patients and healthy controls in the caudate nucleus, as well as an association between radial distance in the head of the caudate nucleus and the Mini-Mental State Examination score in PD patients [25]. Similarly, Sterling and colleagues found shape difference in the caudate nucleus between PD patients and healthy controls. They also found significant differences in the putamen, as well as an association between a test measuring general cognitive ability (Montreal Cognitive Assessment) and shape in the right putamen [28]. The studies of Apostolova and Sterling relied on manual or semiautomatic approach for segmentation and shape creation. Although manual segmentation is regarded as the gold standard, it is largely operator dependent and time consuming and thus is not suitable

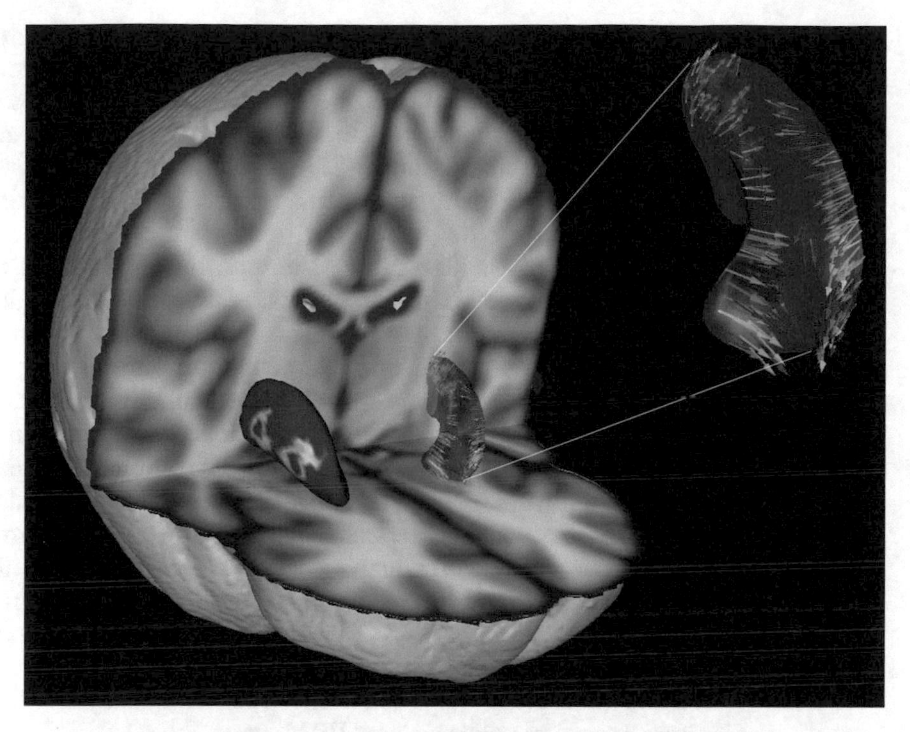

Fig. 2 Results for the vertex-wise group difference analysis of the surface displacement data for the left and right putamen structures in the PD patients versus controls. Blue color represents zone with higher difference between groups. At top right, vectors (arrows shown on the surface) indicate the direction of change [27]

for a very large study. Therefore, the validation of fully automatic approach is of extreme interest, since automatic segmentation method could allow for bigger sample size while minimizing the time needed to perform the study. Two studies have applied a fully automatic segmentation and shape creation method in order to study subcortical nuclei in PD patients, the one by Menke and colleagues (2014) and the one by Nemmi and colleagues (2015). Both these studies have used the FIRST streamline [24], part of the widely used FSL suite [29]. While Menke and colleagues only found shape differences between PD patients and healthy controls in the pallidum (a difference that was not found using standard VBM or volumetric method) [26], the results of the study of Nemmi and colleagues showed more nuanced results (Fig. 2).

In their study, these authors found that PD patients, relative to healthy controls, had lower local volume in the medial aspect of the left caudate nucleus, as well as in the medial aspect of the bilateral putamen. Moreover, Nemmi and colleagues were the first to show an association between shape and motor symptoms as measured with the Unified Parkinson's Disease Rating Scale (UPDRS): they found that shape values in the left putamen and thalamus correlated with the motor symptoms in the PD samples [27]. Noteworthy, in the same study the authors compared the discriminant power of volumetric measures of the striatum to discriminant power of striatal shape. The discriminant power (as assessed using a linear discriminant analysis) was always better when shape values were used, compared to volumetric measures [27]. Although the best accuracy obtained using shape values (83% of correctly classified subjects) was not as high as in other studies that have used multimodal information from the subcortical nuclei (e.g., 98% of correctly classified subjects in [10]), it was significantly higher than the highest accuracy obtained using volumetric data (65%) and higher than in other studies that have used cortical thickness [30], iron deposition in the striatum [31], and magnetization transfer [32].

Taken together, the results of the study presented in this section point out to the fact that PD patients clearly show morphometric abnormalities in the subcortical nuclei and especially in the striatum. However, these abnormalities seem to be of small magnitude and localized in specific subregions, rather than being widespread. It is noteworthy that the localized morphometry differences seem to reflect the pattern of greatest dopamine depletion in the striatum [17, 33]. The very mild and localized nature of these morphometric differences suggests that future studies would benefit from using shape-based morphometric analysis rather than more canonical volumetric or VBM ones.

4 Cortex Morphometry and PD

Two MRI quantitative neuroanatomical approaches have been used to evaluate cortical morphological alterations in neurodegenerative diseases: VBM and surface-based morphometry (SBM), such as cortical folding or cortical thickness.

For whole brain volumetric analysis, VBM studies have shown gray matter reductions in frontal, temporal associative, limbic, and paralimbic areas [20, 34–36] in PD patients relative to healthy controls. Moreover, these gray matter alterations have been related to cognitive impairment in these patients such as visuospatial and visuoperceptual deficits [37]. Indeed, some VBM investigations failed to find significant differences between non-demented PD patients and healthy subjects [38]. SBM methods seem to be more sensitive than VBM to identify regional cortical changes (i.e., decrease of cortical folding measures in the frontal sulcus or decrease of cortical thinning in lateral occipital cortex [37] and decrease of cortical gyrification in the inferior parietal, lateral orbitofrontal, and entorhinal cortices [39] associated to non-demented PD patients). However, further investigations are required to corroborate these results, because some studies did not found cortical gray thickness differences between PD patients with normal cognition and healthy controls [40, 41]. According to neuropathological studies of PD [3, 42], in the earliest disease stage, neural degeneration occurs in motor nuclei and limbic areas that propagate to the inferior temporal and paralimbic cortex and further reach associative areas like the prefrontal lobes. At this stage, patients become symptomatic, manifesting the typical motor features of PD; and the prefrontal cortical changes are essentially functional, expressing a consequence of dopaminergic level (*functional deafferentation*). At the later stage, the *neuronal degeneration* spreads at cortical level, leading to greater cortical atrophy and consequent cognitive decline [30, 43, 44]. This may explain why there has been a lack of reliable results from magnetic resonance cortical atrophy studies in non-demented patients with PD. Regional cortical thinning may be already present also at the time of the diagnosis in patients with early, untreated PD who do not meet the criteria for mild cognitive impairment [37], suggesting that cortical thinning can serve as a marker for initial cognitive decline in early PD.

Moreover, it is noteworthy to underline that the surface-based techniques using volumetric acquisitions seem to be promising also to better understand the underlying pathophysiology of certain motor complications (levodopa-induced dyskinesias) or non-motor signs (depression and impulse control disorders [ICDs]) in PD, as well as the modality of cognitive impairment progression. Increase of cortical thickness in the right inferior frontal sulcus and in the orbitofrontal regions and insula has been found, respectively, in PD patients with levodopa-induced dyskinesia and those with depression compared to PD patients without levodopa-induced dyskinesia [45, 46] and those without depression [47]. With regard to cortical thickness changes in patients with PD and ICDs, two studies have shown seemingly discordant results. The first one [48] revealed cortical thinning in many regions involved in reward pathways, such as the superior orbitofrontal, rostral middle frontal, and caudal middle frontal regions, in patients with PD and multiple ICDs. The second study [49] demonstrated that PD patients with ICDs have an increased cortical thickness in limbic regions when compared with those without ICDs. The discrepancies between the two studies may stem from intergroup clinical differences, such as age, disease duration, or the levodopa equivalent daily dose. Another study using a cortical folding method [50] has revealed a reduced orbitofrontal cortex local gyrification index (which quantifies the cortex folding and is strictly associated to the surface area width) in PD patients with pathological gambling.

Although future investigations are needed to verify these observations, the overall results of these recent studies suggest that the corticometry has certainly provided further insights in the progression of cognitive decline and in the pathophysiology of motor and non-motor symptoms in PD.

5 Conclusion

Although the contribution of morphometric indexes in PD pathophysiology understanding was sometimes unsatisfactory and sometimes contradictory, the latest methodological advances are very promising. In addition, MRI is able to provide brain morphometric indexes for the three levels (brainstem, basal ganglia, and cortex) simultaneously which is a great advantage to characterize changes in PD patient brains. Previous neuroimaging studies have generally focused on only one of these three levels using only one MRI modality. Recently, Péran and colleagues have developed a multimodal MRI (mMRI) approach to study such disorders. The concept is based on combination of MRI parameters from different MRI sequences sensitive to different tissue characteristics (e.g., volume atrophy, iron deposition, and microstructural damage). mMRI is able to characterize the physiological aging in healthy subjects [51] demonstrating that physiological aging does not influence the same parameters in the functioning anatomical region. Our group demonstrated also that mMRI is able to discriminate PD patients from healthy control subjects with high accuracy thanks to multiparametric nigrostriatal signature [10]. The highly discriminating combinations were composed of markers from three different MRI parameters suggesting that the MRI parameters provide different but complementary information. Among these discriminating markers, the increased substantia nigra iron content played a key role. Numerous histochemical studies and MRI studies have confirmed this specific iron accumulation (for review: [52]). The combination of different MR biomarkers opens new perspectives to investigate pathological changes, such as disease progression and long-term drug impact, detecting non-dopaminergic degeneration and evaluating neuroprotective effects.

References

1. Hughes AJ, Daniel SE, Kilford L, Lees AJ (1992) Accuracy of clinical diagnosis of idiopathic Parkinson's disease: a clinico-pathological study of 100 cases. J Neurol Neurosurg Psychiatry 55:181–184
2. Alexander GE, DeLong MR, Strick PL (1986) Parallel organization of functionally segregated circuits linking basal ganglia and cortex. Annu Rev Neurosci 9:357–381. https://doi.org/10.1146/annurev.ne.09.030186.002041
3. Braak H, Del Tredici K, Rüb U et al (2003) Staging of brain pathology related to sporadic Parkinson's disease. Neurobiol Aging 24:197–211. https://doi.org/10.1016/S0197-4580(02)00065-9
4. Grinberg LT, Rueb U, Alho AT di L, Heinsen H (2010) Brainstem pathology and non-motor symptoms in PD. J Neurol Sci 289:81–88. https://doi.org/10.1016/j.jns.2009.08.021
5. Jellinger KA (2011) Synuclein deposition and non-motor symptoms in Parkinson disease. J Neurol Sci 310:107–111. https://doi.org/10.1016/j.jns.2011.04.012
6. Jubault T, Brambati SM, Degroot C et al (2009) Regional brain stem atrophy in idiopathic Parkinson's disease detected by

anatomical MRI. PLoS One 4:e8247. https://doi.org/10.1371/journal.pone.0008247

7. Adachi M, Hosoya T, Haku T et al (1999) Evaluation of the substantia nigra in patients with Parkinsonian syndrome accomplished using multishot diffusion-weighted MR imaging. Am J Neuroradiol 20:1500–1506
8. Barbagallo G, Sierra-Peña M, Nemmi F et al (2016) Multimodal MRI assessment of nigro-striatal pathway in multiple system atrophy and Parkinson disease. Mov Disord 31:325–334. https://doi.org/10.1002/mds.26471
9. DA Z, Wonderlick JS, Ashourian P et al (2013) Substantia nigra volume loss before basal forebrain degeneration in early Parkinson disease. JAMA Neurol 70:241–247. https://doi.org/10.1001/jamaneurol.2013.597
10. Péran P, Cherubini A, Assogna F et al (2010) Magnetic resonance imaging markers of Parkinson's disease nigrostriatal signature. Brain 133:3423–3433. https://doi.org/10.1093/brain/awq212
11. Geng D-Y, Li Y-X, Zee C-S (2006) Magnetic resonance imaging-based volumetric analysis of basal ganglia nuclei and substantia nigra in patients with Parkinson's disease. Neurosurgery 58:256–262. https://doi.org/10.1227/01.NEU.0000194845.19462.7B. discussion 256–262
12. Lehéricy S, Bardinet E, Poupon C et al (2014) 7 tesla magnetic resonance imaging: a closer look at substantia nigra anatomy in Parkinson's disease. Mov Disord 29:1574–1581. https://doi.org/10.1002/mds.26043
13. Hotter A, Esterhammer R, Schocke MFH, Seppi K (2009) Potential of advanced MR imaging techniques in the differential diagnosis of parkinsonism. Mov Disord 24:S711–S720. https://doi.org/10.1002/mds.22648
14. Seppi K, Schocke MFH, Prennschuetz-Schuetzenau K et al (2006) Topography of putaminal degeneration in multiple system atrophy: a diffusion magnetic resonance study. Mov Disord 21:847–865. https://doi.org/10.1002/mds.20843
15. Kosta P, Argyropoulou MI, Markoula S, Konitsiotis S (2006) MRI evaluation of the basal ganglia size and iron content in patients with Parkinson's disease. J Neurol 253:26–32. https://doi.org/10.1007/s00415-005-0914-9
16. Lee SH, Kim SS, Tae WS et al (2011) Regional volume analysis of the Parkinson disease brain in early disease stage: gray matter, white matter, striatum, and thalamus. Am J Neuroradiol 32:682–687. https://doi.org/10.3174/ajnr.A2372
17. Pitcher TL, Melzer TR, MacAskill MR et al (2012) Reduced striatal volumes in Parkinson's disease: a magnetic resonance imaging study. Transl Neurodegener 1:17. https://doi.org/10.1186/2047-9158-1-17
18. Messina D, Cerasa A, Condino F et al (2011) Patterns of brain atrophy in Parkinson's disease, progressive supranuclear palsy and multiple system atrophy. Parkinsonism Relat Disord 17:172–176. https://doi.org/10.1016/j.parkreldis.2010.12.010
19. Ashburner J, Friston KJ (2000) Voxel-based morphometry—the methods. NeuroImage 11:805–821. https://doi.org/10.1006/nimg.2000.0582
20. Summerfield C, Junque C, Tolosa E et al (2005) Structural brain changes in Parkinson disease with dementia. Arch Neurol 62:281–285. https://doi.org/10.1001/archneur.62.2.281
21. Brenneis C, Seppi K, Schocke MF et al (2003) Voxel-based morphometry detects cortical atrophy in the Parkinson variant of multiple system atrophy. Mov Disord 18:1132–1138. https://doi.org/10.1002/mds.10502
22. Pan PL, Shi HC, Zhong JG et al (2013) Gray matter atrophy in Parkinson's disease with dementia: evidence from meta-analysis of voxel-based morphometry studies. Neurol Sci 34:613–619. https://doi.org/10.1007/s10072-012-1250-3
23. Tessitore A, Amboni M, Cirillo G et al (2012) Regional gray matter atrophy in patients with Parkinson disease and freezing of gait. Am J Neuroradiol 33(9):1804. https://doi.org/10.3174/ajnr.A3066
24. Patenaude B, Smith SM, Kennedy DN, Jenkinson M (2011) A Bayesian model of shape and appearance for subcortical brain segmentation. NeuroImage 56:907–922. https://doi.org/10.1016/j.neuroimage.2011.02.046
25. Apostolova LG, Beyer M, Green AE et al (2010) Hippocampal, caudate, and ventricular changes in Parkinson's disease with and without dementia. Mov Disord 25:687–695. https://doi.org/10.1002/mds.22799
26. Menke RAL, Szewczyk-Krolikowski K, Jbabdi S et al (2014) Comprehensive morphometry of subcortical grey matter structures in early-stage Parkinson's disease. Hum Brain Mapp 35:1681–1690. https://doi.org/10.1002/hbm.22282
27. Nemmi F, Sabatini U, Rascol O, Péran P (2015) Parkinson's disease and local atrophy in subcortical nuclei: insight from shape analysis. Neurobiol Aging 36:424–433. https://doi.org/10.1016/j.neurobiolaging.2014.07.010

28. Sterling NW, Du G, Lewis MM et al (2013) Striatal shape in Parkinson's disease. Neurobiol Aging 34:2510–2516. https://doi.org/10.1016/j.neurobiolaging.2013.05.017
29. Jenkinson M, Beckmann CF, Behrens TEJ et al (2012) FSL. NeuroImage 62:782–790. https://doi.org/10.1016/j.neuroimage.2011.09.015
30. Zarei M, Ibarretxe-Bilbao N, Compta Y et al (2013) Cortical thinning is associated with disease stages and dementia in Parkinson's disease. J Neurol Neurosurg Psychiatry 84:875–881. https://doi.org/10.1136/jnnp-2012-304126
31. Boelmans K, Holst B, Hackius M et al (2012) Brain iron deposition fingerprints in Parkinson's disease and progressive supranuclear palsy. Mov Disord 27:421–427. https://doi.org/10.1002/mds.24926
32. Eckert T, Sailer M, Kaufmann J et al (2004) Differentiation of idiopathic Parkinson's disease, multiple system atrophy, progressive supranuclear palsy, and healthy controls using magnetization transfer imaging. NeuroImage 21:229–235. https://doi.org/10.1016/j.neuroimage.2003.08.028
33. Smith Y, Villalba R (2008) Striatal and extrastriatal dopamine in the basal ganglia: an overview of its anatomical organization in normal and Parkinsonian brains. Mov Disord 23:534–547. https://doi.org/10.1002/mds.22027
34. Burton EJ, McKeith IG, Burn DJ et al (2004) Cerebral atrophy in Parkinson's disease with and without dementia: a comparison with Alzheimer's disease, dementia with Lewy bodies and controls. Brain 127:791–800. https://doi.org/10.1093/brain/awh088
35. Nagano-Saito A, Washimi Y, Arahata Y et al (2005) Cerebral atrophy and its relation to cognitive impairment in Parkinson disease. Neurology 64:224–229. https://doi.org/10.1212/01.WNL.0000149510.41793.50
36. Ramírez-Ruiz B, Martí MJ, Tolosa E et al (2005) Longitudinal evaluation of cerebral morphological changes in Parkinson's disease with and without dementia. J Neurol 252:1345–1352. https://doi.org/10.1007/s00415-005-0864-2
37. Pereira JB, Ibarretxe-Bilbao N, Marti MJ et al (2012) Assessment of cortical degeneration in patients with Parkinson's disease by voxel-based morphometry, cortical folding, and cortical thickness. Hum Brain Mapp 33:2521–2534. https://doi.org/10.1002/hbm.21378
38. Price S, Paviour D, Scahill R et al (2004) Voxel-based morphometry detects patterns of atrophy that help differentiate progressive supranuclear palsy and Parkinson's disease. NeuroImage 23:663–669. https://doi.org/10.1016/j.neuroimage.2004.06.013
39. Zhang Y, Zhang J, Xu J et al (2014) Cortical gyrification reductions and subcortical atrophy in Parkinson's disease. Mov Disord 29:122–126. https://doi.org/10.1002/mds.25680
40. Price CC, Tanner J, Nguyen PT et al (2016) Gray and white matter contributions to cognitive frontostriatal deficits in non-demented Parkinson's disease. PLoS One 11:1–21. https://doi.org/10.1371/journal.pone.0147332
41. Danti S, Toschi N, Diciotti S et al (2015) Cortical thickness in de novo patients with Parkinson disease and mild cognitive impairment with consideration of clinical phenotype and motor laterality. Eur J Neurol 22:1564–1572. https://doi.org/10.1111/ene.12785
42. Braak H, Ghebremedhin E, Rüb U et al (2004) Stages in the development of Parkinson's disease-related pathology. Cell Tissue Res 318:121–134. https://doi.org/10.1007/s00441-004-0956-9
43. Pagonabarraga J, Corcuera-Solano I, Vives-Gilabert Y et al (2013) Pattern of regional cortical thinning associated with cognitive deterioration in Parkinson's disease. PLoS One 8(1): e54980. https://doi.org/10.1371/journal.pone.0054980
44. Hwang KS, Beyer MK, Green AE et al (2013) Mapping cortical atrophy in Parkinson's disease patients with dementia. J Parkinsons Dis 3:69–76. https://doi.org/10.3233/JPD-120151
45. Cerasa A, Morelli M, Augimeri A et al (2013) Prefrontal thickening in PD with levodopa-induced dyskinesias: new evidence from cortical thickness measurement. Park Relat Disord 19:123–125. https://doi.org/10.1016/j.parkreldis.2012.06.003
46. Cerasa A, Salsone M, Morelli M et al (2013) Age at onset influences neurodegenerative processes underlying PD with levodopa-induced dyskinesias. Park Relat Disord 19:883–888. https://doi.org/10.1016/j.parkreldis.2013.05.015
47. Huang P, Lou Y, Xuan M et al (2016) Cortical abnormalities in Parkinson's disease patients and relationship to depression: a surface-based morphometry study. Psychiatry Res 250:24–28. https://doi.org/10.1016/j.pscychresns.2016.03.002
48. Biundo R, Weis L, Facchini S et al (2015) Patterns of cortical thickness associated with

impulse control disorders in Parkinson's disease. Mov Disord 30:688–695. https://doi.org/10.1002/mds.26154

49. Tessitore A, Santangelo G, De Micco R et al (2016) Cortical thickness changes in patients with Parkinson's disease and impulse control disorders. Parkinsonism Relat Disord 24:119–125. https://doi.org/10.1016/j.parkreldis.2015.10.013
50. Cerasa A, Salsone M, Nigro S et al (2014) Cortical volume and folding abnormalities in Parkinson's disease patients with pathological gambling. Park Relat Disord 20:1209–1214. https://doi.org/10.1016/j.parkreldis.2014.09.001
51. Cherubini A, Péran P, Caltagirone C et al (2009) Aging of subcortical nuclei: microstructural, mineralization and atrophy modifications measured in vivo using MRI. NeuroImage 48:29–36. https://doi.org/10.1016/j.neuroimage.2009.06.035
52. Ward RJ, Zucca FA, Duyn JH et al (2014) The role of iron in brain ageing and neurodegenerative disorders. Lancet Neurol 13:1045–1060. https://doi.org/10.1016/S1474-4422(14)70117-6

Check for updates

Chapter 17

Brain Morphometry in Multiple Sclerosis

Ilona Lipp, Nils Muhlert, and Valentina Tomassini

Abstract

Changes in the volume of brain gray matter in people with multiple sclerosis (MS) are a core MRI feature of the disease. Alterations in gray matter volume have been found to reflect the severity of the disease but also the emergence of cognitive impairment in people with MS. These findings have led to increasing research into the neuropathological factors affecting gray matter volume and outcomes associated with gray matter volume changes. In this chapter, we discuss the methods used to assess gray matter volume on MRI, links to brain lesions, and other MS-related MRI features and the potential underlying mechanisms that may lead to gray matter atrophy. We consider the relationships between altered regional gray matter volume and both physical and cognitive disability. Last, we discuss how a better understanding of brain volume changes and their implications in MS may help us to improve the use of this metric in clinical trials and in estimating prognosis in individual patients.

Key words Multiple sclerosis, MRI, Atrophy, Cortical thickness, Brain volume, Cognition, Disability

Abbreviations

BMS	Benign multiple sclerosis
CNS	Central nervous system
CSF	Cerebrospinal fluid
DMT	Disease-modifying treatment
DTI	Diffusion tensor imaging
EDSS	Expanded disability status scale
GM	Gray matter
MRI	Magnetic resonance imaging
MS	Multiple sclerosis
MSFC	Multiple sclerosis functional composite
RRMS	Relapsing-remitting multiple sclerosis
PPMS	Primary progressive multiple sclerosis

Ilona Lipp and Nils Muhlert contributed equally to this work.

Gianfranco Spalletta et al. (eds.), *Brain Morphometry*, Neuromethods, vol. 136,
https://doi.org/10.1007/978-1-4939-7647-8_17, © Springer Science+Business Media, LLC 2018

SPMS	Secondary progressive multiple sclerosis
VBM	Voxel-based morphometry
WM	White matter

1 Introduction

Multiple sclerosis (MS) is an autoimmune condition of the central nervous system (CNS) characterized by recurrent episodes of inflammation with demyelination and neurodegeneration [1]. These episodes manifest themselves through a variety of symptoms, most commonly affecting motor, sensory, visual, and cognitive functions [2]. MS usually onsets in young adulthood (aged 20–40 years) and is more common in women. In the majority of cases, after a clinically isolated syndrome, MS evolves in a relapsing-remitting form, followed by secondary progressive multiple sclerosis (SPMS); a progressive worsening of clinical disability over time, with an insidious onset, characterizes the primary progressive multiple sclerosis (PPMS) form of the disease [3].

MS is characterized by an immune attack against the myelin sheaths surrounding the axons of the neurons [4]. This inflammation can lead to axonal damage [5]. Evidence of neurodegeneration that extends beyond the areas of inflammation suggested that there is a primary neurodegenerative component in MS, followed by secondary inflammation [6–8].

Inflammatory damage in MS can be detected with magnetic resonance imaging (MRI) [9], which has been used both as a diagnostic and a prognostic tool. MS lesions in the white matter (WM) can be seen as hyperintense on T2-weighted acquisitions. Active lesions are detected as hyperintense areas on T1-weighted scans acquired after the injection of the contrast agent gadolinium-DTPA and indicate a breakdown in the blood-brain barrier. Hypointense lesions visible on post-contrast T1-weighted scans reflect severe tissue damage and are called "chronic black holes" [10, 11]. Lesions also occur in the gray matter (GM) [12] and are characterized by transection of axons and dendrites as well as cell death [13]. GM lesions histopathologically differ from WM lesions, as they do not involve breakdown of the blood-brain barrier and show significantly less inflammation [14–16].

MS lesions have traditionally been the main focus of disease diagnosis [17], prognosis [18] and response to treatment [19, 20]. More recently, MS research has turned its focus to GM abnormalities and brain volume loss not only as prognostic factors [21] but also as outcome measures of clinical trials assessing the efficacy of disease-modifying treatments (DMTs) [22]. In this chapter, we discuss the available methods for measuring MS pathology in the brain GM using MRI. We then discuss potential causes for GM changes and the relationship with damage in WM. Last, we outline the relationship between GM damage and disability and its potential applicability to monitor and predict disease evolution. This chapter will not cover global or tract-specific WM atrophy or atrophy of the spinal cord and optic nerve, despite their undisputed contribution to symptom and disease severity [23–25]. The term *atrophy* will be used to describe a decrease in GM volume over time. In contrast, we will use the term *volume loss* to describe a mean difference between patients with MS and healthy controls.

2 Measuring Atrophy in MS: Common Methods and Challenges

Brain volume is commonly measured using high-resolution T1-weighted MRI scans. Isotropic resolutions of, or around, 1 mm can be acquired within clinically feasible times (<10 min) and have sufficient contrast for distinguishing separate tissue classes (e.g., GM, WM, cerebrospinal fluid [CSF]). Delineation of specific structures is occasionally carried out using manual tracing, but the vast majority of studies use automated analytic techniques for estimating volume. These automated techniques benefit from less user bias, are highly reproducible, and demonstrate comparable results to manually defined region-of-interest approaches [26].

Segmentation algorithms first separate the brain images into different tissue classes. Tissue fractions can be generated by dividing each tissue volume by the summed GM, WM, and CSF volumes, called the intracranial volume. Regional brain volumes can also be estimated by voxel-based morphometry (VBM) or shape-based parcellation algorithms. The former relies on the registration of each voxel within the image to a common template, with the amount of warping needed to move voxels from their original, native space into the common space serving as an index of regional volume. Common analysis routines used for VBM are implemented in the statistical parametric mapping (SPM; Wellcome Trust Centre for Neuroimaging, UCL Institute of Neurology, http://www.fil.ion.ucl.ac.uk) [27] software and in FSL (the FMRIB Software Library, where FMRIB is the Oxford Centre for Functional MRI of the Brain [28]). Shape-based parcellation of subcortical structures (such as the hippocampus, thalamus, and caudate nucleus) can be carried out using FreeSurfer [29, 30] or FSL's FIRST (FMRIB's Integrated Registration and Segmentation Tool [31]), which exploits the probabilistic relationships between the intensity and shape of these structures. The use of shape-based parcellation routines is particularly advantageous for determining volumes of striated structures (i.e., those containing both GM and WM), like the thalamus and globus pallidus, which can be challenging to segment using standard techniques. Alternatively, surface-based analysis methods can be used to measure the cortical surface area or cortical thickness, using the FreeSurfer software [32]. The variety of software tools available provides a number of choices for measuring brain tissue volumes in people with MS. These methods are not free from confounds, as the algorithms used by each software package can affect the final results. The results of VBM analysis can differ depending on whether it is carried out in SPM or FSL [33, 34]. This can be relatively small differences introduced by segmentation algorithms, to large differences introduced by registration algorithms and statistical approaches. The lack of a gold standard for measuring volume precludes calibration of algorithms or even an assessment of which software offers the most accurate results.

A number of potential confounds are known to affect findings in studies of brain volume and cortical thickness. VBM results may be influenced by the field strength, with regional differences seen when comparing images acquired on 1.5 and 3 T MRI scanners [35]. Many clinical studies are conducted on hospital scanners with 1.5 or 1 T field strengths. The choice of smoothing kernel can also influence findings [36], as well as the use of modulation, a scaling method, which maintains local volume size during registration to standard space [37]. Additional factors specific to MS should also be considered. In MS, the presence of WM lesions can affect both estimates of GM volumes [38] and cortical thickness [39]. This bias can occur due to an inappropriate assignment of lesion signal intensities into WM signal intensity distributions, affecting the estimation of the boundary between different tissue classes, and subsequent tissue volumes. The bias caused by WM lesions can be minimized by marking and "filling" lesions using mean WM intensities from the whole brain [38] or from the surrounding neighboring voxels of lesions [39–42].

3 Regional Distribution of GM Volume Changes

Most studies on GM volume in MS have examined global whole brain or GM volume. Most commonly used measures have been brain parenchymal fraction (BPF; e.g., Bermel et al. [43]) or the width of the third ventricle (a surrogate marker of brain atrophy). Both suggest decreased total brain volume in MS patients relative to healthy controls [44–48], and ventricle size is highly correlated with BPF [49]. Enlarged ventricles are commonly observed in MS brains with long disease duration (Fig. 1).

MS-related damage is not randomly distributed over the brain [1]. In the WM, lesions are particularly common within periventricular regions, such as those neighboring the lateral ventricle and in the superior and posterior corona radiata. This observation has

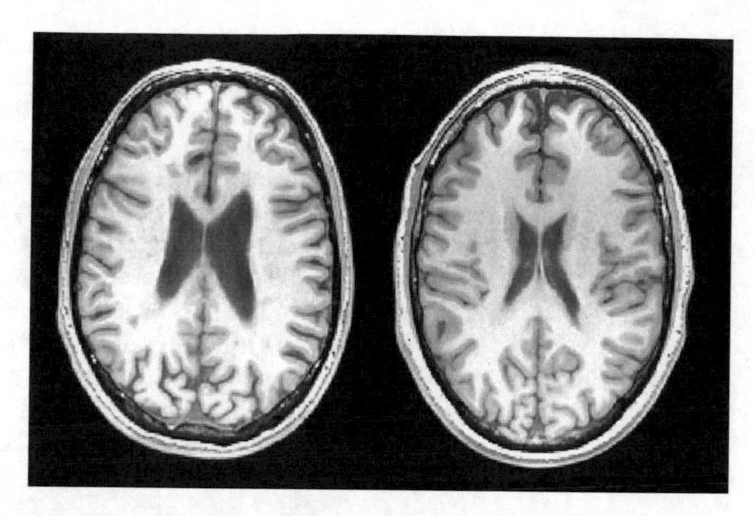

Fig. 1 Left: T1-weighted axial brain slice of a 36-year-old male with RRMS and a disease duration of 12 years. Right: scan acquired with the same parameters in similar location of a male of the same age

been quantitatively confirmed by demonstrating highly consistent lesion probability maps across studies [50–53]. Similarly, within GM, demyelinated lesions, which are more frequently subpial, are seen along the entire cortical ribbon, but their incidence and size are significantly larger in cortical sulci and in deep invaginations of the brain surface [1]. GM and WM damage do not happen completely independently, and histochemistry shows that GM neurodegenerative processes are mostly pronounced in the cortex overlaying subcortical demyelinated lesions [1]. Histopathological [16, 54, 55] and neuroimaging studies [56] show the greatest number of cortical lesions in the frontal and temporal lobes. However, comparing patients with healthy controls, GM volume loss has been reported throughout cortical and subcortical areas. Individual studies have reported significant loss within the cingulate cortex [57, 58], temporal lobe [57, 59], insula [59, 60], and cerebellum [57, 60, 61]. A meta-analysis of 19 studies [62] has concluded that GM loss in RRMS and CIS is most pronounced in the thalamus and basal ganglia, within the cingulate cortex and around the central gyrus. Cortical thinning in MS patients compared to healthy controls has been found in frontal, parietal, occipital, temporal, and insular lobes [63–66]. The volume loss in MS patients compared to healthy controls is likely a result of higher atrophy rates. In a longitudinal study conducted with patients with PPMS, Eshaghi et al. [67] showed regional differences in GM atrophy, with the greatest atrophy observed in the cingulate gyri. Regional differences in GM atrophy studies may be influenced by methodological factors, as discussed above. Despite this caution, it is clear that atrophy in MS occurs in both cortical and subcortical GM.

The basal ganglia form complex circuits with cortical and subcortical structures and are involved in motor and cognitive functions [68]. In RRMS, focal atrophy within basal ganglia structures has been reported, with up to 20% volume loss within the caudate nuclei [49, 69]. Many basal ganglia circuits also involve the thalamus, which coordinates several sensory-motor pathways and has been viewed as a hub between cortical and subcortical regions. Thalamic volume changes are frequently reported in MS [70], with neuronal volume loss at levels up to 30% [45, 71]. Thalamic atrophy impacts significantly upon cortical functional networks in MS [72]. Indeed, given the role of the thalamus as a hub region in the brain, it has become clear that GM volume within this region may be one of the most important predictors of clinical and cognitive dysfunction in MS [73–75].

4 Interplay Between GM and WM Pathology

The processes that characterize MS occur primarily through the loss of myelin, followed by axonal degeneration as a result of demyelination and subsequent metabolic changes [76]. MRI research in MS was initially dominated by a focus on WM changes. Despite increasingly common investigations of GM changes, at a theoretical level, these effects are persistently attributed mainly to prior WM changes [66]. This attribution is partly supported by evidence that the distribution of GM damage throughout the brain is in part related to damage to connected WM. This relationship has been shown using WM lesions and also more subtle changes in "normal-appearing" WM. For example, the ventricular enlargement seen in the lateral and third ventricles is associated with the total volume of WM lesions [77]. Also, regional GM atrophy in MS can be statistically predicted by diffusion tensor imaging (DTI) metrics (probing microstructural alterations) in connected WM tracts [66]. This relationship between GM and WM damage was observed in both deep and cortical GM in those with RRMS and with deep GM in those with SPMS. These findings provide a clear link between WM and GM changes but also suggest that pathology in different subtypes of MS may be driven by separate pathophysiological mechanisms [1]. We do not yet know the order in which different forms of pathology occur or what drives each form of pathology, but well-designed longitudinal imaging studies in MS can help to address these questions.

The thalamus has also been well studied in MS, with evidence suggesting that it may be both directly affected by GM pathology and indirectly affected by WM pathology in connected tracts. A correlation between total lesion volume and GM volume in the thalamus has been reported repeatedly [78–81]. In people with clinically isolated syndrome (CIS), thalamocortical tracts show a tenfold higher lesion density than other tracts [82]. In the same study, lesion volume specifically within these thalamocortical tracts was found to significantly predict thalamic volume. Similarly in MS, thalamic volume has been found to be associated with WM lesion volume and diffusion MRI metrics (fractional anisotropy) in its connected tracts [66]. At histopathology, substantial deep GM pathology is seen in MS, with focal demyelinating lesions, inflammation and diffuse neurodegeneration [83]. However, a recent MRI study has identified substantial GM abnormalities on magnetization transfer imaging in GM regions' neighboring ventricles, suggesting the role of CSF factors in driving pathology [84]. This interesting hypothesis is supported by the general finding that deep GM lesions tend to follow ventricles and CSF, providing an alternative pathophysiological mechanism for atrophy in deep GM structures [16].

5 Atrophy and the Influence of Disease Stage

Throughout adulthood, the normal brain shrinks, and the rate of shrinkage accelerates throughout the lifespan. In MS, this shrinking happens faster, at a rate of about 0.5% per year compared to 0.3% per year in healthy controls [85–87]. Some studies report even higher atrophy rates of up to 2% [88, 89], and atrophy rates may be higher still in those with faster disease evolution [90]. This increase in atrophy rate can explain the, often substantial, brain volume differences between MS patients and healthy controls [90, 91]. Volume loss compared to controls has been reported in RRMS [62, 92], as well as in progressive forms [67, 93, 94].

It has been argued that MS is a neurodegenerative disease and it enters the progressive stage when the brain's capacity to compensate for damage has exceeded its limit [6]. With this in mind, one would expect GM damage to be more extensive and accelerated in progressive stages, which is supported by histological studies [83, 95], but which is not always found in MRI studies [46, 96, 97]. Some studies suggest that the spatial distribution of atrophy differs between disease subtypes and could be responsible for the different clinical phenotypes [1, 77].

GM volume loss has been observed at the earliest stages of the disease, with changes apparent in patients with CIS, i.e., those who have experienced a single inflammatory event and in whom MS has not yet been diagnosed. Cross-sectional studies report GM volume loss in the whole-brain GM [98], thalamus, basal ganglia, and brain stem in CIS [99] and within both deep GM and cortical areas [60]. However, these findings are not consistently seen, with many failing to show GM alterations [98, 100, 101]. Differences in the nature of samples could help to explain these inconsistencies.

For example, CIS patients with higher lesion load have been found to have lower subcortical GM volumes [102]. Additionally, there are likely to be differences between those with CIS who eventually convert to MS and those who do not. "Converters" have been found to show lower GM volume before conversion compared to "non-converters" [61], as well as higher GM atrophy rates within the first few years after the clinical event [59, 103].

Ceccarelli et al. [100] found significantly different rates of GM volume loss among clinical phenotypes (CIS, RRMS, SPMS, and PPMS), with it being highest in SPMS patients. Similarly, lower GM volume has been reported for RRMS when compared to patients with progressive MS [98, 104, 105]. Such comparisons may sometimes be confounded by the typically longer disease duration and older age of SPMS compared to RRMS patients, as disease duration influences brain volumes in both RRMS and SPMS [89]. However, other factors are likely to have a greater influence, as a longitudinal study has demonstrated that RRMS patients who later convert to progressive course show lower baseline BPF and GM and greater decreases in these metrics than those who do not convert within the subsequent 4 years [87]. Therefore, it is likely that pathophysiological mechanisms relating to the progressive pathology of MS accelerate volume changes over and above the effects of age or disease duration. Additionally, the regional distribution of volume loss seems to differ between early MS and later stages of progressive MS [106].

One approach to study the influence of disease severity on GM volume is by comparing patients with benign MS (BMS) [107] to MS patients with a similar disease duration but more severe disease. Patients with BMS show lower GM volumes than healthy controls [79] but also lower cortical lesion volume [108] and less severe GM volume changes than patients with RRMS [98, 109] or SPMS [79]. Some evidence indicates that it may be the spatial distribution of GM changes, rather than the extent, that differs between patients with BMS and other patients [53]. Additionally, patients with clinically BMS but cognitive deficits do have GM volume loss similar to more severe forms of MS [110].

Only a few longitudinal studies have directly compared atrophy rates between disease subtypes. Ge et al. [89] compared rates between RRMS and SPMS, reporting a 2% annual decrease in brain volume in SPMS and a 1.5% decrease in RRMS, but this difference was not statistically significant. Similarly, Kalkers et al. [111] and De Stefano et al. [85] showed no differences in annualized atrophy rate between MS subtypes. Korteweg et al. [112] suggest that patients with higher baseline lesion volume have a greater atrophy rate over the subsequent 2 years. Looking at cortical thickness, Calabrese et al. [106] demonstrated a faster change over time in SPMS and late RRMS than in early RRMS. Studies examining BMS suggest that it may be characterized by slower rates of atrophy than patients with more active early MS [109, 113]. Possible confounds for studies comparing atrophy rates across different disease subtypes are the baseline characteristics of the cohorts [114]. Larger-scale studies are needed to identify predictors of brain atrophy rates.

6 Mechanisms Leading to GM Atrophy

As GM volume loss seems to be a hallmark in MS and relates to both disease severity and disease prognosis [115], identifying the mechanisms underlying GM atrophy in MS could offer useful tools for both monitoring the disease and measuring the efficacy of therapeutic interventions [116]. In order to identify and understand the molecular and cellular processes underlying the GM changes visible on MRI, a combination of histopathological evidence and advanced brain imaging tools is crucial.

Initial views on disease mechanisms in MS focused on axonal loss as the cause for degeneration of connected GM tissue (i.e., Wallerian degeneration). A recent line of thought is that MS is not initially triggered by the immune system malfunctioning (outside-in hypothesis [117]) but that demyelination may be the underlying disease mechanism, with myelin debris then triggering the immune system to react, amplifying the tissue damage (inside-out hypothesis [6]). This view was bolstered by Stys et al. [8], amongst others, based on evidence that demyelination also occurs in areas with low levels of inflammation (such as the GM [13]), which argues against a purely inflammation-driven disease process; note that cortical inflammation has been reported in early MS [118, 119] and meningeal inflammation might be involved in cortical neuronal loss [120]. In either case, the demyelination that co-occurs with inflammation triggers several mechanisms that can affect the GM.

Recently, attention has been drawn to the involvement of mitochondria and energy metabolism following demyelinating events [121]. These biochemical processes can also trigger neuronal loss in MS tissue. Active microglia and macrophages, which are present at inflammatory sites, release radical oxygen species and nitric oxide, which can oxidize macromolecules in the neurons, inhibit their function, and promote their degeneration [122]. In particular, radical oxygen and nitric oxygen species can impact upon mitochondrial function as they directly inhibit parts of the respiratory chain [123]. Additionally, due to demyelination, ion channels are rearranged along the axons of the neurons, and more energy is needed for ion transport. This can lead to an increase in nonenergy-demanding ion exchange mechanisms and to a Ca^{2+} overload in the cell, further triggering mechanisms of neurodegeneration [123]. Iron is stored as ferritin in myelin, and if myelin is destructed, iron starts to move freely and contributes to oxidative stress [76]. An increased demand in energy coupled with energy deficiency can lead to cell death and axonal transection. These findings suggest that axonal loss and demyelination can affect GM damage in both direct and indirect ways.

In a recent paper, Haider et al. [1] distinguish two main patterns of neurodegeneration. The first is characterized by retrograde degeneration and the pattern of damage that follows WM lesions, as discussed above. The second is related to oxidative stress and happens across the entire cortex. Combining ex vivo MRI scanning with histology in post-mortem studies can help to understand which of these molecular processes underlie MRI measures of GM volume. For example, Popescu et al. [124] found that neuronal density, neuronal size, and axonal density all contribute to GM volume measures. Therefore, axonal degeneration, as well as demyelination, could contribute to in vivo measures of GM atrophy. In deep GM, there is reduced neuronal density in MS patients compared to healthy controls, with lower neuronal density in demyelinated MS brain tissue than in non-demyelinated tissue and lower neuronal density in non-demyelinated MS brain tissue compared to healthy control tissue [16, 45]. These findings indirectly suggest that the GM volume loss detected with MRI might be at least influenced by neuronal loss triggered by demyelination.

As with most studies of MRI changes in MS, imaging studies investigating GM changes are challenging to carry out. For example, it is estimated that only a small subset of existing cortical lesions can be detected with conventional MRI methods [125, 126]. Therefore, abnormalities and changes in GM volume measures may reflect the presence of undetected GM lesions [14]. In support of this view, non-lesional MS cortical tissue shows similar cell density (neuronal, glial, and synaptic density) to tissue in healthy volunteers [127]. A recent MRI study investigated the spatial co-localization of GM lesions and atrophy and found only weak correspondence [128]. In a longitudinal study [106], the appearance of new cortical lesion correlated with cortical atrophy early in the disease course, but only when averaged across the whole GM and not on a region by region basis. Even if undetected GM lesions contribute to findings of GM atrophy in MS, it still needs to be resolved whether this is a methodological effect (e.g., different relaxation time in lesioned GM) or whether the cellular changes within lesions (e.g., cell death) compromise GM volume. Another challenge concerns the registration between MRI data and histological slices. Newer methods such as CLARITY, which permits histological assessment in intact tissue, will allow more direct links and comparisons between MRI images and histology [129].

7 Application and Predictive Value of GM Volume Measures

Whilst the mechanisms underlying GM atrophy are not fully understood, GM volume measures might be useful as markers for disease evolution and prognosis [22]. WM lesion count and volume remain the most commonly used clinical outcome measures for pharmaceutical studies, as DMTs aim to reduce the risk of developing new lesions and new symptoms. However, there are substantial drawbacks to an exclusive focus on WM pathology. The correlation between WM lesion load and disability scores is inconsistent, a phenomenon often referred to as the "clinico-radiological paradox" [130]. It is unclear whether this paradox is caused by methodological limitations, such as missed lesional tissue [131, 132], or whether parameters other than lesions are better predictors of MS damage. Increasingly, researchers have focused on the relationship between atrophy and disability. It has been proposed that GM volume can explain some of the remaining variance and may even be a better indicator of disease progression and clinical outcome than WM damage [133, 134]. Only recently, studies have started looking at the effects of drugs on GM atrophy, with DMTs having been shown to slow atrophy rates [22].

7.1 Clinical Disability Correlates with Whole-Brain GM Volume and Regional Volume Loss

The clinical relevance of whole-brain and regional GM volumes can be established by examining their associations with disability scores. The heterogeneity of disability in MS creates obvious challenges when assessing this relationship. The symptoms cover a wide range of sensory, motor, and cognitive dysfunctions, which creates difficulties when summarizing overall disability levels. The most commonly used measure of disability is the Expanded Disability Status Scale (EDSS [135]), which is based on the neurological examination of various functional systems but is primarily affected by the mobility of a patient. The second most frequently used measure of disability is the multiple sclerosis functional composite (MSFC [136]), which consists of tests of mobility, fine motor skills, and cognition. The range of symptoms covered by the EDSS and MSFC makes pinpointing of regional associations between disability and GM atrophy challenging. Additionally, the EDSS is not an interval scale, which makes it statistically more challenging to detect clinical-MRI associations [137]. Despite these limitations, a number of studies have reported negative correlations between disability scores and GM volume measures in the form of either whole-brain volume, GM volume, ventricular volume, or cortical thickness [46, 49, 87, 98, 105, 138–141].

More regionally specific correlations have been demonstrated for cortical as well as deep GM. Correlations between EDSS and GM volume and thickness were shown in the precentral, medial and superior frontal cortices, cingulate, insula and other cortical regions [61, 65, 142], thalamus, putamen, and cerebellum [61, 102]. In a more symptom-specific analysis approach, Calabrese et al. [63] showed that the severity of motor symptoms was correlated with cortical thickness in the precentral gyrus, while visual symptoms were correlated with thickness in the occipital cortex. Similarly, Henry et al. [99] related cerebellar system symptoms with atrophy in the cerebellum and disability measured with the MSFC with atrophy in the cerebellum, caudate, and putamen.

Due to the variety of methods used, results across studies are far from consistent with some failing to find an association between disability and GM volumes or cortical thickness [64, 143]. In an approach to combine existing evidence, Lansley et al. [62] conducted a meta-analysis of 19 VBM studies that had looked at the relationship between EDSS and GM volume. While widespread brain volume loss in patients compared to healthy controls was confirmed, the analysis revealed only a single cluster, in which GM volume was related to EDSS, encompassing the left pre- and postcentral cortex.

7.2 The Relationship Between GM Damage and Cognition

A burgeoning area of MS research is the one exploring links between GM changes and cognitive impairment [144]. This work has become of increasing importance as the drugs used to treat MS have proven effective in slowing disability worsening, but as yet have had limited impact on slowing cognitive decline [144]. GM lesions and volume are significant predictors of cognitive impairment, both in cross-sectional [145] and longitudinal studies [146]. GM lesions and atrophy are independent predictors of cognitive function [128], emphasizing the need to consider each separately. In addition, other MRI metrics have proved useful for understanding cognitive changes, such as alterations in the complexity of diffusion in GM [147, 148] or abnormally low levels of glutamate in relevant GM regions [149]. These changes offered similar or greater predictive value than measures of regional brain volume alone, so complement the use of conventional imaging techniques when trying to understand the role of subtle structural GM damage on cognitive impairment.

7.3 Atrophy Rate Correlates with Disability Change

Even stronger evidence that GM volume and MS disability influence each other comes from studies that show how both of these variables change in synchrony. The change of GM volume per year has been shown to correlate with change of disability as measured by EDSS and MSFC [67, 77, 87, 138, 150–154]. This is the case even when correlating the short-term change of GM volume with long-term change in disability [138], suggesting that brain atrophy may not spontaneously slow down. Even GM volume at baseline can predict disease progression. Most of this evidence comes from studies comparing CIS patients who convert to MS with patients who do not convert. In particular, lower baseline GM volume and more pronounced GM atrophy are reported in "converters" [61, 87, 103].

Recent studies by Filippi et al. [155] and Popescu et al. [156] showed that baseline GM atrophy can predict EDSS up to 13 years later even when controlling for baseline EDSS [156].

The sizes of these reported effects are all small but statistically significant at a group level. This makes it difficult to predict an individual patient's prognosis based on their MRI scans. Additionally, the effects seem to be widespread across the brain with regional differences between studies, making it challenging to judge which GM measures are the most useful brain imaging markers and predictors of disability progression. In order to identify markers suitable for assessing an individual's prognosis, we need further information about "normal" and "pathological" rates of atrophy. De Stefano et al. [85] attempted to find a cutoff in annualized percent brain volume change that differentiates patients from controls, as well as patients with strong disability incline from patients with weak disability incline. As the effects of GM atrophy seem to be regionally dependent, region-specific markers might allow even higher specificity and sensitivity. This needs large, multicenter longitudinal studies with clear validation samples.

7.4 Atrophy as an Outcome Measure in Clinical Trials

Due to the increasing evidence of the importance of GM pathology in MS, GM volume measures are starting to be used as outcome measures in clinical trials. DMTs such as laquinimod [157], fingolimod [158, 159], interferon beta [161–164] and glatiramer acetate [161, 163] have been shown to slow down GM atrophy [160–162] (for a review, see De Stefano et al. [22]), but no effect [164, 165] or the opposite effect has also been reported [166]. A recent trial based meta analysis has indicated that DMTs may impact upon disability through independent effects on both lesions and atrophy [167].

Clinical trials using GM atrophy as an outcome measure have also had to contend with treatment related decreases in volume. This "pseudoatrophy" can be seen as a paradoxical acceleration in brain volume loss following the initiation of therapy. It is thought to be caused by the resolution of inflammation and may reflect a decrease in edema [168] or changes in the volume of inflammatory cells, such as glial cells [169]. Studies of regional brain volume indicate that pseudoatrophy may be confined to WM regions [170], suggesting that VBM studies of GM volume may be less affected.

8 Summary and Conclusions

Substantial evidence supports the existence and clinical relevance of both cortical and subcortical GM volume loss and atrophy in all MS subtypes. Neuropathology [1] and neuroimaging [171] are progressively clarifying the nature of GM pathology in MS and its relationship with WM damage [172]. Improving further our understanding of the mechanisms underlying GM damage and the relationship between GM and WM damage remains a goal of future research. The clinical relevance of GM damage is supported by the association between clinical characteristics of MS and GM pathology. There is relationship between GM damage and the development of clinical disability, especially cognitive dysfunction [144] . GM plays a role in predicting the evolution of the disease, i.e., the conversion from CIS to MS and from RRMS to progressive MS [61, 87] . GM damage is becoming a relevant outcome measure for immunomodulatory [22] and neuroprotective strategies [173] . Despite this emerging role of GM pathology in MS, more work remains to be done in order to translate the application of GM atrophy measurements in individual patients [174].

References

1. Haider L, Zrzavy T, Hametner S et al (2016) The topography of demyelination and neurodegeneration in the multiple sclerosis brain. Brain 139:807–815. https://doi.org/10.1093/brain/awv398
2. Weiner H, Stankiewicz J (2012) Multiple sclerosis: diagnosis and therapy. In Weiner H, Stankiewicz J (eds) 1st edn. Wiley-Blackwell, Oxford
3. Lublin FD, Reingold SC, Cohen JA et al (2014) Defining the clinical course of multiple sclerosis the 2013 revisions. Neurology 83:278–286
4. Steinman L (1996) Multiple sclerosis: a coordinated immunological attack against myelin in the central nervous system. Cell 85:299–302
5. Bjartmar C, Trapp BD (2001) Axonal and neuronal degeneration in multiple sclerosis: mechanisms and functional consequences. Curr Opin Neurol 14:271–278
6. Trapp BD, Nave K (2008) Multiple sclerosis: an immune or neurodegenerative disorder? Ann Rev Neurosci 31:247–269. https://doi.org/10.1146/annurev.neuro.30.051606.094313
7. Hauser SL, Oksenberg JR (2006) The neurobiology of multiple sclerosis: genes, inflammation, and neurodegeneration. Neuron 52:61–76. https://doi.org/10.1016/j.neuron.2006.09.011
8. Stys PK, Zamponi GW, Van Minnen J, Geurts JJG (2012) Will the real multiple sclerosis please stand up? Nature 13(507):514
9. Tomassini V, Palace J (2009) Multiple sclerosis lesions: insights from imaging techniques. Expert Rev Neurother 9(9):1341–1359
10. van Walderveen M, Kamphorst W, Scheltens P et al (1998) Histopathologic correlate of hypointense lesions on Tl-weighted spin-echo MRI in multiple sclerosis. Neurology 50(95):1282–1288
11. Sahraian M et al (2010) Black holes in multiple sclerosis: definition, evolution, and clinical correlations. Acta Neurol Scand 122:1–8. https://doi.org/10.1111/j.1600-0404.2009.01221.x
12. Kidd D, Barkhof F, Mcconnell R, Algra PR, Allen IV, Revesz T (1999) Cortical lesions in multiple sclerosis. Brain 122:17–26
13. Peterson JW, Bö L, Mörk S, Chang A, Trapp B (2001) Transected neurites, apoptotic neurons, and reduced inflammation in cortical multiple sclerosis lesions. Ann Neurol 50:389–400. https://doi.org/10.1002/ana.1123
14. Geurts JJ, Barkhof F (2008) Grey matter pathology in multiple sclerosis. Lancet Neurol 7(9):841–851. https://doi.org/10.1016/S1474-4422(08)70191-1
15. Klaver R, De Vries HE, Schenk GJ, Geurts JJG (2013) Grey matter damage in multiple sclerosis: a pathology perspective. Prion 7(1):66–75. https://doi.org/10.4161/pri.23499
16. Vercellino M, Masera S, Lorenzatti M et al (2009) Demyelination, inflammation, and neurodegeneration in multiple sclerosis deep gray matter. J Neuropathol Exp Neurol 68(5):489–502. https://doi.org/10.1097/NEN.0b013e3181a19a5a
17. Polman CH, Reingold SC, Banwell B et al (2011) Diagnostic criteria for multiple sclerosis: 2010 revisions to the McDonald criteria.

Ann Neurol 69:292–302. https://doi.org/10.1002/ana.22366
18. Pestalozza IF, Pozzilli C, Di Legge S et al (2005) Monthly brain magnetic resonance imaging scans in patients with clinically isolated syndrome. Mult Scler 11(4):390–394
19. Havrdova E, Galetta S, Hutchinson M et al (2009) Effect of natalizumab on clinical and radiological disease activity in multiple sclerosis: a retrospective analysis of the Natalizumab Safety and Efficacy in Relapsing-Remitting Multiple Sclerosis (AFFIRM) study. Lancet Neurol 8(3):254–260. https://doi.org/10.1016/S1474-4422(09)70021-3
20. Tomassini V, Paolillo A, Russo P et al (2006) Predictors of long-term clinical response to interferon beta therapy in relapsing multiple sclerosis. J Neurol 253:287–293. https://doi.org/10.1007/s00415-005-0979-5
21. Sbardella E, Tomassini V, Stromillo ML et al (2011) Pronounced focal and diffuse brain damage predicts short-term disease evolution in patients with clinically isolated syndrome suggestive of multiple sclerosis. Mult Scler J 17(12):1432–1440. https://doi.org/10.1177/1352458511414602
22. De Stefano N, Airas L, Grigoriadis N et al (2014) Clinical relevance of brain volume measures in multiple sclerosis. CNS Drugs 28(2):147–156. https://doi.org/10.1007/s40263-014-0140-z
23. Audoin B, Ibarrola D, Cozzone PJ, Pelletier J, Ranjeva J (2007) Onset and underpinnings of white matter atrophy at the very early stage of multiple sclerosis a two-year longitudinal MRI/MRSI study of corpus callosum. Mult Scler 13:41–51
24. Bernitsas E, Bao F, Seraji-bozorgzad N et al (2015) Spinal cord atrophy in multiple sclerosis and relationship with disability across clinical phenotypes. Mult Scler Relat Disord 4 (1):47–51. https://doi.org/10.1016/j.msard.2014.11.002
25. Schlaeger R et al (2014) Spinal cord gray matter atrophy correlates with multiple sclerosis disability. Ann Neurol 76(5):568–580
26. Bergouignan L, Chupin M, Czechowska Y et al (2009) NeuroImage can voxel based morphometry, manual segmentation and automated segmentation equally detect hippocampal volume differences in acute depression? NeuroImage 45(1):29–37. https://doi.org/10.1016/j.neuroimage.2008.11.006.
27. Ashburner J (2007) A fast diffeomorphic image registration algorithm. NeuroImage 38:95–113. https://doi.org/10.1016/j.neuroimage.2007.07.007
28. Douaud G et al (2007) Anatomically related grey and white matter abnormalities in adolescent-onset schizophrenia. Brain 130:2375–2386. https://doi.org/10.1093/brain/awm184
29. Fischl B, Salat DH, Busa E et al (2002) Whole brain segmentation: neurotechnique automated labeling of neuroanatomical structures in the human brain. Neuron 33:341–355
30. Fischl B, Salat DH, Kouwe JW, Van der Kouwe AJ, Makris N, Quinn BT, Dale AM (2004) Sequence-independent segmentation of magnetic resonance images. NeuroImage 23:69–84. https://doi.org/10.1016/j.neuroimage.2004.07.016.
31. Patenaude B, Smith SM, Kennedy DN, Jenkinson M, NeuroImage A (2011) Bayesian model of shape and appearance for subcortical brain segmentation. NeuroImage 56 (3):907–922. https://doi.org/10.1016/j.neuroimage.2011.02.046.
32. Gogtay N, Giedd JN, Lusk L et al (2004) Dynamic mapping of human cortical development during childhood through early adulthood. PNAS 101(21):8174–8179
33. Rajagopalan V, Yue GH, Pioro EP (2014) Do preprocessing algorithms and statistical models influence voxel-based morphometry (VBM) results in amyotrophic lateral sclerosis patients? A systematic comparison of popular VBM analytical methods. J Magn Reson Imaging 667:662–667. https://doi.org/10.1002/jmri.24415
34. Muhlert N, Ridgway GR. Discussion forum. Failed replications, contributing factors and careful interpretations: commentary on " A purely confirmatory replication study of structural brain-behaviour correlations " by Boekel et al., 2015. Cortex. 2015:1–5. doi:https://doi.org/10.1016/j.cortex.2015.02.019.
35. Tardif CL, Collins DL, Pike GB (2010) Regional impact of field strength on voxel-based morphometry results. Hum Brain Mapp 957:943–957. https://doi.org/10.1002/hbm.20908.
36. Shen S, Sterr A (2013) Is DARTEL-based voxel-based morphometry affected by width of smoothing kernel and group size? A study using simulated atrophy. J Magn Reson Imaging 1475:1468–1475. https://doi.org/10.1002/jmri.23927
37. Radua J, Canales-rodríguez EJ, Pomarol-clotet E, Salvador R (2014) NeuroImage validity of modulation and optimal settings for advanced voxel-based morphometry. NeuroImage 86:81–90. https://doi.org/10.1016/j.neuroimage.2013.07.084.

38. Chard DT, Jackson JS, Miller DH, Wheeler-kingshott CAM (2010) Reducing the impact of white matter lesions on automated measures of brain gray and white matter volumes. J Magn Reson Imaging 228:223–228. https://doi.org/10.1002/jmri.22214
39. Magon S, Gaetano L, Chakravarty MM et al (2014) White matter lesion filling improves the accuracy of cortical thickness measurements in multiple sclerosis patients: a longitudinal study. BCM Neurosci 15(1):1–10. https://doi.org/10.1186/1471-2202-15-106
40. Sdika M, Pelletier D (2009) Nonrigid registration of multiple sclerosis brain images using lesion in painting for morphometry or lesion mapping. Hum Brain Mapp 1067:1060–1067. https://doi.org/10.1002/hbm.20566
41. Battaglini M, Jenkinson M, De Stefano N (2012) Evaluating and reducing the impact of white matter lesions on brain volume measurements. Hum Brain Mapp 33:2062–2071. https://doi.org/10.1002/hbm.21344
42. Gelineau-morel R, Tomassini V, Jenkinson M, Johansen-berg H, Matthews PM, Palace J (2012) The effect of hypointense white matter lesions on automated gray matter segmentation in multiple sclerosis. Hum Brain Mapp 33:2802–2814. https://doi.org/10.1002/hbm.21402
43. Bermel RA, Sharma J, Tjoa CW, Puli SR, Bakshi R (2003) A semiautomated measure of whole-brain atrophy in multiple sclerosis. J Neurol Sci 208(1–2):57–65. https://doi.org/10.1016/S0022-510X(02)00425-2
44. Berg D, Mäurer M, Warmuth-Metz M, Rieckmann P, Becker G (2000) The correlation between ventricular diameter measured by transcranial sonography and clinical disability and cognitive dysfunction in patients with multiple sclerosis. Arch Neurol 57(9):1289–1292. https://doi.org/10.1001/archneur.57.9.1289.
45. Cifelli A, Arridge M, Jezzard P, Esiri MM, Palace J, Matthews PM (2002) Thalamic neurodegeneration in multiple sclerosis. Ann Neurol 52(5):650–653. https://doi.org/10.1002/ana.10326
46. De Stefano N, Matthews PM, Filippi M et al (2003) Evidence of early cortical atrophy in MS: relevance to white matter changes and disability. Neurology 60(7):1157–1162. https://doi.org/10.1212/01.WNL.0000055926.69643.03
47. Lin X, Blumhardt LD, Constantinescu CS (2003) The relationship of brain and cervical cord volume to disability in clinical subtypes of multiple sclerosis: a three-dimensional MRI study. Acta Neurol Scand 108(6):401–406. https://doi.org/10.1046/j.1600-0404.2003.00160.x
48. Ramasamy DP, Benedict RHB, Cox JL et al (2009) Journal of the neurological sciences extent of cerebellum, subcortical and cortical atrophy in patients with MS a case-control study. J Neurol Sci 282(1–2):47–54. https://doi.org/10.1016/j.jns.2008.12.034.
49. Bermel RA, Innus MD, Tjoa CW, Bakshi R (2003) Selective caudate atrophy in multiple sclerosis: a 3D MRI parcellation study. Neuroreport 14(3):335–339. https://doi.org/10.1097/00001756-200303030-00008.
50. Kincses Z (2011) Lesion probability mapping to explain clinical deficits and cognitive performance in multiple sclerosis. Mult Scler 17:681–689. https://doi.org/10.1177/1352458510391342
51. Bodini B, Battaglini M, De Stefano N et al (2011) T2 lesion location really matters: a 10 year follow-up study in primary progressive multiple sclerosis. J Neurol Neurosurg Psychiatry 82:72–77. https://doi.org/10.1136/jnnp.2009.201574
52. Di Perri C, Battaglini M, Stromillo ML, Bartolozzi ML (2008) Voxel-based assessment of differences in damage and distribution of white matter lesions between patients with primary progressive and relapsing-remitting multiple sclerosis. Arch Neurol 65(2):236–243
53. Ceccarelli A, Rocca MA, Pagani E et al (2008) The topographical distribution of tissue injury in benign MS: a 3T multiparametric MRI study. NeuroImage 39(4):1499–1509. https://doi.org/10.1016/j.neuroimage.2007.11.002.
54. Brownell B, Hughes JT (1962) The distribution of plaques in the cerebrum in multiple sclerosis. J Neurol Neurosurg Psychiatry 25:315
55. Bo L et al (2003) Subpial demyelination in the cerebral cortex of multiple sclerosis patients. J Neuropathol Exp Neurol 62(7):723–732
56. Calabrese M et al (2010) Imaging distribution and frequency of cortical lesions in patients with multiple sclerosis. Neurology 75:1234–1240
57. Bendfeldt K, Kuster P, Traud S et al (2009) Association of regional gray matter volume loss and progression of white matter lesions in multiple sclerosis—a longitudinal voxel-based morphometry study. NeuroImage 45

(1):60–67. https://doi.org/10.1016/j.neuroimage.2008.10.006
58. Prinster A, Quarantelli M, Orefice G et al (2006) Grey matter loss in relapsing-remitting multiple sclerosis: a voxel-based morphometry study. NeuroImage 29 (3):859–867. https://doi.org/10.1016/j.neuroimage.2005.08.034
59. Raz E, Cercignani M, Sbardella E et al (2010) Gray- and white-matter changes 1 year after first clinical episode of multiple sclerosis: MR imaging. Radiology 257(2):448–454. https://doi.org/10.1148/radiol.10100626
60. Audoin B, Zaaraoui W, Reuter F et al (2010) Atrophy mainly affects the limbic system and the deep grey matter at the first stage of multiple sclerosis. J Neurol Neurosurg Psychiatry 81(6):690–695. https://doi.org/10.1136/jnnp.2009.188748
61. Calabrese M, Rinaldi F, Mattisi I et al (2011) The predictive value of gray matter atrophy in clinically isolated syndromes. Neurology 77:257–263
62. Lansley J, Mataix-Cols D, Grau M, Radua J, Sastre-Garriga J (2013) Localized grey matter atrophy in multiple sclerosis: a meta-analysis of voxel-based morphometry studies and associations with functional disability. Neurosci Biobehav Rev 37(5):819–830. https://doi.org/10.1016/j.neubiorev.2013.03.006
63. Calabrese M, Atzori M, Bernardi V et al (2007) Cortical atrophy is relevant in multiple sclerosis at clinical onset. J Neurol 254 (9):1212–1220. https://doi.org/10.1007/s00415-006-0503-6
64. Narayana PA, Govindarajan KA, Goel P et al (2013) NeuroImage: clinical regional cortical thickness in relapsing remitting multiple sclerosis: a multi center study. NeuroImage Clin 2:120–131. https://doi.org/10.1016/j.nicl.2012.11.009.
65. Matsushita T, Madireddy L, Sprenger T et al (2015) Genetic associations with brain cortical thickness in multiple sclerosis. Genes Brain Behav 14:217–227. https://doi.org/10.1111/gbb.12190
66. Steenwijk MD, Daams M, Pouwels PJW et al (2015) Unraveling the relationship between regional gray matter atrophy and pathology in connected white matter tracts in long-standing multiple sclerosis. Hum Brain Mapp 36:1796–1807. https://doi.org/10.1002/hbm.22738
67. Eshaghi A, Bodini B, Ridgway GR et al (2014) Temporal and spatial evolution of grey matter atrophy in primary progressive multiple sclerosis. NeuroImage 86:257–264. https://doi.org/10.1016/j.neuroimage.2013.09.059
68. Tziortzi AC, Haber SN, Searle GE et al (2014) Connectivity-based functional analysis of dopamine release in the striatum using diffusion-weighted MRI and positron emission tomography. Cereb Cortex 24 (5):1165–1177. https://doi.org/10.1093/cercor/bhs397
69. Hasan KM, Halphen C, Kamali A, Nelson FM, Wolinsky JS, PA N (2009) Caudate nuclei volume, diffusion tensor metrics, and T 2 relaxation in healthy adults and relapsing-remitting multiple sclerosis patients: implications for understanding gray matter degeneration. J Magn Reson Imaging 29(1):70–77. https://doi.org/10.1002/jmri.21648
70. Minagar A, Barnett MH, Benedict RHB et al (2013) The thalamus and multiple sclerosis: modern views on pathologic, imaging, and clinical aspects. Neurology 80(2):210–219. https://doi.org/10.1212/WNL.0b013e31827b910b
71. Wylezinska M, Cifelli A, Jezzard P, Palace J, Alecci M (2003) Thalamic neurodegeneration in relapsing-remitting multiple sclerosis. Neurology 60:1949–1954
72. Tewarie P, Schoonheim MM, Schouten DI et al (2015) Functional brain networks: linking thalamic atrophy to clinical disability in multiple sclerosis, a multimodal fMRI and MEG study. Hum Brain Mapp 618:603–618. https://doi.org/10.1002/hbm.22650
73. Batista S, Zivadinov R, Hoogs M, Bergsland N, Michael MH, Benedict BWRHB (2012) Basal ganglia, thalamus and neocortical atrophy predicting slowed cognitive processing in multiple sclerosis. J Neurol 259:139–146. https://doi.org/10.1007/s00415-011-6147-1
74. Gamboa OL, Tagliazucchi E, von Wegner F et al (2014) Working memory performance of early MS patients correlates inversely with modularity increases in resting state functional connectivity networks. NeuroImage 94:385–395. https://doi.org/10.1016/j.neuroimage.2013.12.008
75. Houtchens MK, Killiany R (2007) Thalamic atrophy and cognition in multiple sclerosis. Neurology 69:1213–1223
76. Haider L (2015) Oxidant stress in the pathogenesis of multiple sclerosis. Neurosci Behav Physiol 37(3):209–213
77. Pagani E, Rocca MA, Gallo A et al (2005) Regional brain atrophy evolves differently in patients with multiple sclerosis according to

clinical phenotype. Am J Neuroradiol 26:341–346

78. Mesaros S, Rocca MA, Absinta M et al (2008) Evidence of thalamic gray matter loss in pediatric multiple sclerosis. Neurology 70 (13 PART 2):1107–1112. https://doi.org/10.1212/01.wnl.0000291010.54692.85
79. Mesaros S, Rovaris M, Pagani E et al (2008) A magnetic resonance imaging voxel-based morphometry study of regional gray matter atrophy in patients with benign multiple sclerosis. Arch Neurol 65(9):1223–1230. https://doi.org/10.1001/archneur.65.9.1223.
80. Duan Y, Liu Y, Liang P et al (2012) Comparison of grey matter atrophy between patients with neuromyelitis optica and multiple sclerosis: a voxel-based morphometry study. Eur J Radiol 81(2):e110–e114. https://doi.org/10.1016/j.ejrad.2011.01.065.
81. Audoin B, Davies GR, Thompson AJ, Miller DH, Davies GR, Thompson AJ (2006) Localization of grey matter atrophy in early RRMS a longitudinal study. J Neurol 253:1495–1501. https://doi.org/10.1007/s00415-006-0264-2.
82. Henry RG, Shieh M, Amirbekian B, Chung S, Okuda DT, Pelletier D (2009) Connecting white matter injury and thalamic atrophy in clinically isolated syndromes. J Neurol Sci 282 (1–2):61–66. https://doi.org/10.1016/j.jns.2009.02.379
83. Haider L, Simeonidou C, Steinberger G et al (2014) Multiple sclerosis deep grey matter: the relation between demyelination, neurodegeneration, inflammation and iron. J Neurol Neurosurg Psychiatry 85(12):1386–1395. https://doi.org/10.1136/jnnp-2014-307712
84. Pardini M et al. (2015) Periventricular gradient in thalamic abnormalities in MS: a magnetisation transfer ratio imaging study. In: European Committee for Research and Treatments in MS, Barcelona, Spain
85. De Stefano N, Stromillo ML, Giorgio A et al (2016) Establishing pathological cut-offs of brain atrophy rates in multiple sclerosis. J Neurol Neurosurg Psychiatry 87:93–99. https://doi.org/10.1136/jnnp-2014-309903
86. Filippi M, Preziosa P, Rocca MA (2014) Magnetic resonance outcome measures in multiple sclerosis trials: time to rethink? Curr Opin Neurol 27(3):290–299. https://doi.org/10.1097/WCO.0000000000000095
87. Fisher E, Lee JC, Nakamura K, RA R (2008) Gray matter atrophy in multiple sclerosis: a longitudinal study. Ann Neurol 64 (3):255–265. https://doi.org/10.1002/ana.21436
88. Hardmeier M, Wagenpfeil S, Freitag P et al (2003) Atrophy is detectable within a 3-month period in untreated patients with active relapsing remitting multiple sclerosis. Arch Neurol 60(12):1736–1739. https://doi.org/10.1001/archneur.60.12.1736
89. Ge Y, Grossman RI, Udupa JK et al (2000) Brain atrophy in relapsing-remitting multiple sclerosis and secondary progressive multiple sclerosis: longitudinal quantitative analysis. Radiology 214(3):665–670. https://doi.org/10.1148/radiology.214.3.r00mr30665
90. Chard DT, Griffin CM, Rashid W et al (2004) Progressive grey matter atrophy in clinically early relapsing-remitting multiple sclerosis. Mult Scler 10(4):387–391. https://doi.org/10.1191/1352458504ms1050oa
91. Nygaard GO, Walhovd KB, Sowa P et al (2015) Cortical thickness and surface area relate to specific symptoms in early relapsing—remitting multiple sclerosis. Mult Scler J 21(4):402–414. https://doi.org/10.1177/1352458514543811
92. Battaglini M, Giorgio A, Stromillo ML et al (2009) Voxel-wise assessment of progression of regional brain atrophy in relapsing-remitting multiple sclerosis. J Neurol Sci 282 (1–2):55–60. https://doi.org/10.1016/j.jns.2009.02.322
93. Selpucre J et al (2008) Regional gray matter atrophy in early primary progressive multiple sclerosis. Arch Neurol 63:1175–1180
94. Khaleeli Z, Cercignani M, Audoin B, Ciccarelli O, Miller DH, Thompson AJ (2007) Localized grey matter damage in early primary progressive multiple sclerosis contributes to disability. NeuroImage 37 (1):253–261. https://doi.org/10.1016/j.neuroimage.2007.04.056
95. Kutzelnigg A, Lucchinetti CF, Stadelmann C et al (2005) Cortical demyelination and diffuse white matter injury in multiple sclerosis. Brain 128:2705–2712. https://doi.org/10.1093/brain/awh641
96. Redmond I, Barbosa S, Blumhardt L, Roberts N (2000) Short-term ventricular volume changes on serial MRI in multiple sclerosis. Acta Neurol Scand 102(2):99–105
97. Turner B, Lin X, Calmon G, Roberts N, Blumhardt LD (2003) Cerebral atrophy and disability in relapsing remitting and secondary progressive multiple sclerosis over four years. Mult Scler 9:21–28

98. Fisniku LK, Chard DT, Jackson JS et al (2008) Gray matter atrophy is related to long-term disability in multiple sclerosis. Ann Neurol 64(3):247–254. https://doi.org/10.1002/ana.21423
99. Henry RG, Shieh M, Okuda DT, Evangelista A, Pelletier D (2008) Regional grey matter atrophy in clinically isolated syndromes at presentation. J Neurol Neurosurg Psychiatry 79(11):1236–1244. https://doi.org/10.1136/jnnp.2007.134825
100. Ceccarelli A, Rocca MA, Pagani E et al (2008) A voxel-based morphometry study of grey matter loss in MS patients with different clinical phenotypes. NeuroImage 42 (1):315–322. https://doi.org/10.1016/j.neuroimage.2008.04.173.
101. Raz E, Matter G, Cercignani M, Pozzilli C, Bozzali M, Pantano P (2010) Clinically isolated syndrome suggestive of multiple sclerosis: voxelwise regional investigation of white and gray. Radiology 254(1):227–234
102. Bergsland N, Horakova D, Dwyer MG et al (2012) Subcortical and cortical gray matter atrophy in a large sample of patients with clinically isolated syndrome and early relapsing-remitting multiple sclerosis. Am J Neuroradiol 33(8):1573–1578. https://doi.org/10.3174/ajnr.A3086
103. Dalton CM, Chard DT, Davies GR et al (2004) Early development of multiple sclerosis is associated with progressive grey matter atrophy in patients presenting with clinically isolated syndromes. Brain 127 (5):1101–1107. https://doi.org/10.1093/brain/awh126
104. Lin X, Blumhardt LD (2001) Inflammation and atrophy in multiple sclerosis: MRI associations with disease course. J Neurol Sci 189:99–104
105. Kalkers NF, Bergers E, Castelijns JA, Polman CH, Barkhof F (2001) Optimizing the association between disability and biological markers in MS. Neurology 57:1253–1258
106. Calabrese M, Reynolds R, Magliozzi R et al (2015) Regional distribution and evolution of gray matter damage in different populations of multiple sclerosis patients. PLoS One 10: e0135428. https://doi.org/10.1371/journal.pone.0135428
107. Rovaris M, Barkhof F, Calabrese M et al (2009) MRI features of benign multiple sclerosis. Neurology 72:1693–1701
108. Calabrese M, Filippi M, Rovaris M et al (2009) Evidence for relative cortical sparing in benign multiple sclerosis: a longitudinal magnetic resonance imaging study. Mult Scler J 15:36–41
109. Calabrese M et al (2012) Low degree of cortical pathology is associated with benign course of multiple sclerosis. Mult Scler J 19 (7):904–911. https://doi.org/10.1177/1352458512463767
110. Rovaris M (2001) Cognitive impairment and structural brain damage in benign multiple sclerosis. Neurology 71:1521–1526
111. Kalkers F, Polman C, Barkhof F (2002) Longitudinal brain volume measurement in multiple sclerosis. Arch Neurol 59:1572–1576. http://scholar.google.com/scholar?hl=en&btnG=Search&q=intitle:Longitudinal+Brain+Volume+Measurement+in+Multiple+Sclerosis#0
112. Korteweg T, Rovaris M, Neacsu V et al (2009) Can rate of brain atrophy in multiple sclerosis be explained by clinical and MRI characteristics? Mult Scler 15(4):465–471
113. Gauthier SA, Berger AM, Liptak Z et al (2009) Rate of brain atrophy in benign vs early. Mult Scler 66(2):234–238
114. De Stefano N, Giorgio A (2010) Assessing brain atrophy rates in a large population of untreated multiple sclerosis subtypes. Neurology 74:1868–1876
115. RA B, Bakshi R (2006) The measurement and clinical relevance of brain atrophy in multiple sclerosis. Lancet Neurol 5(2):158–170. https://doi.org/10.1016/S1474-4422(06)70349-0
116. Geurts JJG, Calabrese M, Fisher E, Rudick RA (2012) Measurement and clinical effect of grey matter pathology in multiple sclerosis. Lancet Neurol 11(12):1082–1092. https://doi.org/10.1016/S1474-4422(12)70230-2
117. Hohlfeld R, Wekerle H (2004) Autoimmune concepts of multiple sclerosis as a basis for selective immunotherapy: from pipe dreams to (therapeutic) pipelines. Proc Natl Acad Sci U S A 101:14599–14606
118. Lucchinetti CF, Popescu BFG, Bunyan RF et al (2011) Inflammatory cortical demyelination in early multiple sclerosis. N Engl J Med 365(23):2188–2197. https://doi.org/10.1056/NEJMoa1100648
119. Popescu BFG, Bunyan RF, Parisi JE, Ransohoff RM, Lucchinetti CF (2011) A case of multiple sclerosis presenting with inflammatory cortical demyelination. Neurology 76 (20):1705–1710. https://doi.org/10.1212/WNL.0b013e31821a44f1
120. Magliozzi R, Howell OW, Reeves C et al (2010) A gradient of neuronal loss and

meningeal inflammation in multiple sclerosis. Ann Neurol 68:477–493. https://doi.org/10.1002/ana.22230

121. Paling D, Golay X, Miller D (2011) Energy failure in multiple sclerosis and its investigation using MR techniques. J Neurol 258:2113–2127. https://doi.org/10.1007/s00415-011-6117-7

122. Lassmann H, Van Horssen J (2016) Oxidative stress and its impact on neurons and glia in multiple sclerosis lesions. Biochim Biophys Acta 1862(3):506–510. https://doi.org/10.1016/j.bbadis.2015.09.018

123. Witte ME, Mahad DJ, Lassmann H, van Horssen J (2014) Mitochondrial dysfunction contributes to neurodegeneration in multiple sclerosis. Trends Mol Med 20(3):179–187. https://doi.org/10.1016/j.molmed.2013.11.007

124. Popescu V, Klaver R, Voorn P et al (2015) What drives MRI-measured cortical atrophy in multiple sclerosis? Mult Scler J 21:1280–1290. https://doi.org/10.1177/1352458514562440

125. Geurts JJG, Bo L, Pouwels PJW, Castelijns JA, Polman CH, Barkhof F (2005) Cortical lesions in multiple sclerosis: combined postmortem MR imaging and histopathology. Am J Neuroradiol 26:572–577

126. Seewann A, Roosendaal SD, Wattjes MP, Van Der Valk P, Barkhof F, Polman CH (2012) Postmortem verification of MS cortical lesion detection with 3D DIR. Neurology 78(09):302–308

127. Wegner C, Esiri MM, Chance SA, Palace J (2006) Neocortical neuronal, synaptic, and glial loss in multiple sclerosis. Neurology 67:960–967. https://doi.org/10.1212/01.wnl.0000237551.26858.39

128. Van De Pavert SHP, Muhlert N, Sethi V et al (2016) DIR-visible grey matter lesions and atrophy in multiple sclerosis: partners in crime? J Neurol Neurosurg Psychiatry 87:461–467. https://doi.org/10.1136/jnnp-2014-310142

129. Spence RD, Kurth F, Itoh N et al (2014) NeuroImage bringing CLARITY to gray matter atrophy. NeuroImage 101:625–632. https://doi.org/10.1016/j.neuroimage.2014.07.017.

130. Barkhof F (1999) MRI in multiple sclerosis: correlation with expanded disability status scale (EDSS). Mult Scler 5(4):283–286. https://doi.org/10.1191/135245899678846221

131. Hackmack K, Weygandt M, Wuerfel J, Pfueller CF, Friedemann JB, Haynes PJ (2012) Can we overcome the " clinico-radiological paradox" in multiple sclerosis ? J Neurol 259:2151–2160. https://doi.org/10.1007/s00415-012-6475-9

132. Mistry N, Tallantyre EC, Dixon JE et al (2011) Focal multiple sclerosis lesions abound in "normal appearing white matter". Mult Scler J 17(11):1313–1323. https://doi.org/10.1177/1352458511415305

133. Horakova D, Dwyer MG, Havrdova E et al (2009) Gray matter atrophy and disability progression in patients with early relapsing-remitting multiple sclerosis. A 5-year longitudinal study. J Neurol Sci 282(1–2):112–119. https://doi.org/10.1016/j.jns.2008.12.005.

134. Roosendaal SD, Bendfeldt K, Vrenken H et al (2011) Grey matter volume in a large cohort of MS patients: relation to MRI parameters and disability. Mult Scler 17(9):1098–1106. https://doi.org/10.1177/1352458511404916

135. Kurtzke JF (1983) Rating neurologic impairment in multiple sclerosis: an expanded disability status scale (EDSS). Neurology 33(11):1444–1453

136. Fischer JS, Rudick RA, Cutter GR, Reingold SC, Ms N, Clinical S (1999) The multiple sclerosis functional composite measure (MSFC): an integrated approach to MS clinical outcome assessment. Mult Scler 5:244–250

137. Twork S, Wiesmeth S, Spindler M et al (2010) Disability status and quality of life in multiple sclerosis: non-linearity of the Expanded Disability Status Scale (EDSS). Health Qual Life Outcomes 8(55):8–13

138. Fisher E, Rudick RA, Simon JH et al (2002) Eight-year follow-up study of brain atrophy in patients with MS. Neurology 59:1412–1420

139. Sailer M, Fischl B, Salat D et al (2003) Focal thinning of the cerebral cortex in multiple sclerosis. Brain 126(8):1734–1744. https://doi.org/10.1093/brain/awg175

140. Tedeschi G, Lavorgna L, Russo P et al (2005) Brain atrophy and lesion load in a large population of patients with. Neurology 65:280–285

141. Sanfilipo MP, Benedict RHB, Sharma J, Weinstock-guttman B, Bakshi R (2005) The relationship between whole brain volume and disability in multiple sclerosis: a comparison of normalized gray vs. white matter with misclassification correction. NeuroImage 26:1068–1077. https://doi.org/10.1016/j.neuroimage.2005.03.008

142. Charil A, Dagher A, Lerch JP, Zijdenbos AP, Worsley KJ, Evans AC (2007) Focal cortical atrophy in multiple sclerosis: relation to lesion load and disability. NeuroImage 34 (2):509–517. https://doi.org/10.1016/j.neuroimage.2006.10.006
143. Prinster A, Quarantelli M, Lanzillo R et al (2010) A voxel-based morphometry study of disease severity correlates in relapsing—remitting multiple sclerosis. Mult Scler 16 (1):45–54. https://doi.org/10.1177/1352458509351896
144. Rocca MA, Amato MP, De SN et al (2015) Clinical and imaging assessment of cognitive dysfunction in multiple sclerosis. Lancet Neurol 14(3):302–317. https://doi.org/10.1016/S1474-4422(14)70250-9
145. Calabrese M (2009) Cortical lesions and atrophy associated with cognitive impairment in relapsing-remitting multiple sclerosis. Arch Neurol 66(9):1144–1150
146. Nygaard GO, Celius EG, Benavent SADR, Sowa P (2015) A longitudinal study of disability, cognition and gray matter atrophy in early multiple sclerosis patients according to evidence of disease activity. PLoS One 10: e0135974. https://doi.org/10.1371/journal.pone.0135974
147. Muhlert N, Sethi V, Schneider T et al (2013) Diffusion MRI-based cortical complexity alterations associated with executive function in multiple sclerosis. J Magn Reson Imaging 63:54–63. https://doi.org/10.1002/jmri.23970
148. Muhlert N, Sethi V, Cipolotti L et al (2015) The grey matter correlates of impaired decision-making in multiple sclerosis. J Neurol Neurosurg Psychiatry 86:530–536. https://doi.org/10.1136/jnnp-2014-308169
149. Muhlert N, Atzori M, De Vita E et al (2014) Memory in multiple sclerosis is linked to glutamate concentration in grey matter regions. J Neurol Neurosurg Psychiatry 85:833–839. https://doi.org/10.1136/jnnp-2013-306662
150. Losseff NA, Wang L, Lai HM et al (1996) Progressive cerebral atrophy in multiple sclerosis a serial MRI study. Brain 119:2009–2019
151. Minneboo A, Jasperse B, Barkhof F et al (2008) Predicting short-term disability progression in early multiple sclerosis: added value of MRI parameters. J Neurol Neurosurg Psychiatry 79:917–924. https://doi.org/10.1136/jnnp.2007.124123
152. Rudick RA, Cutter G, Baier M et al (2001) Use of the multiple sclerosis functional composite to predict disability in relapsing MS. Neurology 56(10):1324–1330. https://doi.org/10.1212/WNL.56.10.1324
153. Chen JT, Narayanan S, Collins DL, Smith SM, Matthews PM, Arnold DL (2004) Relating neocortical pathology to disability progression in multiple sclerosis using MRI. NeuroImage 23(3):1168–1175. https://doi.org/10.1016/j.neuroimage.2004.07.046
154. Rudick RA, Lee JC, Nakamura K, Fisher E (2009) Gray matter atrophy correlates with MS disability progression measured with MSFC but not EDSS. J Neurol Sci 282 (1–2):106–111. https://doi.org/10.1016/j.jns.2008.11.018
155. Filippi M, Preziosa P, Copetti M, Riccitelli G, Horsfield MA, Martinelli V (2013) Gray matter damage predicts the accumulation of disability 13 years later in MS. Neurology 81:1759–1767
156. Popescu V, Agosta F, Hulst HE et al (2013) Brain atrophy and lesion load predict long term disability in multiple sclerosis. J Neurol Neurosurg Psychiatry 84(10):1082–1091. https://doi.org/10.1136/jnnp-2012-304094
157. Filippi M, Rocca MA, Pagani E et al (2014) Placebo-controlled trial of oral laquinimod in multiple sclerosis: MRI evidence of an effect on brain tissue damage. J Neurol Neurosurg Psychiatry 85:851–858. https://doi.org/10.1136/jnnp-2013-306132
158. Kappos L, Radue E, O'Connor P et al (2011) A placebo-controlled trial of oral fingolimod in relapsing multiple sclerosis. N Engl J Med 362(5):387–401
159. Radue E, O'Connor P, Polman C, Hohlfeld R, Calabresi P, Selmaj K et al (2012) Impact of fingolimod therapy on magnetic resonance imaging outcomes in patients with multiple sclerosis. Arch Neurol 69 (10):1259–1269. https://doi.org/10.1001/archneurol.2012.1051
160. Zivadinov R, Locatelli L, Cookfair D et al (2007) Interferon beta-1a slows progression of brain atrophy in relapsing-remitting multiple sclerosis predominantly by reducing gray matter atrophy. Mult Scler 13(4):490–501
161. Calabrese M, Bernardi V, Atzori M et al (2012) Effect of disease-modifying drugs on cortical lesions and atrophy in relapsing—remitting multiple sclerosis. Mult Scler J 18 (4):418–424. https://doi.org/10.1177/1352458510394702

162. Rudick RA, Fisher E, Lee J, Simon J, Jacobs L (1999) Use of the brain parenchymal fraction to measure whole brain atrophy in relapsing-remitting MS. Neurology 53:1698–1704
163. Rinaldi F, Perini P, Atzori M, Favaretto A, Seppi D, Gallo P (2015) Disease-modifying drugs reduce cortical lesion accumulation and atrophy progression in relapsing-remitting multiple sclerosis: results from a 48-month extension study. Mult Scler Int 2015:Article 369348
164. De Stefano N, Pia M, Stubinski B et al (2012) Efficacy and safety of subcutaneous interferon beta-1a in relapsing—remitting multiple sclerosis: further outcomes from the IMPROVE study. J Neurol Sci 312(1–2):97–101. https://doi.org/10.1016/j.jns.2011.08.013
165. Rovaris M, Comi G, Rocca MA, Wolinsky JS, Filippi M (2001) Short-term brain volume change in relapsing—remitting multiple sclerosis effect of glatiramer acetate and implications. Brain 124:1803–1812
166. Bendfeldt K, Egger H, Nichols TE et al (2010) Effect of immunomodulatory medication on regional gray matter loss in relapsing—remitting multiple sclerosis—a longitudinal MRI study. Brain Res 1325:174–182. https://doi.org/10.1016/j.brainres.2010.02.035
167. Sormani MP, Arnold DL, De Stefano N (2014) Treatment effect on brain atrophy correlates with treatment effect on disability in multiple sclerosis. Ann Neurol 75:43–49. https://doi.org/10.1002/ana.24018
168. Zivadinov R, Stu O (2008) Mechanisms of action of disease-modifying agents and brain volume changes in multiple sclerosis. Neurology 71:136–144
169. De Stefano N, Arnold DL (2015) Towards a better understanding of pseudoatrophy in the brain of multiple sclerosis patients. Mult Scler J 21(6):675–676. https://doi.org/10.1177/1352458514564494
170. Vidal-Jordana A, Sastre-garriga J, Pérez-Miralles F et al (2013) Early brain pseudoatrophy while on natalizumab therapy is due to white matter volume changes. Mult Scler J 19:1175–1181. https://doi.org/10.1177/1352458512473190
171. Calabrese M, Magliozzi R, Ciccarelli O et al (2015) Exploring the origins of grey matter damage in multiple sclerosis. Nat Rev Neurosci 16(3):147–158. https://doi.org/10.1038/nrn3900
172. Bodini B, Altmann DR, Tozer D, Miller DH, Wheeler-Kingshott C (2016) White and gray matter damage in primary progressive MS: the chicken or the egg? Neurology 86:170–176
173. Barkhof F, Calabresi P, Miller D, Reingold S (2009) Imaging outcomes for neuroprotection and repair in multiple sclerosis trials. Nat Rev Neurol 5:256–266
174. Wattjes M, Rovira À, Miller D et al (2015) MAGNIMS consensus guidelines on the use of MRI in multiple sclerosis—establishing disease prognosis and monitoring patients. Nat Rev Neurol 11(10):597–606. https://doi.org/10.1038/nrneurol.2015.157

Chapter 18

Brain Morphometry: Epilepsy

Dewi S. Schrader, Neda Bernasconi, and Andrea Bernasconi

Abstract

Magnetic resonance imaging (MRI) is essential in the presurgical investigation of patients with epilepsy because of its unmatched ability in visualizing structural brain pathology. In many cases, however, best-practice MRI is unremarkable and thus unable to reveal the potential surgical target. This chapter considers the application of advanced imaging techniques in temporal lobe epilepsy and neocortical epilepsy and discusses various techniques, such as MRI morphometry of the mesiotemporal lobe structures and the neocortex, hippocampal shape analysis, and computational modeling of epileptogenic lesions. By revealing subtle lesions that escape visual inspection, image processing of structural MRI demonstrates increased sensitivity compared to conventional techniques. Aside from diagnostics, ongoing developments provide unprecedented opportunities for the design of new biomarkers to stage and monitor the disease.

Key words Magnetic resonance imaging, Epilepsy, Temporal lobe epilepsy, Focal cortical dysplasia, MRI volumetry, Shape analysis, Cortical thickness, Machine learning

1 Introduction

Approximately 50 million people worldwide have epilepsy, making it one of the most common neurological diseases. Individuals with epilepsy have an underlying structural, metabolic, or genetic abnormality of the brain that causes excessive neuronal activity that leads to a transient change in behavior termed an epileptic seizure. In clinical practice, epilepsy is defined as two or more unprovoked seizures [1]. About a third of patients with epilepsy do not respond to medication despite the current availability of over two-dozen antiepileptic drugs. Drug-resistant epilepsy is a serious condition associated with high risk for psychosocial impairment, cognitive decline, and mortality.

In patients with focal onset of seizures, when a structural brain abnormality is identified, it can be surgically resected, offering patients the chance of a complete cure [2]. The two most common forms of surgically remediable syndromes are temporal lobe epilepsy secondary to mesiotemporal sclerosis [3] and extra-temporal lobe neocortical epilepsy secondary to cortical dysplasia [4]. Timely

Gianfranco Spalletta et al. (eds.), *Brain Morphometry*, Neuromethods, vol. 136,
https://doi.org/10.1007/978-1-4939-7647-8_18,

identification and treatment of these lesions reduce the long-term negative effects of seizures and medications and have a positive effect on cognition [5].

Magnetic resonance imaging (MRI) is essential in the presurgical investigation of patients with drug-resistant epilepsy because of its unmatched ability in visualizing structural brain pathology. Despite technical improvements in MRI hardware and sequences, however, best-practice MRI is unremarkable and thus unable to reveal the potential surgical target in more than 50% of patients [6]. The lack of an identifiable lesion on MRI is currently one of the biggest barriers to epilepsy surgery. Over the years, this clinical difficulty has motivated the development of computer-aided methods aimed at analyzing brain morphology and signal intensity. The purpose of this chapter is to give an overview of image processing methods that have significantly improved the detection of brain lesions in the most common forms of drug-resistant epilepsy.

2 Temporal Lobe Epilepsy

In temporal lobe epilepsy (TLE), the most common drug-resistant epilepsy in adults, seizures arise primarily from the mesial temporal structures. The histopathological hallmark is mesial temporal sclerosis, which is characterized by cell loss and astrocytic gliosis [7]. These pathological changes often involve the hippocampus but also the amygdala, the entorhinal cortex, the temporopolar cortex, and the gray and white matter of the temporal lobe [8]. The availability of surgical specimen has allowed to devise specific classification schemes for hippocampal sclerosis based on patterns and degree of anomalies across various subfields [7]. On conventional MRI, mesiotemporal sclerosis is classically characterized by changes in the hippocampus including atrophy, loss of internal structure, and decreased T1 signal and increased T2 signal intensity. Additional features on MRI include atrophy of the ipsilateral fornix and mammillary body and atrophy of temporal lobe gray and white matter. Besides atrophy, about 40% of TLE patients present with shape and positioning anomalies of the hippocampus and surrounding structures [9]. This so-called hippocampal malrotation is characterized by an abnormally rounded vertically orientated hippocampus accompanied by a deep collateral sulcus [10, 11]. Hippocampal malrotation is likely a marker of an underlying neurodevelopmental anomaly that may contribute to temporal lobe epileptogenesis.

2.1 Hippocampal Volumetry

The analysis of the morphological characteristics of the hippocampus through visual inspection is subjective, as it is based mainly on the experience of the observer. The development of quantitative post-processing techniques has provided objective indicators of symmetry and atrophy. In this context, manual volumetry performed on T1-weighted anatomical MRI has been the first technique to assess morphology of the hippocampus and other mesiotemporal lobe structures (Fig. 1); as such, it has shown

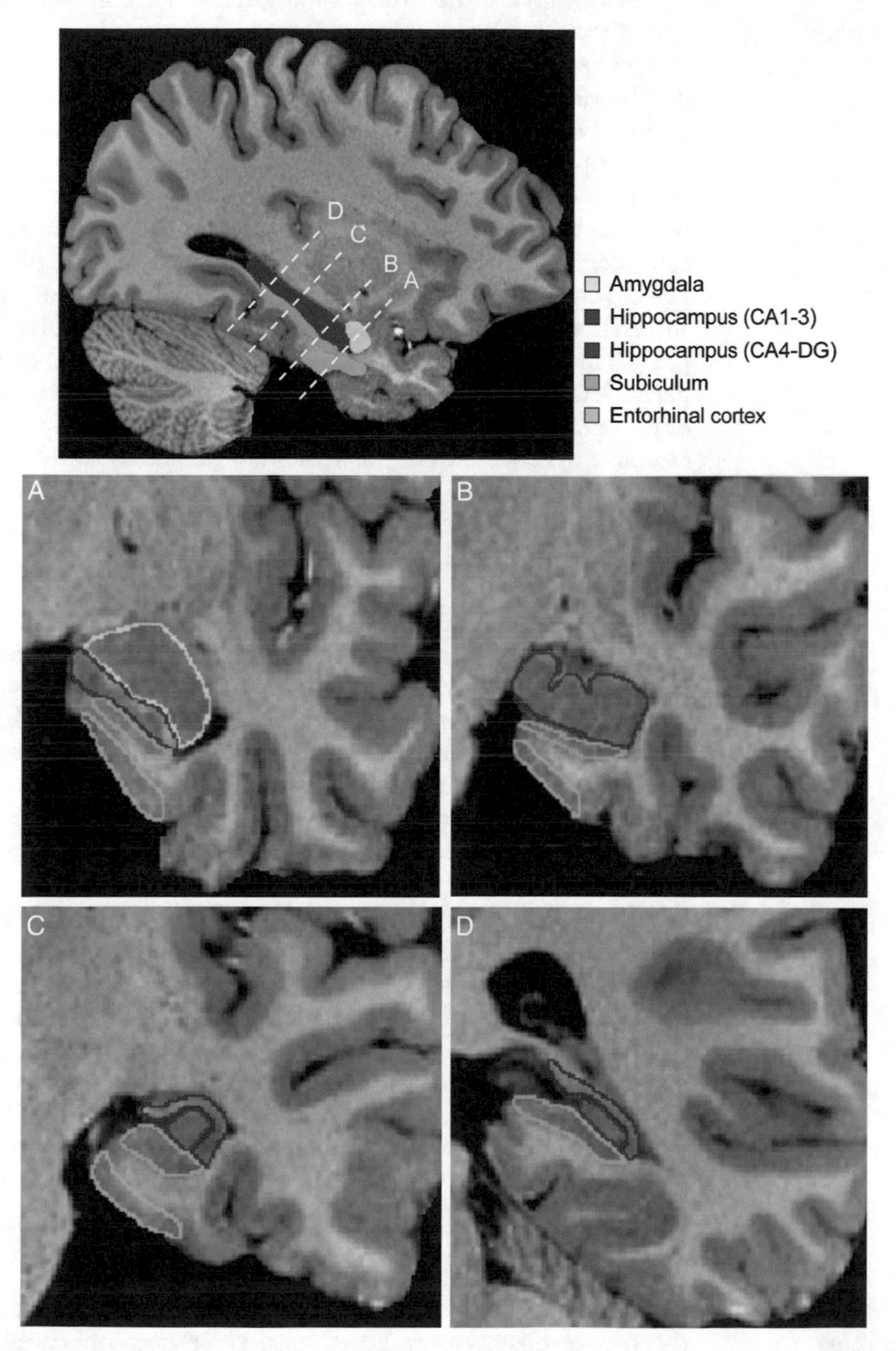

Fig. 1 Mesial temporal lobe manual segmentation on 3 T micrometric MRI. Sagittal T1-weighted MRI at 600 μm isotropic resolution of a control overlaid with manual labels of MTL structures. Dotted lines refer to coronal sections in A, B, C, and D at various levels along the temporal lobe

increased sensitivity of hippocampal atrophy compared to visual MRI analysis [12–18]. Indeed, hippocampal volumetry has been repeatedly shown to accurately lateralize the seizure focus in >90% of cases [19].

In agreement with histopathological data, a considerable number of studies have established that structural brain abnormalities extend beyond the hippocampus to involve the adjacent mesiotemporal structures, namely, the amygdala and the parahippocampal region. Volumetry of the entorhinal cortex [12, 13], amygdala, and temporopolar region [16], as well as the thalamus [20], also has a role in lateralization of the seizure focus. For instance, in patients with normal hippocampal volumes, entorhinal cortex volumetry can lateralize the seizure focus in an additional 25% of patients [14]. Importantly, the degree of MRI volume loss has been shown to correlate with the degree of cell loss on surgical specimens [21]. Thus, quantification of mesiotemporal structural changes is particularly important, and strongly recommended, when considering epilepsy surgery in order to detect subtle atrophy and to establish whether the contralateral mesiotemporal lobe is structurally normal. Indeed, bilateral mesial temporal lobe atrophy raises concerns of markedly reduced chance of seizure freedom after selective amygdalohippocampectomy or anterior temporal lobe resection and an increased risk of memory impairment.

Despite its clinical utility, manual volumetry is time consuming. Ongoing developments in image processing since the early 1990s have led to increasingly sophisticated automatic hippocampal segmentation techniques. Most algorithms have used atlas-based (single or multiple, or probabilistic) and deformable models techniques [see [22] for review]. Up until recently, results in epilepsy had been quite disappointing [23–29]. Recent methods rely on a multi-template framework to account for interindividual local anatomical variability. While the majority employed a purely voxel-based strategy, adopting a surface-based library improves flexibility to model shape deformations often seen in disease but also in 10–15% of healthy subjects. We have shown that the relatively poor performance of previous methods was likely due to the fact that they did not account for hippocampal malrotation [9]. To overcome this challenge, our group has developed an automated hippocampal segmentation algorithm that integrates deformable parametric surfaces and multiple templates in a unified framework [30]. Compared to other state-of-the-art algorithms, our method maintains a high level of performance regardless of atrophy and variations in shape.

2.2 Hippocampal Subfield Volumetry

Developments in high-field MRI at 3 T and beyond, together with the use of phased-array head coils, offer new opportunities to appraise hippocampal internal structure by unveiling strata rich in white matter and improved identification of the hippocampal sulcus, which separates cornu ammonis (CA) and subiculum from the dentate gyrus (DG). Paralleling advances in hardware, a number of studies have provided MRI-based guidelines to manually segment hippocampal subfields [31–38].

While substantial progress has been made, challenges remain, particularly when attempting to separate individual CA subfields from one another, which compromises reliability within and across analysts. Opting for high reliability, we recently developed a segmentation protocol that divides the hippocampal formation into consistently identifiable subregions, guided by intensity and morphology of the densely myelinated molecular layer, together with few geometry-based boundaries flexible to overall mesiotemporal anatomy [39]. Specifically, we combined presubiculum, parasubiculum, and subiculum proper into a single label (subiculum); joined CA1, 2, and 3 (CA1–3); and merged CA4 with the DG (CA4-DG). While segmentation relied primarily on T1-weighted data, T2-weighted images offered additional guidance. We made available online the full set of multispectral images in high-resolution native and stereotaxic (MNI152) space, the manual labels, together with a probabilistic atlas that can inform functional and structural imaging assessments of the hippocampal formation.

From a practical perspective, manual subfield segmentation requires anatomical expertise and is prohibitively time-consuming, precluding its use in a clinical setting. Few automated methods have been proposed, relying on multiple templates together with Bayesian inference and fusion of ex vivo/in vivo landmarks [40, 41], label propagation to intermediate templates [42], or combinations of label fusion (taking inter-template similarity into account) and post-hoc segmentation correction [43]. These methods operate either on anisotropic T2-weighted images, or T1-weighted images with standard millimetric, or submillimetric resolution. Only one study so far [42] calculated Dice overlaps of manual and automated labels in millimetric T1-weighted images, with modest performance (Dice: 0.56–0.65). We have devised a novel approach for hippocampal subfield segmentation operating on T1-weighted images, which combines multi-template feature matching with deformable parametric surfaces and vertex-wise patch sampling, relying on point-wise correspondence across the template library [44]. We used patches; they compactly represent shape, anatomy, texture, and intensity [45]. In controls, accuracy was excellent, with Dice overlap >80% when using either submillimetric or millimetric images. Performance remained robust when reducing the size of the template library, an advantageous feature given high demands on expertise/time for the generation of subfield-specific atlases. Our results were superior to the widely used *FreeSurfer* algorithm [40], both with respect to the segmentation accuracy in controls (80% vs. 68%) and seizure focus lateralization in patients (93% vs. 79%).

2.3 Hippocampal Shape Analysis

In patients with drug-resistant TLE, hippocampal volumes may be normal in up to 40% of cases, despite hippocampal histopathology showing subtle mesiotemporal sclerosis. This is likely due to the fact that MRI volumetry provides a global estimate of atrophy, thus limiting the sensitivity to detect subtle focal anomalies. In contrast, statistical surface shape modeling allows the localization of structural alterations at a submillimetric level. Statistical surface shape modeling has been applied to TLE [46–48]. However, this method does not allow differentiating volume variations from shifting and/or bending. To overcome this issue, we proposed a new approach to quantify local volume changes by computing the surface-based Jacobian determinant [49]. In brief, mesiotemporal labels are first converted to surface meshes and parameterized using spherical harmonics with point distribution model (SPHARM-PDM) guaranteeing anatomical correspondence across subjects [50], a requirement for sensitive and reliable statistics. For each mesiotemporal structure, SPHARM-PDM surfaces of each individual are rigidly aligned to a template (constructed from the mean surface of controls and patients) with respect to the centroid and the longitudinal axis of the first order ellipsoid. To correct for differences in head size, surfaces are inversely scaled with respect to intracranial volume. In our original method, to compute volume changes, we applied the heat equation to interpolate the vertex-wise displacement vectors within the volume enclosed by the SPHARM-PDM surface boundary [49]. Data simulation showed that this approach enables accurate quantification of local volume changes without interference of shifting/bending.

Combined cross-sectional modeling of disease duration in a large cohort of patients and longitudinal analysis in a subset of patients that delayed surgery, we localized consistent patterns of progressive atrophy in hippocampal CA1, anterolateral entorhinal, and amygdalar laterobasal group bilaterally [51]. Using 3D surface modeling of high-resolution MRI, we statistically mapped thalamic atrophy, located mainly in the ipsilateral medial divisions. We showed that atrophic changes in the thalamus covary with thickness of frontal, parahippocampal, and lateral temporal cortices [52], suggesting a role of the thalamocortical network in the pathophysiology of TLE.

The contemporary shift toward evidence-based practice promotes computational neuroimaging as a means to provide biomarkers for diagnostics and prognostics. In this context, data mining is a method of choice to extract critical features, patterns, and relationships from high-dimensional datasets that might otherwise be missed. In TLE MRI-based pattern, learning has been used mainly to discern patients from controls and lateralize the seizure focus [53–56]. Combining shape analysis with machine learning, we recently identified subregional variations across mesiotemporal structures, not captured by global MRI volumetry [57]. Indeed, four classes optimally partitioned a large cohort of drug-resistant TLE with a unilateral seizure focus. Anomalies were not restricted to the hippocampus alone in any given class; furthermore, except in one class, alterations were bilaterally distributed. Importantly, the majority of patients diagnosed with unilateral hippocampal atrophy by means of conventional MRI volumetry fell into two classes: half of them presented with strictly ipsilateral mesiotemporal atrophy, while the others showed bilateral asymmetric atrophy. Similarly, most patients with normal global hippocampal volumetry were divided into those with subtle bilateral symmetric atrophy in all mesiotemporal structures, and a class with a paradoxical makeup, i.e., focal increases in hippocampal and amygdala volumes. Class-wise surface-based classifiers accurately predicted outcome in 92 ± 1% of patients, outperforming classifiers based on conventional MRI volumetry. Class membership was also associated with distinct patterns of histopathological damage, emphasizing the ability of machine learning to disentangle the differential contribution of morphology to patient phenotypes, ultimately refining the prognosis of epilepsy surgery.

Lateralization of the seizure focus remains a core challenge in the management of drug-resistant TLE. In patients with unclear lateralization and/or normal hippocampal MRI findings, epileptologists perform invasive EEG investigation with intracerebral electrodes implantation; in these cases, resective epilepsy surgery is often delayed or not performed because the location of the epileptic focus remains elusive. We recently proposed a multispectral MRI-based clinical decision support approach to carry out automated seizure focus lateralization [58]. Based on high-resolution, submillimetric 3 T MRI with hippocampal subfield segmentations, this approach samples MRI features along the medial sheet of each subfield to minimize partial volume effects. To establish correspondence of sampling points across subjects, we propagate a spherical harmonic parameterization derived from the hippocampal boundary along a Laplacian gradient field toward the medial sheet. The medial sheet analysis can integrate multiple MRI parameters (volume, T2-intensity, and diffusion parameters) in a common surface-based reference frame (Fig. 2).

Applying statistical pattern learners that operate on subregional surface data achieved excellent accuracy for individualized focus lateralization (based on T2 intensity) and postsurgical seizure outcome prediction (based on columnar volume), emphasizing complimentary yield of different features across clinical applications [59]. Collectively, our findings recommend multiparametric MRI analysis of subregional data as a comprehensive in vivo phenotyping of hippocampal pathology in TLE.

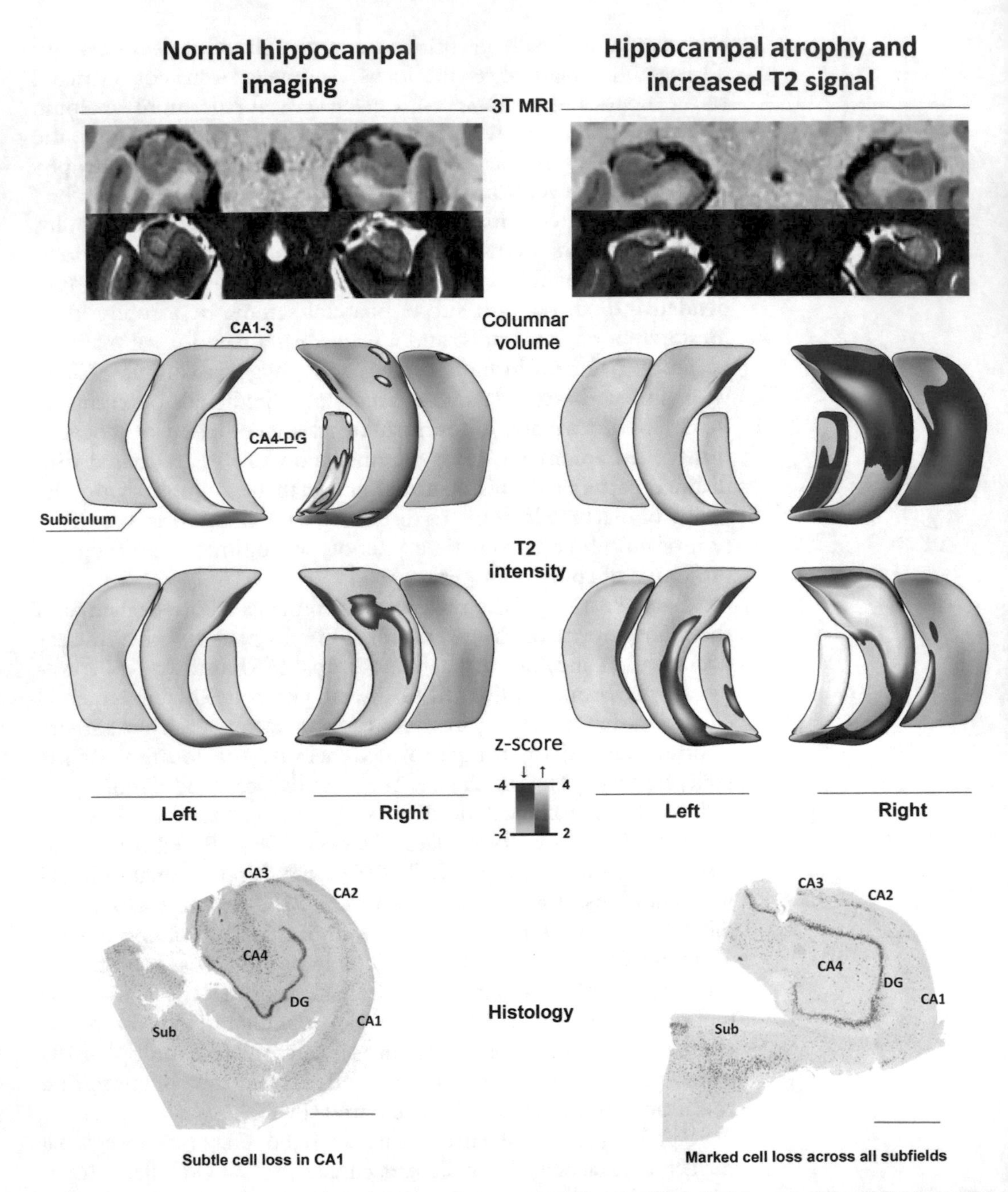

Fig. 2 Laminar analysis of hippocampal subfields. Coronal 3 T T1w and T2w MRI (top panels), subfields medial sheet (laminar) feature maps *z*-normalized with respect to controls (center), and histopathological NeuN stain of the resected tissue (bottom) in two patients with right TLE. In the patient with normal hippocampal imaging (left), sheet-based analysis revealed subtle volume increases in CA4-dentate (dentate gyrus, DG) and CA1 ipsilateral to the focus. Histology revealed a minimal neuronal loss in CA1, with gliosis (stain not shown). In the case with MRI-visible atrophy and increased hippocampal T2 signal (right panels), sheet-based analysis confirmed the visual findings, with extensive laminar atrophy and T2 signal anomalies in CA1, CA4-DG, and subiculum. Marked cell loss and gliosis across all subfields were also seen in the specimen

2.4 Whole-Brain Morphometry

Contrary to methods restricted to selected brain areas, voxel-based morphometry (VBM) detects structural differences in GM and WM "density" throughout the brain, without a priori assumptions [60, 61]. A detailed technical description, as well as discussion regarding the validity of VBM, can be found elsewhere [62–64]. Briefly, it involves the spatial normalization of T1 high-resolution scans (with good contrast between GM and the other tissues) from all individuals into a common stereotactic space, extraction of a particular tissue type (usually GM and WM) from the normalized scan of each subject, smoothing of the extracted tissue maps, and, finally, performing statistical analysis to localize and make group inferences. Usually all spatial processing and statistical analysis are performed with the freely available statistical parametric software (SPM) package (http://www.fil.ion.uck.ac.uk/spm) and VBM toolboxes (http://dbm.neuro.uni-jena.de).

Most VBM studies in TLE have focused on refractory patients (including children and adults) [15, 65, 66]. Although most of the selected patients in these studies presented with mesiotemporal sclerosis, many failed to detect reduced hippocampal GM density [67–69]. Conversely, most studies have consistently shown GM reduction in other limbic areas including the cingulum and thalamus, as well as extra-limbic areas including the frontal lobe [15, 70–75]. The consistency of widespread cortical anomalies was shown in a meta-analysis of six studies using activation likelihood estimation [76].

Voxel-based morphometry may not be the most appropriate approach to assess cortical pathology, since its smoothing step neglects anatomical relationships across a folded surface [77]. Furthermore, anatomical variability in gyrification and sulcation may reduce the sensitivity to detect significant effects, even after nonlinear warping and smoothing [62, 78, 79]. Measuring cortical thickness is a more direct and biologically meaningful technique to quantify atrophy as it allows studying continuous changes with respect to the anatomy of the folded cortical surface. Advanced image processing methods enable automatic and reliable measurement of cortical thickness by calculating the distance between corresponding points on both the GM and WM surfaces across the entire cortical mantle [80–83]. Cortical thickness provides a biologically meaningful in vivo index of morphology reflecting cell size, density, and arrangement [84].

Studies from independent groups have shown that TLE is associated with widespread atrophy that is particularly marked in the fronto-central regions [15, 85–89] (Fig. 3). Bootstrap methods showed that at least 20 subjects are necessary to reliably observe these patterns of atrophy [85].

By measuring cortical thickness in patients with TLE and controls in a combined cross-sectional and longitudinal analysis, we demonstrated that patients suffering from epilepsy for longer than

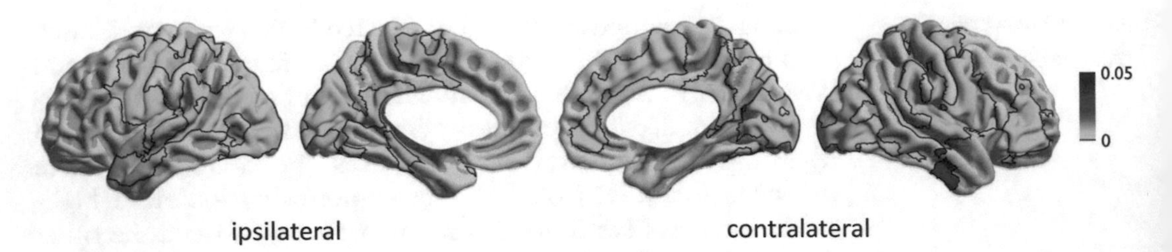

Fig. 3 Whole-brain morphometry in temporal lobe epilepsy. Compared to healthy controls, patients with drug-resistant temporal lobe epilepsy present with widespread neocortical thinning ipsi- and contralateral to the seizure focus. Cortical thinning is found not only in the temporal lobes but also in fronto-central regions. Significant clusters are thresholded using random field theory

14 years have a more rapid progression of neocortical atrophy compared to those with shorter disease duration [88]. Progressive neocortical atrophy was distinct from normal aging and likely reflects seizure-induced damage. Our findings are in line with functional data that show progressive cognitive decline in temporal lobe epilepsy [90] and extension of the epileptogenic network with recurrent EEG discharges [91]. Taken together, they provide compelling evidence that TLE is a progressive disorder that warrants serious consideration of early surgery when seizures are medication-resistant.

3 Neocortical Epilepsy

Neocortical epilepsy related to focal cortical dysplasia (FCD), a congenital malformation of cortical development due to abnormal neuronal proliferation [92], accounts for more than half of pediatric and a quarter of adult patients [4]. The histopathological spectrum includes lesions characterized only by disruptions of the hexalaminar cortical structure (FCD type I) to those associating dyslamination with dysmorphic neurons or large balloon cells (FCD type II); gliosis is generally found across the lesional tissue, and demyelination is present at the gray-white matter junction as well as in the subcortical white matter [93]. Main FCD features on structural MRI include abnormally thick cortical gray matter (50–92% of cases) and blurring of the gray-white matter interface (60–80%) [4, 94]. Analysis of T2-weighted images reveals gray matter hyperintensity in 46–92% of lesions, and sensitivity of FLAIR images (Fig. 4) is even higher (71–100%). The prominence of MRI findings generally reflects the degree of the underlying histopathological abnormality [95]. In many patients, however, the dysplastic features may be very subtle, and the MRI, consequently, was reported as unremarkable [6, 94, 96, 97].

This clinical challenge has motivated the development of computer-aided methods aimed at modeling in vivo characteristics of dysplasias.

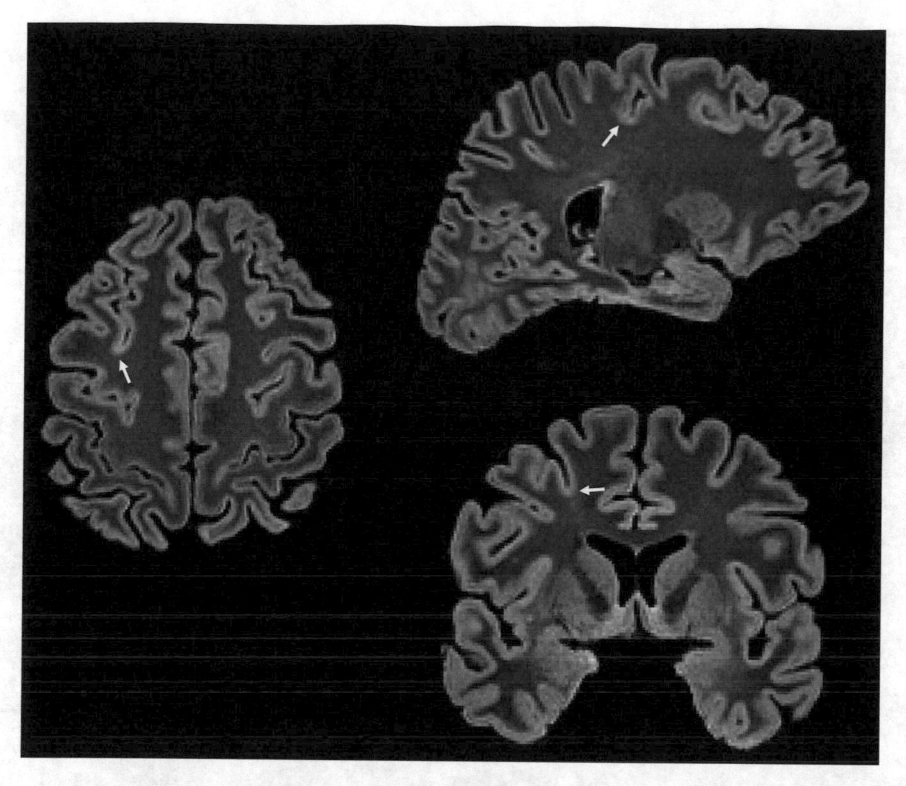

Fig. 4 Focal cortical dysplasia with non-diagnostic routine MRI. Representative example of a patient with drug-resistant frontal lobe epilepsy in whom routine MRI was initially reported as unremarkable. Detailed inspection of the FLAIR images, however, revealed a subtle blurring at the bottom of a deep frontal sulcus. The histopathological examination of the surgical specimen demonstrated a focal cortical dysplasia Type-II with dysplastic neurons

3.1 Computational Models of Focal Cortical Dysplasia

Several groups have applied VBM of T1-weighted images to detect structural abnormalities related to MRI-visible FCD type II in single patients. These studies have reported increases in GM concentration co-localizing with the lesion in 63–86% of cases [98–102]. Notably, the marked hyperintensity of some FCD lesions leads to GM misclassification, reducing considerably the sensitivity of the technique [103]. MRI intensities or quantitative MRI contrasts such as T2 relaxometry, double inversion recovery, and magnetization transfer ratio imaging are also amenable to voxel-based comparison. These approaches have a sensitivity of 87–100% in detecting obvious FCD lesions [104–107] but have low yield in cases with negative conventional MRI [108, 109].

Our group pioneered the design of computer-based algorithms modeling the distinctive characteristics of FCD type II [100, 110–112]. Our initial methods targeted cortical thickening and blurring of the GM-WM boundary on T1-weighted MRI, enhancing visual detection rates by 30% relative to standard evaluation; others have replicated our techniques with similar results [102, 113–116]. The evaluation of multiple maps generated

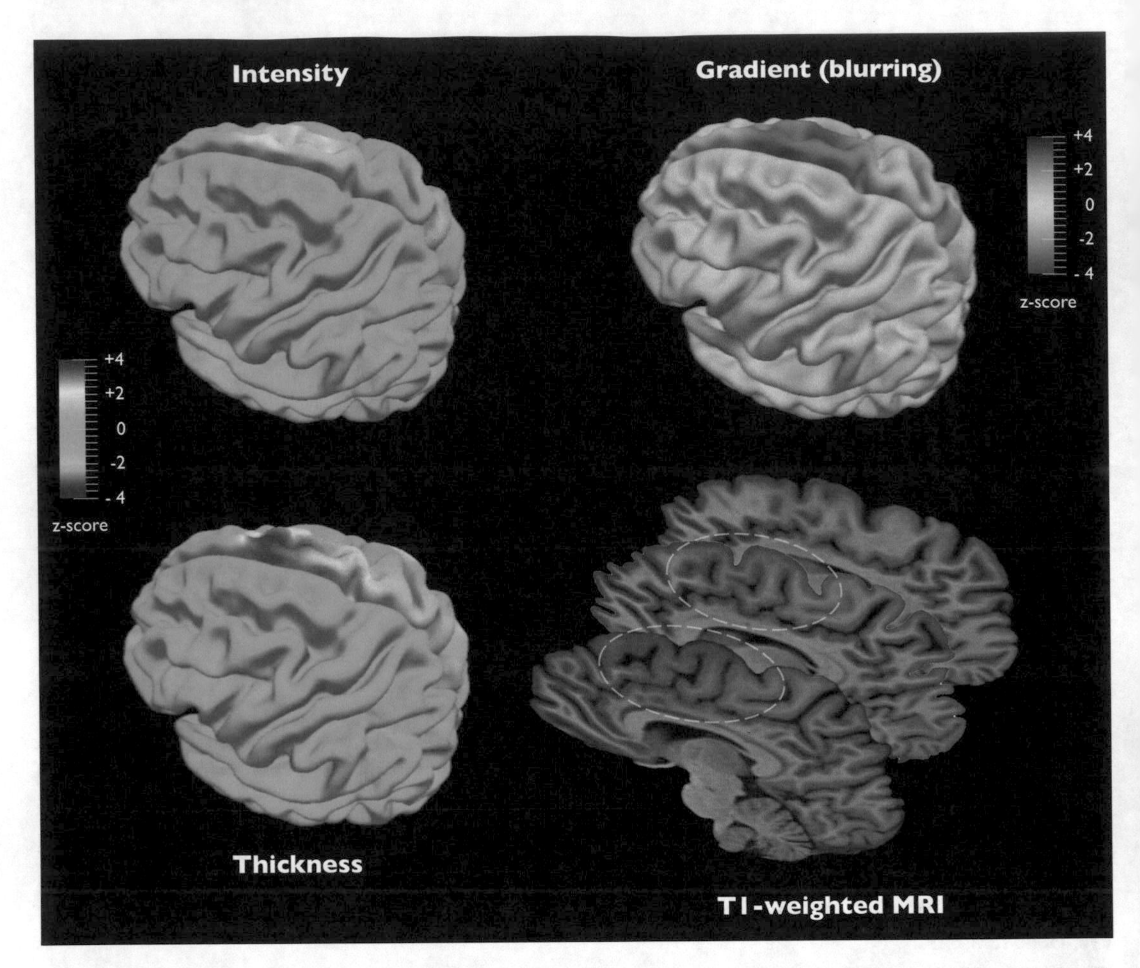

Fig. 5 Surface-based computational models of focal cortical dysplasia with subtle features. In this patient with right frontal lobe epilepsy, high-resolution 3 T MRI was reported as normal. The three color-coded maps display computational models of cortical dysplasia derived from T1-weighted MRI and normalized with respect to healthy controls (*z*-scores). The pre-central and the cingulate regions present with a widespread increase in cortical thickness, decreased gradient (blurring) and subtle increased intensity. Histopathology of the surgical specimen confirmed the presence of focal cortical dysplasia ILAE Type-IIA. The dotted ellipsoids on the T1-weighted MRI sagittal sections surround the dysplastic cortex. This relatively large lesion was initially overlooked likely due to its subtle characteristics

through these quantitative techniques is done visually, so that the yield and diagnostic confidence depend on the reader's familiarity with the algorithm (Fig. 5).

Limited generalizability also stems from the fact that these approaches have been validated with lesions recognized on routine radiological evaluation [100, 102, 114, 116, 117]. Moreover, integrating the diverse and complex information embedded in the various maps requires expertise that few centers have developed so far. Another source of difficulty arises from the paucity or lack of localizing clinical semiology and surface EEG findings to direct the search for FCD, particularly in patients with frontal lobe epilepsy.

We previously developed a series of algorithms for automatic FCD detection on 1.5 T MRI, which relied on voxel-based texture analysis combined with a Bayesian classifier [118]. They were validated with mid- to large-sized lesions visible on routine radiological inspection; nevertheless, classification failed in up to 20% of cases. Such performance is unsatisfactory given current referral patterns to epilepsy surgery centers, with an increasing number of patients with non-diagnostic clinical MRI, even at 3.0 T. Notably, the absence of a visible lesion is one of the greatest challenges in epilepsy surgery and has led to an increase in invasive EEG studies with implanted intracranial electrodes. Yet, without informed, image-guided implantation, even with widespread coverage, EEG sampling errors may occur in up to 40% of cases; consequently, the target cannot be defined, and the outcome of surgery, if considered, is poorer [2]. In these patients, lesions are subtle, with morphological characteristics that may differ only slightly from normal tissue.

Compared to voxel-based techniques, a surface-based approach preserves cortical topology and quantifies sulco-gyral anomalies, at times the only sign of dysgenesis [119]. In our more recent work, we thus opted for a surface-based framework and combined various morphological (cortical thickness, curvature, and sulcal depth) and intensity features (relative intensity and intensity gradients) taking advantage of their covariance to unveil subthreshold tissue properties not readily identified on a single modality. For automated lesion detection, we chose a linear discriminant classifier, a supervised technique that is mathematically robust and simple to interpret [120], and developed a two-step procedure. The first vertex-wise classifier recognized vertices with the highest detection rate; the second was designed to remove false-positive clusters using additional texture features and spatial priors. We cross-validated our findings using a leave-one-out strategy and obtained a sensitivity of 74% (Fig. 6).

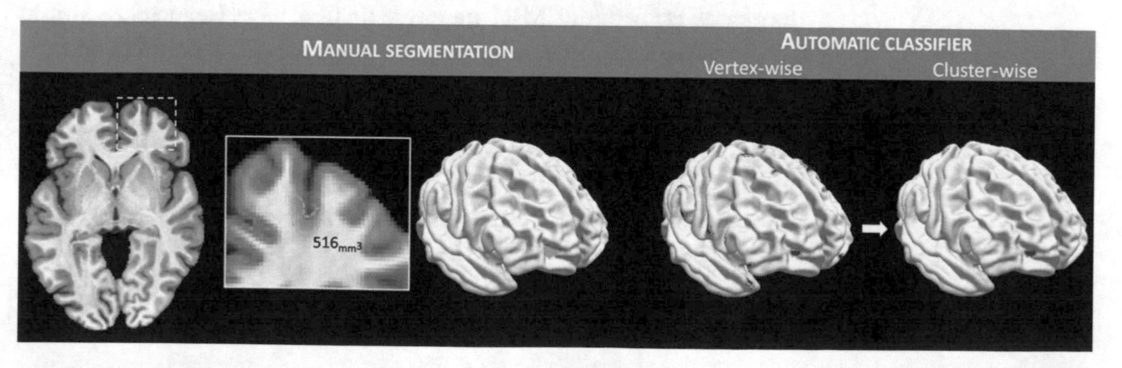

Fig. 6 Automated FCD type II detection. The axial T1-weighted MRI sections show the region containing the FCD (dashed square). The magnified panel displays the manually segmented FCD label (dotted green line) and its volume; the label (green) is projected onto the surface template. In these three examples, the vertex-wise classification identified several putative lesions (in red), whereas the subsequent cluster-wise classification discarded all false positives except the cluster co-localizing with the manual label (in blue)

Given that all subjects were initially diagnosed as MRI negative, this fully automated approach offered a substantial gain in sensitivity over standard radiological assessment. When applying our classifier trained on 3.0 T images to an independent dataset of patients with histologically proven FCD acquired at 1.5 T, we maintained high sensitivity (71%), supporting generalizability.

3.2 Whole-Brain Morphometry

Because of the crucial role of MRI in defining the surgical target, imaging studies in FCD have been primarily dedicated to lesion detection in single patients [121]. Whole-brain cohort-specific structural brain anomalies remain largely unknown [122]. We recently examined whole-brain MRI morphology in dysplasia-related frontal lobe epilepsy. Relative to controls, FCD type I displayed multilobar cortical thinning that was most marked in ipsilateral frontal cortices. Conversely, type II showed thickening in temporal and postcentral cortices. Cortical folding also diverged, with increased complexity in prefrontal cortices in type I and decreases in type II. We hypothesized that cortical thickening in type II may indicate delayed pruning, while a thin cortex in type I likely results from combined effects of seizure excitotoxicity and the primary malformation. In addition, group-level patterns had a translational value for individual diagnostics as they successfully guided automated subtype classification (type I: 100%; type II: 96%), seizure focus lateralization (type I, 92%; type II, 86%), and outcome prediction (type I, 92%; type II, 82%).

4 Conclusion

By revealing subtle lesions that previously eluded visual inspection, quantitative image processing of structural MRI has clearly demonstrated increased sensitivity compared to conventional techniques. Overall, the most significant clinical impact of post-processing is that cases considered MRI negative at first have been increasingly become MRI positive, offering the life-changing benefits of epilepsy surgery to more patients. Importantly, as advanced image post-processing can be performed on clinical MRI yielding 3D millimetric multicontrast images, they offer a substantial cost-benefit. Aside from diagnostics, ongoing developments provide unprecedented opportunities for the design of new biomarkers to stage and monitor the disease.

References

1. Fisher RS, Acevedo C, Arzimanoglou A, Bogacz A, Cross JH, Elger CE, Engel J Jr, Forsgren L, French JA, Glynn M, Hesdorffer DC, Lee BI, Mathern GW, Moshe SL, Perucca E, Scheffer IE, Tomson T, Watanabe M, Wiebe S (2014) ILAE official report: a practical clinical definition of epilepsy. Epilepsia 55(4):475–482. https://doi.org/10.1111/epi.12550
2. Tellez-Zenteno JF, Hernandez Ronquillo L, Moien-Afshari F, Wiebe S (2010) Surgical outcomes in lesional and non-lesional epilepsy: a systematic review and meta-analysis. Epilepsy Res 89(2–3):310–318. https://doi.org/10.1016/j.eplepsyres.2010.02.007. S0920-1211(10)00041-0 [pii]
3. Spencer S, Huh L (2008) Outcomes of epilepsy surgery in adults and children. Lancet Neurol 7(6):525–537. https://doi.org/10.1016/S1474-4422(08)70109-1
4. Lerner JT, Salamon N, Hauptman JS, Velasco TR, Hemb M, Wu JY, Sankar R, Donald Shields W, Engel J Jr, Fried I, Cepeda C, Andre VM, Levine MS, Miyata H, Yong WH, Vinters HV, Mathern GW (2009) Assessment and surgical outcomes for mild type I and severe type II cortical dysplasia: a critical review and the UCLA experience. Epilepsia 50(6):1310–1335. https://doi.org/10.1111/j.1528-1167.2008.01998.x. EPI1998 [pii]
5. Skirrow C, Cross JH, Cormack F, Harkness W, Vargha-Khadem F, Baldeweg T (2011) Long-term intellectual outcome after temporal lobe surgery in childhood. Neurology 76(15):1330–1337. https://doi.org/10.1212/WNL.0b013e31821527f0. 76/15/1330 [pii]
6. Bernasconi A, Bernasconi N, Bernhardt BC, Schrader D (2011) Advances in MRI for 'cryptogenic' epilepsies. Nat Rev Neurol 7(2):99–108. https://doi.org/10.1038/nrneurol.2010.199
7. Blumcke I, Coras R, Miyata H, Ozkara C (2012) Defining clinico-neuropathological subtypes of mesial temporal lobe epilepsy with hippocampal sclerosis. Brain Pathol 22(3):402–411. https://doi.org/10.1111/j.1750-3639.2012.00583.x
8. Thom M (2014) Review: hippocampal sclerosis in epilepsy: a neuropathology review. Neuropathol Appl Neurobiol 40(5):520–543. https://doi.org/10.1111/nan.12150
9. Kim H, Chupin M, Colliot O, Bernhardt BC, Bernasconi N, Bernasconi A (2012) Automatic hippocampal segmentation in temporal lobe epilepsy: impact of developmental abnormalities. NeuroImage 59(4):3178–3186. https://doi.org/10.1016/j.neuroimage.2011.11.040
10. Bernasconi N, Kinay D, Andermann F, Antel S, Bernasconi A (2005) Analysis of shape and positioning of the hippocampal formation: an MRI study in patients with partial epilepsy and healthy controls. Brain 128(Pt 10):2442–2452
11. Kim H, Mansi T, Bernasconi A, Bernasconi N (2013) Disentangling hippocampal shape anomalies in epilepsy. Front Neurol 4:131. https://doi.org/10.3389/fneur.2013.00131
12. Bernasconi N, Bernasconi A, Andermann F, Dubeau F, Feindel W, Reutens DC (1999) Entorhinal cortex in temporal lobe epilepsy: a quantitative MRI study. Neurology 52(9):1870–1876
13. Bernasconi N, Bernasconi A, Caramanos Z, Antel SB, Andermann F, Arnold DL (2003) Mesial temporal damage in temporal lobe epilepsy: a volumetric MRI study of the hippocampus, amygdala and parahippocampal region. Brain 126(Pt 2):462–469
14. Bernasconi N, Bernasconi A, Caramanos Z, Dubeau F, Richardson J, Andermann F, Arnold DL (2001) Entorhinal cortex atrophy in epilepsy patients exhibiting normal hippocampal volumes. Neurology 56(10):1335–1339
15. Bernasconi N, Duchesne S, Janke A, Lerch J, Collins DL, Bernasconi A (2004) Whole-brain voxel-based statistical analysis of gray matter and white matter in temporal lobe epilepsy. NeuroImage 23(2):717–723
16. Sankar T, Bernasconi N, Kim H, Bernasconi A (2008) Temporal lobe epilepsy: differential pattern of damage in temporopolar cortex and white matter. Hum Brain Mapp 29(8):931–944
17. Natsume J, Bernasconi N, Andermann F, Bernasconi A (2003) MRI volumetry of the thalamus in temporal, extratemporal, and idiopathic generalized epilepsy. Neurology 60(8):1296–1300
18. Lee JW, Andermann F, Dubeau F, Bernasconi A, MacDonald D, Evans A, Reutens DC (1998) Morphometric analysis of the temporal lobe in temporal lobe epilepsy. Epilepsia 39(7):727–736
19. Duncan JS, Winston GP, Koepp MJ, Ourselin S (2016) Brain imaging in the assessment for epilepsy surgery. Lancet Neurol 15

(4):420–433. https://doi.org/10.1016/S1474-4422(15)00383-X

20. Bernasconi A, Bernasconi N, Natsume J, Antel SB, Andermann F, Arnold DL (2003) Magnetic resonance spectroscopy and imaging of the thalamus in idiopathic generalized epilepsy. Brain 126:2447–2454
21. Goubran M, Bernhardt BC, Cantor-Rivera D, Lau JC, Blinston C, Hammond RR, de Ribaupierre S, Burneo JG, Mirsattari SM, Steven DA, Parrent AG, Bernasconi A, Bernasconi N, Peters TM, Khan AR (2016) In vivo MRI signatures of hippocampal subfield pathology in intractable epilepsy. Hum Brain Mapp 37(3):1103–1119. https://doi.org/10.1002/hbm.23090
22. Dill V, Franco AR, Pinho MS (2015) Automated methods for hippocampus segmentation: the evolution and a review of the state of the art. Neuroinformatics 13(2):133–150. https://doi.org/10.1007/s12021-014-9243-4
23. McDonald CR, Hagler DJ Jr, Ahmadi ME, Tecoma E, Iragui V, Dale AM, Halgren E (2008) Subcortical and cerebellar atrophy in mesial temporal lobe epilepsy revealed by automatic segmentation. Epilepsy Res 79(2–3):130–138. https://doi.org/10.1016/j.eplepsyres.2008.01.006. S0920-1211(08)00036-3 [pii]
24. Pardoe HR, Pell GS, Abbott DF, Jackson GD (2009) Hippocampal volume assessment in temporal lobe epilepsy: how good is automated segmentation? Epilepsia 50(12):2586–2592. https://doi.org/10.1111/j.1528-1167.2009.02243.x. EPI2243 [pii]
25. Hammers A, Heckemann R, Koepp MJ, Duncan JS, Hajnal JV, Rueckert D, Aljabar P (2007) Automatic detection and quantification of hippocampal atrophy on MRI in temporal lobe epilepsy: a proof-of-principle study. NeuroImage 36(1):38–47
26. Avants BB, Yushkevich P, Pluta J, Minkoff D, Korczykowski M, Detre J, Gee JC (2010) The optimal template effect in hippocampus studies of diseased populations. NeuroImage 49(3):2457–2466
27. Chupin M, Hammers A, Liu RS, Colliot O, Burdett J, Bardinet E, Duncan JS, Garnero L, Lemieux L (2009) Automatic segmentation of the hippocampus and the amygdala driven by hybrid constraints: method and validation. NeuroImage 46(3):749–761. https://doi.org/10.1016/j.neuroimage.2009.02.013. S1053-8119(09)00152-9 [pii]
28. Heckemann RA, Keihaninejad S, Aljabar P, Rueckert D, Hajnal JV, Hammers A (2010) Improving intersubject image registration using tissue-class information benefits robustness and accuracy of multi-atlas based anatomical segmentation. NeuroImage 51(1):221–227. https://doi.org/10.1016/j.neuroimage.2010.01.072. S1053-8119(10)00094-7 [pii]
29. Akhondi-Asl A, Jafari-Khouzani K, Elisevich K, Soltanian-Zadeh H (2011) Hippocampal volumetry for lateralization of temporal lobe epilepsy: automated versus manual methods. Neuroimage 54(Supplement 1):S218–S226
30. Kim H, Mansi T, Bernasconi N, Bernasconi A (2012) Surface-based multi-template automated hippocampal segmentation: application to temporal lobe epilepsy. Med Image Anal 16(7):1445–1455. https://doi.org/10.1016/j.media.2012.04.008
31. Mueller SG, Stables L, AT D, Schuff N, Truran D, Cashdollar N, Weiner MW (2007) Measurement of hippocampal subfields and age-related changes with high resolution MRI at 4T. Neurobiol Aging 28(5):719–726
32. La Joie R, Fouquet M, Mezenge F, Landeau B, Villain N, Mevel K, Pelerin A, Eustache F, Desgranges B, Chetelat G (2010) Differential effect of age on hippocampal subfields assessed using a new high-resolution 3T MR sequence. NeuroImage 53(2):506–514. https://doi.org/10.1016/j.neuroimage.2010.06.024
33. Malykhin NV, Lebel RM, Coupland NJ, Wilman AH, Carter R (2010) In vivo quantification of hippocampal subfields using 4.7 T fast spin echo imaging. NeuroImage 49(2):1224–1230. https://doi.org/10.1016/j.neuroimage.2009.09.042
34. Henry TR, Chupin M, Lehericy S, Strupp JP, Sikora MA, Sha ZY, Ugurbil K, Van de Moortele PF (2011) Hippocampal sclerosis in temporal lobe epilepsy: findings at 7 T. Radiology 261(1):199–209. https://doi.org/10.1148/radiol.11101651
35. Wisse LE, Gerritsen L, Zwanenburg JJ, Kuijf HJ, Luijten PR, Biessels GJ, Geerlings MI (2012) Subfields of the hippocampal formation at 7 T MRI: in vivo volumetric assessment. NeuroImage 61(4):1043–1049. https://doi.org/10.1016/j.neuroimage.2012.03.023
36. Winterburn JL, Pruessner JC, Chavez S, Schira MM, Lobaugh NJ, Voineskos AN, Chakravarty MM (2013) A novel in vivo atlas of human hippocampal subfields using high-resolution 3 T magnetic resonance imaging. NeuroImage 74:254–265. https://doi.org/10.1016/j.neuroimage.2013.02.003

37. Goubran M, Rudko DA, Santyr B, Gati J, Szekeres T, Peters TM, Khan AR (2014) In vivo normative atlas of the hippocampal subfields using multi-echo susceptibility imaging at 7 Tesla. Hum Brain Mapp 35 (8):3588–3601. https://doi.org/10.1002/hbm.22423
38. Rhindress K, Ikuta T, Wellington R, Malhotra AK, Szeszko PR (2015) Delineation of hippocampal subregions using T1-weighted magnetic resonance images at 3 Tesla. Brain Struct Funct 220(6):3259–3272. https://doi.org/10.1007/s00429-014-0854-1
39. Kulaga-Yoskovitz J, Bernhardt BC, Hong SJ, Mansi T, Liang KE, van der Kouwe AJ, Smallwood J, Bernasconi A, Bernasconi N (2015) Multi-contrast submillimetric 3 Tesla hippocampal subfield segmentation protocol and dataset. Sci Data 2:150059. https://doi.org/10.1038/sdata.2015.59
40. Van Leemput K, Bakkour A, Benner T, Wiggins G, Wald LL, Augustinack J, Dickerson BC, Golland P, Fischl B (2009) Automated segmentation of hippocampal subfields from ultra-high resolution in vivo MRI. Hippocampus 19(6):549–557. https://doi.org/10.1002/hipo.20615
41. Iglesias JE, Sabuncu MR, Van Leemput K, Alzheimer's Disease Neuroimaging I (2013) Improved inference in Bayesian segmentation using Monte Carlo sampling: application to hippocampal subfield volumetry. Med Image Anal 17(7):766–778. https://doi.org/10.1016/j.media.2013.04.005
42. Pipitone J, Park MT, Winterburn J, Lett TA, Lerch JP, Pruessner JC, Lepage M, Voineskos AN, Chakravarty MM, Alzheimer's Disease Neuroimaging I (2014) Multi-atlas segmentation of the whole hippocampus and subfields using multiple automatically generated templates. NeuroImage 101:494–512. https://doi.org/10.1016/j.neuroimage.2014.04.054
43. Yushkevich PA, Amaral RS, Augustinack JC, Bender AR, Bernstein JD, Boccardi M, Bocchetta M, Burggren AC, Carr VA, Chakravarty MM, Chetelat G, Daugherty AM, Davachi L, Ding SL, Ekstrom A, Geerlings MI, Hassan A, Huang Y, Iglesias JE, La Joie R, Kerchner GA, LaRocque KF, Libby LA, Malykhin N, Mueller SG, Olsen RK, Palombo DJ, Parekh MB, Pluta JB, Preston AR, Pruessner JC, Ranganath C, Raz N, Schlichting ML, Schoemaker D, Singh S, Stark CE, Suthana N, Tompary A, Turowski MM, Van Leemput K, Wagner AD, Wang L, Winterburn JL, Wisse LE, Yassa MA, Zeineh MM, Hippocampal Subfields G (2015) Quantitative comparison of 21 protocols for labeling hippocampal subfields and parahippocampal subregions in in vivo MRI: towards a harmonized segmentation protocol. NeuroImage 111:526–541. https://doi.org/10.1016/j.neuroimage.2015.01.004
44. Caldairou B, Bernhardt BC, Kulaga-Yoskovitz J, Kim H, Bernasconi N, Bernasconi A. (2016) A Surface Patch-Based Segmentation Method for Hippocampal Subfields. In: Ourselin S, Joskowicz L, Sabuncu M, Unal G, Wells W. (eds) Medical Image Computing and Computer-Assisted Intervention – MICCAI 2016. MICCAI 2016. Lecture Notes in Computer Science, vol 9901. Springer, Cham. https://doi.org/10.1007/978-3-319-46723-8_44
45. Giraud R, Ta VT, Papadakis N, Manjon JV, Collins DL, Coupe P, Alzheimer's Disease Neuroimaging I (2016) An optimized patchmatch for multi-scale and multi-feature label fusion. Neuroimage 124(Pt A):770–782. https://doi.org/10.1016/j.neuroimage.2015.07.076
46. Hogan RE, Wang L, Bertrand ME, Willmore LJ, Bucholz RD, Nassif AS, Csernansky JG (2004) MRI-based high-dimensional hippocampal mapping in mesial temporal lobe epilepsy. Brain 127(Pt 8):1731–1740. https://doi.org/10.1093/brain/awh197. awh197 [pii]
47. Hogan RE, Carne RP, Kilpatrick CJ, Cook MJ, Patel A, King L, O'Brien TJ (2008) Hippocampal deformation mapping in MRI negative PET positive temporal lobe epilepsy. J Neurol Neurosurg Psychiatry 79 (6):636–640. https://doi.org/10.1136/jnnp.2007.123406. jnnp.2007.123406 [pii]
48. Maccotta L, Moseley ED, Benzinger TL, Hogan RE (2015) Beyond the CA1 subfield: local hippocampal shape changes in MRI-negative temporal lobe epilepsy. Epilepsia 56(5):780–788. https://doi.org/10.1111/epi.12955
49. Kim H, Besson P, Colliot O, Bernasconi A, Bernasconi N (2008) Surface-based vector analysis using heat equation interpolation: a new approach to quantify local hippocampal volume changes. Med Image Comput Comput Assist Interv 11(Pt 1):1008–1015
50. Styner M, Oguz I, Xu S, Brechbühler C, Pantazis D, Levitt JJ, Shenton ME, Gerig G (2006) Framework for the Statistical Shape Analysis of Brain Structures using SPHARM-PDM. The insight journal:242–250
51. Bernhardt BC, Kim H, Bernasconi A, Bernasconi N (2013) Patterns of subregional mesiotemporal disease progression in temporal lobe

epilepsy. Neurology 81(21):1840–1847. https://doi.org/10.1212/01.wnl.0000436069.20513.92

52. Bernhardt BC, Bernasconi N, Kim H, Bernasconi A (2012) Mapping thalamocortical network pathology in temporal lobe epilepsy. Neurology 78(2):129–136. https://doi.org/10.1212/WNL.0b013e31823efd0d
53. Keihaninejad S, Heckemann RA, Gousias IS, Hajnal JV, Duncan JS, Aljabar P, Rueckert D, Hammers A (2012) Classification and lateralization of temporal lobe epilepsies with and without hippocampal atrophy based on whole-brain automatic MRI segmentation. PLoS One 7(4):e33096. https://doi.org/10.1371/journal.pone.0033096
54. Duchesne S, Bernasconi N, Bernasconi A, Collins DL (2006) MR-based neurological disease classification methodology: application to lateralization of seizure focus in temporal lobe epilepsy. NeuroImage 29(2):557–566
55. Focke NK, Yogarajah M, Symms MR, Gruber O, Paulus W, Duncan JS (2012) Automated MR image classification in temporal lobe epilepsy. NeuroImage 59(1):356–362. https://doi.org/10.1016/j.neuroimage.2011.07.068
56. Mueller SG, Young K, Hartig M, Barakos J, Garcia P, Laxer KD (2013) A two-level multimodality imaging Bayesian network approach for classification of partial epilepsy: preliminary data. NeuroImage 71:224–232. https://doi.org/10.1016/j.neuroimage.2013.01.014
57. Bernhardt BC, Hong SJ, Bernasconi A, Bernasconi N (2015) Magnetic resonance imaging pattern learning in temporal lobe epilepsy: classification and prognostics. Ann Neurol 77(3):436–446. https://doi.org/10.1002/ana.24341
58. Kim H, Bernhardt BC, Kulaga-Yoskovitz J, Caldairou B, Bernasconi A, Bernasconi N (2014) Multivariate hippocampal subfield analysis of local MRI intensity and volume: application to temporal lobe epilepsy. Med Image Comput Comput Assist Interv 17(Pt 2):170–178
59. Bernhardt BC, Bernasconi A, Liu M, Hong SJ, Caldairou B, Goubran M, Guiot MC, Hall J, Bernasconi N (2016) The spectrum of structural and functional imaging abnormalities in temporal lobe epilepsy. Ann Neurol 80(1):142–153. https://doi.org/10.1002/ana.24691
60. Salmond CH, Ashburner J, Vargha-Khadem-F, Connelly A, Gadian DG, Friston KJ (2002) Distributional assumptions in voxel-based morphometry. NeuroImage 17(2):1027–1030
61. Ashburner J, Friston KJ (2000) Voxel-based morphometry—the methods. NeuroImage 11(6 Pt 1):805–821. https://doi.org/10.1006/nimg.2000.0582
62. Ashburner J, Friston KJ (2001) Why voxel-based morphometry should be used. NeuroImage 14(6):1238–1243
63. Good CD, Johnsrude IS, Ashburner J, Henson RN, Friston KJ, Frackowiak RS (2001) A voxel-based morphometric study of ageing in 465 normal adult human brains. NeuroImage 14(1 Pt 1):21–36
64. Ashburner J (2009) Computational anatomy with the SPM software. Magn Reson Imaging 27(8):1163–1174. https://doi.org/10.1016/j.mri.2009.01.006
65. Yasuda CL, Valise C, Saude AV, Pereira AR, Pereira FR, Ferreira Costa AL, Morita ME, Betting LE, Castellano G, Mantovani Guerreiro CA, Tedeschi H, de Oliveira E, Cendes F (2010) Dynamic changes in white and gray matter volume are associated with outcome of surgical treatment in temporal lobe epilepsy. NeuroImage 49(1):71–79. https://doi.org/10.1016/j.neuroimage.2009.08.014
66. Brazdil M, Marecek R, Fojtikova D, Mikl M, Kuba R, Krupa P, Rektor I (2009) Correlation study of optimized voxel-based morphometry and (1)H MRS in patients with mesial temporal lobe epilepsy and hippocampal sclerosis. Hum Brain Mapp 30(4):1226–1235. https://doi.org/10.1002/hbm.20589
67. Eriksson SH, Thom M, Symms MR, Focke NK, Martinian L, Sisodiya SM, Duncan JS (2009) Cortical neuronal loss and hippocampal sclerosis are not detected by voxel-based morphometry in individual epilepsy surgery patients. Hum Brain Mapp 30(10):3351–3360. https://doi.org/10.1002/hbm.20757
68. Woermann FG, Free SL, Koepp MJ, Ashburner J, Duncan JS (1999) Voxel-by-voxel comparison of automatically segmented cerebral gray matter—a rater-independent comparison of structural MRI in patients with epilepsy. NeuroImage 10(4):373–384. https://doi.org/10.1006/nimg.1999.0481
69. Keller SS, Wieshmann UC, Mackay CE, Denby CE, Webb J, Roberts N (2002) Voxel based morphometry of grey matter abnormalities in patients with medically intractable temporal lobe epilepsy: effects of side of

seizure onset and epilepsy duration. J Neurol Neurosurg Psychiatry 73(6):648–655

70. Bonilha L, Rorden C, Castellano G, Pereira F, Rio PA, Cendes F, Li LM (2004) Voxel-based morphometry reveals gray matter network atrophy in refractory medial temporal lobe epilepsy. Arch Neurol 61(9):1379–1384
71. Mueller SG, Laxer KD, Cashdollar N, Buckley S, Paul C, Weiner MW (2006) Voxel-based optimized morphometry (VBM) of gray and white matter in temporal lobe epilepsy (TLE) with and without mesial temporal sclerosis. Epilepsia 47(5):900–907. https://doi.org/10.1111/j.1528-1167.2006.00512.x
72. Keller SS, Mackay CE, Barrick TR, Wieshmann UC, Howard MA, Roberts N (2002) Voxel-based morphometric comparison of hippocampal and extrahippocampal abnormalities in patients with left and right hippocampal atrophy. NeuroImage 16(1):23–31
73. Bernasconi A (2004) Quantitative MR imaging of the neocortex. Neuroimaging Clin N Am 14(3):425–436
74. Yasuda CL, Betting LE, Cendes F (2010) Voxel-based morphometry and epilepsy. Expert Rev Neurother 10(6):975–984. https://doi.org/10.1586/ern.10.63
75. Keller SS, Roberts N (2008) Voxel-based morphometry of temporal lobe epilepsy: an introduction and review of the literature. Epilepsia 49(5):741–757. https://doi.org/10.1111/j.1528-1167.2007.01485.x
76. Li J, Zhang Z, Shang H (2012) A meta-analysis of voxel-based morphometry studies on unilateral refractory temporal lobe epilepsy. Epilepsy Res 98(2-3):97–103. https://doi.org/10.1016/j.eplepsyres.2011.10.002
77. Singh V, Chertkow H, Lerch JP, Evans AC, Dorr AE, Kabani NJ (2006) Spatial patterns of cortical thinning in mild cognitive impairment and Alzheimer's disease. Brain 129(Pt 11):2885–2893
78. Bookstein FL (2001) "Voxel-based morphometry" should not be used with imperfectly registered images. NeuroImage 14(6):1454–1462
79. Tisserand DJ, Pruessner JC, Sanz Arigita EJ, van Boxtel MP, Evans AC, Jolles J, Uylings HB (2002) Regional frontal cortical volumes decrease differentially in aging: an MRI study to compare volumetric approaches and voxel-based morphometry. NeuroImage 17(2):657–669
80. Dale AM, Fischl B, Sereno MI (1999) Cortical surface-based analysis. I. Segmentation and surface reconstruction. NeuroImage 9(2):179–194
81. MacDonald D, Kabani N, Avis D, Evans AC (2000) Automated 3-D extraction of inner and outer surfaces of cerebral cortex from MRI. NeuroImage 12(3):340–356
82. Kim JS, Singh V, Lee JK, Lerch J, Ad-Dab'bagh Y, MacDonald D, Lee JM, Kim SI, Evans AC (2005) Automated 3-D extraction and evaluation of the inner and outer cortical surfaces using a Laplacian map and partial volume effect classification. NeuroImage 27(1):210–221
83. Thompson PM, Hayashi KM, Simon SL, Geaga JA, Hong MS, Sui Y, Lee JY, Toga AW, Ling W, London ED (2004) Structural abnormalities in the brains of human subjects who use methamphetamine. J Neurosci 24(26):6028–6036
84. Scholtens LH, de Reus MA, van den Heuvel MP (2015) Linking contemporary high resolution magnetic resonance imaging to the von Economo legacy: a study on the comparison of MRI cortical thickness and histological measurements of cortical structure. Hum Brain Mapp 36(8):3038–3046. https://doi.org/10.1002/hbm.22826
85. Bernhardt BC, Bernasconi N, Concha L, Bernasconi A (2010) Cortical thickness analysis in temporal lobe epilepsy: reproducibility and relation to outcome. Neurology 74(22):1776–1784. https://doi.org/10.1212/WNL.0b013e3181e0f80a. 74/22/1776 [pii]
86. Mueller SG, Laxer KD, Barakos J, Cheong I, Garcia P, Weiner MW (2009) Widespread neocortical abnormalities in temporal lobe epilepsy with and without mesial sclerosis. NeuroImage 46(2):353–359. https://doi.org/10.1016/j.neuroimage.2009.02.020
87. Bernhardt BC, Worsley KJ, Besson P, Concha L, Lerch JP, Evans AC, Bernasconi N (2008) Mapping limbic network organization in temporal lobe epilepsy using morphometric correlations: insights on the relation between mesiotemporal connectivity and cortical atrophy. NeuroImage 42(2):515–524. https://doi.org/10.1016/j.neuroimage.2008.04.261
88. Bernhardt BC, Worsley KJ, Kim H, Evans AC, Bernasconi A, Bernasconi N (2009) Longitudinal and cross-sectional analysis of atrophy in pharmacoresistant temporal lobe epilepsy. Neurology 72(20):1747–1754. https://doi.org/10.1212/01.wnl.0000345969.57574.f5. 01.wnl.0000345969.57574.f5 [pii]
89. Lin JJ, Salamon N, Lee AD, Dutton RA, Geaga JA, Hayashi KM, Luders E, Toga AW, Engel J Jr, Thompson PM (2007) Reduced neocortical thickness and complexity mapped in mesial temporal lobe epilepsy with

hippocampal sclerosis. Cereb Cortex 17 (9):2007–2018. https://doi.org/10.1093/cercor/bhl109. bhl109 [pii]

90. Helmstaedter C, Kurthen M, Lux S, Reuber M, Elger CE (2003) Chronic epilepsy and cognition: a longitudinal study in temporal lobe epilepsy. Ann Neurol 54(4):425–432
91. Bartolomei F, Chauvel P, Wendling F (2008) Epileptogenicity of brain structures in human temporal lobe epilepsy: a quantified study from intracerebral EEG. Brain 131 (Pt 7):1818–1830. https://doi.org/10.1093/brain/awn111. awn111 [pii]
92. Guerrini R, Dobyns WB (2014) Malformations of cortical development: clinical features and genetic causes. Lancet Neurol 13 (7):710–726. https://doi.org/10.1016/S1474-4422(14)70040-7
93. Blumcke I, Thom M, Aronica E, Armstrong DD, Vinters HV, Palmini A, Jacques TS, Avanzini G, Barkovich AJ, Battaglia G, Becker A, Cepeda C, Cendes F, Colombo N, Crino P, Cross JH, Delalande O, Dubeau F, Duncan J, Guerrini R, Kahane P, Mathern G, Najm I, Ozkara C, Raybaud C, Represa A, Roper SN, Salamon N, Schulze-Bonhage A, Tassi L, Vezzani A, Spreafico R (2011) The clinicopathologic spectrum of focal cortical dysplasias: a consensus classification proposed by an ad hoc Task Force of the ILAE Diagnostic Methods Commission. Epilepsia 52 (1):158–174. https://doi.org/10.1111/j.1528-1167.2010.02777.x
94. Cohen-Gadol AA, Ozduman K, Bronen RA, Kim JH, Spencer DD (2004) Long-term outcome after epilepsy surgery for focal cortical dysplasia. J Neurosurg 101(1):55–65
95. Muhlebner A, Coras R, Kobow K, Feucht M, Czech T, Stefan H, Weigel D, Buchfelder M, Holthausen H, Pieper T, Kudernatsch M, Blumcke I (2012) Neuropathologic measurements in focal cortical dysplasias: validation of the ILAE 2011 classification system and diagnostic implications for MRI. Acta Neuropathol 123(2):259–272. https://doi.org/10.1007/s00401-011-0920-1
96. Krsek P, Maton B, Jayakar P, Dean P, Korman B, Rey G, Dunoyer C, Pacheco-Jacome E, Morrison G, Ragheb J, Vinters HV, Resnick T, Duchowny M (2009) Incomplete resection of focal cortical dysplasia is the main predictor of poor postsurgical outcome. Neurology 72(3):217–223. https://doi.org/10.1212/01.wnl.0000334365.22854.d3
97. Bernasconi A, Bernasconi N (2011) Unveiling epileptogenic lesions: the contribution of image processing. Epilepsia 52(Suppl 4):20–24. https://doi.org/10.1111/j.1528-1167.2011.03146.x
98. Bonilha L, Montenegro MA, Rorden C, Castellano G, Guerreiro MM, Cendes F, Li LM (2006) Voxel-based morphometry reveals excess gray matter concentration in patients with focal cortical dysplasia. Epilepsia 47 (5):908–915. https://doi.org/10.1111/j.1528-1167.2006.00548.x. EPI548 [pii]
99. Bruggemann JM, Wilke M, Som SS, Bye AM, Bleasel A, Lawson JA (2007) Voxel-based morphometry in the detection of dysplasia and neoplasia in childhood epilepsy: combined grey/white matter analysis augments detection. Epilepsy Res 77(2–3):93–101
100. Colliot O, Bernasconi N, Khalili N, Antel SB, Naessens V, Bernasconi A (2006) Individual voxel-based analysis of gray matter in focal cortical dysplasia. NeuroImage 29 (1):162–171
101. Merschhemke M, Mitchell TN, Free SL, Hammers A, Kinton L, Siddiqui A, Stevens J, Kendall B, Meencke HJ, Duncan JS (2003) Quantitative MRI detects abnormalities in relatives of patients with epilepsy and malformations of cortical development. NeuroImage 18(3):642–649
102. Wilke M, Kassubek J, Ziyeh S, Schulze-Bonhage A, Huppertz HJ (2003) Automated detection of gray matter malformations using optimized voxel-based morphometry: a systematic approach. NeuroImage 20 (1):330–343
103. Colliot O, Antel SB, Naessens VB, Bernasconi N, Bernasconi A (2006) In vivo profiling of focal cortical dysplasia on high-resolution MRI with computational models. Epilepsia 47(1):134–142
104. Rugg-Gunn FJ, Boulby PA, Symms MR, Barker GJ, Duncan JS (2006) Imaging the neocortex in epilepsy with double inversion recovery imaging. NeuroImage 31(1):39–50. https://doi.org/10.1016/j.neuroimage.2005.11.034. S1053-8119(05)02486-9 [pii]
105. Focke NK, Symms MR, Burdett JL, Duncan JS (2008) Voxel-based analysis of whole brain FLAIR at 3T detects focal cortical dysplasia. Epilepsia 49(5):786–793
106. Rugg-Gunn FJ, Boulby PA, Symms MR, Barker GJ, Duncan JS (2005) Whole-brain T2 mapping demonstrates occult abnormalities in focal epilepsy. Neurology 64 (2):318–325
107. Rugg-Gunn FJ, Eriksson SH, Boulby PA, Symms MR, Barker GJ, Duncan JS (2003) Magnetization transfer imaging in focal epilepsy. Neurology 60(10):1638–1645

108. Focke NK, Bonelli SB, Yogarajah M, Scott C, Symms MR, Duncan JS (2009) Automated normalized FLAIR imaging in MRI-negative patients with refractory focal epilepsy. Epilepsia 50(6):1484–1490. https://doi.org/10.1111/j.1528-1167.2009.02022.x. EPI2022 [pii]
109. Salmenpera TM, Symms MR, Rugg-Gunn FJ, Boulby PA, Free SL, Barker GJ, Yousry TA, Duncan JS (2007) Evaluation of quantitative magnetic resonance imaging contrasts in MRI-negative refractory focal epilepsy. Epilepsia 48(2):229–237
110. Bernasconi A, Antel SB, Collins DL, Bernasconi N, Olivier A, Dubeau F, Pike GB, Andermann F, Arnold DL (2001) Texture analysis and morphological processing of magnetic resonance imaging assist detection of focal cortical dysplasia in extra-temporal partial epilepsy. Ann Neurol 49(6):770–775
111. Antel SB, Bernasconi A, Bernasconi N, Collins DL, Kearney RE, Shinghal R, Arnold DL (2002) Computational models of MRI characteristics of focal cortical dysplasia improve lesion detection. NeuroImage 17 (4):1755–1760
112. Colliot O, Mansi T, Bernasconi N, Naessens V, Klironomos D, Bernasconi A (2006) Improved segmentation of focal cortical dysplasia lesions on MRI using expansion towards cortical boundaries. In: IEEE-ISBI Macro to Nano, pp. 323–326
113. Huppertz HJ, Grimm C, Fauser S, Kassubek J, Mader I, Hochmuth A, Spreer J, Schulze-Bonhage A (2005) Enhanced visualization of blurred gray-white matter junctions in focal cortical dysplasia by voxel-based 3D MRI analysis. Epilepsy Res 67(1-2):35–50. https://doi.org/10.1016/j.eplepsyres.2005.07.009. S0920-1211(05)00164-6 [pii]
114. Huppertz HJ, Kurthen M, Kassubek J (2009) Voxel-based 3D MRI analysis for the detection of epileptogenic lesions at single subject level. Epilepsia 50(1):155–156. https://doi.org/10.1111/j.1528-1167.2008.01734.x. EPI1734 [pii]
115. Bien CG, Szinay M, Wagner J, Clusmann H, Becker AJ, Urbach H (2009) Characteristics and surgical outcomes of patients with refractory magnetic resonance imaging-negative epilepsies. Arch Neurol 66 (12):1491–1499. https://doi.org/10.1001/archneurol.2009.283. 66/12/1491 [pii]
116. Kassubek J, Huppertz HJ, Spreer J, Schulze-Bonhage A (2002) Detection and localization of focal cortical dysplasia by voxel-based 3-D MRI analysis. Epilepsia 43(6):596–602
117. Thesen T, Quinn BT, Carlson C, Devinsky O, DuBois J, McDonald CR, French J, Leventer R, Felsovalyi O, Wang X, Halgren E, Kuzniecky R (2011) Detection of epileptogenic cortical malformations with surface-based MRI morphometry. PLoS One 6(2):e16430. https://doi.org/10.1371/journal.pone.0016430
118. Antel SB, Collins DL, Bernasconi N, Andermann F, Shinghal R, Kearney RE, Arnold DL, Bernasconi A (2003) Automated detection of focal cortical dysplasia lesions using computational models of their MRI characteristics and texture analysis. NeuroImage 19(4):1748–1759
119. Raymond AA, Fish DR, Sisodiya SM, Alsanjari N, Stevens JM, Shorvon SD (1995) Abnormalities of gyration, heterotopias, tuberous sclerosis, focal cortical dysplasia, microdysgenesis, dysembryoplastic neuroepithelial tumor and dysgenesis of archicortex in epilepsy. Clinical, EEG and neuroimaging features in 100 adult patients. Brain 118:629–660
120. Wang S, Summers RM (2012) Machine learning and radiology. Med Image Anal 16 (5):933–951. https://doi.org/10.1016/j.media.2012.02.005
121. Hong SJ, Kim H, Schrader D, Bernasconi N, Bernhardt BC, Bernasconi A (2014) Automated detection of cortical dysplasia type II in MRI-negative epilepsy. Neurology 83 (1):48–55. https://doi.org/10.1212/WNL.0000000000000543
122. Fonseca Vde C, Yasuda CL, Tedeschi GG, Betting LE, Cendes F (2012) White matter abnormalities in patients with focal cortical dysplasia revealed by diffusion tensor imaging analysis in a voxelwise approach. Front Neurol 3:121. https://doi.org/10.3389/fneur.2012.00121

Chapter 19

Brain Morphometry: Schizophrenia

Chiara Chiapponi, Pietro De Rossi, Fabrizio Piras, Tommaso Gili, and Gianfranco Spalletta

Abstract

During the last three decades, neuroimaging techniques have emerged as an essential tool for the noninvasive examination of subtle brain dysfunctions in psychiatric patient populations. These techniques helped clarify in vivo some of the principal neurobiological characteristic of the quintessential major mental disorder, schizophrenia, providing evidence for potential biomarkers that might improve the diagnostic process, therapeutic monitoring, and outcomes. Here, we describe the main brain morphometric techniques (including volumetric, shape analysis, and microstructural techniques) currently used in research on schizophrenia and summarize the results of the most important studies in the field. This chapter aims at providing an exhaustive description of the state-of-the-art in vivo brain morphology in schizophrenia in order to better characterize the disorder from a neurobiological point of view, thus providing a comprehensive background for further research on this topic. A general picture emerges in which schizophrenia is characterized by widespread brain cortical and subcortical, structural, and microstructural anomalies since the early phases of the course of illness. Volumetric, shape, and microstructural disruptions in structures of the fronto-temporo-parietal network are predominant both cross-sectionally and in terms of altered developmental trajectories and outcome prediction. Future studies on brain morphometric indices in schizophrenia should focus on their reliability as predictors of treatment response through longitudinal designs in large samples.

Key words Schizophrenia, Brain morphometry, Region of interest, VBM, Cortical thickness, Shape analysis, Microstructure, Structural covariance

1 Introduction

One major goal of biomedical research is to find efficient biomarkers to improve the accuracy of diagnosis and, in turn, improve patient outcomes. Identification of a valid biomarker is based on observations that the biomarker is detected in patients with a specific disorder and not in healthy controls. This approach has been successful for some disorders such as quantification of amyloid deposition, tau proteinopathy, and neuronal degeneration in Alzheimer's disease. However, for psychiatric disorders in general and schizophrenia (SZ) in particular, it poses particular challenges.

Gianfranco Spalletta et al. (eds.), *Brain Morphometry*, Neuromethods, vol. 136,
https://doi.org/10.1007/978-1-4939-7647-8_19,

SZ is a chronic and highly disabling disorder with a lifetime prevalence of about 0.7% [1]. The symptoms of SZ include positive symptoms (i.e., delusions and hallucinations), negative symptoms (e.g., alogia, blunted affect, etc.), behavioral and/or conceptual disorganization, and catatonic symptoms [2]. The clinical diagnosis of SZ, as happens for any other psychiatric disorder, is based on subjective (distress) and/or behavioral (dysfunction) criteria. Therefore, it is difficult to see how using a biomarker to help decide whether a person does or does not have a disorder can improve diagnosis and treatment. More refined employment of biomarkers might be beneficial to drive the definition of homogeneous disorder subtypes that may cut across traditional boundaries of DSM-defined disorders [3] or to obtain an estimate of the likelihood of a specific outcome, such as response to pharmacological treatments [4].

A further issue connected to the search for biomarkers in SZ is clarifying their role in the development of the disease. Indeed, clinical and pathophysiological features differ among high-risk subjects, those in a prodromal stage, patients with a recent onset, and those in a chronic phase. Therefore it is fundamental to link potential biomarkers to the specific phase of the disease they refer to.

Biomarkers can be found from manifold techniques such as neuroimaging, neuropathology, neuroimmunology, neuropsychology, and genetics. Here, we will focus on neuroimaging to highlight the major and up to date morphometric features of SZ patients. In particular, magnetic resonance imaging (MRI) represents the first-choice technique to obtain a wide range of noninvasive morphometric information. Indeed, although radiological inspection of MRI brain scans cannot be used as a diagnostic tool in SZ, research studies have shown that it is a promising technique to better understand the pathophysiology of the disorder.

The crucial MRI-based work of Pantelis and colleagues [5] showed that subjects at high risk of developing SZ have smaller volumes of frontal and temporal cortices, which are the same brain areas that have been reported as smaller in individuals with an established diagnosis of SZ. Together with these gray matter (GM) changes, high-risk individuals also have white matter (WM) alterations similar to first-episode patients. The affected WM regions are widespread and include associative fibers connecting fronto-parieto-temporal (superior longitudinal fasciculus) and fronto-parieto-occipital (inferior fronto-occipital fasciculus) regions, commissural fibers (corpus callosum), and cortico-subcortical pathways (corona radiata, corticospinal tract, and corticopontine tract) [6]. Then, the actual state of the art on neuroimaging-derived features of SZ suggests that brain alterations are already present in the prodromal stage of the illness, become more extensive at the time of a first full psychotic episode, and worsen over time, when the illness becomes established. Around this general evidence, the last decades have seen hundreds of publications on brain morphometry in SZ. The importance of neuroanatomical description of SZ patients has been highlighted in the different stages of the illness [7], during aging [8], in the different subtypes of patients [3], etc. Depending on the focus of the study, different morphometric parameters have been taken into consideration. In order to cover a range of parameters as wide as possible, in the following sections we will review the main morphometric evidence found in subjects diagnosed with SZ according to the different methods adopted to post-process and analyze structural images. We will in turn highlight the specificity and importance of each methodological approach adopted and the different morphometric parameters obtained to describe the brain in SZ. In particular we will first focus on macroscopic morphometric features detected by region of interest (ROI) and voxel-based approach as well as by investigating cortical thickness, shape of subcortical structures, and the covariance between different brain regions. We will then explore microscopic morphometric characteristics of SZ overviewing the main findings obtained by diffusion tensor imaging (DTI) methods.

2 Macroscopic Morphometry

2.1 ROI-Based Findings

ROI approach starts from manual delineation of brain anatomical regions chosen according to a priori hypothesis. Methods for tracing ROIs and the choice of anatomical landmarks refer to internationally agreed criteria and are based on human brain atlases. Manual tracing has remained the gold standard in neuroimaging for psychiatric research, particularly when segmenting regions with great sulcal variability. However, its intrinsic limits (error proneness, inter- and intra-observer variability, time consumed, etc.) encouraged the development of semi- and fully automatic tracing methods, in particular to isolate deep gray matter nuclei, to identify hippocampal subfields, or to divide brain cortex in distinct compartments. Nevertheless, since automatism needs special mathematical and computational skills not always available and because of the lack of universal segmentation algorithm that can be applied satisfactorily for all structures, manual tracing still lingers.

Cross-sectional ROI-based studies on SZ highlighted important pathological changes in patients compared to healthy subjects, in both GM and WM [9]. In chronic patients, the most replicated findings are ventricular enlargement [10] and general cerebral atrophy [11], particularly located in the frontal lobe and limbic and paralimbic regions [12, 13]. Evidence from studies on first-episode patients and from longitudinal investigations suggest that volumetric abnormalities in the abovementioned areas are often already detectable at illness outbreak and may progress over time, depending on a combination of factors as age of onset, illness duration, pharmacological treatment, and clinical outcome [14]. One of the major limitations of ROI studies is that they focus on specific, a priori defined, brain regions, neglecting the rest of the brain. As an alternative, a whole brain approach, without any a priori hypothesis, has been encouraged by the development of voxel-based morphometry (VBM) technique.

2.2 VBM Findings

VBM is a technique born in the 1990s and developed by Ashburner and Friston [15]. Depending upon the use of modulated or unmodulated images of the tissue considered (GM or WM), it allows comparison of voxel-by-voxel tissue volume or density, respectively [15, 16]. Group differences are generally reported in terms of T or Z score statistics rather than volume units. Clusters in which group comparisons survive a pre-specified statistical threshold are graphically represented through statistical maps and tables reporting the coordinate and P value of the statistical peak. General limitations of VBM technique include confounds in normalization and segmentation during the preprocessing steps, accuracy of spatial localization of results, and validity of statistical inference when using data not normally distributed [17]. However, VBM represents a unique approach to test brain abnormalities in clinical populations with no a priori hypothesis, and it is still one of the most used tools to investigate brains in SZ [9]. The majority of VBM studies refer to GM, but the technique has been used to evaluate also WM. As evidenced by recent meta-analyses [18, 19], the most robust and replicated findings in VBM studies on SZ patients implicate pathological reduction of volume (or density) distributed bilaterally over the fronto-insular cortex, anterior cingulate cortex, superior temporal gyrus, thalamus, hippocampus, and parahippocampal region. In most of these brain areas, voxel-based abnormalities refer to both GM and WM tissue. Specifically, GM alterations are already present in the prodromal and early stage of the illness [20], they are larger in male, chronic, with negative symptoms and poor outcome SZ patients. A meta-analysis distinguishing GM densityration studies from those considering GM volume showed that reductions in concentration are stronger and more consistent than reductions in volume, and authors concluded that GM changes observed with MRI in SZ may not necessarily result from a unitary pathological process [18]. VBM studies on WM reveal that volume is less sensitive to WM abnormalities in SZ than microstructural measures of WM integrity, and this might be related to the fact that T1-weighted contrast is less sensitive to reveal damage in this tissue. Differently from GM evidences, WM findings are not influenced by gender and negative symptoms.

Currently available VBM longitudinal studies in SZ have been recently meta-analyzed, showing progressive gray matter volume decreases and lateral ventricular enlargements in patients over time, with gray matter volumes inversely correlated with cumulative exposure to antipsychotic treatments but neither with duration of illness nor with illness severity [21]. Nonetheless, a large longitudinal study following up 202 patients over a 7-year period allows to make subtler considerations about this aspect [22]. In fact, the study shows that relapse duration (and not the number of relapses) in SZ patients is associated with significant volume decrease both globally and locally (frontal regions). At the same time, cumulative exposure to antipsychotic drugs seems to be associated with the development of significant anomalies in several generalized brain volumetric measures over time, thus suggesting that an optimal management of antipsychotic dosage is needed in order to prevent relapses and preserve brain tissue. The future of VBM in SZ should go in the direction of further longitudinal studies, examining GM and WM simultaneously at multiple time points to deepen the understanding of the nature and the development of structural abnormalities in the disorder.

2.3 Cortical Thickness Findings

Along with surface area (SA), cortical thickness (CT) is one of the two dimensions of the cortical sheet determining cortical volume (CV). Other parameters carrying unique information on cortical development are the convex hull area (CHA, i.e., the exposed cortical surface) and the gyrification index (GI), resulting from the ratio between total SA and CHA [23]. Of all these indices, CT is indeed the one which was studied in more detail, especially with respect to longitudinal studies on cortical development in healthy humans [24]. Since the late 1990s, CT has been studied both cross-sectionally and longitudinally in SZ, including its most severe early-onset variants, casting some light on the peculiar way the brain develops and changes over time under this profoundly disabling condition.

In fact, VBM might be influenced by the sulcal widening in SZ [10, 25, 26]. Furthermore, a surface-based approach is more appropriate for shape analysis, and, finally, it can avoid noise generated from resampling in the voxel-based analysis.

In terms of pathogenetic significance, the decrease in CT which is hypothesized and often found in SZ may reflect several neurobiologically relevant phenomena such as synaptic over pruning, neurodegeneration, changes in myelination, development of neurotransmitter systems, and disrupted regional gene expression [27–29].

The first study on CT in SZ showed decreased CT metrics in inferior-medial portions of frontal and temporal regions [30], and this predominantly frontotemporal CT disruption has been more recently replicated in larger samples including both chronic and first-episode patients [31–33].

CT was also used in a principal component analysis, identifying a cortical thinning pattern composed of bilateral superior, middle, inferior, and medial frontal gyri, precentral gyrus, bilateral superior and medial temporal cortices, the postcentral gyrus, the bilateral precuneus, and the bilateral cingulate gyrus as significantly discriminative for SZ [34]. In particular, cortical thinning in the aforementioned regions compared to healthy controls showed the best diagnostic discriminative validity for SZ.

Furthermore, there is evidence that genetic factors might have an impact on cortical thinning. Studies in siblings or relatives of patients yielded mixed results, with some supporting a decreased CT in temporal and/or frontal areas in relatives [35, 36] and others not supporting the same finding [37, 38]. However, there is evidence that cortical thinning is associated with specific SZ risk genes such as the neurogranin gene, GSK-3beta, ZNF804A, DISC1, and risk variants of the glutamatergic regulatory genes [39–43].

Disease progression seems to have an impact on cortical thickness as well. Findings based on longitudinal data seem to confirm the central role of a frontotemporal network's disruption in SZ. In fact, the largest longitudinal CT study comparing SZ patients and healthy subjects so far showed thinner baseline left orbitofrontal, right parahippocampal, and bilateral superior temporal cortices in SZ, with an excessive widespread cortical thinning over time [44]. The aforementioned thinning was particularly pronounced in the temporal cortex bilaterally and in the left frontal area. Furthermore, cortical thinning over time was more pronounced in patients with poor outcome and higher cumulative exposition to typical antipsychotic drugs, while exposition to atypical antipsychotics was associated with less pronounced cortical thinning. However, it is not clear whether abnormal cortical thinning in SZ merely represents a neurobiological phenomenon paralleling the course of illness independently of treatments and whether it necessarily reflects neurotoxicity and/or negative neuroplasticity. In fact, in a recent longitudinal study on treatment-resistant SZ patients, Ahmed and colleagues measured CT before and 6–9 months after clozapine treatment [45]. They found that, despite most patients improved both symptomatically and functionally on clozapine, switching to the new treatment was not significantly associated with arrest or reversal of widespread and frontotemporal thinning.

CT studies on childhood and adolescence cortical development in subjects who display symptoms of SZ in young adulthood (i.e., typical onset of symptoms in SZ) are lacking. However, although childhood-onset SZ (COS) is a rare and particularly severe form of the disorder beginning before the age of 13, it provides the opportunity to study early brain development in psychotic disorders. A recent, large, case-control longitudinal study of CT in COS showed that in this particular condition the vast majority of brain cortical regions displays a trait-like CT deficit persisting throughout the whole follow-up, while areas within the cingulo-frontotemporal module (including right inferior frontal gyrus, right orbital cortex, right gyrus rectus, and left posterior cingulate) display altered maturational trajectories over time [46].

A general picture emerges where CT alterations in SZ are mainly represented by widespread cortical thinning, which is particularly pronounced in frontotemporal areas since the earlier phases of the disorder. In this context, several studies tested the hypothesis that subgroups of patients, with different symptom profiles, show different degrees of CT changes in frontotemporal areas and/or distinct CT anomalies in different brain regions.

Other studies focused on whether CT alterations seen in SZ are specific for this disorder or they overlap, at least to some extent, with CT anomalies peculiar to other major psychiatric disorders such as bipolar disorder.

As for studies focusing on specific symptomatic dimensions peculiar to SZ, it is possible that patients characterized by more severe negative symptoms display a generalized cortical thinning [33, 47], whereas patients with more pronounced disorganized and paranoid dimensions seem to have a more localized prefrontotemporal cortical thinning mostly converging in the orbitofrontal and dorsolateral prefrontal cortices [47]. However, a cortical thinning pattern similar to the one found by Nenadic and colleagues is not seen in deficit SZ, the subtype of the disorder which is characterized by severe, primary, and stable negative symptoms [48]. In fact, a recent study found no cortical thinning pattern differences between deficit SZ patients and non-deficit SZ patients with the exception of the anterior cingulate cortex, which displayed reduced CT in deficit patients [49]. Furthermore, another study found no CT reduction differences at all between deficit patients and their non-deficit counterparts [50].

As for the association between CT and more specific symptoms, a recent example is provided by an interesting study by Chen and colleagues on first-episode SZ patients [51]. They showed that auditory verbal hallucinations are associated with lower CT in the right Heschl's gyrus (i.e., the transverse temporal gyrus), and the degree of cortical thinning was correlated with the severity of hallucinations.

As previously mentioned, studies on the specificity of CT alterations observed in SZ focused on the comparison between SZ and bipolar disorders. In particular, SZ and bipolar disorder type 1 showed similar patterns of cortical thinning in the frontal lobes and in the temporo-parietal junction. However, in SZ CT was specifically reduced in the inferior frontal lobes, inferior-ventral temporal lobes, and the left anterior cingulate cortex [52]. A more recent study on a smaller sample substantially confirmed the previously described results and found a significant association between cortical thinning in the inferior frontal gyrus and psychomotor speed and executive functioning in SZ but not in bipolar disorder [53].

2.4 Shape Analysis

Recent developments in shape analysis methodology have led to morphometric measures that can enrich data derived from volumetric and cortical thickness analyses [54]. Interestingly, investigations on shape of subcortical structures in psychiatric patients revealed group alterations where investigations on volume did not, indicating that shape, more than size, is a sensitive parameter to identify subtle structural anomalies [55]. Many are the computational methods available to probe shape of selected brain structures (see Mamah et al. [56] for a complete list and description). Among the most popularly used, FSL [57] provides the tool FIRST which calculates vertex-wise statistics to investigate localized shape differences [58]. FIRST allows for a model-based segmentation and registration of anatomical images, where volumetric labels are parameterized as surface meshes. Vertex locations from each participant are projected onto the surface of the average shape transformed to Montreal Neurological Institute space. Scalar projection values are processed with univariate permutation methods using FSL's *randomize* tool, corrected at the cluster level using threshold-free cluster enhancement [59]. Alternatively, FreeSurfer (http://surfer.nmr.mgh.harvard.edu/) is often used to segment subcortical structures and create surface mesh models. When FreeSurfer segmentations are not suitable for shape analysis (often because of image noise), smooth surfaces can be created incorporating large deformation diffeomorphic metric mapping (LDDMM) [60] into the FreeSurfer pipeline. LDDMM computes the velocity vectors that transform one binary image to another giving the metric distance that ensures smoothness in the space of velocity vector fields that are generated by the group of infinite dimensional diffeomorphisms (which is the generalization of rotations, translations and scale group), the necessary group for studying shape. These distances give a precise mathematical description of what shapes are similar and different.

The most popular structures for shape investigations in SZ are hippocampus [61–64] and thalamus [54, 65, 66]. The importance of fine shape analysis for both structures resides in the chance to highlight morphometric features of specific hippocampal or thalamic subfields, each with specific role and function. To date, beside a general agreement on the correlation between the intensity of shape deformation and severity of symptoms and antipsychotic dosage [67–69], the findings on the localization of abnormalities in the shape of these structures in SZ are still discordant. Possible causes of heterogeneity can be ascribed to the computational methodology adopted, to the study design, or to the diverse nature of patients involved. Indeed, due to its central role in the different stages of SZ, hippocampus has been investigated both in chronically ill and in first-episode or ultrahigh risk patients. Studies on chronic SZ patients showed hippocampal shape alterations either in anterior, posterior, and lateral hippocampal regions [61, 63,

70–72], as well as no overall differences in hippocampal shape with respect to healthy control subjects [64]. Studies on first-episode or ultrahigh risk patients are still a minority and revealed either hippocampal surface contraction on areas of the anterior hippocampus and lateral parts of its body and tail [67], either a significant inversion in the left ventral posterior hippocampus [69].

As for the thalamus, while some authors showed the absence of contour deformations [73], the majority highlighted a diffuse pattern of shape changes in SZ, showing involvement of nuclei at anterior and posterior extremes of the thalamic complex [65, 66] as well as of the pulvinar and ventral lateral nuclei [54, 74]. This is consistent with deficits commonly observed in SZ patients such as planning, self-monitoring, sensorial integration, attention, and modulation of alertness aimed at learning and episodic memory.

2.5 Interregional Coupling and Structural Covariance

The wide distribution of local brain abnormalities in SZ, as evidenced by the above-presented techniques, suggested that the understanding of SZ pathophysiology could be improved by looking at brain morphometry from a more general perspective. Indeed, thinking at brain regions as elements of a complex and multilevel network evidenced that alteration in the connections between the different nodes of such net may be considered as one of the causes of SZ [4, 75]. While alteration in WM connectivity will be explored in the next paragraph, here we will deepen the issue of GM structural covariance, i.e., interregional correlations of regional GM morphology [76]. Such an approach examines whether an anatomical change in one brain area correlates with changes in other areas. As underlined by Evans [77], significant interregional morphometric correlation can exist independently from the presence of a direct fiber connection between regions. Indeed, anatomical correlation, as well as functional one, can result from indirect connection mediated by third party.

Structural covariance is particularly intriguing for researches in psychiatry and particularly in SZ, which has long been seen as a connectivity disorder [78]. Moreover, structural covariance has been shown to be sensitive to harmonized developmental brain modifications [76] and to be suitable to study diffuse pathological brain processes [79].

Straightforward covariance analyses looking at the relationship between volume and cortical thickness in different brain regions showed that SZ is associated with altered bilateral symmetry of these regions. In particular, SZ patients turned out to have exaggeration of leftward thalamic asymmetry [65], reduced laterality in planum temporale [80] and hippocampus [81], and reversal of the normal asymmetry of the inferior parietal cortex [82]. Moreover, researches considering the relationship between volumetric measure variance of the prefrontal and temporal cortex revealed both enhanced [83–85] and decreased [86] covariance in SZ with respect to healthy subjects, while stronger frontoparietal associations emerged in patients [82, 87]. The latter also showed altered corticothalamic coupling [88, 89] and increased interhemispheric prefrontal association [90].

With a more sophisticated approach, structural covariance can be explored using complex network analysis methods, such as graph theory [91]. Standard structural images can be processed to obtain information on network structure and topology through parameters such as clustering coefficient, path length, and network efficiency or to measure the importance of specific nodes of the graph through parameters such as betweenness, closeness, and eigenvector centrality [92]. Bassett et al. used a priori hub definitions and found a loss of frontal hubs, increased connection distance, and reduced hierarchy in SZ [93]. More recently, Zhang et al. showed that SZ patients had decreased betweenness centrality in associative regions of the cortex and increased centrality in limbic and paralimbic areas [94].

To date, an agreed interpretation of altered connections between brain regions does not exist yet. An increase in correlations in patients may be related to over connectivity or contemporary GM loss, while a decrease in correlations may suggest disconnectivity or confined degeneration.

3 Microscopic Morphometry

To complete the description of brain morphometry in SZ, investigation on WM tracts is of key importance, particularly aiming at clarifying SZ pathophysiology. The election technique to quantify organization of WM microstructure is DTI. This relatively new MRI-based technique measures the diffusivity of water molecules in brain tissue, which depends on barriers and obstacles imposed by microstructure. Such barriers slow down the diffusing particles or even impose an upper limit on their overall displacement, resulting in a hindered and restricted diffusion. In WM in particular, the barriers are represented by myelin sheaths and microtubules [95], and their integrity can be evaluated using DTI microstructural indexes, e.g., Fractional Anisotropy (FA), axial diffusivity (AD), radial diffusivity (RD), and mean diffusivity (MD). Although over-interpretation without strong theoretical foundations or additional data from other sources should be discouraged [96], these parameters are fundamental for the description of WM microstructure. FA is the most commonly investigated index to estimate diffusion anisotropy in WM, and it is used as a proxy for the extent of alignment and organization of cellular microstructure within WM projections and for the integrity of structural connectivity. AD represents diffusion along fiber direction and may be related to axonal integrity; RD refers to diffusion perpendicular to fiber direction and can give information on myelin sheaths around axons [97, 98], while MD describes the rotationally invariant magnitude of the average water diffusion. Such microstructural parameters can be analyzed using many approaches. As well as macroscopic morphometry, DTI metrics can also be evaluated on a voxel-by-voxel basis using VBM, normalizing the images onto a stereotaxic brain atlas. Alternatively, in case of strong a priori hypotheses, statistical comparisons between diagnostic populations can be performed on specific ROIs. More recently, tract-based spatial statistics (TBSS) has been developed, and it allows projecting voxel-based diffusion metrics onto an alignment-invariant tract representation, the so-called skeleton [99]. Moreover, novel tractographic methods have now emerged to assess structural connectivity in major WM tracts that connect different regions in the brain. To date, independently from the technique adapted to analyze DTI-derived parameters over single voxels, ROIs or WM tracts, a massive number of studies on WM microstructural morphometry in SZ have been published. Two recent meta-analyses considering studies on clinically heterogeneous populations formed by patients at different stages of SZ, with different treatment exposure, and different clinical diagnosis agreed on a general pattern of alteration in WM microstructural morphometry. In 2009, Ellison-Wright and Bullmore meta-analyzed 15 studies investigating FA with a voxel-based approach on a total of 407 patients affected by SZ, schizoaffective disorder, or first-episode psychosis [100]. They showed that 12 of the 15 studies examined evidenced two major clusters of altered WM integrity in the left frontal and temporal lobes. Such result suggest that two brain networks might be altered in SZ, namely, the cerebello-thalamocortical circuit and the temporal network interconnecting the frontal lobe, insula, hippocampus/amygdala, and occipital lobe. Concordantly, the meta-analysis performed by Bora and colleagues showed that both treated and untreated SZ patients have decreased FA in the bilateral genu of the corpus callosum, anterior cingulate cortex/medial frontal WM, and the right anterior limb of internal capsule and the right external capsule/corona radiata [19].

More recently, Canu and coauthors performed an updated review dividing evidences on (i) high-risk individuals, (ii) first-episode and drug-naive patients, and (iii) chronic, treated patients and looked at the relationship between WM alterations and clinical manifestations [75]. Authors showed that SZ is characterized by WM alterations from the preclinical to the chronic stages, and the causes of such alterations are likely found in abnormal neurodevelopmental processes. The most affected WM tracts are the corticospinal tract, interhemispheric connections, long association WM tracts, cerebello-thalamocortical circuit, and limbic system. WM alterations in SZ are related to positive and negative symptoms and cognitive impairment and can predict clinical outcome and response to treatment. Studies using the most recent graph analysis techniques suggest that abnormal connectivity of brain hubs and disrupted integration of information seem to be a core aspect of the SZ. Again, a shortcoming of WM connectivity investigation is again the small number of longitudinal studies that would be a key step for understanding the disease progression.

4 Notes

All the different techniques and analysis approaches above described are giving valuable contributions in the description of brain morphometry in SZ. However, at the actual state of technological development in MRI and accuracy in software analysis, each morphometric index has a strong potential, but it is not accurate enough to be used as a biomarker for SZ. A feasible step forward in the definition of a neuroimaging profile of SZ patients can be performed using a multimodal approach. Indeed, whereas using a single data processing method can capture a particular aspect of SZ pathology, the combined application of multiple imaging modalities with multiple image processing protocols may elucidate a far more comprehensive understanding of its nature. The feasibility of this approach resides in the fact that a complex and multilevel set of information on brain morphometry can be gained from a single MRI acquisition.

Another issue limiting the impact of the morphometric indices described is that they are mainly used in cross-sectional studies. Indeed, more and more efforts should be put in longitudinal studies, particularly in those focusing on ultrahigh-risk populations. Such an approach would give a fundamental understanding of the neurodevelopmental processes involved in SZ pathology and offer the chance to investigate the reliability of eventual neuroanatomical predictors of treatment response.

References

1. Saha S, Chant D, Welham J, McGrath J (2005) A systematic review of the prevalence of schizophrenia. PLoS Med 2:0413–0433
2. American Psychiatric Association (2000) Diagnostic and statistical manual of mental disorders, vol XLIV. American Psychiatric Association, Washington, DC, p 947 S
3. Spalletta G, De Rossi P, Piras F, Iorio M, Dacquino C et al (2015) Brain white matter microstructure in deficit and non-deficit subtypes of schizophrenia. Psychiatry Res Neuroimaging 231:252–261
4. Dazzan P (2014) Neuroimaging biomarkers to predict treatment response in schizophrenia: the end of 30 years of solitude? Dialogues Clin Neurosci 16:491–503
5. Pantelis C, Velakoulis D, McGorry PD, Wood SJ, Suckling J et al (2003) Neuroanatomical abnormalities before and after onset of psychosis: a cross-sectional and longitudinal MRI comparison. Lancet 361:281–288
6. Carletti F, Woolley JB, Bhattacharyya S, Perez-Iglesias R, Fusar-Poli P et al (2012) Alterations in white matter evident before the onset of psychosis. Schizophr Bull 38:1170–1179
7. Ellison-Wright I, Glahn DC, Laird AR, Thelen SM, Bullmore E (2008) The anatomy of

first-episode and chronic schizophrenia: an anatomical likelihood estimation meta-analysis. Am J Psychiatry 165:1015–1023

8. Chiapponi C, Piras F, Fagioli S, Piras F, Caltagirone C et al (2013) Age-related brain trajectories in schizophrenia: a systematic review of structural MRI studies. Psychiatry Res 214:83–93
9. Perlini C, Bellani M, Brambilla P (2012) Structural imaging techniques in schizophrenia. Acta Psychiatr Scand 126:235–242
10. Shenton ME, Dickey CC, Frumin M, McCarley RW (2001) A review of MRI findings in schizophrenia. Schizophr Res 49:1–52
11. Andreone N, Tansella M, Cerini R, Versace A, Rambaldelli G et al (2007) Cortical white-matter microstructure in schizophrenia diffusion imaging study. Br J Psychiatry 191:113–119
12. Levitt JJ, Bobrow L, Lucia D, Srinivasan P (2010) A selective review of volumetric and morphometric imaging in schizophrenia. Curr Top Behav Neurosci 4:243–281
13. Baiano M, David A, Versace A, Churchill R, Balestrieri M et al (2007) Anterior cingulate volumes in schizophrenia: a systematic review and a meta-analysis of MRI studies. Schizophr Res 93:1–12
14. Olabi B, Ellison-Wright I, McIntosh AM, Wood SJ, Bullmore E et al (2011) Are there progressive brain changes in schizophrenia? A meta-analysis of structural magnetic resonance imaging studies. Biol Psychiatry 70:88–96
15. Ashburner J, Friston KJ (2005) Unified segmentation. NeuroImage 26:839–851
16. Keller SS, Wilke M, Wieshmann UC, Sluming VA, Roberts N (2004) Comparison of standard and optimized voxel-based morphometry for analysis of brain changes associated with temporal lobe epilepsy. NeuroImage 23:860–868
17. Mechelli A, Price C, Friston KJ, Ashburner J (2005) Voxel-based morphometry of the human brain: methods and applications. Curr Med Imaging Rev 1:105–113
18. Fornito A, Yücel M, Patti J, Wood SJ, Pantelis C (2009) Mapping grey matter reductions in schizophrenia: an anatomical likelihood estimation analysis of voxel-based morphometry studies. Schizophr Res 108:104–113
19. Bora E, Fornito A, Radua J, Walterfang M, Seal M et al (2011) Neuroanatomical abnormalities in schizophrenia: a multimodal voxelwise meta-analysis and meta-regression analysis. Schizophr Res 127:46–57
20. Fusar-Poli P, Radua J, McGuire P, Borgwardt S (2012) Neuroanatomical maps of psychosis onset: voxel-wise meta-analysis of antipsychotic-naive VBM studies. Schizophr Bull 38:1297–1307
21. Fusar-Poli P, Smieskova R, Kempton MJ, Ho BC, Andreasen NC et al (2013) Progressive brain changes in schizophrenia related to antipsychotic treatment? A meta-analysis of longitudinal MRI studies. Neurosci Biobehav Rev 37:1680–1691
22. Andreasen NC, Liu D, Ziebell S, Vora A, Ho B-C (2013) Relapse duration, treatment intensity, and brain tissue loss in schizophrenia: a prospective longitudinal MRI study. Am J Psychiatry 170:609–615
23. Van Essen DC, Drury HA (1997) Structural and functional analyses of human cerebral cortex using a surface-based atlas. J Neurosci 17:7079–7102
24. Raznahan A, Shaw P, Lalonde F, Stockman M, Wallace GL et al (2011) How does your cortex grow? J Neurosci 31:7174–7177
25. Ashburner J, Friston KJ (2000) Voxel-based morphometry--the methods. NeuroImage 11:805–821
26. Narr KL, Bilder RM, Toga AW, Woods RP, Rex DE et al (2005) Mapping cortical thickness and gray matter concentration in first episode schizophrenia. Cereb Cortex 15:708–719
27. Lewis DA (1997) Development of the prefrontal cortex during adolescence: insights into vulnerable neural circuits in schizophrenia. Neuropsychopharmacology 16:385–398
28. McGlashan TH, Hoffman RE (2000) Schizophrenia as a disorder of developmentally reduced synaptic connectivity. Arch Gen Psychiatry 57:637–648
29. Choi KH, Zepp ME, Higgs BW, Weickert CS, Webster MJ (2009) Expression profiles of schizophrenia susceptibility genes during human prefrontal cortical development. J Psychiatry Neurosci 34:450–458
30. Kuperberg GR, Broome MR, McGuire PK, David AS, Eddy M et al (2003) Regionally localized thinning of the cerebral cortex in schizophrenia. Arch Gen Psychiatry 60:878–888
31. Goldman AL, Pezawas L, Mattay VS, Fischl B, Verchinski BA et al (2009) Widespread reductions of cortical thickness in schizophrenia and spectrum disorders and evidence of heritability. Arch Gen Psychiatry 66:467–477
32. Crespo-Facorro B, Roiz-Santiáñez R, Pérez-Iglesias R, Rodriguez-Sanchez JM, Mata I

et al (2011) Global and regional cortical thinning in first-episode psychosis patients: relationships with clinical and cognitive features. Psychol Med 41:1449–1460

33. Rimol LM, Nesvåg R, Hagler DJ, Bergmann O, Fennema-Notestine C et al (2012) Cortical volume, surface area, and thickness in schizophrenia and bipolar disorder. Biol Psychiatry 71:552–560
34. Yoon U, Lee J-M, Im K, Shin Y-W, Cho BH et al (2007) Pattern classification using principal components of cortical thickness and its discriminative pattern in schizophrenia. NeuroImage 34:1405–1415
35. Byun MS, Kim JS, Jung WH, Jang JH, Choi J-S et al (2012) Regional cortical thinning in subjects with high genetic loading for schizophrenia. Schizophr Res 141:197–203
36. Sprooten E, Papmeyer M, Smyth AM, Vincenz D, Honold S et al (2013) Cortical thickness in first-episode schizophrenia patients and individuals at high familial risk: a cross-sectional comparison. Schizophr Res 151:259–264
37. Yang Y, Nuechterlein KH, Phillips O, Hamilton LS, Subotnik KL et al (2010) The contributions of disease and genetic factors towards regional cortical thinning in schizophrenia: the UCLA family study. Schizophr Res 123:116–125
38. Boos HBM, Cahn W, van Haren NEM, Derks EM, Brouwer RM et al (2012) Focal and global brain measurements in siblings of patients with schizophrenia. Schizophr Bull 38:814–825
39. Brauns S, Gollub RL, Roffman JL, Yendiki A, Ho B-C et al (2011) DISC1 is associated with cortical thickness and neural efficiency. NeuroImage 57:1591–1600
40. Schultz CC, Nenadic I, Koch K, Wagner G, Roebel M et al (2011) Reduced cortical thickness is associated with the glutamatergic regulatory gene risk variant DAOA Arg30Lys in schizophrenia. Neuropsychopharmacology 36:1747–1753
41. Bergmann O, Haukvik UK, Brown AA, Rimol LM, Hartberg CB et al (2013) ZNF804A and cortical thickness in schizophrenia and bipolar disorder. Psychiatry Res 212:154–157
42. Blasi G, Napolitano F, Ursini G, Di Giorgio A, Caforio G et al (2013) Association of GSK-3β genetic variation with GSK-3β expression, prefrontal cortical thickness, prefrontal physiology, and schizophrenia. Am J Psychiatry 170:868–876
43. Walton E, Geisler D, Hass J, Liu J, Turner J et al (2013) The impact of genome-wide supported schizophrenia risk variants in the neurogranin gene on brain structure and function. PLoS One 8:e76815
44. van Haren NEM, Schnack HG, Cahn W, van den Heuvel MP, Lepage C et al (2011) Changes in cortical thickness during the course of illness in schizophrenia. Arch Gen Psychiatry 68:871–880
45. Ahmed M, Cannon DM, Scanlon C, Holleran L, Schmidt H et al (2015) Progressive brain atrophy and cortical thinning in schizophrenia after commencing clozapine treatment. Neuropsychopharmacology 40:2409–2417
46. Alexander-Bloch AF, Reiss PT, Rapoport J, McAdams H, Giedd JN et al (2014) Abnormal cortical growth in schizophrenia targets normative modules of synchronized development. Biol Psychiatry 76:438–446
47. Nenadic I, Yotter RA, Sauer H, Gaser C (2015) Patterns of cortical thinning in different subgroups of schizophrenia. Br J Psychiatry 206:479–483
48. Kirkpatrick B, Buchanan RW (1990) The neural basis of the deficit syndrome of schizophrenia. J Nerv Ment Dis 178:545–555
49. Takayanagi M, Wentz J, Takayanagi Y, Schretlen DJ, Ceyhan E et al (2013) Reduced anterior cingulate gray matter volume and thickness in subjects with deficit schizophrenia. Schizophr Res 150:484–490
50. Voineskos AN, Foussias G, Lerch J, Felsky D, Remington G et al (2013) Neuroimaging evidence for the deficit subtype of schizophrenia. JAMA Psychiat 70:472–480
51. Chen X, Liang S, Pu W, Song Y, Mwansisya TE et al (2015) Reduced cortical thickness in right Heschl's gyrus associated with auditory verbal hallucinations severity in first-episode schizophrenia. BMC Psychiatry 15:152
52. Rimol LM, Hartberg CB, Nesvåg R, Fennema-Notestine C, Hagler DJ et al (2010) Cortical thickness and subcortical volumes in schizophrenia and bipolar disorder. Biol Psychiatry 68:41–50
53. Knöchel C, Reuter J, Reinke B, Stäblein M, Marbach K et al (2016) Cortical thinning in bipolar disorder and schizophrenia. Schizophr Res 172:78–85
54. Mamah D, Alpert KI, Barch DM, Csernansky JG, Wang L (2016) Subcortical neuromorphometry in schizophrenia spectrum and bipolar disorders. NeuroImage Clin 11:276–286
55. Wang L, Mamah D, Harms MP, Karnik M, Price JL et al (2008) Progressive deformation of deep brain nuclei and hippocampal-

amygdala formation in schizophrenia. Biol Psychiatry 64:1060–1068

56. Mamah D, Barch DM, Csernansky JG (2009) Neuromorphometric measures as endophenotypes of schizophrenia spectrum disorders. In: Ritsner MS (ed) The handbook of neuropsychiatric biomarkers, endophenotypes and genes. Springer, Amsterdam, pp 87–122
57. Smith SM, Jenkinson M, Woolrich MW, Beckmann CF, Behrens TEJ et al (2004) Advances in functional and structural MR image analysis and implementation as FSL. NeuroImage 23(Suppl 1):S208–S219
58. Patenaude B, Smith SM, Kennedy DN, Jenkinson M (2011) A Bayesian model of shape and appearance for subcortical brain segmentation. NeuroImage 56:907–922
59. Smith SM, Nichols TE (2009) Threshold-free cluster enhancement: addressing problems of smoothing, threshold dependence and localisation in cluster inference. NeuroImage 44:83–98
60. Beg M, Miller M, Trouve A, Younes L (2005) Computing large deformation metric mapping via geodesic flows of diffeomorphisms. Int J Comput Vis 61:139–157
61. Csernansky JG, Wang L, Jones D, Rastogi-Cruz D, Posener JA et al (2002) Hippocampal deformities in schizophrenia characterized by high dimensional brain mapping. Am J Psychiatry 159:2000–2006
62. Qiu A, Tuan T, Woon P, Abdul-Rahman M, Graham S et al (2010) Hippocampal-cortical structural connectivity disruptions in schizophrenia: an integrated perspective from hippocampal shape, cortical thickness, and integrity of white matter bundles. NeuroImage 52:1181–1189
63. Styner M, Lieberman JA, Pantazis D, Gerig G (2004) Boundary and medial shape analysis of the hippocampus in schizophrenia. Med Image Anal 8:197–203
64. Shenton ME, Gerig G, McCarley RW, Székely G, Kikinis R (2002) Amygdala-hippocampal shape differences in schizophrenia: the application of 3D shape models to volumetric MR data. Psychiatry Res 115:15–35
65. Csernansky JG, Schindler MK, Splinter NR, Wang L, Gado M et al (2004) Abnormalities of thalamic volume and shape in schizophrenia. Am J Psychiatry 161:896–902
66. Smith MJ, Wang L, Cronenwett W, Mamah D, Barch DM et al (2011) Thalamic morphology in schizophrenia and schizoaffective disorder. J Psychiatr Res 45:378–385
67. Mamah D, Harms MP, Barch D, Styner M, Lieberman JA et al (2012) Hippocampal shape and volume changes with antipsychotics in early stage psychotic illness. Front Psych 3:96
68. Zierhut KC, Graßmann R, Kaufmann J, Steiner J, Bogerts B et al (2013) Hippocampal CA1 deformity is related to symptom severity and antipsychotic dosage in schizophrenia. Brain 136:804–814
69. Dean DJ, Orr JM, Bernard JA, Gupta T, Pelletier-Baldelli A et al (2016) Hippocampal shape abnormalities predict symptom progression in neuroleptic-free youth at ultrahigh risk for psychosis. Schizophr Bull 42:161–169
70. Haller JW, Christensen GE, Joshi SC, Newcomer JW, Miller MI et al (1996) Hippocampal MR imaging morphometry by means of general pattern matching. Radiology 199:787–791
71. Csernansky JG, Joshi S, Wang L, Haller JW, Gado M et al (1998) Hippocampal morphometry in schizophrenia by high dimensional brain mapping. Proc Natl Acad Sci U S A 95:11406–11411
72. Wang L, Joshi SC, Miller MI, Csernansky JG (2001) Statistical analysis of hippocampal asymmetry in schizophrenia. NeuroImage 14:531–545
73. Womer FY, Wang L, Alpert KI, Smith MJ, Csernansky JG et al (2014) Basal ganglia and thalamic morphology in schizophrenia and bipolar disorder. Psychiatry Res 223:75–83
74. Danivas V, Kalmady SV, Venkatasubramanian G, Gangadhar BN (2013) Thalamic shape abnormalities in antipsychotic naïve schizophrenia. Indian J Psychol Med 35:34–38
75. Canu E, Agosta F, Filippi M (2015) A selective review of structural connectivity abnormalities of schizophrenic patients at different stages of the disease. Schizophr Res 161:19–28
76. Alexander-Bloch A, Giedd JN, Bullmore E (2013) Imaging structural co-variance between human brain regions. Nat Rev Neurosci 14:322–336
77. Evans AC (2013) Networks of anatomical covariance. NeuroImage 80:489–504
78. Zugman A, Assunção I, Vieira G, Gadelha A, White TP et al (2015) Structural covariance in schizophrenia and first-episode psychosis: an approach based on graph analysis. J Psychiatr Res 71:89–96
79. Fornito A, Bullmore ET (2015) Reconciling abnormalities of brain network structure and

function in schizophrenia. Curr Opin Neurobiol 30:44–50

80. Ratnanather JT, Poynton CB, Pisano DV, Crocker B, Postell E et al (2013) Morphometry of superior temporal gyrus and planum temporale in schizophrenia and psychotic bipolar disorder. Schizophr Res 150:476–483
81. Kim SH, Lee J-M, Kim H-P, Jang DP, Shin Y-W et al (2005) Asymmetry analysis of deformable hippocampal model using the principal component in schizophrenia. Hum Brain Mapp 25:361–369
82. Buchanan RW, Francis A, Arango C, Miller K, Lefkowitz DM et al (2004) Morphometric assessment of the heteromodal association cortex in schizophrenia. Am J Psychiatry 161:322–331
83. Breier A, Buchanan RW, Elkashef A, Munson RC, Kirkpatrick B et al (1992) Brain morphology and schizophrenia a magnetic resonance imaging study of limbic, prefrontal cortex, and caudate structures. Arch Gen Psychiatry 49:921–926
84. Mitelman SA, Buchsbaum MS, Brickman AM, Shihabuddin L (2005) Cortical intercorrelations of frontal area volumes in schizophrenia. NeuroImage 27:753–770
85. Mitelman SA, Shihabuddin L, Brickman AM, Buchsbaum MS (2005) Cortical intercorrelations of temporal area volumes in schizophrenia. Schizophr Res 76:207–229
86. Woodruff PW, Wright IC, Shuriquie N, Russouw H, Rushe T et al (1997) Structural brain abnormalities in male schizophrenics reflect fronto-temporal dissociation. Psychol Med 27:1257–1266
87. Abbs B, Liang L, Makris N, Tsuang M, Seidman LJ et al (2011) Covariance modeling of MRI brain volumes in memory circuitry in schizophrenia: sex differences are critical. NeuroImage 56:1865–1874
88. Mitelman SA, Brickman AM, Shihabuddin L, Newmark R, Chu KW et al (2005) Correlations between MRI-assessed volumes of the thalamus and cortical Brodmann's areas in schizophrenia. Schizophr Res 75:265–281
89. Mitelman SA, Byne W, Kemether EM, Hazlett EA, Buchsbaum MS (2006) Correlations between volumes of the pulvinar, centromedian, and mediodorsal nuclei and cortical Brodmann's areas in schizophrenia. Neurosci Lett 392:16–21
90. Wheeler AL, Voineskos AN (2014) A review of structural neuroimaging in schizophrenia: from connectivity to connectomics. Front Hum Neurosci 8:653
91. Bullmore E, Sporns O (2009) Complex brain networks: graph theoretical analysis of structural and functional systems. Nat Rev Neurosci 10:186–198
92. Telesford QK, Simpson SL, Burdette JH, Hayasaka S, Laurienti PJ (2011) The brain as a complex system: using network science as a tool for understanding the brain. Brain Connect 1:295–308
93. Bassett DS, Bullmore E, Verchinski BA, Mattay VS, Weinberger DR et al (2008) Hierarchical organization of human cortical networks in health and schizophrenia. J Neurosci 28:9239–9248
94. Zhang Y, Lin L, Lin C-P, Zhou Y, Chou K-H et al (2012) Abnormal topological organization of structural brain networks in schizophrenia. Schizophr Res 141:109–118
95. Beaulieu C (2002) The basis of anisotropic water diffusion in the nervous system—a technical review. NMR Biomed 15:435–455
96. Jones DK, Knösche TR, Turner R (2013) White matter integrity, fiber count, and other fallacies: the do's and don'ts of diffusion MRI. NeuroImage 73:239–254
97. Marner L, Nyengaard JR, Tang Y, Pakkenberg B (2003) Marked loss of myelinated nerve fibers in the human brain with age. J Comp Neurol 462:144–152
98. Peters A, Sethares C (2002) Aging and the myelinated fibers in prefrontal cortex and corpus callosum of the monkey. J Comp Neurol 442:277–291
99. Smith SM, Jenkinson M, Johansen-Berg H, Rueckert D, Nichols TE et al (2006) Tract-based spatial statistics: voxelwise analysis of multi-subject diffusion data. NeuroImage 31:1487–1505
100. Ellison-Wright I, Bullmore E (2009) Meta-analysis of diffusion tensor imaging studies in schizophrenia. Schizophr Res 108:3–10

Chapter 20

Bipolar Disorders

Delfina Janiri, Elisa Ambrosi, Emanuela Danese, Isabella Panaccione, Alessio Simonetti, and Gabriele Sani

Abstract

In the last decades, researchers put their efforts in searching for specific biomarkers to make the diagnosis of psychiatric disorders more accurate, to better profile psychiatric patients, and, essentially, to improve their response to treatment and their outcome. This is particularly important in the field of bipolar disorders (BDs), which are severe and chronic disorders with polymorphic clinical manifestations that may be the expression of different underlying neurobiological alterations.

Currently, neuroimaging techniques are considered an essential tool for the noninvasive evaluation of putative brain dysfunctions of these patients. Neuroimaging studies aim to find brain alterations associated with the diagnosis of BD or with some psychopathological dimensions (such as mood lability, emotional instability, etc.) specific of BD patients. These findings should improve the understanding of the mechanisms underlying the disorder and may become potential targets of pharmacological and non-pharmacological treatment (i.e., neurocognitive rehabilitation).

This chapter aims at providing an exhaustive review of the studies on brain morphology in bipolar disorder in order to better characterize it from a neurobiological point of view, thus providing a comprehensive background for further research on this topic.

Structural abnormalities were found in regions pertaining to specific networks postulated as vulnerable in BD. Specifically, alterations in the neocortical areas of the brain, in particular the frontotemporal circuit, and in subcortical areas of the brain have been identified.

In general, the findings suggest that it is necessary to consider large-scale neural circuitries more than specific brain structures to understand the neurobiological mechanisms underlying BD. Moreover, our review points up the need for longitudinal studies that might better investigate, and cast light on, these issues.

Key words Bipolar disorder, Brain morphometry, Region of interest, VBM, Cortical thickness, Shape analysis, Microstructure

1 Introduction

Bipolar disorder (BD) is a severe, chronic disorder characterized by interchanging episodes of depression and mania or hypomania, with some relatively asymptomatic periods (euthymia). Epidemiological studies estimate lifetime prevalence rate of 0.6–1% for bipolar I disorder (BDI), 0.4–1.1% for bipolar II disorder (BDII), and

Gianfranco Spalletta et al. (eds.), *Brain Morphometry*, Neuromethods, vol. 136,
https://doi.org/10.1007/978-1-4939-7647-8_20,

2.4–5.1% for subthreshold BD [1, 2]. Depending on the presence of manic or hypomanic episodes, there are two major types of BD: type I, if at least one episode of mania is present, and type II, if at least one episode of hypomania and one of depression occur. This oversimplified distinction does not take into account the many different clinical presentations of the disease observed in patients. For example, it is possible to identify different cycle patterns, depending on the sequence of the mood alterations (i.e., mania-depression-euthymia, depression-mania-euthymia, irregular cycle pattern) [3]; different longitudinal courses, according to the frequency of cyclicity (i.e., continuous long cycle vs. rapid cyclicity) [4] or to the predominant polarity [5]; different manifestations of mood episodes (i.e. pure vs. mixed) [6]; and so on. These clinical aspects, even more than the diagnosis in itself, concur to dramatically modify the outcome of patients in terms of response to treatment, functioning level, or risk of suicide. Therefore, it would not be unlikely if distinct BD patients, sharing the same diagnosis but showing different clinical features, also differed from a neurobiological point of view.

In the last decades, researchers put their efforts in searching for specific biomarkers to make the diagnosis more accurate, to better profile patients, and, essentially, to improve their response to treatment and outcome. These studies are essential, also, to verify how reliable pure clinically based diagnoses, such as the *Diagnostic and Statistical Manual of Mental Disorders* (DSM)-based diagnosis, are.

The study of the neurobiology of BD involves different areas, including neuropathology, genetics, immunology, neuropsychology, and magnetic resonance imaging (MRI). The latter is the object of this chapter.

Neuroimaging studies aim to find brain alterations associated with the diagnosis of BD or with some psychopathological dimensions (such as mood lability, emotional instability, etc.) specific of BD patients. These findings should improve the understanding of the mechanisms underlying the disorder and may become potential targets of pharmacological and non-pharmacological treatment (i.e., neurocognitive rehabilitation).

Although these studies are intriguing, there are some issues to deal with. The vast majority of the studies are cross-sectional, while there are just few longitudinal studies [7]. This is a critical issue in a long-lasting disorder like BD, because it is practically impossible to establish whether brain modifications observed in patients precede the onset of the disorder or occur as a consequence of it. In other words, it is not possible to find a predictive marker of the disease before the onset of the disease itself, and it's still uncertain whether brain alterations observed in individuals at high risk for BD can actually be considered as proper risk factors for developing the disease and used in preventive interventions [8]. Longitudinal studies would be crucial in clarifying this matter, but to date they are still quite scarce, being time-consuming, expensive, and burdened with a high rate of dropouts. Also, patients enrolled in these studies are usually undergoing pharmacological treatments, and some of the drugs commonly prescribed have been demonstrates to somehow affect brain morphology. For all of these reasons, neuroimaging studies often give contrasting results.

In this chapter, we reviewed neuroimaging studies aiming to identify macro- and microstructural brain changes in BD patients, in order to identify alterations of specific brain regions, along with definite cerebral systems, associated with the diagnosis of BD.

2 Macrostructural Findings

2.1 Frontal Region

Alongside psychiatric symptoms, a considerable extent of BD patients presents cognitive deficits, especially in three core domains: attention, executive functions, and emotional processing. Brain frontal regions, with their multiple connections, play a key role in modulating all these three functions, and abnormalities in these areas have been consequently considered as a potential biomarker of BD. The majority of studies reporting structural changes in the frontal cortex of BD patients utilize voxel-based morphometry (VBM) rather than region of interest (ROI) techniques. Indeed, differently from ROI-based meta-analyses which reported non-specific changes in the brain structure of patient with BD, [9, 10], two VBM meta-analyses, using T- and Z- maps and anatomical likelihood estimation approach, demonstrated gray matter reduction in the right ventrolateral prefrontal cortex (VLPFC) (Brodmann areas (BA) 44, 45, and 47) [11] and in the anterior cingulate cortex (ACC) (BA 24,32, and 33) [12], respectively.

Studies that analyzed cortical thickness confirmed alterations mainly in the ACC [13–17] and VLPFC [13, 16, 18, 19]; however, volumetric thickness reduction in other brain areas, such as the dorsolateral prefrontal cortex [17, 20, 21], medial orbitofrontal cortex [13, 16–18, 22], and superior frontal cortex [13, 16, 17, 20, 22–24], was also reported.

Studies that tried to define specific neuroanatomical markers related to the "core" clinical dimensions and subcategories of BD frequently reported inconsistent results. However, a significant correlation between psychotic symptoms and decreased cortical thickness in the left VLPFC (BA44), left dorsolateral prefrontal cortex, and left temporal pole was reported [13]. Furthermore, only one study reported smaller gray matter volume within the right medial orbitofrontal region and reduced cortical thickness in the right medial orbitofrontal region and in the left superior temporal region in BD patients; these alterations were more severe in BDI than in BDII patients [24]. Rapid cyclers (i.e., patients with at least four mood episodes per year) exhibited smaller ventromedial prefrontal cortex volumes than non-rapid cyclers [25], and in female subjects with history of suicide attempt, an inverted correlation between the volume of the anterior genu corpus callosum and impulsivity was found [26].

The majority of investigators didn't find a relationship between age, age at onset, illness duration, or number of affective episodes and prefrontal volumes [27–30]. Nonetheless, an interaction between VLPFC volume and age was found in a cross-sectional study in adolescents and young adults, suggesting that there may be an acceleration of normal age-related volume loss in late adolescence and early adulthood in patients with BD [25].

Studies on at-risk subjects reported very few replicated findings, the most important of which being increased volumes in the VLPFC of the healthy offspring of BD parents compared with age-matched subjects with no family history of BD [31]. Taken together, these findings led some investigators to suggest that VLPFC may be characterized by an initial increase in volume early in the course of illness, only later followed by long-term anatomic changes [32].

It's important to point out that some of the inconsistencies reported in the literature may be related to variability of the age of the sample, mood state, history of substance use, or treatment. This is specifically true for the ACC. In fact, one major environmental influence that emerges in studies of the ACC is the impact of medication [33], in particular in terms of neurotrophic effects. The majority of data demonstrated this effect is provided by lithium [29, 34, 35], even if evidences for valproate and quetiapine are present [36].

These data show that, despite the substantial literature, very few structural brain abnormalities were consistently found in the frontal lobes in BD patients. This may suggest that findings are spurious and that lack of consistency might be due to several reasons, including methodological issues, e.g., small sample sizes. However, it is important to note that some of the few replicated findings, mostly regarding VLPFC and ACC, involve alterations in brain structures that are part of the fronto-limbic network postulated to be abnormal in BD [32]. In this network, VLPFC and ACC have extensive connections with the amygdala and other subcortical areas related to emotions and mood [37] and mediate voluntary and automatic emotion regulation subprocesses, respectively [38]. A loss of prefrontal modulation of limbic system may underlie some symptoms in BD, and, conversely, dysfunction within limbic areas may disrupt the activity of cortical regions, resulting in bipolar symptoms [39]. Therefore, the alterations reported in these two areas could be interpreted as the structural correlates of the failure in cortical modulatory activity.

Moreover, structural studies suggest, even with the limitations described above, that some abnormalities in prefrontal cortical areas may be already present in the early stages of BD or even predate the onset of the illness [39]. Functional and psychological data appear to confirm this theory [40] and led researchers to redefine the nature of BD. In this view, BD could be seen as a pattern of abnormalities in the neural development that gives way to progressive structural and functional changes. These changes drive clinical symptomatology and are in turn exacerbated by the consequences of these symptoms [40]. Even if the amount of data that support the neurodevelopmental model is growing, more evidences are needed to confirm the reliability of this theory.

2.2 Temporal Region

Temporal lobe could be considered as another vulnerable area in the pathophysiology of BD. A meta-analysis of VBM studies, in fact, found gray matter volumetric reductions in the right temporal cortex of patients with BD (middle temporal gyrus, superior temporal gyrus, temporopolar area, and insula) [11], and one of the most replicated cortical thickness findings is a reduction of the left temporal lobe. Specifically, two ROI [41, 42] and four whole brain [16, 19, 20, 24] studies reported cortical thinning of the left superior temporal gyrus in BD patients compared to controls, even if more widespread cortical thinning across the temporal lobes, including regions of the temporal pole (BA 38), middle temporal gyrus (BA 21) [16, 20], and inferior temporal gyrus (BA 20) [16, 19, 20], was also reported.

Some studies aimed at finding a link between volumetric changes in temporal cortex and clinical variables. Two VBM studies showed an inverse correlation between volume of the superior temporal gyrus and age but failed in finding an association with the duration of illness or number of affective episodes [43, 44]. In contrast, another study reported increased volumes in this area over time in patients but not in controls [45]. No correlation was found between temporal lobe alterations and psychotic features, rapid cycling, or suicide attempts [46–49]. When the distinction between the two main subtypes of BD was considered, smaller volumes in left superior temporal regions were observed in BDII compared with BDI [24].

In young patients with BD, overall temporal cortical findings have been mixed. Several studies reported decreased superior temporal gyrus volume in children and adolescents with BD [50–52]. However, those who examined correlations with age [50] and duration of illness [51] didn't find any association. One study failed to detect any difference in temporal lobe volume between healthy adolescents and those with BD [50], while another reported increased left temporal lobe volumes in the BD group [53].

Taken together, these findings demonstrate that temporal cortex alterations are frequently present in BD and suggest that temporal cortex dysfunction could play a role in the pathophysiology of this disorder [54]. Temporal cortex holds complex functions related to mood regulation, and distinct areas within the temporal cortex probably underlie different aspects of mood regulation. The middle temporal gyrus, for example, is particularly important in the experience of emotions [55], while superior temporal gyrus plays a major role in social cognition and perception [56]. Therefore, the volume and thickness reductions found in these areas can be related to impairments in the theory of mind observed in BD [54]. The insula has been implicated in language processing [57], unsuccessful response inhibition [58], and emotion processing [59, 60]. A volumetric reduction in this region consequently may explain the impaired inhibitory function in BD patients or their inability to appreciate complex emotions, such as disgust [59, 61]. Moreover, it is important to differentiate between the anterior and the posterior areas of temporal cortex because they were found to have distinct developmental patterns [62]. Thus, differences in the development in these areas can partially explain the structural discrepancies found in the brain of children and adolescents with BD.

2.3 Parietal Region

The parietal cortex, separated from the frontal lobe by the central sulcus and from the occipital lobe by the parieto-occipital sulcus, is an associative structure that plays a critical role in sensorimotor integration processes. Because of its functions, alterations in this area might be responsible for symptoms such as hallucinations, thought disorders, bizarre behavior, volition and attentional deficits, social withdrawal, flat affect, and aggressiveness. All these symptoms can be found in many psychiatric disorders, including schizophrenia (SZ) and BD [63]. Brain volumetric deficits of patients with schizophrenia and BD have been compared, showing overlapping clusters of white matter (WM) losses in parietal regions. In particular WM reductions were described in parietal lobe in both first-episode BD patients and SZ patients, compared to healthy subjects [64]. Moreover, the same study showed gray matter (GM) volume reduction in the right precuneus in patients with BD. Cui and colleagues in 2011 compared a group of patients with paranoid schizophrenia with a group of patients with psychotic type I BD. Both paranoid schizophrenia and bipolar mania groups showed reduction of gray matter volume in the inferior parietal lobule, but different hemispheres were affected in both groups (right hemisphere in schizophrenia and left hemisphere in BD) [65].

In accordance with these results, more studies described abnormalities in the inferior parietal lobule and precuneus in patients with BD [28, 66–68]. Therefore, it is possible to hypothesize that these two regions could be directly involved in the neuropathophysiological process underpinning BD. Inferior parietal lobule is traditionally related to attention, and attention deficits, particularly spatial attention, have been frequently described in patients with BD. Precuneus, on the other hand, has been implicated in normal responses to emotional stimuli that could be altered in BD.

Not only reduced parietal cortical volumes but also increased volumes have been reported in patients with BD. Adler et al. found an increase of gray matter density in the inferior parietal lobule, superior parietal lobule, and precuneus in first-episode BD and suggested that changes in parietal activity might represent a central abnormality in this disease [66], indirectly altering prefrontal cortex functioning [69]. They proposed that the increase of gray matter could be explained as preapoptotic osmotic changes occurring in damaged brain cells or, conversely, as cellular hypertrophy, neuronal overgrowth, or increased pruning. Overall, these alterations might mark areas of neuronal degeneration occurring early in the course of the disease. The same group also described increased gray matter in primary and secondary motor cortex [70], and, even if motor dysfunction is not considered a typical feature of BD, several studies have reported a functional impairment of fine motor control, as well as other vague neurological symptoms [71].

Other studies focused on cortical thickness in BD and found cortical thinning in primary sensory regions in the parietal lobe [17], in the superior and inferior parietal gyri, and in the supramarginal gyrus [19]. Comparing cortical thickness in depressed BD patients and patients with major depressive disorder (MDD), Lan et al. found that the left inferior parietal and the right precuneus regions were thinner in BD than in both MDD and healthy controls (HC) [21].

In conclusion, the parietal lobe, and particularly the inferior parietal lobule and the precuneus, is probably involved in the physiopathology of BD since the earlier stages. However, findings are often inconsistent. Again, the differences observed might be due to the heterogeneity of subjects and the methodological differences. Additional studies are needed to clarify the role played by parietal regions in BD.

2.4 Occipital Region

Sensory cortex, usually considered just as a vehicle of information to other areas engaged in emotion processing, has been recently involved in the neuropathophysiology of BD. In particular, abnormalities in BD were found in the occipital cortex, which is the visual processing center of the human brain.

Macrostructural studies mainly showed reduced GM density and lower cortical thickness in the occipital cortex of patients with BD [17, 72, 73]. These findings are in line with previously reported deficits in visual processing and perception in these patients [74]. Also, reported deficits in working memory tasks, which partly engage visual areas of the occipital cortex, might be related to neuroanatomical abnormalities in occipital brain regions [75]. A recent study by Abé et al. [72] showed that both patients with

BDI and with BDII had lower cortical thickness in the medial occipital lobe (containing the lingual, pericalcarine, and cuneus cortices) than HC. The same study, when comparing patients with BD who took lithium or antiepileptic drugs with those who did not, found specific differences in medial occipital regions. Specifically, the effect of the two types of medication had opposite directionality: lithium use was associated with larger and antiepileptic drugs with lower cortical volume/thickness in the medial occipital cortex. In this context it is interesting that the use of antiepileptic drugs has been associated with visual impairments [76]. Results on lithium are in line with a recent study detecting thinner occipital cortices in lithium-free, but not in lithium-treated, BD patients, compared to HC [77]. Moreover, the same study showed that lithium-treated BD patients consistently exhibited nonsignificant trends for greater cortical thickness than lithium-free patients. Another study found that long-term lithium treatment was associated with increased GM volumes in occipital cortex [46], highlighting the importance of considering not only the mere exposure to lithium in BD but primarily the length of treatment. This is in line with a recent study that found that long-term, but not short-term, lithium treatment may exert neuroprotective effects on specific hippocampal subfields [78].

Some studies focused on the effect of specific clinical characteristics on occipital cortex in BD, suggesting that it could modulate specific aspects of the disorder. Hallucinations were negatively correlated with GM volume within the occipital cortex in patients with BD [79], and BD patients that attempted suicide during lifetime showed reduced GM volumes in several brain areas including occipital cortex [46]. Furthermore, a study on adolescents with BD showed that patients with co-occurring cannabis use demonstrated increased gray matter volume in the right middle occipital cortex compared to BD adolescents free of any substance use [80].

Occipital cortex recently has been indicated as a possible marker of risk for BD. Hanford et al. [81], to test this speculation, compared GM volumes between symptomatic and asymptomatic children of parents with BD and healthy children of healthy parents. Children were defined as "symptomatic" when a psychiatric diagnosis could be made. Interestingly, they found that when comparing symptomatic and asymptomatic children of parents with BD, GM volumes were comparable in all regions except the lateral occipital cortex. Other studies, aiming at verifying genetic and environmental influences on cortical thickness in BD, showed that thinner occipital cortices were associated with environmental factors related to BD [18, 82]. Taken together, these results suggest that environmental factors could play a role in driving the liability to develop BD via occipital cortex and that this liability could be expressed in reduced GM volumes in occipital cortex.

In summary, occipital cortex could provide an objective neurobiological marker to increase diagnostic precision in identifying BD. Future studies should continue to focus more thoroughly on this relationship, considering in particular the link between occipital cortex, specific psychopathological aspects of BD, response to treatment, and risk to develop affective disorders.

2.5 Cerebellum

The cerebellum has traditionally been considered as only involved in motor control and programming. However, an increasing number of studies are now highlighting the role of the cerebellum in regulating different functions, such as emotion processing, decision-making, attention, working memory, and impulsive behaviors. It has been suggested that these functions are mediated by multiple connections to various regions of the brain, including the motor areas, the prefrontal cortex, and the limbic system. Specifically, two pathways of connections have been identified, the cortico-ponto-cerebellar and the cerebello-thalamo-cortical, which allow the cerebellum to play an important role in cognitive and emotional regulation [83]. If this ability is damaged, cognitive deficits and psychopathological manifestations could come to light. Schmahmann proposed the construct of "dysmetria of thought," which is the inability of the cerebellum to regulate consistency and appropriateness of cognitive process, resulting in an unpredictability of social interaction [84]. Structural abnormalities of the cerebellum have been associated with many neuropsychiatric disorders, including BD [85]. Current knowledge of the neural underpinnings of BD provides a map of neural circuits involving the cerebellum, and many studies have demonstrated structural cerebellar changes associated with BD. In particular, cerebellar atrophy is indicated as a potential, albeit aspecific, biological marker for BD. Though cerebellar GM reduction is the most common finding in patients with BD [86, 87], also reductions of WM [88] and both GM and WM were found [7], as well as no structural abnormalities [89, 90]. These contrasting findings may be due to medication and in particular to the use of lithium. Lithium, in fact, has a supposed neuroprotective effect, already suggested for other cerebral regions such as hippocampus, which could eventually impact on cerebellum [83]. A voxel-based MRI study found increased volumes in the cerebellum, bilaterally, in first-episode BD subjects [66]. This result may be due to the specific population tested by the authors and suggests that increased GM density in first-episode bipolar patients may indicate a primary abnormality in BD, later followed by a progressive decline of GM volume. In support of this hypothesis, a longitudinal VBM study over 4 years showed a more rapid decrease in cerebellar GM in patients with BD than healthy controls [91]. Other studies confirm that the damage to the cerebellum is progressive over the course of the illness and showed an association between reduced GM volume and number of episodes [87, 92].

Nevertheless, it is not yet clear if the reduction of cerebellar areas is related to the progression of the illness. A recent study found a bilateral reduction of cerebellar volume not related to the course of BD, suggesting that cerebellar alterations in BD patients might be present from the very beginning of the illness and remain stable over time [7].

An important question to address in studies on cerebellar volumes and BD is whether the reduction involved the whole cerebellum or only specific areas. Anatomically the cerebellum can be divided into the anterior lobe (lobules I–V), the posterior lobe (lobules VI–IX), and the flocculonodular lobe (lobule X); the first two can be further divided into vermis (the median portion) and hemisphere (the lateral portion). Baldacara et al. found that the left and right cerebellum and vermis were significantly smaller in BDI compared to healthy controls [86]. Other studies, based on the anterior-posterior dichotomy of motor vs. cognitive functions of cerebellum, identified the posterior cerebellum as a vulnerable area in BD. Specifically, two studies on patients with BD found a reduction in GM of the right crus (lobules VII), which seems critical for impulsive behaviors, projecting to dorsolateral prefrontal cortex and parietal association areas. Moreover, MRI studies using ROI approach found an inverse correlation between number of episodes and volumes of vermal area V3 (lobules VIII–X) [92] and V2 (lobules VI–VII) [87], which are connected, respectively, with the limbic system and thalamus. Nevertheless, a recent study found decreased volumes of the anterior cerebellar lobules, primarily involved in the motor control system, and focused on the importance of psychomotor alterations in BD [7].

Cerebellar volume changes have been also indicated as a potential neuroanatomical correlate of genetic risk for BD. Specifically, Eker et al. found GM volume deficits in the right cerebellum of both BD patients and their non-BD siblings, compared to healthy controls. Conversely, other studies identified group differences in the left cerebellum between patients with BD, healthy first-degree relatives, and controls [93].

In conclusion findings indicate the cerebellum as an interesting area of study for pathophysiology of BD. Reduced volume, in particular, could be a potential biological marker for BD. It was not clear, however, which parts of the cerebellum are primarily involved and how atrophy is related to the course of the illness. Moreover, there are no studies investigating cerebellar structural differences between type I and type II BD.

2.6 Deep Gray Matter

Current views in neuroscience highlight that many cognitive and affective functions of the brain rely on neural networks including subcortical gray matter structures. Neocortical areas of the brain are anatomically connected – either directly or through the thalamus – with the basal ganglia (putamen, caudate, nucleus accumbens, and globus pallidus), amygdala, and hippocampus. Therefore, these subcortical structures can modulate the activity of the cortex, and abnormalities in these areas can diffusely impact brain functioning. Deep gray matter alterations have been demonstrated in many psychiatric disorders; in particular, subcortical structures are involved in neural circuitry underlying emotion processing and regulation, which seems to be compromised in BD [38]. Neuroimaging studies have showed altered volumes of deep gray matter in patients with BD, and abnormalities in the shape of these structures were also reported in BD patients.

Reduced volumes of hippocampus, amygdala, thalamus, caudate, putamen, pallidus, and accumbens were found in patients with BD compared to healthy controls [19, 94]. The predominant pattern of reduced subcortical gray matter volumes may follow an elevated activity in these structures, then resulting in a neurotoxic effect [38]. Nevertheless, other studies reported enlarged subcortical volumes in patients with BD [30, 94], whereas retrospective meta-analyses found no changes in deep gray matter when comparing BD patients and HC [10]. A recent study by the Enhancing Neuroimaging Genetics Through Meta-Analysis (ENIGMA) consortium on 1710 subjects with BD attempted to compensate for the substantial heterogeneity across previous volumetric studies on subcortical structures [95]. They quantified case-control differences in subcortical structures automatically segmented using a specific software and found consistent volumetric reductions in patients with BD, compared to healthy controls, in the bilateral hippocampus and thalamus and a trending significant reduction in the amygdala. They did not find volumetric differences comparing patients with BDI and with BDII; however, the magnitude of differences in brain volumes was larger in subjects with BDI. Moreover, they found that patients taking lithium at the time of assessment presented larger thalamic volumes compared with patients not taking lithium.

When studying subcortical structures, it is relevant to note that some of these areas are not uniform structures and that it is important to consider this heterogeneity. The hippocampus, in particular, consists in subfields with distinct morphology: the cornu ammonis (CA) subfields CA1-4, the dentate gyrus (DG), the fimbria, the subiculum, and the presubiculum. To better examine the hippocampal volume reduction observed in patients with BD, some studies focused on hippocampal subfields and found differences between clinical groups and healthy controls. The largest study on this topic analyzed hippocampal subfield volumes segmented with FreeSurfer and total hippocampal volumes in 192 subjects with BD compared to 300 healthy controls [96]. The BD group had smaller bilateral cornu ammonis subfields CA2/3, CA4/dentate gyrus, presubiculum, subiculum, and right CA1 volumes than healthy control subjects. No differences between patients with BDI and

BDII were found. Moreover, smaller subiculum volumes were related to poorer verbal memory in patients.

Structural studies on deep gray matter in BD focused not only on volume changes but also on abnormalities in the surface of subcortical structures. In recent years, in fact, developments in shape analysis methodology have shown that investigation of the three-dimensional surfaces of brain subcortical structures can supplement data derived from volumetric analysis. Specifically, deformations were found in the thalamus [97], hippocampus [98], caudate, and putamen [99]. Interestingly, a study found shape differences in the striatum of drug-naïve patients with BD, more prominent for the right side, and did not find the same results for drug-treated subjects with BD [99]. This result suggests that shape differences may be modulated by treatment. Nevertheless, shape analysis studies on subcortical structures are still too scarce and based on samples too small to yield an overall consensus on the topic.

Summarizing, volumetric studies have shown that reduced subcortical gray matter volumes are probably involved in the neurobiological mechanisms underlying BD. No differences between BDI and BDII were found, and this suggests a similar origin between BD subtypes. However, it is not still clear whether this loss of volumes is related to a loss of specific functions. Moreover, biological mechanisms mediating the reduction of subcortical structures have not yet been explained. Shape analysis may provide a useful method to identify subtle anatomical abnormalities in BD and could eventually adduce additional information to traditional volumetric studies.

3 Microstructural Findings

Over the last decade, the application of diffusion tensor imaging (DTI) has allowed the in vivo study of WM microstructure. In normal conditions, water molecules tend to diffuse parallel (axial diffusivity, AD) rather than perpendicular (radial diffusivity, RD) to the main axis of fibers, given the effect of myelin on restricting the diffusivity. The variance of the direct measures of the diffusion magnitude in these directions (fractional anisotropy, FA) reflects the integrity of axons and myelin sheaths.

Recent studies provide strong evidence for the presence of WM microstructural alterations in all major commissural, association, and projection tracts in BD. Meta-analytic findings show that compared to HC, BD patients have reduced FA in the left posterior cingulum, right anterior superior longitudinal fasciculus (SLF), right inferior longitudinal fasciculus (ILF), right inferior fronto-occipital fasciculus (IFOF), left genu of corpus callosum (CC), and anterior thalamic radiation [100]. The involvement of these WM tracts suggests a pattern of altered connectivity mainly in the frontal limbic network, which plays a central role in the proposed emotional dysregulation theory [37]. Indeed, an imbalance between the automatic-ventral and voluntary-dorsal systems implicated in emotional regulation may explain mood instability in BD. Moreover, morphometric and functional studies reported decreased GM and reduced activation in regions subserving emotional regulation [101], such that altered WM connectivity may be responsible for a deficient communication between neurons.

However, the exact cause of variations in FA is still unclear. The interpretation of FA changes rely on the concurrent changes in both radial diffusivity (RD) and axial diffusivity (AD), respectively, expression of myelin, and axonal damage, but only few studies report these values [102, 103]. Interestingly, FA decreases have been more frequently associated with RD increases suggesting that demyelination might contribute to the loss of WM integrity.

Whether or not the abovementioned abnormalities reflect the phenotypic heterogeneity of BD at a clinical level is debated. The under-recruitment of BDII in neuroimaging studies led to inconsistent conclusions. Compared to BDI, BDII showed lower FA in the inferior longitudinal fasciculus (ILF), superior longitudinal fasciculus (SLF), corona radiate, and internal capsule [102], higher FA in the uncinate fasciculus and right temporal WM [48], or no FA differences [24].

Although brain structure is generally considered stable, dynamic changes of WM microstructure may contribute to the different illness phases (e.g., depression, mania/hypomania, euthymia) and different impairments of cognitive domains across BD subtypes. Interestingly, depressed BD showed larger load of alterations compared to both euthymic [104] and manic patients [103]. Particularly, reduced FA was reported in the left superior frontal WM, right SLF, and posterior limb of internal capsule [104]. The release of pro-inflammatory factors, which is increased in prolonged stress system activation such as depression, might contribute to state-dependent WM changes [105]. Furthermore, positive association between FA reductions in all major tracts (e.g., corticospinal, corpus callosum, ILF, SLF, IFOF, uncinate fasciculus, anterior thalamic radiation) and cognitive impairment in working memory, processed speed performance, and attention has been reported in BD patients [103, 106]. The reduced fiber coherence together with a slowing down in the rate conduction along fibers might be hypothesized to progressively lead to deficits in cognitive functions.

Finally, the potential contribution of pharmacologic treatment in relation to these findings should be considered. Although some studies fail to detect an effect of medication [106], few studies have shown normalizing effects of lithium on DTI parameters. Indeed, lithium-treated subjects showed increased FA [107] and AD [108] in the body of CC and several WM tracts including forceps major, cingulum, SLF, ILF, posterior thalamic radiation, corticospinal, and corona radiata.

Overall, findings are consistent with the involvement of WM abnormalities in the pathophysiology of BD. However, further investigations are needed to clarify the pathogenetic mechanism underlying the loss of WM integrity.

4 Conclusion

In this chapter, we reviewed the major structural brain alterations in BD patients observed in a large number of neuroimaging studies, and we noticed that different macro- and micro-alterations have been described in numerous brain structures. Structural abnormalities were found in regions pertaining to specific networks postulated as vulnerable in BD. Specifically, it is possible to identify a parallel alteration in the neocortical areas of the brain, in particular the frontotemporal circuit, and in subcortical areas of the brain. These areas are reciprocally connected, and they influence each other. Accordingly, BD can be conceptualized, in neural circuitry terms, as the result of alterations in the networks modulating the emotional processing and regulation, and in the reward-processing circuitry. Nevertheless, alterations were found not only in the areas traditionally considered involved in the affective disorders but also in some areas of the brain that are usually associated with other functions. Intriguingly, the cerebellum, implicated in motor control and programming, and the occipital cortex, the center of the visual processing of the human brain, could be interesting areas of study for the pathophysiology of BD, modulating affective and cognitive functions. Therefore, an observation that stands out when analyzing these studies is that, despite the high variability of results, most of the anatomical structures that appear to be constantly modified in BD belong to complex brain circuitry underlying major affective and cognitive functions, whose alterations represent the clinical core of the disease. In this perspective, it seems important to switch the attention of the researchers from the single brain structure to the whole system that might be affected in BD, bearing in mind that it is still a matter of debate whether or not structural modifications are associated to functional alterations of neural activity and how this affects the circuitry in its whole (Fig. 1). In addition, virtually none of the alterations observed so far is specific for BD; rather, many of them are shared with other psychiatric diseases, such as schizophrenia or major depressive disorder. For example, both BD and schizophrenia patients show reduced hippocampus, amygdala ,and thalamus volumes, although these alterations are more marked in schizophrenia, and lower hippocampal volumes have been described in

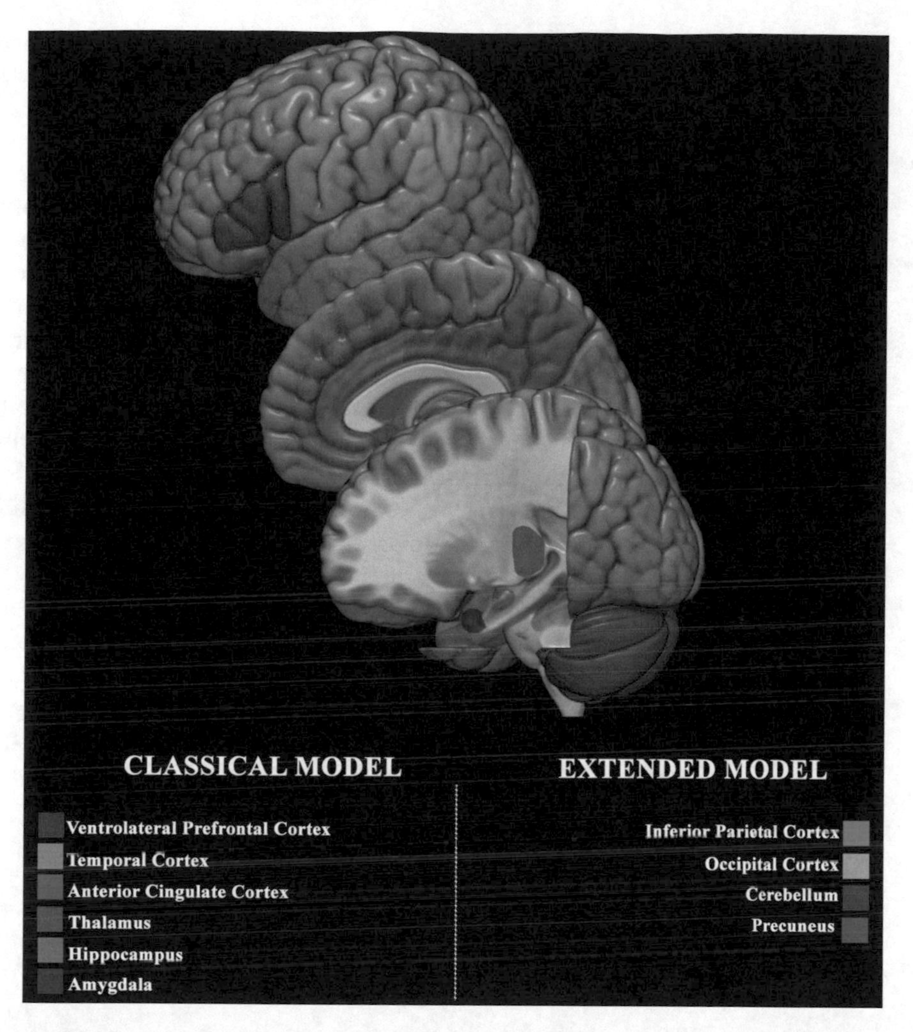

Fig. 1 Cerebral basis of bipolar disorder. The classical model of brain areas traditionally related to bipolar disorders (left box) and the extended one, including areas recently considered as part of large-scale neural circuitries underlying bipolar disorders

MDD patients, albeit to a lesser extent. Therefore, it is possible that structural abnormalities observed in BD are the expression of a "common vulnerability" for mental disease that involves complex brain circuitries. Accordingly, single regions of these circuitries could be differently involved in different mental disorders, rather than being specific target of specific diseases.

The above inferences aside, there are still some limitations of neuroimaging studies in BD.

As repeatedly pointed out, contrasting results might be due to several reasons, including methodological differences in sample selection and numerosity, diagnostic instruments, volumetric assessment, statistical data analysis, and so on. In addition, it seems important to highlight the influence that different clinical features displayed by patients might be a confounding factor when observing the results. Indeed, many studies do not even differentiate between BDI and BDII in their selected sample, and it cannot be ruled out that various clinical manifestations of the same phase, depressive or manic, could be underlied by different neurobiological alterations. The role of medications is another important factor to be taken into account when studying the modifications in brain morphology, since many of the drugs most commonly used in the treatment of BD may differently affect size and shape of different brain structures, in particular when administered long term. This aspect is of particular relevance if we consider that most of the available data on brain structural modifications in BD are obtained by cross-sectional studies which, given to intrinsic methodological limitations, cannot be of any use in clarifying whether the described alterations are a cause or a consequence of the disease. Additionally, in such studies the impact of medications in determining the variations is impossible to be quantified.

In conclusion, our chapter suggests that it is necessary to consider large-scale neural circuitries more than specific brain structures to understand neurobiological mechanisms underlying BD. Moreover, our review points up the need for longitudinal studies that might better investigate and finally cast light on these issues (Tables 1, 2, 3, 4 and 5).

Table 1
Macrostructural alterations in frontal and temporal regions of patients with bipolar disorders

Study	Sample	Comparison group	Technique	Areas analyzed	Major findings
Arnone et al. (2009)	1223 BD; 670 S; 29 SA	1940 HC	Random-effect meta-analysis of ROI-based studies	ROI	BD patients showed enlarged lateral ventricles and increased volumes of bilateral globus pallidus than HC; BD showed increased right amygdala volume than S
Atmaca et al. (2007)	30 BD	10 HC	Morphometry (ROI approach)	ROI (cingulate cortex)	Drug-free patients had significantly smaller left anterior cingulate cortex and left posterior CC volumes compared with patients treated with valproate and valproate plus quetiapine and HC. In addition, a trend toward significant difference was found between valproate plus quetiapine group and valproate group in regard to only left anterior CC
Bansal et al. (2013)	38 BDI	58 HC	Cortical thickness	Whole brain	BD showed decreased CT in the bilateral superior frontal and left parahippocampal regions and general brain-wide increase in CT including lateral frontal, temporal, lateral parietal, and lateral occipital regions than HC
Benedetti et al. (2011)	59 BD; depressed	–	VBM	Whole brain	Smaller GM volumes in dorsolateral prefrontal cortex, orbitofrontal cortex, anterior cingulate cortex, superior temporal cortex parieto-occipital cortex, and basal ganglia in BD with than without substance abuse. No effect of illness duration was found

(continued)

Table 1 (continued)

Study	Sample	Comparison group	Technique	Areas analyzed	Major findings
Bearden et al. (2007)	28 BD	28 HC	3D parametric mesh model	Whole brain	Lithium-treated bipolar patients had significantly greater overall gray matter volumes compared with HC although patients not on lithium did not differ from normal control subjects for left and right hemisphere, respectively; lobar gray matter volumes were significantly greater in lithium-treated bipolar patients versus HC in frontal, temporal, and parietal lobes, but this difference was not significant in the occipital lobe. Gray matter density was significantly greater in Li + bipolar patients for both whole brain and frontal and anterior cingulate ROIs. In non-lithium-treated patients, gray matter density did not differ from normal control subjects for both whole brain and frontal and anterior cingulate ROIs
Blumberg et al. (2006)	37 BD	56 HC	Morphometry (ROI approach)	ROI (VLPFC)	Smaller ventral PFC volume in rapid cycler than in non-rapid cycler BD. The volumes of the other temporal lobe structures did not differ significantly between the two groups. No significant differences were found for any anatomical measures between drug-free and lithium-treated bipolar patients bipolar type I and bipolar type II subjects or depressed and euthymic bipolar individuals

Brambilla et al. (2003)	18 BDI; 6 BDII, 13 euthymic, 10 depressed, 1 hypomanic	36 HC	Morphometry (ROI approach)	ROI: hippocampus, amygdala, superior temporal gyrus	Bipolar patients had significantly larger left amygdala volumes compared with controls
Chen et al. (2004)	13 BDI; 3 BDII; 1 BD NOS	21 HC	Morphometry (ROI approach)	Amygdala, hippocampus, temporal gray matter, temporal lobe	There was a trend to smaller left amygdala volumes in BD versus HC. BD did not show significant differences in right or left hippocampus temporal lobe gray matter, temporal lobe, or right amygdala volumes compared with HC. Furthermore, there was a direct correlation between left amygdala volumes and age in BD, whereas in HC there was an inverse correlation
Ekman et al. (2010)	55 BDI	–	VBM	Whole brain	Smaller gray matter volume in the inferior frontal gyri of the dorsolateral prefrontal cortices correlated significantly to the lifetime number of manic episodes. No association between local gray matter volume and the lifetime number of depression episodes or illness duration was found
Ellison-Wright and Bullmore (2010)	2058 S, 366 BD	2131 HC (for S), 497 (for BD)	Meta-analysis of VBM studies using anatomical likelihood estimation approach	Whole brain	BD showed gray matter reductions in the anterior cingulate and bilateral insula as compared with HC
Elvsåshagen et al. (2013)	36 BDII	42 HC	Cortical thickness	Whole brain	BDII showed decreased CT in two clusters comprised of the left ventromedial prefrontal (BA 10/32), left subgenual anterior cingulate (BA 24/25), bilateral dorsal medial prefrontal (BA 8/9/10/32), and

(continued)

Table 1 (continued)

Study	Sample	Comparison group	Technique	Areas analyzed	Major findings
					bilateral dorsolateral prefrontal (BA 9/10/46), and one cluster comprised of the left superior, middle, and inferior temporal (BA 22/38, 21, 20) regions than HC. No significant correlations between CT and illness duration, mood state, or family history were found in BDII. Those BD taking medications has less severe CT deficits in prefrontal regions than those BD not taking medications
Ha et al. (2009)	3 BDI; 23 BDII; depressed	23 HC	VBM	Whole brain	No difference between BDI and BDII
Hajek et al. (2013)	50 unaffected and 36 affected relatives of BD	49 HC	VBM	Whole brain	Five groups, including the unaffected and affected relatives of BD probands from each center as well as participants early in the course of BD, showed larger right inferior frontal gyrus volumes than control subjects. The rIFG volume correlated negatively with illness duration and, relative to the controls, was smaller among BD individuals with long-term illness burden and minimal lifetime lithium exposure. Li-treated subjects had normal rIFG volumes despite substantial illness burden

Hegarty et al. (2012)	17 BDI; 19 ADHD;18 BDI + ADHD. Euthymic	31 HC	Cortical thickness	Whole brain	BD showed decreased cortical thickness (CT) in bilateral orbitofrontal (BA11); left ventrolateral prefrontal (BA 44) ($p = 0.05$); bilateral frontal pole (BA10); bilateral dorsomedial prefrontal (BA8, left BA 9); bilateral superior frontal (BA 6); left anterior cingulate (BA 24/32); bilateral superior, middle, and inferior temporal (BA 38, 21, 20); left angular (BA39); left supramarginal (BA40); and right occipital (BA 18/19) regions than HC. BD showed decreased CT in the left lateral orbitofrontal (BA 47) and a trend for an increased CT in the anterior cingulate (BA25) cortex than patients with BD + ADHD
Huang et al. (2011)	44 BD	–	VBM	Whole brain	Larger left caudate nucleus and left middle frontal gyrus volumes and smaller right posterior cingulate gyrus volumes in BD with late-onset mania than with early-onset mania. No effect of medication status; effects of mood symptoms severity and illness duration unknown
Hulshoff et al. (2012)	49 twin pairs with BD, 26 twin pairs with S	83 HC twin pairs	Cortical thickness	Whole brain	BD reported decreased CT in the right orbitofrontal, bilateral parahippocampal, and right medial occipital calcarine and increased CT in the left supramarginal, left fusiform, left precentral, and right postcentral regions than HC. No significant differences were found between BD and S

(continued)

Table 1 (continued)

Study	Sample	Comparison group	Technique	Areas analyzed	Major findings
Foland-Ross et al. (2013)	31 BDI	32 HC	Cortical thickness	Whole brain	BD showed decreased CT in the bilateral orbitofrontal (BA 11), bilateral dorsomedial prefrontal (BA 8, left BA 9), left ventrolateral prefrontal (BA 44) ($p = 0.038$), left frontal pole (BA 10), and left anterior cingulate (BA 24/32) regions than HC. In BD significant negative correlations between duration of illness and CT in the left medial prefrontal (BA 24, 43, 8); between interval, illness onset, and treatment onset and CT in the left paracingulate/medial prefrontal (BA 24, 32, 8); and between the number of depressive episodes and CT in the left orbitofrontal cortex (BA 8, 10, 11) and a positive correlation between the number of hospitalizations for mania and CT in the left subgenual anterior cingulate (BA 25) were found. Significant correlation of psychosis and decreased CT in the left ventrolateral (BA44) and left dorsolateral prefrontal cortex and left temporal pole were also reported

Fornito et al. (2008)	24 BDI	24 HC	Cortical thickness	ROI (ACC and paracingulate cortex (PaC))	BD showed decreased CT in the left rostral and right dorsal paracingulate and trend for decreased CT in the left dorsal paracingulate regions and trend for increased CT in the left rostral anterior cingulate than HC
Fornito et al. (2009)	26 BDI, manic	26 HC	Cortical thickness	ROI (ACC and paracingulate cortex (PaC))	Males with BD reported increased CT in the right subcallosal anterior ACC than HC
Frazier et al. (2005)	32 BDI, manic	15 HC	Morphometry (based on image parcellation)	ROI: neocortex	Relative to controls, the BD had significantly smaller bilateral parietal and left temporal lobes than HC. Analysis of parietal and temporal gyri showed significantly smaller volume in bilateral postcentral gyrus and in left superior temporal and fusiform gyri in BD as compared with HC. The parahippocampal gyri were bilaterally increased in BD as compared with HC. No association between total temporal volume and duration of illness was reported
Gutiérrez-Galve et al. (2012)	25 BDI; 11 BDII; depressed	–	Morphometry (ROI approach)	Frontal and temporal regions	No difference in frontal or temporal cortical thickness or area between BDI and BDII
Lan et al. (2009)	17 BD; 56 MDD	54 HC	Cortical thickness	Whole brain	BD showed decreased CT in the right caudal middle frontal, right posterior cingulate, bilateral inferior parietal, left superior parietal, and right supramarginal regions. BD showed decreased CT in the right caudal middle frontal, left inferior parietal, and right precuneus compared to

(continued)

Table 1 (continued)

Study	Sample	Comparison group	Technique	Areas analyzed	Major findings
					MDD. As compared with MDD, BD showed decreased CT in the left inferior parietal bilateral superior frontal, right superior parietal, right precuneus, right rostral middle frontal, and right fusiform regions. No significant correlations between CT and HAM-D scores or age of onset of illness for the right caudal middle frontal, left inferior parietal, or right precuneus regions were found in BD
Li et al. (2011)	24 BDI	36 HC	VBM	Whole brain	There was no difference in whole-brain gray matter volume between BDI and HC. Optimized vVBM showed that BD had smaller volumes in the left inferior parietal lobule, right superior temporal gyrus, right middle frontal gyrus, and left caudate as compared with HC. Only the volume of the right middle frontal gyrus was correlated with duration of illness and number of episodes in patients
Lisy et al. (2011)	58 BDI	48 HC	VBM	Whole brain	At baseline, BD showed reduced gray matter volume in portions of the frontal cortex including left medial and middle frontal gyri [Brodmann area (BA 6)], left insula (BA 13), and left precentral gyrus (BA 6) than HC. Gray matter volume was also

					reduced in temporal structures including the right superior temporal gyrus (BA 13), bilateral inferior temporal gyrus (BA 37/47), and right amygdala as well as the right precuneus (BA 7). BD showed evidence of increasing gray matter volume over time in the right medial temporal gyrus (BA 21), left inferior and superior temporal gyri (BA 20/38), right VLPFC (BA 10) and parahippocampal gyrus (BA 28), and right caudate. There were no areas of significantly decreased gray matter volume. Conversely, healthy subjects showed significantly decreased subcortical gray matter bilaterally, without any areas of significantly increased volume. When compared with healthy subjects, bipolar disorder subjects showed significantly greater increases in gray matter volumes in several brain regions, specifically in the left superior temporal gyrus
Lyoo et al. (2006)	25 BD	21 HC	Cortical thickness	Whole brain	BD showed decreased CT in the left dorsolateral prefrontal (BA 46), right orbitofrontal (BA10), left anterior (BA 24/32) and posterior (BA 23) cingulate, left middle occipital (BA 18), right angular (BA39), right fusiform (BA 19), and bilateral postcentral (BA 3/4) regions than HC. No significant differences were found between BDI and BDII. Correlation of decreased CT in the left middle frontal cortex and right post central cortex with increased duration of illness in BD

(continued)

Table 1
(continued)

Study	Sample	Comparison group	Technique	Areas analyzed	Major findings
Makris et al. (2012)	18 BD; 31 BD + ADHD	23 HC	Cortical thickness	ROI (32 ROIs); whole brain	BD showed increased CT in the left posterior cingulate (BA 23), right middle temporal, right angular (BA 39), bilateral fusiform(BA 37), right posterior insula, and right lateral occipital (BA 18/19) regions and decreased CT in the right prefrontal (BA 9/10/11) region than HC. No difference were found between BDn and BD+ ADHD
Maller et al. (2014)	25 BDI; 14 BDII, depressed	31 HC	Cortical thickness	Whole brain	BD showed decreased CT in the right superior frontal, left superior temporal ($p = 0.016$), left superior ($p = 1.0e{-}4$) and inferior parietal, right supramarginal, right precuneus, and right pars opercularis regions than HC. BDI showed decreased CT in the right isthmus cingulate and left inferior parietal and increased CT in the left pars triangularis regions than HC. Decreased CT in the right superior frontal, right superior temporal, right transverse temporal, and right precentral regions were found in BDII as compared with HC. BDI showed decreased CT in the right medial orbitofrontal and left superior temporal regions than BDII). Negative correlations between CT and

					HAM-D scores in the right superior frontal and right superior temporal were found in the BD group
Matsuo et al. (2010)	20 BD	27 HC	Morphometry (ROI approach)	ROI: CC	Anterior CC genu area inversely correlated with impulsivity in BD with SA history
McDonald et al. (2004)	404 BD	HC	Random-effect meta-analysis of ROI-based studies	ROI	BD patients had enlargement of the right lateral ventricle than HC
Qiu et al. (2008)	20 HC; 20 BD	20 HC	Cortical thickness	Whole brain	BD showed decreased CT in the left lingual and increased CT in the right middle temporal cortex than HC. BD showed increased CT in left inferior frontal, right middle and superior temporal, and left cuneus regions than S. Correlations of decreased CT in the middle temporal gyrus with lower GAF was found in BD
Ratnanather et al. (2014)	36 BDI; 31 S	27 HC	Morphometry (ROI approach)	ROI: superior temporal sulcus, planum temporale	S showed decreased CT in the left planum temporale than BDI and HC. BDI showed decreased CT in the left planum temporale than HC
Rimol et al. (2010)	139 BD, 173 S	207 HC	Cortical thickness	Whole brain	No significant differences were found between BD and HC and between BD and S. Decreased CT in the left orbitofrontal, right superior frontal, left posterior superior temporal, right inferior temporal, right parahippocampal, bilateral inferior parietal, right superior parietal, and right supramarginal regions was found in BDI as compared with HC. No significant correlations between CT

(continued)

Table 1 (continued)

Study	Sample	Comparison group	Technique	Areas analyzed	Major findings
					and duration of illness or symptom severity were found in BD
Sassi et al. (2004)	27 BD	39 HC	Morphometry (ROI approach)	ROI: CC	Untreated BD had decreased left anterior cingulate volumes compared with healthy control subjects, respectively, and compared with lithium-treated patients. The cingulate volumes in lithium-treated patients were not significantly different from those of HC subjects. Lithium-treated patients also presented significant inverse correlation between age at onset of illness and right anterior cingulate volumes. No significant correlations were found between age, age at onset of illness, or length of illness and cingulate volumes among untreated patients. No other significant correlations between age and any other cingulate measurements were found in patients or HC. Moreover, we did not find any significant correlation between number of previous affective episodes
Sarnicola et al. (2009)	71 BDI	82 HC	VBM	Whole brain	In BD negative correlations between age and gray matter were observed in the medial frontal gyrus and insula on the

					right and superior temporal gyrus and cerebellum on the left. In HC, negative correlations between age and gray matter volumes were only found in the cerebellum bilaterally. There was no differential effect of age on the two diagnostic groups with respect to cognitive task performance
Selvaraj et al. (2012)	239 BD	281 HC	Meta-analysis of VBM studies	Whole brain	BD patients had reduced gray matter in the right ventral prefrontal cortex, insula, temporal cortex, and claustrum than HC
Strakowski et al. (1999)	24 BDI, manic	22 HC	Morphometry (ROI approach)	ROI (cerebrum, prefrontal lobe, striatum, globus pallidus, hippocampus, amygdala)	BD demonstrated a significant overall difference in structural volumes in these regions compared with HC. In particular, the amygdala was enlarged in BD. Brain structural volumes were not significantly associated with duration of illness, prior medication exposure, number of previous hospital admissions, or duration of substance abuse. Separating patients into first-episode and multiple-episode subgroups revealed no significant differences in any structure
Wilke et al. (2004)	10 BDI; mixed, manic	52 HC	VBM	Whole brain	BD showed localized gray matter deficits in the medial temporal lobe, the orbitofrontal cortex, and the anterior cingulate and greater bilateral basal ganglia volumes than HC

ACC anterior cingulate cortex, *ADHD* attention deficit and hyperactivity disorder, *BD* bipolar disorder, *CC* cingulate cortex, *CT* cortical thickness, *HC* healthy controls, *MDD* major depressive disorder, *SA* schizoaffective disorder, *S* schizophrenia, *PaC* paracingulate cortex, *VLPFC* ventrolateral prefrontal cortex, *VBM* voxel-brain morphometry

Table 2
Macrostructural alterations in parietal regions of patients with bipolar disorders

Study	Sample	Comparison group	Technique	Area analyzed	Major findings
Nenadic et al. (2015)	17 psychotic BDI, 34 SZ	34 HC	VBM	Whole brain	Comparison of the BD group with HC for GM VBM analyses ($p < 0{,}001$ uncorrected) showed two significant clusters in the right supramarginal gyrus and the left parieto-occipital cortex
Watson et al. (2012)	24 BDI, 25 SZ	88 HC	VBM	Whole brain	Compared to HC, BD showed a GM reduction in the right precuneus and a WM reduction in the left inferior parietal lobe
Li et al. (2011)	24 BDI	36 HC	VBM	Whole brain	Compared to HC, BDI showed a GM reduction in the left inferior parietal lobule
Cui et al. (2011)	23 SZ, 24 BDI	36 HC	VBM	Whole brain	Compared to HC, SZ showed reduced GM volume in the right inferior parietal lobule and BDI in the left inferior parietal lobule
Adler et al. (2007)	33 BDI	33 HC	VBM	Whole brain	Compared to HC, BDI showed increased GM volume in the paracentral lobule and increased GM density in the inferior parietal lobule, superior parietal lobule, and precuneus
Adler et al. (2005)	32 BDI	27 HC	VBM	Whole brain	Compared to HC, BD showed increased GM volume in the supplemental motor cortex, bilaterally, and decreased GM volume in the right superior parietal lobule
Rimol et al. (2010)	139 BD, 173 SZ	207 HC	FreeSurfer	Whole brain	The BDI group showed cortical thinning in the inferior parietal gyrus, the superior parietal gyrus, and the right supramarginal gyrus
Lyoo et al. (2006)	25 BD	21 HC	FreeSurfer	Whole brain	BD patients show cortical thinning in primary sensory regions in the parietal lobe
Lan et al. (2014)	18 BD, 56 MDD	54 HC	FreeSurfer	Whole brain	Regions thinner in BD patients compared to HC: left and right inferior parietal, left superior parietal, and right supramarginal. Regions located in the left inferior parietal and the right precuneus are thinner in BD patients compared to MDD
Nugent et al. (2006)	36 BD	65 HC	VBM	Whole brain	Reduction of GM in the bilateral inferior parietal lobule in treatment-naive BD patients compared to HC

BD bipolar disorder, *HC* healthy controls, *MDD* major depressive disorder, *SA* schizoaffective disorder, *VBM* voxel-brain morphometry

Table 3
Macrostructural alterations in cerebellum of patients with bipolar disorders

Study	Sample	Comparison group	Technique	Area analyzed	Major findings
Kim et al. (2013)	49 BD (29 BDI and 19 BDII, 24 medication-naïve and 25 medication-treated)	50 HC	Voxel-based morphometry (VBM)	ROI (cerebellum)	Compared to HC, BD showed a greater reduction in gray matter density of the posterior cerebellar regions, including the bilateral vermis and the right crus. Positive correlations of longer duration of illness with bilateral vermal gray matter deficits were observed only in medication-naive BD
Baldaçara et al. (2011)	40 BDI (20 with and 20 without history of suicide attempt)	22 HC	VBM	ROI (cerebellum)	Compared to HC, BD showed a greater reduction in gray matter density of the left cerebellum, right cerebellum, and vermis. No volumetric differences between the BD with and without suicidal attempt
Mills et al. (2005)	39 BD (18 with first-episode bipolar disorder and 21 with multiple-episode)	32 HC	VBM	ROI (cerebellum)	Vermal subregion V2 volume was significantly smaller in multiple-episode BD than in first-episode patients and HC. Vermal subregion V3 was significantly smaller in multiple-episode BD than in healthy subjects
Eker et al. (2014)	28 BDI and 28 healthy siblings	30 HC	VBM	Whole-brain and additional ROI analyses (cerebellum)	Compared to HC, BD and their healthy siblings presented volume deficits in the right cerebellum
Monkul et al. (2008)	16 young BD (12 BDI and 4BDII)	21 HC	VBM	ROI (cerebellum)	The number of previous affective episodes and vermis area V2 were inversely correlated in the male BD group. There was a trend to smaller vermis V2 areas in BD than in HC

(continued)

Table 3 (continued)

Study	Sample	Comparison group	Technique	Area analyzed	Major findings
Redlich et al. (2014)	58 BDI depressed, 58 unipolar depressed patients	58 HC	VBM	Whole brain	Reductions in white matter volume within the cerebellum and hippocampus were found in BD compared to HC
Sani et al. (2016)	78 BD (49 BDI and 29 BDII)	78 HC	VBM	Whole brain	A bilateral reduction of cerebellar volume, not related to the course of BD, was found in BD
McDonald et al. (2006)	37 BDI with psychotic symptoms, 25 SZ	52 HC	VBM	Whole brain	Individuals with SZ presented gray matter deficits in the cerebellum. BD had no significant regions of gray matter abnormality
Yüksell et al. (2012)	28,BDI with psychotic symptoms, 58 SZ	43 HC	VBM	Whole brain	GM volume was increased in the right posterior cerebellum in SZ compared to HC. BDI did not show significant GM deficits compared to HC or SZ
Adler et al. (2007)	First-episode 33 BDI	33 HC	VBM	Whole brain	Increased volume in cerebellum bilaterally was observed in BD
Moorhead et al. (2007)	21 BDI	21 HC	VBM	Whole brain	Patients with BD showed a longitudinal decline in cerebellar gray matter density over 4 years than HC
Del Bello et al. (1999)	30 BDI (16 with a first manic episode and 14 with prior manic episodes)	15 HC	VBM	ROI (cerebellum)	V3 area was significantly smaller in multiple-episode BDI than in first-episode BDI or HC

BD bipolar disorder, *HC* healthy controls, *MDD* major depressive disorder, *SA* schizoaffective disorder, *VBM* voxel-brain morphometry

Table 4
Macrostructural alterations in the deep gray matter of patients with bipolar disorders

Study	Sample	Comparison group	Technique	Area analyzed	Major findings
Rimol et al. (2014)	173 SZ, 139 BD (87 BDI and 52 BDII)	HC 207	Automatic segmentation (FreeSurfer)	Cortical thickness, subcortical volumes	Patient groups showed subcortical volume reductions bilaterally in the hippocampus, the left thalamus, the right nucleus accumbens, the left cerebellar cortex, and the brainstem, along with substantial ventricular enlargements
Sacchet et al. (2015)	40 BDI, 57 MDD, 35 remitted MDD	61 HC	Automatic segmentation (FreeSurfer)	Subcortical volumes	For the caudate, both the BD and the MDD participants had smaller volumes than did the HC
Womer et al. (2014)	33PBDI (12 without history of psychotic features,21 with history of psychotic features) 32 SZ	27 HC	Automatic segmentation (FreeSurfer) (volumes and shape)	Basal ganglia and thalamus	Significant volume differences were found in the caudate and globus pallidus, with volumes smallest in the BD group without history of psychotic features. Shape abnormalities showing inward deformation of superior regions of the caudate were observed in BD compared with HC. Shape differences were also found in the globus pallidus and putamen when comparing the BD and SZ groups
MacMaster et al. (2014)	32 unipolar depressed subjects,14 BD	22 HC	VBM	ROI (hippocampus, dorsolateral prefrontal cortex, anterior cingulate cortex, caudate, putamen and thalamus)	BD displayed reduced left hippocampal and right/left putamen volumes compared to HC

(continued)

Table 4 (continued)

Study	Sample	Comparison group	Technique	Area analyzed	Major findings
DelBello et al. (2004)	23 BDI adolescents	20 HC	VBM	Whole brain and ROI (amygdala, globus pallidus, caudate, putamen, thalamus)	Adolescents with BD exhibited smaller amygdala and enlarged putamen compared with HC
Hibar et al. (2016)	1.710 BD (1.394 PBDI and 361 BDII)	2.594 HC	Automatic segmentation (FreeSurfer)	Subcortical volumes	BD, compared to HC, showed volumetric reductions in the bilateral hippocampus and thalamus and a trending significant reduction in the amygdala. No differences between BDI and BDII. BD taking lithium exhibited larger thalamic volumes compared with BD not taking lithium
Hauvik et al. (2015)	210 SZ, 192 BD (117 BDI and 75 BDII)	HC 300	Automatic segmentation (FreeSurfer)	Hippocampal subfields	Volume differences between groups were found in hippocampal subfields cornu ammonis (CA)2/3, CA4/dentate gyrus, presubiculum, and subiculum bilaterally and right CA1; patients with SZ or BD had smaller volumes than HC subjects for all subfields except the presubiculum, where only patients with SZ had smaller volumes

Mamah et al. (2016)	52 SZ, 12 schizotypal personality disorder, 49 psychotic BD 24 and nonpsychotic BD	40 HC	Automatic segmentation (FreeSurfer) (volumes and shape)	Hippocampus, amygdala, caudate, nucleus accumbens, putamen, globus pallidus and thalamus	Inward deformation was present in the posterior thalamus in SZ and BD compared to HC
Quigley et al. (2015)	60 BDI	60 HC	Automatic segmentation (FreeSurfer) (volumes and shape)	Subcortical volumes	BDI displayed significantly smaller left hippocampal volumes and significantly larger left lateral ventricle volumes compared with HC. Shape analysis revealed an area of contraction in the anterior head and medial border of the left hippocampus, as well as expansion in the right hippocampal tail medially, in patients compared with HC
Hwang et al. (2006)	49 BD (35 BDI and 14 BDII)	37 HC	Automatic segmentation (volumes and shape)	Caudate and putamen	Shape differences, more prominent for the right side, were found for drug-naive BD, relative to HC, but not for drug-treated BD

BD bipolar disorder, *HC* healthy controls, *MDD* major depressive disorder, *SA* schizoaffective disorder, *SZ* schizophrenia, *VBM* voxel-brain morphometry

Table 5
Microstructural brain alterations of patients with bipolar disorders

Study	Sample	Comparison group	Technique	Areas analyzed	Major findings
Ajilore et al. (2015)	24 BDI euthymic	23 HC	DTI	ROI (corpus callosum (CC))	Positive correlation was found between FA values in the body, genu, and splenium of corpus callosum and processing speed ability
Ambrosi et al. (2016)	25 BDI, 25 BDII euthymic	50 HC	DTI	Whole brain	Compared to BDI and HC, BDII showed decreased FA in the right inferior longitudinal fasciculus (ILF) with unchanged AD and RD. Compared to HC, BDI and BDII showed decreased AD and RD in the left internal capsule, lower AD in the left ILF, the right corticospinal tract, and bilateral cerebellum
Benedetti et al. (2011)	15 MDD, 15 BDI euthymic	21 HC	DTI	ROI (bilateral amygdala, orbitofrontal cortex, subgenual cingulate cortex, supragenual cingulate cortex, lateral prefrontal cortex, insula)	Compared to HC and MDD, BD had significantly decreased FA and increased RD and AD in the majority of WM fiber bundles connecting structures of the anterior limbic network including the uncinate fasciculus, cingulum, and prefrontal cortico-cortical WM. Lithium was associated with normal diffusivity values in tracts connecting the amygdala with the subgenual cingulate cortex

Benedetti et al. (2013)	70 BDI depressed	–	DTI	Whole brain	Long-term exposure to lithium was associated with increases in AD in several WM tracts including CC, forceps major, anterior and posterior cingulum bundle, left superior longitudinal fasciculus (SLF), left ILF, left posterior thalamic radiation, bilateral superior and posterior corona radiata, and bilateral corticospinal tract
Caseras et al. (2015)	17 BDI, 15 BDII euthymic	–	DTI	ROI (uncinate fasciculus, corticospinal tract)	Compared to BDII, BDI showed reduced FA in the right uncinate fasciculus associated with increased RD
Chan et al. (2010)	16 BDI remitted first-episode mania	16 HC	DTI	Whole brain	Compared to HC, BD patients showed decreased FA and increased RD in the left anterior frontal WM, left insula, right posterior thalamic radiation, left posterior cingulum, and genu of corpus callosum
Ha et al. (2011)	12 BDI, 13 BDII euthymic	–	DTI	Whole brain	Compared to BDII, BDI showed lower FA in the right ILF
Liu et al. (2010)	14 BDI, 13 BDII euthymic	–	DTI	Whole brain	Compared to BDI, BDII showed lower FA in the right precuneus, right inferior frontal gyrus, and left inferior prefrontal areas
Lu et al. (2011)	13 BDI and 21 SZ at untreated first-episode psychosis	18 HC	DTI	Whole brain	Compared to HC, BD showed decreased FA in several WM tracts. Compared to SZ, BD showed significantly lower FA in the cingulum, internal capsule, posterior corpus callosum, ILF, and IFOF. Lower FA was associated with increased RD

(continued)

Table 5 (continued)

Study	Sample	Comparison group	Technique	Areas analyzed	Major findings
Macritchie et al. (2010)	20 BDI, 8 BDII euthymic	28 HC	DTI	ROI (genu, body, and splenium of CC)	Compared to HC, the BD showed reduced FA in all the regions of CC. Lithium-treated group showed increased FA in the body of CC compared to the non-lithium-treated group
Magioncalda et al. (2016)	21 BDI depressed, 20 BDI manic, 20 BDI euthymic	42 HC	DTI	Whole brain	Compared to HC, BD showed FA decrease and RD increase in the anterior thalamic radiation, cingulate gyrus, corticospinal tract, forceps major and minor, ILF, inferior fronto-occipital fasciculus (IFOF), SLF, and uncinate fasciculus. The global load of WM abnormalities was larger in depression, intermediate in mania, and smaller in euthymia. Considering the BD as whole sample, cognitive deficits at the continuous performance test were positively associated with FA value and negatively associated with RD value
Maller et al. (2014)	16 BDI, 15 BDII depressed	31 HC	DTI	Whole brain	No differences between BDI and BDII
McKenna et al. (2015)	26 BDI euthymic	36 HC	DTI	ROI (dorsolateral prefrontal cortex, supramarginal gyri, uncinate fasciculus and bilateral superior SLF)	For processing speed, the genu and splenium of CC and right SLF from DTI were significant predictors of cognitive

					performance selectively for BD. No effect of medication status was found
Nortje et al. (2013)	390 BD	354 HC	Voxel-based meta-analysis of DTI studies using anisotropic effect size-signed differential mapping (ES-SDM)	Whole brain	Compared to HC, BD showed decreased FA in the right ILF, IFOF, left posterior cingulate
Oertel-Knochel et al. (2015)	30 BDI euthymic	32 HC	DTI	Whole brain and ROI (anterior thalamic radiation, fornix, CC)	Compared to HC, BD showed lower FA in the right middle frontal gyrus, the left parahippocampal gyrus, the fornix, and the medial thalamus. The lower FA was associated with higher AD and RD in the fornix and higher RD in the corpus callosum. Positive correlation was found between problem-solving ability and RD in the fornix and in the right thalamic radiation. No effect of medication status was found
Poletti et al. (2015)	78 BDI depressed	–	DTI	Whole brain	Attention and information processing were positively associated with FA in several WM tracts with signal peak in forceps major and anterior thalamic radiation. Working memory was positively associated with AD in several WM tracts with signal peak in bilateral superior corona radiata and left ILF. Executive functions were negatively associated with RD in several WM tracts with signal peak in the

(continued)

Table 5 (continued)

Study	Sample	Comparison group	Technique	Areas analyzed	Major findings
					left ILF, right posterior thalamic radiation, and left corona radiata
Wise et al. (2016)	536 BD	489 HC	Voxel-based meta-analysis of DTI studies using anisotropic effect size-signed differential mapping (ES-SDM)	Whole brain	Compared to HC, BD showed decreased FA in the left posterior cingulum, the right anterior SLF, and the left genu of CC
Zanetti et al. (2009)	16 BDI depressed, 21 BDI euthymic	26 HC	DTI	Whole brain	Compared to HC, overall BD showed decreased FA in the external capsule bilaterally, right SLF, and right ILF. Compared to euthymic BD, depressed BD showed reduced FA in the superior frontal WM (next to the dorsal cingulate gyrus), posterior limb of internal capsule, and right SLF

AD axial diffusivity, *BD* bipolar disorder, *CC* corpus callosum, *DTI* diffusion tensor imaging, *FA* fractional anisotropy, *HC* healthy controls, *IFOF* inferior fronto-occipital fasciculus, *ILF* inferior longitudinal fasciculus, *MDD* major depressive disorder, *RD* radial diffusivity, *SA* schizoaffective disorder, *SLF* superior longitudinal fasciculus, *SZ* schizophrenia, *VBM* voxel-brain morphometry

References

1. Judd LL, Akiskal HS (2003) The prevalence and disability of bipolar spectrum disorders in the US population: re-analysis of the ECA database taking into account subthreshold cases. J Affect Disord 73:123–131
2. Merikangas KR, Akiskal HS, Angst J, Greenberg PE, Hirschfeld RM, Petukhova M et al (2007) Lifetime and 12-month prevalence of bipolar spectrum disorder in the National Comorbidity Survey replication. Arch Gen Psychiatry 64:543–552
3. Koukopoulos A, Reginaldi D, Tondo L, Visioli C, Baldessarini RJ (2013) Course sequences in bipolar disorder: depressions preceding or following manias or hypomanias. J Affect Disord 151:105–110
4. Koukopoulos A, Sani G, Koukopoulos AE, Minnai GP, Girardi P, Pani L et al (2003) Duration and stability of the rapid-cycling course: a long-term personal follow-up of 109 patients. J Affect Disord 73:75–85
5. Colom F, Vieta E, Daban C, Pacchiarotti I, Sánchez-Moreno J (2006) Clinical and therapeutic implications of predominant polarity in bipolar disorder. J Affect Disord 93:13–17
6. Sani G, Napoletano F, Vöhringer PA, Sullivan M, Simonetti A, Koukopoulos A et al (2014) Mixed depression: clinical features and predictors of its onset associated with antidepressant use. Psychother Psychosom 83:213–221
7. Sani G, Chiapponi C, Piras F, Ambrosi E, Simonetti A, Danese E et al (2016) Gray and white matter trajectories in patients with bipolar disorder. Bipolar Disord 18:52–62
8. Fusar-Poli P, Howes O, Bechdolf A, Borgwardt S (2012) Mapping vulnerability to bipolar disorder: a systematic review and meta-analysis of neuroimaging studies. J Psychiatry Neurosci 37:170–184
9. Arnone D, Cavanagh J, Gerber D, Lawrie SM, Ebmeier KP, McIntosh AM (2009) Magnetic resonance imaging studies in bipolar disorder and schizophrenia: meta-analysis. Br J Psychiatry 195:194–201
10. McDonald C, Zanelli J, Rabe-Hesketh S, Ellison-Wright I, Sham P, Kalidindi S et al (2004) Meta-analysis of magnetic resonance imaging brain morphometry studies in bipolar disorder. Biol Psychiatry 56:411–417
11. Selvaraj S, Arnone D, Job D, Stanfield A, Farrow TF, Nugent AC et al (2012) Grey matter differences in bipolar disorder: a meta-analysis of voxel-based morphometry studies. Bipolar Disord 14:135–145
12. Ellison-Wright I, Bullmore E (2010) Anatomy of bipolar disorder and schizophrenia: a meta-analysis. Schizophr Res 117:1–12
13. Foland-Ross LC, Thompson PM, Sugar CA, Madsen SK, Shen JK, Penfold C et al (2011) Investigation of cortical thickness abnormalities in lithium-free adults with bipolar I disorder using cortical pattern matching. Am J Psychiatry 168:530–539
14. Fornito A, Malhi GS, Lagopoulos J, Ivanovski B, Wood SJ, Saling MM et al (2008) Anatomical abnormalities of the anterior cingulate and paracingulate cortex in patients with bipolar I disorder. Psychiatry Res 162:123–132
15. Fornito A, Yücel M, Wood SJ, Bechdolf A, Carter S, Adamson C et al (2009) Anterior cingulate cortex abnormalities associated with a first psychotic episode in bipolar disorder. Br J Psychiatry 194:426–433
16. Hegarty CE, Foland-Ross LC, Narr KL, Sugar CA, McGough JJ, Thompson PM et al (2012) ADHD comorbidity can matter when assessing cortical thickness abnormalities in patients with bipolar disorder. Bipolar Disord 14:843–855
17. Lyoo IK, Sung YH, Dager SR, Friedman SD, Lee JY, Kim SJ et al (2006) Regional cerebral cortical thinning in bipolar disorder. Bipolar Disord 8:65–74
18. Hulshoff Pol HE, van Baal GC, Schnack HG, Brans RG, van der Schot AC, Brouwer RM et al (2012) Overlapping and segregating structural brain abnormalities in twins with schizophrenia or bipolar disorder. Arch Gen Psychiatry 69:349–359
19. Rimol LM, Hartberg CB, Nesvåg R, Fennema-Notestine C, Hagler DJ Jr, Pung CJ et al (2010) Cortical thickness and subcortical volumes in schizophrenia and bipolar disorder. Biol Psychiatry 68:41–50
20. Elvsåshagen T, Westlye LT, Bøen E, Hol PK, Andreassen OA, Boye B et al (2013) Bipolar II disorder is associated with thinning of prefrontal and temporal cortices involved in affect regulation. Bipolar Disord 15:855–864
21. Lan MJ, Chhetry BT, Oquendo MA, Sublette ME, Sullivan G, Mann JJ et al (2014) Cortical thickness differences between bipolar depression and major depressive disorder. Bipolar Disord 16:378–388
22. Makris N, Seidman LJ, Brown A, Valera EM, Kaiser JR, Petty CR et al (2012) Further understanding of the comorbidity between attention-deficit/hyperactivity disorder and

bipolar disorder in adults: an MRI study of cortical thickness. Psychiatry Res 202:1–11

23. Bansal R, Hao X, Liu F, Xu D, Liu J, Peterson BS (2013) The effects of changing water content, relaxation times, and tissue contrast on tissue segmentation and measures of cortical anatomy in MR images. Magn Reson Imaging 31:1709–1730
24. Maller JJ, P T, RH T, S MQ, PB F (2014) Volumetric, cortical thickness and white matter integrity alterations in bipolar disorder type I and II. J Affect Disord 169:118–127
25. Blumberg HP, Krystal JH, Bansal R, Martin A, Dziura J, Durkin K et al (2006) Age, rapid-cycling, and pharmacotherapy effects on ventral prefrontal cortex in bipolar disorder: a cross-sectional study. Biol Psychiatry 59:611–618
26. Matsuo K, Nielsen N, Nicoletti MA, Hatch JP, Monkul ES, Watanabe Y et al (2010) Anterior genu corpus callosum and impulsivity in suicidal patients with bipolar disorder. Neurosci Lett 469:75–80
27. Ekman CJ, Lind J, Rydén E, Ingvar M, Landén M (2010) Manic episodes are associated with grey matter volume reduction - a voxel-based morphometry brain analysis. Acta Psychiatr Scand 122:507–515
28. Li M, Cui L, Deng W, Ma X, Huang C, Jiang L et al (2011) Voxel-based morphometric analysis on the volume of gray matter in bipolar I disorder. Psychiatry Res 191:92–97
29. Sassi RB, Brambilla P, Hatch JP, Nicoletti MA, Mallinger AG, Frank E et al (2004) Reduced left anterior cingulate volumes in untreated bipolar patients. Biol Psychiatry 56:467–475
30. Strakowski SM, DelBello MP, Sax KW, Zimmerman ME, Shear PK, Hawkins JM et al (1999) Brain magnetic resonance imaging of structural abnormalities in bipolar disorder. Arch Gen Psychiatry 56:254–260
31. Hajek T, Cullis J, Novak T, Kopecek M, Blagdon R, Propper L et al (2013) Brain structural signature of familial predisposition for bipolar disorder: replicable evidence for involvement of the right inferior frontal gyrus. Biol Psychiatry 73:144–152
32. Nery FG, Monkul ES, Lafer B (2013) Gray matter abnormalities as brain structural vulnerability factors for bipolar disorder: a review of neuroimaging studies of individuals at high genetic risk for bipolar disorder. Aust N Z J Psychiatry 47:1124–1135
33. Emsell L, McDonald C (2009) The structural neuroimaging of bipolar disorder. Int Rev Psychiatry 21:297–313
34. Bearden CE, Thompson PM, Dutton RA, Frey BN, Peluso MA, Nicoletti M et al (2008) Three-dimensional mapping of hippocampal anatomy in unmedicated and lithium-treated patients with bipolar disorder. Neuropsychopharmacology 33:1229–1238
35. Manji HK, Moore GJ, Chen G (2000) Clinical and preclinical evidence for the neurotrophic effects of mood stabilizers: implications for the pathophysiology and treatment of manic-depressive illness. Biol Psychiatry 48:740–754
36. Atmaca M, Ozdemir H, Cetinkaya S, Parmaksiz S, Belli H, Poyraz AK et al (2007) Cingulate gyrus volumetry in drug free bipolar patients and patients treated with valproate or valproate and quetiapine. J Psychiatr Res 41:821–827
37. Phillips ML, Ladouceur CD, Drevets WC (2008) A neural model of voluntary and automatic emotion regulation: implications for understanding the pathophysiology and neurodevelopment of bipolar disorder. Mol Psychiatry 13(829):833–857
38. Phillips ML, Swartz HA (2014) A critical appraisal of neuroimaging studies of bipolar disorder: toward a new conceptualization of underlying neural circuitry and a road map for future research. Am J Psychiatry 171:829–843
39. Strakowski SM, Delbello MP, Adler CM (2005) The functional neuroanatomy of bipolar disorder: a review of neuroimaging findings. Mol Psychiatry 10:105–116
40. Schneider MR, DelBello MP, McNamara RK, Strakowski SM, Adler CM (2012) Neuroprogression in bipolar disorder. Bipolar Disord 14:356–374
41. Qiu A, Vaillant M, Barta P, Ratnanather JT, Miller MI (2008) Region-of-interest-based analysis with application of cortical thickness variation of left planum temporale in schizophrenia and psychotic bipolar disorder. Hum Brain Mapp 29:973–985
42. Ratnanather JT, Cebron S, Ceyhan E, Postell E, Pisano DV, Poynton CB et al (2014) Morphometric differences in planum temporale in schizophrenia and bipolar disorder revealed by statistical analysis of labeled cortical depth maps. Front Psychol 5:94
43. Brambilla P, Harenski K, Nicoletti M, Sassi RB, Mallinger AG, Frank E et al (2003) MRI investigation of temporal lobe structures in bipolar patients. J Psychiatr Res 37:287–295
44. Sarnicola A, Kempton M, Germanà C, Haldane M, Hadjulis M, Christodoulou T

et al (2009) No differential effect of age on brain matter volume and cognition in bipolar patients and healthy individuals. Bipolar Disord 11:316–322

45. Lisy ME, Jarvis KB, DelBello MP, Mills NP, Weber WA, Fleck D et al (2011) Progressive neurostructural changes in adolescent and adult patients with bipolar disorder. Bipolar Disord 13:396–405
46. Benedetti F, Radaelli D, Poletti S, Locatelli C, Falini A, Colombo C et al (2011) Opposite effects of suicidality and lithium on gray matter volumes in bipolar depression. J Affect Disord 135:139–147
47. Gutiérrez-Galve L, Bruno S, Wheeler-Kingshott CA, Summers M, Cipolotti L, Ron MA (2012) IQ and the fronto-temporal cortex in bipolar disorder. J Int Neuropsychol Soc 18:370–374
48. Ha TH, Ha K, Kim JH, Choi JE (2009) Regional brain gray matter abnormalities in patients with bipolar II disorder: a comparison study with bipolar I patients and healthy controls. Neurosci Lett 456:44–48
49. Huang SH, Tsai SY, Hsu JL, Huang YL (2011) Volumetric reduction in various cortical regions of elderly patients with early-onset and late-onset mania. Int Psychogeriatr 23:149–154
50. Chen HH, Nicoletti MA, Hatch JP, Sassi RB, Axelson D, Brambilla P et al (2004) Abnormal left superior temporal gyrus volumes in children and adolescents with bipolar disorder: a magnetic resonance imaging study. Neurosci Lett 363:65–68
51. Frazier JA, Breeze JL, Makris N, Giuliano AS, Herbert MR, Seidman L et al (2005) Cortical gray matter differences identified by structural magnetic resonance imaging in pediatric bipolar disorder. Bipolar Disord 7:555–569
52. F W, Kalmar JH, Womer FY, Edmiston EE, Chepenik LG, Chen R et al (2011) Olfactocentric paralimbic cortex morphology in adolescents with bipolar disorder. Brain 134 (Pt 7):2005–2012
53. Wilke M, Kowatch RA, DelBello MP, Mills NP, Holland SK (2004) Voxel-based morphometry in adolescents with bipolar disorder: first results. Psychiatry Res 131:57–69
54. Takahashi T, Malhi GS, Wood SJ, Yücel M, Walterfang M, Kawasaki Y et al (2010) Gray matter reduction of the superior temporal gyrus in patients with established bipolar I disorder. J Affect Disord 123:276–282
55. Aalto S, Näätänen P, Wallius E, Metsähonkala L, Stenman H, Niem PM et al (2002) Neuroanatomical substrata of amusement and sadness: a PET activation study using film stimuli. Neuroreport 13:67–73
56. Allison T, Puce A, McCarthy G (2000) Social perception from visual cues: role of the STS region. Trends Cogn Sci 4:267–278
57. Chee MW, Soon CS, Lee HL, Pallier C (2004) Left insula activation: a marker for language attainment in bilinguals. Proc Natl Acad Sci U S A 101:15265–15270
58. Ramautar JR, Slagter HA, Kok A, Ridderinkhof KR (2006) Probability effects in the stop-signal paradigm: the insula and the significance of failed inhibition. Brain Res 1105:143–154
59. Malhi GS, Lagopoulos J, Sachdev PS, Ivanovski B, Shnier R, Ketter T (2007) Is a lack of disgust something to fear? A functional magnetic resonance imaging facial emotion recognition study in euthymic bipolar disorder patients. Bipolar Disord 9:345–357
60. Phillips ML, Young AW, Senior C, Brammer M, Andrew C, Calder AJ et al (1997) A specific neural substrate for perceiving facial expressions of disgust. Nature 389:495–498
61. McIntosh AM, Whalley HC, McKirdy J, Hall J, Sussmann JE, Shankar P et al (2008) Prefrontal function and activation in bipolar disorder and schizophrenia. Am J Psychiatry 165:378–384
62. Gogtay N, Giedd JN, Lusk L, Hayashi KM, Greenstein D, Vaituzis AC et al (2004) Dynamic mapping of human cortical development during childhood through early adulthood. Proc Natl Acad Sci U S A 101:8174–8179
63. Teixeira S, Machado S, Velasques B, Sanfim A, Minc D, Peressutti C et al (2014) Integrative parietal cortex processes: neurological and psychiatric aspects. J Neurol Sci 338:12–22
64. Watson DR, Anderson JME, Bai F, Barrett SL, McGinnity TM, Mulholland CC et al (2012) A voxel based morphometry study investigating brain structural changes in first episode psychosis. Behav Brain Res 227:91–99
65. Cui L, Li M, Deng W, Guo W, Ma X, Huang C et al (2011) Overlapping clusters of grey matter deficits in paranoid schizophrenia and psychotic bipolar mania with family history. Neurosci Lett 489:94–98
66. Adler CM, DelBello MP, Jarvis K, Levine A, Adams J, Strakowski SM (2007) Voxel-based studyof structural changes in first-episode patients with bipolar disorder. Biol Psychiatry 61:776–781

67. Nenadic I, Maitra R, Langbein K, Dietzek M, Lorenz C, Smesny S et al (2015) Brain structure in schizophrenia vs psychotic bipolar I disorder: a VBM study. Schizophr Res 165:212–219
68. Nugent AC, Milham MP, Bain EE, Mah L, Cannon DM, Marrett S et al (2006) Cortical abnormalities in bipolar disorder investigated with MRI and voxel-based morphometry. NeuroImage 30:485–497
69. Asanuma C, Andersen RA, Cowan WM (1985) The thalamic relations of the caudal inferior parietal lobule and the lateral prefrontal cortex in monkeys: divergent cortical projections from cell clusters in the medial pulvinar nucleus. J Comp Neurol 241:357–381
70. Adler CM, Levine AD, DelBello MP, Strakowski SM (2005) Changes in gray matter volume in patients with bipolar disorder. Biol Psychiatry 58:151–157
71. Negash A, Kebede D, Alem A, Melaku Z, Deywssa N, Shibire T et al (2004) Neurological soft signs in bipolar disorder. J Affect Disord 80:221–230
72. Abé C, Ekman C-J, Sellgren C, Petrovic P, Ingvar M, Landén M (2015) Cortical thickness, volume and surface area in patients with bipolar disorder types I and II. J Psychiatry Neurosci 41:150093
73. James A, Hough M, James S, Burge L, Winmill L, Nijhawan S et al (2011) Structural brain and neuropsychometric changes associated with pediatric bipolar disorder with psychosis. Bipolar Disord 13:16–27
74. O'Bryan RA, Brenner CA, Hetrick WP, O'Donnell BF (2014) Disturbances of visual motion perception in bipolar disorder. Bipolar Disord 16:354–365
75. Allen DN, Randall C, Bello D, Armstrong C, Frantom L, Cross C et al (2010) Are working memory deficits in bipolar disorder markers for psychosis? Neuropsychology 24:244–254
76. Roff Hilton EJ, Hosking SL, Betts T (2004) The effect of antiepileptic drugs on visual performance. Seizure 13:113–128
77. Giakoumatos CI, Nanda P, Mathew IT, Tandon N, Shah J, Bishop JR et al (2015) Effects of lithium on cortical thickness and hippocampal subfield volumes in psychotic bipolar disorder. J Psychiatr Res 61:180–187
78. Simonetti A, Sani G, Dacquino C, Piras F, De Rossi P, Caltagirone C et al (2016) Hippocampal subfield volumes in short- and long-term lithium-treated patients with bipolar I disorder. Bipolar Disord 18(4):352–362
79. Song J, Han DH, Kim SM, Hong JS, Min KJ, Cheong JH et al (2015) Differences in gray matter volume corresponding to delusion and hallucination in patients with schizophrenia compared with patients who have bipolar disorder. Neuropsychiatr Dis Treat 11:1211–1219
80. Jarvis K, DelBello MP, Mills N, Elman I, Strakowski SM, Adler CM (2008) Neuroanatomic comparison of bipolar adolescents with and without cannabis use disorders. J Child Adolesc Psychopharmacol 18:557–563
81. Hanford LC, Hall GB, Minuzzi L, Sassi RB (2016) Gray matter volumes in symptomatic and asymptomatic offspring of parents diagnosed with bipolar disorder. Eur Child Adolesc Psychiatry 25(9):959–967
82. Bootsman F, Brouwer RM, Schnack HG, van Baal GCM, van der Schot AC, Vonk R et al (2015) Genetic and environmental influences on cortical surface area and cortical thickness in bipolar disorder. Psychol Med 45:193–204
83. Phillips JR, Hewedi DH, Eissa AM, Moustafa AA (2015) The cerebellum and psychiatric. Front Public Health 3:1–8
84. Schmahmann JD (2010) The role of the cerebellum in cognition and emotion: personal reflections since 1982 on the dysmetria of thought hypothesis, and its historical evolution from theory to therapy. Neuropsychol Rev 20:236–260
85. Konarski JZ, McIntyre RS, Grupp LA, Kennedy SH (2005) Is the cerebellum relevant in the circuitry of neuropsychiatric disorders? J Psychiatry Neurosci 30:178–186
86. Kim D, Cho HB, Dager SR, Yurgelun-Todd DA, Yoon S, Lee JH et al (2013) Posterior cerebellar vermal deficits in bipolar disorder. J Affect Disord 150:499–506
87. Monkul ES, Hatch JP, Sassi RB, Axelson D, Brambilla P, Nicoletti MA et al (2008) MRI study of the cerebellum in young bipolar patients. Prog Neuro-Psychopharmacology Biol Psychiatry 32:613–619
88. Redlich R, Almeida JRC, Grotegerd D, Opel N, Kugel H, Heindel W et al (2014) Brain morphometric biomarkers distinguishing unipolar and bipolar depression. A voxel-based morphometry-pattern classification approach. JAMA Psychiat 71:1222–1230
89. McDonald C, Marshall N, Sham PC, Bullmore ET, Schulze K, Chapple B et al (2006) Regional brain morphometry in patients with schizophrenia or bipolar disorder and their unaffected relatives. Am J Psychiatry 163:478–487

90. Yüksell C, McCarthy J, Shinn A, Pfaff DL, Baker JT, Heckers S et al (2012) Gray matter volume in schizophrenia and bipolar disorder with psychotic features. Schizophr Res 138:177–182
91. Moorhead TWJ, McKirdy J, Sussmann JED, Hall J, Lawrie SM, Johnstone EC et al (2007) Progressive gray matter loss in patients with bipolar disorder. Biol Psychiatry 62:894–900
92. DelBello MP, Strakowski SM, Zimmerman ME, Hawkins JM, Sax KW (1999) MRI analysis of the cerebellum in bipolar disorder: a pilot study. Neuropsychopharmacology 21:63–68
93. Sarıçiçek A, Yalın N, Hıdıroğlu C, Çavuşoğlu B, Taş C, Ceylan D et al (2015) Neuroanatomical correlates of genetic risk for bipolar disorder: a voxel-based morphometry study in bipolar type I patients and healthy first degree relatives. J Affect Disord 186:110–118
94. DelBello MP, Zimmerman ME, Mills NP, Getz GE, Strakowski SM (2004) Magnetic resonance imaging analysis of amygdala and other subcortical brain regions in adolescents with bipolar disorder. Bipolar Disord 6:43–52
95. Hibar DP, Westlye LT, van Erp TGM, Rasmussen J, Leonardo CD, Faskowitz J et al (2016) Subcortical volumetric abnormalities in bipolar disorder. Mol Psychiatry 21:1–7. https://doi.org/10.1038/mp.2015.227
96. Haukvik UK, Westlye LT, Mørch-Johnsen L, Jørgensen KN, Lange EH, Dale AM et al (2015) In vivo hippocampal subfield volumes in schizophrenia and bipolar disorder. Biol Psychiatry 77:581–588
97. Mamah D, Alpert KI, Barch DM, Csernansky JG, Wang L (2016) Subcortical neuromorphometry in schizophrenia spectrum and bipolar disorders. NeuroImage Clin 11:276–286
98. Quigley SJ, Scanlon C, Kilmartin L, Emsell L, Langan C, Hallahan B et al (2015) Volume and shape analysis of subcortical brain structures and ventricles in euthymic bipolar I disorder. Psychiatry Res Neuroimaging 233:324–330
99. Hwang J, In KL, Dager SR, Friedman SD, Jung SO, Jun YL et al (2006) Basal ganglia shape alterations in bipolar disorder. Am J Psychiatry 163:276–285
100. Nortje G, Stein DJ, Radua J, Mataix-Cols D, Horn N (2013) Systematic review and voxel-based meta-analysis of diffusion tensor imaging studies in bipolar disorder. J Affect Disord 150:192–200
101. Houenou J, Frommberger J, Carde S, Glasbrenner M, Diener C, Leboyer M et al (2011) Neuroimaging-based markers of bipolar disorder: evidence from two meta-analyses. J Affect Disord 132:344–355
102. Ambrosi E, Chiapponi C, Sani G, Manfredi G, Piras F, Caltagirone C et al (2016) White matter microstructural characteristics in bipolar I and bipolar II disorder: a diffusion tensor imaging study. J Affect Disord 189:176–183
103. Magioncalda P, Martino M, Conio B, Piaggio N, Teodorescu R, Escelsior A et al (2016) Patterns of microstructural white matter abnormalities and their impact on cognitive dysfunction in the various phases of type I bipolar disorder. J Affect Disord 193:39–50
104. Zanetti MV, Jackowski MP, Versace A, Almeida JRC, Hassel S, Duran FLS et al (2009) State-dependent microstructural white matter changes in bipolar I depression. Eur Arch Psychiatry Clin Neurosci 259:316–328
105. Miralbell J, Soriano JJ, Spulber G, López-Cancio E, Arenillas JF, Bargalló N et al (2012) Structural brain changes and cognition in relation to markers of vascular dysfunction. Neurobiol Aging 33:1003.e9–1003.e17
106. McKenna BS, Theilmann RJ, Sutherland AN, Eyler LT (2015) Fusing functional MRI and diffusion tensor imaging measures of brain function and structure to predict working memory and processing speed performance among inter-episode bipolar patients. J Int Neuropsychol Soc 21:330–341
107. Macritchie KA, Lloyd AJ, Bastin ME, Vasudev K, Gallagher P, Eyre R et al (2010) White matter microstructural abnormalities in euthymic bipolar disorder. Br J Psychiatry 196:52–58
108. Benedetti F, Bollettini I, Barberi I, Radaelli D, Poletti S, Locatelli C et al (2013) Lithium and GSK3-β promoter gene variants influence white matter microstructure in bipolar disorder. Neuropsychopharmacology 38:313–327

Chapter 21

Voxel-Based Morphometry Imaging Studies in Major Depression

Nicola Dusi, Giuseppe Delvecchio, Chiara Rovera, Carlo A. Altamura, and Paolo Brambilla

Abstract

Major depressive disorder is a frequent psychiatric illness with increasing incidence and heterogeneous course, which affects all the adults' lifespan. Clinical outcome and treatment response can vary a lot among patients; therefore, there has been an increasing demand for biological markers for early identification, clinical course prediction, and treatment response to this disease. In this context, imaging techniques offer a valid tool for the biological characterization of the disease. In particular magnetic resonance imaging with voxel-based morphometry analysis performs wide investigations of whole-brain alterations with an atheoretical approach. This technique allowed the identification of an altered network involving medial prefrontal cortex, anterior cingulate, insula, and limbic areas, such as hippocampus and amygdala. This network is involved in emotional processing, cognitive control on affectiveness and mnesic abilities. These areas, particularly limbic ones, have been observed by imaging studies to be responsive to pharmacological treatment; also, preserved volumes in these areas are indicators of better clinical response. According to these observations, imaging technique can be considered a valid application for a thorough comprehension of MDD features and to follow-up clinical response in this group of patients.

Key words Magnetic resonance imaging, Prefrontal cortex, Cingulate, Insula, Limbic areas, Hippocampus, Amygdala

1 Introduction

Major depressive disorder (MDD) is a common psychiatric illness, which frequently affects people across the lifespan and determines severe psychosocial functioning limitation. The lifetime prevalence of MDD is about 16%, being one of the leading causes of disease burden worldwide; according to the World Health Organization (WHO) estimates, its prevalence and disabling effects are expected to grow in the next years. It should be mentioned that late-life MDD is a form of depression that affects people in their mature life, usually over 65 years old. In this context, the constant rate of aging of the world population predicts an increase of late-life MDD.

Gianfranco Spalletta et al. (eds.), *Brain Morphometry*, Neuromethods, vol. 136,
https://doi.org/10.1007/978-1-4939-7647-8_21,

Interestingly, clinically relevant depressive symptoms affect older people in a range between 10 and 15%, whereas 1–5% of people reach the criteria for MDD. The syndrome itself can show different clinical symptoms and illness course. Several individuals undergo a benign course, which remit after few months of treatment and do not have relapses, and a more pervasive one, characterized by poor prognosis with recurrent forms of the disease and treatment failures [1].

Although antidepressant pharmacological treatment is the main therapeutic approach to MDD, clinical response to the first pharmacological choice is about 50% [2]. In general, antidepressants' selection does not rely on objective and quantitative indicators but on clinical impression, and successful treatment is often reached after subsequent pharmacological trials. As antidepressant prescription is rising constantly, there is an increasing need for clear decision-making procedures that guarantee a faster and reliable cure. Indeed, time spent waiting for clinical response and the adoption of iterative therapeutic interventions raise enormously the disease burden for the patients and the direct and indirect costs for healthcare systems [3]. Therefore, there is an increasing interest for the identification of the neurobiological markers of MDD, which can improve our understanding of the pathophysiological processes involved in this disease. In this perspective, magnetic resonance imaging (MRI) allows noninvasive assessment of the neural underpinnings of MDD and treatment response and clinical outcome [4]. However, current research has not yet managed to translate into the clinical field validated indicators of brain disrupted circuitries implicated in MDD. Indeed, MDD is a heterogeneous illness, which may be in principle characterized by specific clinical manifestations and biological substrates. Therefore, featuring homogenous patient groups based on specific brain markers would be useful to delineate clear diagnostic dimensions and tailored intervention strategies [5].

To date, most of the structural MRI studies in MDD found fronto-temporo-limbic gray matter (GM) volume reductions [6], and white matter (WM) hyperintensities, particularly in late-life MDD [7]. These observations supported the contribution of an altered emotional processing network, which involve prefrontal regions, hippocampus, and amygdala, in MDD [8]. This theory has been confirmed by converging evidence from different imaging methodologies, both functional and structural, such as MRI, positron-emission tomography (PET), and single-photon emission computed tomography (SPECT). Specifically, reduced cortical thickness and cellular composition have been described in frontal cortex GM coupled with altered WM structure and histology, by diffusion tensor imaging (DTI) as well as by task-based and resting state functional MRI (fMRI) studies.

2 Imaging Findings on MDD by Region of Interest (ROI) Studies

Region of interest (ROI) analysis is a hypothesis-driven technique that assesses brain morphology on specific a priori selected areas, with manual, rather than computerized, post-processing of images [9]. The ROI approach usually focuses on areas that are anatomically well described, with clear landmarks and already considered as relevant for the given disease. This technique requires long and strict training of operators and rigid rater reliability procedures and therefore is a time-consuming and expensive technique.

Whereas whole-brain volume and total GM volumes were not reported to be reduced in depressed patients, either in older or adult age [10–15], larger third [16, 17] and lateral ventricles [18, 19] were observed, although not consistently [20–23]. Moreover, WM hyperintensities have often been observed especially in geriatric population [24, 25].

ROI studies have also extensively focused on the limbic system, since it includes structures taking part to emotional processes as well as to memory functions and stress response, such as the hippocampus [26, 27]. Smaller hippocampal volumes have been reported in MDD in several cross-sectional and longitudinal studies including chronic, remitted, not medicated, and treatment-resistant populations; these finding have been confirmed by two independent meta-analyses [28, 29].

Another key limbic structure is the amygdala, involved in fear modulation, phobic reactions, and reward processes. Smaller amygdala volumes were observed in chronic [30–32], recurrent [33–35], and not medicated MDD patients compared to healthy controls [36], although not in all studies [37–39]. Enlarged amygdala volumes have also been reported [40–43]. These findings have been explained by sampling biases, differences in the samples employed, and by tracing biases, difficulties in delineating clear thresholds between amygdala and hippocampus, which do not have distinct anatomical boundaries.

Increasing interest has been driven toward the exploration of cortical alterations in MDD, which presented GM volume reductions in dorsolateral prefrontal cortex (DLPFC), ventrolateral prefrontal cortex (VLPFC) [44, 45], orbitofrontal cortex (OFC) [46], and anterior cingulate cortex (ACC) [12, 47]. Moreover, several reports are found in MDD lower caudate and putamen volumes [47–49] and putamen/globus pallidus hyperintensities [50, 51].

All together, these findings sustain the hypothesis of reduced volumes in MDD in prefronto-limbic areas with higher hyperintensities in cerebral white matter and basal ganglia, particularly in older MDD patients.

3 Voxel-Based Morphometry (VBM) Findings in MDD

The vastness of depressive manifestations, even those strictly included under the clinical label of MDD, gives reason of the inconsistency of findings, which reflect discrepancies in samples, MDD severity, age, medication status, illness phase, and duration, as well as methodological differences related to imaging procedures, such as images acquisition, machinery applied, or analyses' techniques [52]. Therefore, in order to develop sharper ways to distinguish between state and trait markers of the disease, an increasing interest for the identification of intermediate phenotypes of MDD has lately been raised. Intermediate phenotypes are considered as phenotypic features of the population that expresses a certain disease, which do not pertain to the core symptoms of the disease, but are inherited with the disease itself, are state independent, and are closely related to the disease genotype. In order to be considered as an endophenotype, a measure has to be associated with illness, be heritable, be apparent in a subject regardless of whether the illness is active, and co-segregate with illness within families. Thus, intermediate phenotypes are more closely related to the core functions of the genes implicated in a certain disease than the phenotype of the disease itself and are helpful tools in isolating those genes, which are relevant to the illness manifestations. In this field, wide brain morphological analyses, such as VBM, can be applied to test whether certain brain areas' alteration are eligible consistent endophenotipical markers of psychiatric illness. For these reasons, several studies have used neuroimaging measures as endophenotypes in order to identify possible generic risk factors for psychiatric disorders.

VBM consists in a technique that allows a wider investigation of the brain, which is not biased by the a priori driven investigation of the ROI approach. VBM can detect, in a whole-brain analysis, the component of white and gray matter that takes part to a certain brain circuitry. VBM have reached a comparable accuracy to the ROI method by overcoming manual tracing limitations and biases.

Therefore, VBM is the technique of choice when alterations are supposed to involve the entire brain and if we want to exclude operator biases. However, the two techniques can be seen as complementary tools, which offer different but convergent types of information. VBM has the advantage of a faster, exploratory approach that can search the entire brain for anomalies in an atheoretical way, whereas the ROI approach is a detailed and elegant confirmatory analysis that can validate VBM findings.

Several MRI studies employing a VBM approach assessed brain alterations in MDD, mostly with a cross-sectional design. In the next paragraphs, the main alterations concerning single brain areas or circuitries associated with different MDD populations, illness outcome, and treatment response will be addressed. VBM studies consistently found hippocampal and parahippocampal abnormalities in MDD [53–60]. Opel and colleagues (2016) found smaller hippocampal volumes in MDD patients, but not in their first degree relatives, and in healthy controls which were exposed to childhood maltreatment [61]. This finding is not surprising, especially because it has been shown that early traumatic experiences have a high incidence on people who then develop MDD. Additionally, smaller hippocampal volumes have been found in subjects with positive familial anamnesis for MDD, compared to healthy controls without predisposition to MDD and MDD patients [54]. The lack of a positive finding in MDD patients has been explained by the authors by considering that these patients were on antidepressants treatment at the time of the assessment, which might have exerted a neuroprotective effect on this group with a subtle volume enlargement in respect to subjects with genetic predisposition to MDD, but without any clinical manifestation of the disease. Nevertheless, these findings suggested that this structure could be considered not only a genetic substrate but also a marker of early environmental risk factors for developing MDD. Furthermore, smaller hippocampal volumes have been observed in adult population, irrespectively to pharmacological treatment [59, 60] and age [55]. Stratmann and colleagues (2014) found inverse correlation between hippocampal volumes and number of episodes, whereas Egger and colleagues (2008) observed a significant correlation between volume loss and duration of illness, indicating that this region is sensitive to factors related to the illness status, which seems to have a degenerative effect on its tropism [55, 59]. Moreover, the atrophy of the hippocampus has been connected to a specific action of stress-related hormones, to the function of the brain-derived neurotrophic factor (BDNF) gene, or to an enhanced inflammatory response. Also, hippocampus has been reported to have neuromodulatory response to pharmacological treatment, which results in volume expansion. Along with hippocampus, smaller amygdala volumes have been observed, particularly in drug-naïve young females with MDD [62] and in elderly patients [55]. The first study is in line with previous observations, emerging from ROI approaches, exploring MDD in female patients during both adulthood [34] and childhood [63]. These studies, along with the study from Egger and colleagues (2008), postulated that the alteration in amygdala volume could be an early but stable marker of MDD, not influenced by pharmacological treatment [64].

Further, most of the studies that demonstrated a significant reduction in brain regions within the limbic system also revealed an altered prefrontal-anterior cingulate (ACC) pathway in MDD patients [3, 53–57, 61, 64–74]. Specifically, a study by Opel and colleagues (2016), which compared MDD patients and their first-degree relatives, found a widespread prefrontal volume reduction within the OFC, which is an area implicated in the enteroceptive and subjective elaboration of emotional experience and therefore might be associated with the genesis of MDD [61]. Interestingly, the reduction of prefrontal regions is not specific to MDD but a common feature that characterizes also patients with bipolar disorder [65], especially the inferior frontal gyrus, which is an area considered to play a regulatory function on emotion integration and intensity. Additionally, inferior and middle frontal gyrus was found to be smaller both in drug-naïve MDD patients [57, 69, 74] and chronic patients, with smaller volumes associated with longer illness duration [67]. These areas were responsive, in terms of volume enlargement, to pharmacological treatment, along with OFC [66]; on the other hand, smaller frontal and prefrontal volumes were related to drug resistance and illness severity [3, 67, 68]. Specifically, Korgaonkar and colleagues (2015) followed a sample of 74 patients for 8 weeks and tested weather baseline imaging data could differentiate remitters from non-remitters, after 8 weeks of antidepressant treatment. The authors observed lower volumes in the left middle frontal gyrus and right angular gyrus cortex, which turned out to have about 85% accuracy of predicting non-remitting patients [3]. This observation is in line with imaging applications on data-driven classification such as pattern recognition, which are promising instruments on identification of markers of diagnosis or clinical course [75]. On the other hand, lower superior frontal cortex, dorsal prefrontal cortex, and ACC were also observed among patients with geriatric MDD [54–56] who achieved remission [70], indicating that prefrontal pathology could be considered as a stable marker of the disease, not influenced by treatment outcome, aging, or illness duration.

Moreover, a GM volume reduction in ACC was observed in treatment-naïve and treatment undergoing patients in several [54, 57, 64, 76], but not all studies [77]. Indeed, Frodl and colleagues (2008) did not find a reduction in ACC volumes in MDD patients, but, in contrast, they found that larger ACC volumes were a predictor of better clinical outcome and lower number of hospitalizations [77]. According to these findings, frequent depressive episodes might have a degenerating effect on medial brain structures like cingulate cortex, and the lack of volume reduction or even enlargement of this area after treatment administration might be considered as a marker of resilience to illness process.

All together, these observations, which report an altered structural network of anterior medial structures in MDD, represent an interesting finding for two main reasons. First, from a methodological point of view, employing a wider volumetric analysis, such as VBM, which allows to capture brain alterations in different areas that are intercorrelated, suggests that the origin of MDD neuropathology is associated with several brain regions connected with each other rather than single, isolated structures. Second, from a more substantial point of view, these studies identify a core network within prefrontal and limbic regions which are associated with mood disorders, especially because they exert an important role in the modulation of cognitive and emotional processes.

Along with medial temporal, limbic structures and frontal/prefrontal cortices volume alterations, temporal cortex volume reductions have been observed in treatment-resistant [67], suicidal [58], treatment-naïve [78], remitted [70], and geriatric [56, 79] MDD patients. Furthermore, VBM analyses on MDD reported altered insular volume, which has been under-investigated with the ROI approach [56, 58, 59, 61, 66–69, 71, 80, 81]. The majority of VBM studies found lower insular volumes in currently ill [58, 59, 61, 68, 80], treatment-resistant [67], geriatric [56, 71] and remitted patients [80], drug-naïve patients [69], and first-degree relatives [61], whereas larger volumes were observed in drug-naïve patients [66].

Moreover, GM volume reductions in MDD have also been reported in posterior brain structures, such as the posterior cingulate cortex/precuneus [56, 57, 71, 73, 74] and cuneus [71–73], as well as the cerebellum [57, 76, 53, 56, 68], which have not usually considered as core areas for depressive disorder. Lower volumes in cuneus and precuneus have been consistently found in older age patients [71, 73] and, to a lesser extent, in adults [72]. Precuneus, through its anatomical and functional connection with posterior cingulate cortex, is part of a group of medial brain areas, implicated in cognitive and emotional processes, which formed the default mode network (DMN) and has been widely studied in psychiatric disorders. Similarly, smaller cerebellar volumes were observed in treatment ongoing [53, 57, 76], naïve [57], geriatric and suicidal [56] MDD patients. Although this region is not considered as a crucial area for MDD, the cerebellum, along with posterior cingulate/precuneus and insula, is also part of those median structures functionally implicated in the DMN. Therefore, all together, these observations, which include areas traditionally considered of secondary interest in mood disorders, demonstrated a further advantage of whole brain analyses with respect to a priori selected ones.

Late-life MDD deserves ad hoc considerations, particularly because depression among elder people can be particularly disabling, by limiting day life activities and by requiring caregiving and assistance. Also, late-life MDD presents clinical and diagnostic complexity due to high medical comorbidity and the overlap with neurological syndromes, such as cognitive impairments and dementias, which mimic depressive symptoms, at their early stages, and share some clinical features. Indeed, in this group of patients,

MDD is often the early manifestation of Alzheimer's disease, Parkinson's disease, and cerebrovascular strokes. There is growing evidence of correlations between MDD and poor cognitive function or decline in the elderly. Therefore, late-onset depression challenges physicians on its detection and treatment, especially because this population responds slower, poorer than younger population, and has lower treatment options due to side effects' limitations related to poli-pharmacotherapy and concomitant illness. Differences between early and late-onset MDD might reflect different biological processes implicated in the pathophysiology of the disease. The first can be explained by biological models involving stress response hormones and neurotoxicity caused by glucocorticoids. The second would imply a vascular model of etiopathology, which is similar to those proposed for several neurodegenerative diseases. Recent VBM studies reported lower volumes in whole-brain GM, as well as in the right superior frontal cortex, OFC, left postcentral cortex, right middle temporal cortex, and, less frequently, hippocampus and amygdala, in MDD geriatric patients compared to healthy controls [55, 70], though not by all studies [82–84]. The implication of prefrontal cognitive control areas rather than limbic areas, in remitted MDD geriatric patients, might indicate a possible greater impairment in executive functions in older age MDD. With this regard, in an effort to differentiate geriatric MDD patients with or without cognitive impairment, Xie and colleagues 2012 carried out a correlation analysis, which included both imaging and clinical data, and reported that lower GM volumes in inferior frontal gyrus, left medial frontal gyrus, and left insula, together with deficit of episodic memory, uniquely characterized MDD with mild cognitive impairment. However, the authors also found some common alterations, especially in those areas consistently implicated in MDD, such as ventromedial prefrontal cortex, OFC, anterior and posterior cingulate, insula, and hippocampus [71]. Moreover, besides GM alterations, several studies reported widespread WM disruption in geriatric MDD [56, 73]. These observations supported evidence of WM hyperintensities in older age MDD [25, 85] and have been confirmed by different imaging approaches. Thus, WM pathology could be either a marker or a secondary effect of MDD in older age, and its correlation with cognitive deterioration should be object of further investigation.

Imaging studies can also be applied to assess features of response to pharmacological treatments. Unfortunately this topic has been addressed by few cross-sectional studies and even fewer applied a longitudinal design to test brain changes after weeks of pharmacological administration. Nevertheless, this information is of great interest in the research for biological markers linked to clinical outcome [4].

As reported by several studies, treatment-naïve patients presented lower GM volumes in DLPFC, middle prefrontal cortex [66], medial frontal gyrus [57], superior frontal gyrus [69, 74], precentral gyrus [86], anterior cingulate [62], caudate [87], hippocampus [60, 88], amygdala [57, 62, 74], supplementary motor areas [88], cerebellum [57], and insula [69] [66] compared to healthy controls. These data should be considered as preliminary observations, which have to be coupled to the results of those studies reporting findings about treatment effect. Treatment outcome have been mainly addressed by cross-sectional studies which compared acutely ill patients or patients under treatment with patients in remission or with a population of healthy controls. Although the majority of the drugs investigated are serotonin selective reuptake inhibitors (SSRIs), as they are the most prescribed antidepressants nowadays, there is still not a sufficient number of studies allowing to draw imaging profiles of response linked to single molecules.

However, from the available literature, it seems that treatment response or better clinical outcome was associated with larger DLPFC [72, 89], medial prefrontal [3, 72, 90], OFC [91], superior frontal gyrus [92], anterior cingulate [72, 90, 93, 94], temporal lobe [67], insula [94], posterior cingulate/precuneus [72, 90], and angular gyrus [3]. Absence of differences between drug-naïve and treated or remitted MDD patients were usually attributed to heterogeneity of the samples, differences in treatment durations, or lack of thorough information about clinical course of illness of the samples. Few longitudinal VBM studies have assessed MDD patients before and after treatment intervention to test the effective influence of pharmacological treatment on the brain structure. A 12-week administration of SSRI sertraline determined DLPFC enlargement, a core area of top-down regulation of emotion, in MDD patients [79]. Similarly, 6 weeks of treatment with duloxetine, noradrenergic, and serotonergic reuptake inhibitor caused larger GM volumes in the inferior frontal cortex and cerebellum in patients who achieved remission after pharmacological intervention [95]. Frodl and colleagues (2008) followed up a group of MDD patients on several antidepressants treatments for 3 years and found lower decline in dorsomedial prefrontal cortex, DLPFC, ACC, and left hippocampus in remitted MDD patients compared to non-remitted ones [96]. On the other hand, 8 weeks of fluoxetine treatment did not provoke any volume change between baseline and follow-up in a sample of MDD patients [66]. Similarly, 6 months of treatment with tricyclic nortriptyline did not cause structural changes among patients; nevertheless, functional electroencephalographic changes have been observed in prefrontal cortex, which presented lower activity coupled with depressive symptoms amelioration in MDD patients [97]. Moreover, effective electroconvulsive treatment (twice a week until symptoms remission) was accompanied by GM enlargement in the right caudate nucleus, superior and medial temporal lobe, and insula [81]. These data have been interpreted as a possible neuromodulatory effect or as an acute vascular reaction to electrical trauma. Finally, these findings, which have to be considered just as preliminary and sparse observations [66], offer further evidence of an altered network in MDD which encompasses prefrontal areas and medial temporal (insula) and subcortical areas (hippocampus and amygdala) together with median structures such as the anterior cingulate, posterior cingulate/precuneus, and cerebellum.

4 Conclusions

Magnetic resonance imaging (MRI) investigations offer new insights on understanding and treating MDD. Indeed, MRI can help to detect signs of neurobiological alterations that are implicated in the disease, to identify biomarkers for distinguishing specific MDD syndrome subpopulations, and to find features associated with clinical outcome and treatment response. Lower GM volumes in prefrontal areas (including DLPFC, OFC, ventromedial prefrontal cortex), anterior cingulate cortex (ACC), medial temporal areas, insula, limbic areas (hippocampus and amygdala), and precuneus (for a review, see [10]) have been shown. In particular, ACC and insula are crucial in the regulation of emotional processing. Insula integrates information from limbic areas and frontal areas and, particularly, its anterior division is involved in cognitive, affective, and concentration processes. Insula and prefrontal dysfunctions might therefore determine the lack of cognitive inhibition to negative feelings, rumination, and neuropsychological impairments, which are frequently observed in MDD. These data are also supported by fMRI findings, which reported insular cortex involvement during processing of specific emotions, such as disgust and sadness.

Additionally, the neuropathology underlying these alterations might be partially sustained by stress. Indeed, stress response hormones, such as glucocorticoids, would be oversecreted in demanding conditions, such as early traumatic exposure or emotional withdrawal, determining a second messenger pathway activation which, in turn, leads to lower level of neurotrophins, like the BDNF, and, ultimately, to atrophic processes. This hypothesis is supported by the relatively high incidence of traumatic experiences in the anamnesis of patients with MDD, by the lower expression of BDNF in the limbic structures, by preclinical data on hippocampus atrophy in response to glucocorticoids' administration, as well as by the neuroprotective effects of antidepressant treatment on the same areas. Moreover, the alteration in the fronto-limbic network might be differentiated according to the stage of life in which MDD occurs. On the one hand, early-onset MDD could affect more severely the limbic part of the circuit, including mainly the hippocampus, and could be characterized by loss of episodic memory. On the other hand, late-onset MDD could involve more the frontal part of the circuitry and could be characterized by impairments in executive functions. Additionally, in geriatric MDD, besides glucocorticoids' stress response, metabolic and vascular processes may play a role, potentially leading to increased oxidative status [98] and neuronal deterioration.

In conclusion, VBM represents a powerful and reasonably applicable approach for the assessment of structural alterations in MDD, considered as a consistent method for whole brain analyses. Even though this approach has several advantages, it does not come without limitations. Several technical and procedural elements should be taken into account when interpreting the results. Registration accuracy is crucial as errors in this phase may lead to misinterpretation of the localization of the estimates. Mathematical approaches to registration techniques are continuously implemented in order to improve realignment efficiency. Lately, nonlinear registration algorithms with millions of parameters, such as Diffeomorphic Anatomical Registration using Exponentiated Lie (DARTEL) algebra registration method has been considered preferable, with better inter- and intra-operator reliability, whereas previous techniques were less precise than ROI tracing in detecting alterations in certain given areas of small dimensions. Another matter of consideration is that MRI does not allow direct identification of neuropathological alterations related to the results obtained. It is therefore unclear which mechanisms cause volume modifications reported in MDD: they could pertain to different cellular populations, processes, or combined effects of several converging elements, which cannot be distinguished. The application of different MRI modalities on the same samples might help overcoming such limitations. For instance, morphometric analysis could take advantage of the contribution of diffusion-weighted imaging, which can add information about the integrity of the connecting fibers underpinning the cortical structures. Future MRI studies in MDD should therefore aim at integrating different MR techniques, coupled with genetics and cognitive investigations, in order to provide complementary observations and further understand the pathophysiology of the disease (Figs. 1 and 2).

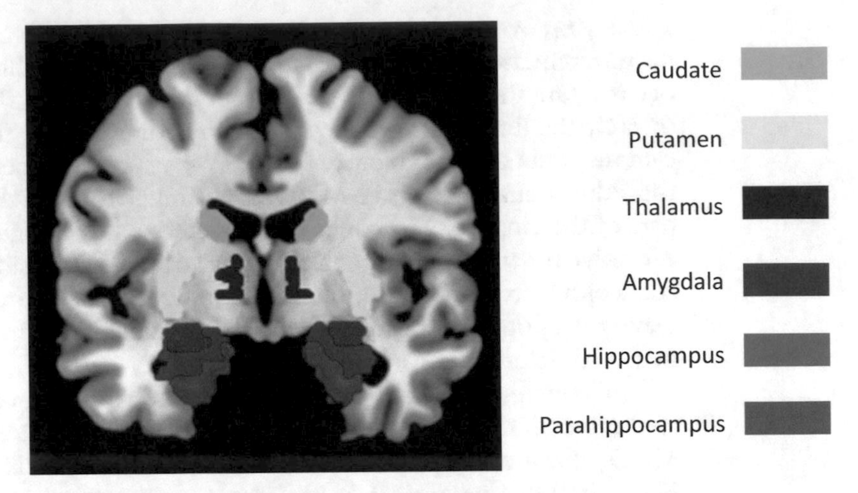

Fig. 1 Subcortical regions consistently found to be involved in major depressive disorder

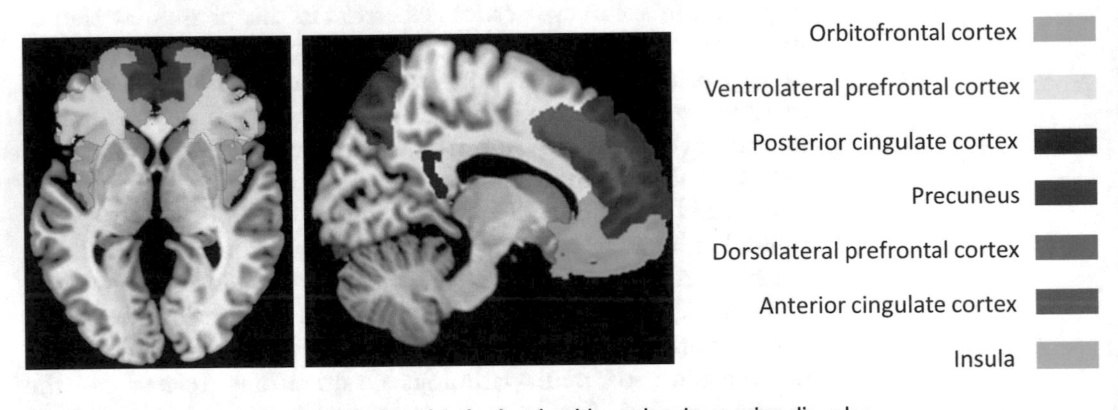

Fig. 2 Cortical regions consistently found to be involved in major depressive disorder

Acknowledgments

GDV and PB were partially supported by a grant from the Italian Ministry of Health (RF-2011-02352308).

References

1. Altamura AC, Percudani M (1993) The use of antidepressants for long-term treatment of recurrent depression: rationale, current methodologies, and future directions. J Clin Psychiatry 54(Suppl):29–37; discussion 38
2. Altamura AC, Mauri M (1985) Plasma concentrations, information and therapy adherence during long-term treatment with antidepressants. Br J Clin Pharmacol 20 (6):714–716
3. Korgaonkar MS, Rekshan W, Gordon E, Rush AJ, Williams LM, Blasey C, Grieve SM (2015) Magnetic resonance imaging measures of brain structure to predict antidepressant treatment outcome in major depressive disorder. EBioMedicine 2(1):37–45. https://doi.org/10.1016/j.ebiom.2014.12.002
4. Dusi N, Perlini C, Bellani M, Brambilla P (2012) Searching for psychosocial endophenotypes in schizophrenia: the innovative role of

brain imaging. Riv Psichiatr 47(2):76–88. https://doi.org/10.1708/1069.11712

5. Bellani M, Dusi N, Brambilla P (2013) Can brain imaging address psychosocial functioning and outcome in schizophrenia? In: Thornicroft G, Ruggeri M, Goldberg D (eds) Improving mental health care: the global challenge. Wiley-Blackwell, NJ, USA, pp 281–290
6. Campbell S, MacQueen G (2006) An update on regional brain volume differences associated with mood disorders. Curr Opin Psychiatry 19 (1):25–33. https://doi.org/10.1097/01.yco.0000194371.47685.f2
7. Videbech P (1997) MRI findings in patients with affective disorder: a meta-analysis. Acta Psychiatr Scand 96(3):157–168
8. Koolschijn PC, van Haren NE, Lensvelt-Mulders GJ, Hulshoff Pol HE, Kahn RS (2009) Brain volume abnormalities in major depressive disorder: a meta-analysis of magnetic resonance imaging studies. Hum Brain Mapp 30(11):3719–3735. https://doi.org/10.1002/hbm.20801
9. Andreone N, Tansella M, Cerini R, Rambaldelli G, Versace A, Marrella G, Perlini C, Dusi N, Pelizza L, Balestrieri M, Barbui C, Nose M, Gasparini A, Brambilla P (2007) Cerebral atrophy and white matter disruption in chronic schizophrenia. Eur Arch Psychiatry Clin Neurosci 257(1):3–11. https://doi.org/10.1007/s00406-006-0675-1. Abas MA, Sahakian BJ, Levy R (1990) Neuropsychological deficits and CT scan changes in elderly depressives. Psychological medicine 20 (3):507-520
10. Arnone D, McIntosh AM, Ebmeier KP, Munafo MR, Anderson IM (2012) Magnetic resonance imaging studies in unipolar depression: systematic review and meta-regression analyses. Eur Neuropsychopharmacol 22 (1):1–16. https://doi.org/10.1016/j.euroneuro.2011.05.003
11. Axelson DA, Doraiswamy PM, McDonald WM, Boyko OB, Tupler LA, Patterson LJ, Nemeroff CB, Ellinwood EH Jr, Krishnan KR (1993) Hypercortisolemia and hippocampal changes in depression. Psychiatry Res 47 (2):163–173
12. Coffey CE, Wilkinson WE, Weiner RD, Parashos IA, Djang WT, Webb MC, Figiel GS, Spritzer CE (1993) Quantitative cerebral anatomy in depression. A controlled magnetic resonance imaging study. Arch Gen Psychiatry 50 (1):7–16
13. Dupont RM, Butters N, Schafer K, Wilson T, Hesselink J, Gillin JC (1995) Diagnostic specificity of focal white matter abnormalities in bipolar and unipolar mood disorder. Biol Psychiatry 38(7):482–486
14. Krishnan KR, McDonald WM, Escalona PR, Doraiswamy PM, Na C, Husain MM, Figiel GS, Boyko OB, Ellinwood EH, Nemeroff CB (1992) Magnetic resonance imaging of the caudate nuclei in depression. Preliminary observations. Arch Gen Psychiatry 49 (7):553–557
15. Lavretsky H, Roybal DJ, Ballmaier M, Toga AW, Kumar A (2005) Antidepressant exposure may protect against decrement in frontal gray matter volumes in geriatric depression. J Clin Psychiatry 66(8):964–967
16. Beats B, Levy R, Forstl H (1991) Ventricular enlargement and caudate hyperdensity in elderly depressives. Biol Psychiatry 30 (5):452–458
17. Rabins PV, Pearlson GD, Aylward E, Kumar AJ, Dowell K (1991) Cortical magnetic resonance imaging changes in elderly inpatients with major depression. Am J Psychiatry 148 (5):617–620
18. Dolan RJ, Calloway SP, Mann AH (1985) Cerebral ventricular size in depressed subjects. Psychol Med 15(4):873–878
19. Shima S, Shikano T, Kitamura T, Masuda Y, Tsukumo T, Kanba S, Asai M (1984) Depression and ventricular enlargement. Acta Psychiatr Scand 70(3):275–277
20. Abas MA, Sahakian BJ, Levy R (1990) Neuropsychological deficits and CT scan changes in elderly depressives. Psychol Med 20 (3):507–520
21. Lesser IM, Miller BL, Boone KB, Hill-Gutierrez E, Mehringer CM, Wong K, Mena I (1991) Brain injury and cognitive function in late-onset psychotic depression. J Neuropsychiatry Clin Neurosci 3(1):33–40
22. Rossi A, Stratta P, di Michele V, Bolino F, Nistico R, de Leonardis R, Sabatini MD, Casacchia M (1989) A computerized tomographic study in patients with depressive disorder: a comparison with schizophrenic patients and controls. Acta Psychiatr Belg 89 (1–2):56–61
23. Van den Bossche B, Maes M, Brussaard C, Schotte C, Cosyns P, De Moor J, De Schepper A (1991) Computed tomography of the brain in unipolar depression. J Affect Disord 21 (1):67–74
24. Salloway S, Correia S, Boyle P, Malloy P, Schneider L, Lavretsky H, Sackheim H, Roose S, Krishnan KRR (2002) MRI subcortical hyperintensities in old and very old depressed outpatients: the important role of age in late-life depression. J Neurol Sci

203–204(0):227–233. https://doi.org/10.1016/S0022-510X(02)00296-4

25. Sassi RB, Brambilla P, Nicoletti M, Mallinger AG, Frank E, Kupfer DJ, Keshavan MS, Soares JC (2003) White matter hyperintensities in bipolar and unipolar patients with relatively mild-to-moderate illness severity. J Affect Disord 77(3):237–245
26. Bellani M, Dusi N, Yeh PH, Soares JC, Brambilla P (2011) The effects of antidepressants on human brain as detected by imaging studies. Focus on major depression. Prog Neuro-Psychopharmacol Biol Psychiatry 35 (7):1544–1552. https://doi.org/10.1016/j.pnpbp.2010.11.040
27. Brambilla P, Barale F, Caverzasi E, Soares JC (2002) Anatomical MRI findings in mood and anxiety disorders. Epidemiol Psichiatr Soc 11 (2):88–99
28. Campbell S, Marriott M, Nahmias C, MacQueen GM (2004) Lower hippocampal volume in patients suffering from depression: a meta-analysis. Am J Psychiatry 161(4):598–607
29. McKinnon MC, Yucel K, Nazarov A, MacQueen GM (2009) A meta-analysis examining clinical predictors of hippocampal volume in patients with major depressive disorder. J Psychiatry Neurosci 34(1):41–54
30. Hickie IB, Naismith SL, Ward PB, Scott EM, Mitchell PB, Schofield PR, Scimone A, Wilhelm K, Parker G (2007) Serotonin transporter gene status predicts caudate nucleus but not amygdala or hippocampal volumes in older persons with major depression. J Affect Disord 98(1–2):137–142. https://doi.org/10.1016/j.jad.2006.07.010
31. Keller J, Shen L, Gomez RG, Garrett A, Solvason HB, Reiss A, Schatzberg AF (2008) Hippocampal and amygdalar volumes in psychotic and nonpsychotic unipolar depression. Am J Psychiatry 165(7):872–880. https://doi.org/10.1176/appi.ajp.2008.07081257
32. Sheline YI, Gado MH, Price JL (1998) Amygdala core nuclei volumes are decreased in recurrent major depression. Neuroreport 9 (9):2023–2028
33. Caetano SC, Hatch JP, Brambilla P, Sassi RB, Nicoletti M, Mallinger AG, Frank E, Kupfer DJ, Keshavan MS, Soares JC (2004) Anatomical MRI study of hippocampus and amygdala in patients with current and remitted major depression. Psychiatry Res Neuroimaging 132 (2):141–147. https://doi.org/10.1016/j.pscychresns.2004.08.002
34. Hastings RS, Parsey RV, Oquendo MA, Arango V, Mann JJ (2004) Volumetric analysis of the prefrontal cortex, amygdala, and hippocampus in major depression. Neuropsychopharmacology 29(5):952–959
35. Lorenzetti V, Allen NB, Whittle S, Yücel M (2010) Amygdala volumes in a sample of current depressed and remitted depressed patients and healthy controls. J Affect Disord 120 (1–3):112–119. https://doi.org/10.1016/j.jad.2009.04.021
36. Kronenberg G, Tebartz van Elst L, Regen F, Deuschle M, Heuser I, Colla M (2009) Reduced amygdala volume in newly admitted psychiatric in-patients with unipolar major depression. J Psychiatr Res 43 (13):1112–1117. https://doi.org/10.1016/j.jpsychires.2009.03.007
37. Mervaala E, Fohr J, Kononen M, Valkonen-Korhonen M, Vainio P, Partanen K, Partanen J, Tiihonen J, Viinamaki H, Karjalainen AK, Lehtonen J (2000) Quantitative MRI of the hippocampus and amygdala in severe depression. Psychol Med 30(1):117–125
38. Monkul ES, Hatch JP, Nicoletti MA, Spence S, Brambilla P, Lacerda AL, Sassi RB, Mallinger AG, Keshavan MS, Soares JC (2007) Fronto-limbic brain structures in suicidal and non-suicidal female patients with major depressive disorder. Mol Psychiatry 12(4):360–366. https://doi.org/10.1038/sj.mp.4001919
39. Munn MA, Alexopoulos J, Nishino T, Babb CM, Flake LA, Singer T, Ratnanather JT, Huang H, Todd RD, Miller MI, Botteron KN (2007) Amygdala volume analysis in female twins with major depression. Biol Psychiatry 62(5):415–422. https://doi.org/10.1016/j.biopsych.2006.11.031
40. Frodl T, Meisenzahl EM, Zetzsche T, Born C, Jager M, Groll C, Bottlender R, Leinsinger G, Moller HJ (2003) Larger amygdala volumes in first depressive episode as compared to recurrent major depression and healthy control subjects. Biol Psychiatry 53(4):338–344
41. Lange C, Irle E (2004) Enlarged amygdala volume and reduced hippocampal volume in young women with major depression. Psychol Med 34(6):1059–1064
42. van Eijndhoven P, van Wingen G, van Oijen K, Rijpkema M, Goraj B, Jan Verkes R, Oude Voshaar R, Fernández G, Buitelaar J, Tendolkar I (2009) Amygdala volume marks the acute state in the early course of depression. Biol Psychiatry 65(9):812–818. https://doi.org/10.1016/j.biopsych.2008.10.027
43. Weniger G, Lange C, Irle E (2006) Abnormal size of the amygdala predicts impaired emotional memory in major depressive disorder. J Affect Disord 94(1–3):219–229. https://doi.org/10.1016/j.jad.2006.04.017

44. Drevets WC, Frank E, Price JC, Kupfer DJ, Holt D, Greer PJ, Huang Y, Gautier C, Mathis C (1999) PET imaging of serotonin 1A receptor binding in depression. Biol Psychiatry 46 (10):1375–1387
45. Hirayasu Y, Shenton ME, Salisbury DF, Kwon JS, Wible CG, Fischer IA, Yurgelun-Todd D, Zarate C, Kikinis R, Jolesz FA, McCarley RW (1999) Subgenual cingulate cortex volume in first-episode psychosis. Am J Psychiatry 156 (7):1091–1093
46. Lai T, Payne ME, Byrum CE, Steffens DC, Krishnan KR (2000) Reduction of orbital frontal cortex volume in geriatric depression. Biol Psychiatry 48(10):971–975
47. Krishnan KR (1993) Neuroanatomic substrates of depression in the elderly. J Geriatr Psychiatry Neurol 6(1):39–58
48. Krishnan KR, McDonald WM, Doraiswamy PM, Tupler LA, Husain M, Boyko OB, Figiel GS, Ellinwood EH Jr (1993) Neuroanatomical substrates of depression in the elderly. Eur Arch Psychiatry Clin Neurosci 243(1):41–46
49. Parashos IA, Tupler LA, Blitchington T, Krishnan KRR (1998) Magnetic-resonance morphometry in patients with major depression. Psychiatry Res Neuroimaging 84(1):7–15. https://doi.org/10.1016/S0925-4927(98)00042-0
50. Greenwald BS, Kramer-Ginsberg E, Krishnan RR, Ashtari M, Aupperle PM, Patel M (1996) MRI signal hyperintensities in geriatric depression. Am J Psychiatry 153(9):1212–1215
51. Iidaka T, Nakajima T, Kawamoto K, Fukuda H, Suzuki Y, Maehara T, Shiraishi H (1996) Signal hyperintensities on brain magnetic resonance imaging in elderly depressed patients. Eur Neurol 36(5):293–299
52. Dusi N, Barlati S, Vita A, Brambilla P (2015) Brain structural effects of antidepressant treatment in major depression. Curr Neuropharmacol 13(4):458–465
53. Abe O, Yamasue H, Kasai K, Yamada H, Aoki S, Inoue H, Takei K, Suga M, Matsuo K, Kato T, Masutani Y, Ohtomo K (2010) Voxel-based analyses of gray/white matter volume and diffusion tensor data in major depression. Psychiatry Res 181(1):64–70. https://doi.org/10.1016/j.pscychresns.2009.07.007
54. Amico F, Meisenzahl E, Koutsouleris N, Reiser M, Moller HJ, Frodl T (2011) Structural MRI correlates for vulnerability and resilience to major depressive disorder. J Psychiatry Neurosci 36(1):15–22. https://doi.org/10.1503/jpn.090186
55. Egger K, Schocke M, Weiss E, Auffinger S, Esterhammer R, Goebel G, Walch T, Mechtcheriakov S, Marksteiner J (2008) Pattern of brain atrophy in elderly patients with depression revealed by voxel-based morphometry. Psychiatry Res 164(3):237–244. https://doi.org/10.1016/j.pscychresns.2007.12.018
56. Hwang JP, Lee TW, Tsai SJ, Chen TJ, Yang CH, Lirng JF, Tsai CF (2010) Cortical and subcortical abnormalities in late-onset depression with history of suicide attempts investigated with MRI and voxel-based morphometry. J Geriatr Psychiatry Neurol 23 (3):171–184. https://doi.org/10.1177/0891988710363713
57. Lai CH, Hsu YY, Wu YT (2010) First episode drug-naive major depressive disorder with panic disorder: gray matter deficits in limbic and default network structures. Eur Neuropsychopharmacol 20(10):676–682. https://doi.org/10.1016/j.euroneuro.2010.06.002
58. Peng H, Wu K, Li J, Qi H, Guo S, Chi M, Wu X, Guo Y, Yang Y, Ning Y (2014) Increased suicide attempts in young depressed patients with abnormal temporal-parietal-limbic gray matter volume. J Affect Disord 165:69–73. https://doi.org/10.1016/j.jad.2014.04.046
59. Stratmann M, Konrad C, Kugel H, Krug A, Schoning S, Ohrmann P, Uhlmann C, Postert C, Suslow T, Heindel W, Arolt V, Kircher T, Dannlowski U (2014) Insular and hippocampal gray matter volume reductions in patients with major depressive disorder. PLoS One 9(7):e102692. https://doi.org/10.1371/journal.pone.0102692
60. Zou K, Deng W, Li T, Zhang B, Jiang L, Huang C, Sun X, Sun X (2010) Changes of brain morphometry in first-episode, drug-naive, non-late-life adult patients with major depression: an optimized voxel-based morphometry study. Biol Psychiatry 67 (2):186–188. https://doi.org/10.1016/j.biopsych.2009.09.014
61. Opel N, Zwanzger P, Redlich R, Grotegerd D, Dohm K, Arolt V, Heindel W, Kugel H, Dannlowski U (2016) Differing brain structural correlates of familial and environmental risk for major depressive disorder revealed by a combined VBM/pattern recognition approach. Psychol Med 46(2):277–290. https://doi.org/10.1017/S0033291715001683
62. Tang Y, Wang F, Xie G, Liu J, Li L, Su L, Liu Y, Hu X, He Z, Blumberg HP (2007b) Reduced ventral anterior cingulate and amygdala volumes in medication-naïve females with major depressive disorder: a voxel-based morphometric magnetic resonance imaging study. Psychiatry Res Neuroimaging 156(1):83–86. https://doi.org/10.1016/j.pscychresns.2007.03.005

63. Rosso IM, Cintron CM, Steingard RJ, Renshaw PF, Young AD, Yurgelun-Todd DA (2005) Amygdala and hippocampus volumes in pediatric major depression. Biol Psychiatry 57 (1):21–26. https://doi.org/10.1016/j.biopsych.2004.10.027
64. Tang CY, Friedman J, Shungu D, Chang L, Ernst T, Stewart D, Hajianpour A, Carpenter D, Ng J, Mao X, Hof PR, Buchsbaum MS, Davis K, Gorman JM (2007a) Correlations between diffusion tensor imaging (DTI) and magnetic resonance spectroscopy (1H MRS) in schizophrenic patients and normal controls. BMC Psychiatry 7:25. https://doi.org/10.1186/1471-244X-7-25
65. Cai Y, Liu J, Zhang L, Liao M, Zhang Y, Wang L, Peng H, He Z, Li Z, Li W, Lu S, Ding Y, Li L (2015) Grey matter volume abnormalities in patients with bipolar I depressive disorder and unipolar depressive disorder: a voxel-based morphometry study. Neurosci Bull 31(1):4–12. https://doi.org/10.1007/s12264-014-1485-5
66. Kong L, Wu F, Tang Y, Ren L, Kong D, Liu Y, Xu K, Wang F (2014) Frontal-subcortical volumetric deficits in single episode, medication-naive depressed patients and the effects of 8 weeks fluoxetine treatment: a VBM-DARTEL study. PLoS One 9(1): e79055. https://doi.org/10.1371/journal.pone.0079055
67. Serra-Blasco M, Portella MJ, Gomez-Anson B, de Diego-Adelino J, Vives-Gilabert Y, Puigdemont D, Granell E, Santos A, Alvarez E, Perez V (2013) Effects of illness duration and treatment resistance on grey matter abnormalities in major depression. Br J Psychiatry J Ment Sci 202:434–440. https://doi.org/10.1192/bjp.bp.112.116228
68. Peng J, Liu J, Nie B, Li Y, Shan B, Wang G, Li K (2011) Cerebral and cerebellar gray matter reduction in first-episode patients with major depressive disorder: a voxel-based morphometry study. Eur J Radiol 80(2):395–399. https://doi.org/10.1016/j.ejrad.2010.04.006
69. Lai CH, Wu YT (2014) Frontal-insula gray matter deficits in first-episode medication-naive patients with major depressive disorder. J Affect Disord 160:74–79. https://doi.org/10.1016/j.jad.2013.12.036
70. Yuan Y, Zhu W, Zhang Z, Bai F, Yu H, Shi Y, Qian Y, Liu W, Jiang T, You J, Liu Z (2008) Regional gray matter changes are associated with cognitive deficits in remitted geriatric depression: an optimized voxel-based Morphometry study. Biol Psychiatry 64 (6):541–544. https://doi.org/10.1016/j.biopsych.2008.04.032
71. Xie C, Li W, Chen G, Douglas Ward B, Franczak MB, Jones JL, Antuono PG, Li SJ, Goveas JS (2012) The co-existence of geriatric depression and amnestic mild cognitive impairment detrimentally affect gray matter volumes: voxel-based morphometry study. Behav Brain Res 235(2):244–250. https://doi.org/10.1016/j.bbr.2012.08.007
72. Salvadore G, Nugent AC, Lemaitre H, Luckenbaugh DA, Tinsley R, Cannon DM, Neumeister A, Zarate CA Jr, Drevets WC (2011) Prefrontal cortical abnormalities in currently depressed versus currently remitted patients with major depressive disorder. NeuroImage 54(4):2643–2651. https://doi.org/10.1016/j.neuroimage.2010.11.011
73. Smith GS, Kramer E, Ma Y, Kingsley P, Dhawan V, Chaly T, Eidelberg D (2009) The functional neuroanatomy of geriatric depression. Int J Geriatr Psychiatry 24(8):798–808. https://doi.org/10.1002/gps.2185
74. Yang X, Ma X, Huang B, Sun G, Zhao L, Lin D, Deng W, Li T, Ma X (2015) Gray matter volume abnormalities were associated with sustained attention in unmedicated major depression. Compr Psychiatry 63:71–79. https://doi.org/10.1016/j.comppsych.2015.09.003
75. Squarcina L, Castellani U, Bellani M, Perlini C, Lasalvia A, Dusi N, Bonetto C, Cristofalo D, Tosato S, Rambaldelli G, Alessandrini F, Zoccatelli G, Pozzi-Mucelli R, Lamonaca D, Ceccato E, Pileggi F, Mazzi F, Santonastaso P, Ruggeri M, Brambilla P, Group GU (2017) Classification of first-episode psychosis in a large cohort of patients using support vector machine and multiple kernel learning techniques. NeuroImage 145 (Pt B):238–245. https://doi.org/10.1016/j.neuroimage.2015.12.007
76. Grieve SM, Korgaonkar MS, Koslow SH, Gordon E, Williams LM (2013) Widespread reductions in gray matter volume in depression. Neuroimage Clin 3(0):332–339. https://doi.org/10.1016/j.nicl.2013.08.016
77. Frodl T, Jager M, Born C, Ritter S, Kraft E, Zetzsche T, Bottlender R, Leinsinger G, Reiser M, Moller HJ, Meisenzahl E (2008a) Anterior cingulate cortex does not differ between patients with major depression and healthy controls, but relatively large anterior cingulate cortex predicts a good clinical course. Psychiatry Res 163(1):76–83. https://doi.org/10.1016/j.pscychresns.2007.04.012
78. Guo W, Liu F, Yu M, Zhang J, Zhang Z, Liu J, Xiao C, Zhao J (2014) Functional and

anatomical brain deficits in drug-naive major depressive disorder. Prog Neuro-Psychopharmacol Biol Psychiatry 54:1–6. https://doi.org/10.1016/j.pnpbp.2014.05.008

79. Smith R, Chen K, Baxter L, Fort C, Lane RD (2013) Antidepressant effects of sertraline associated with volume increases in dorsolateral prefrontal cortex. J Affect Disord 146 (3):414–419. https://doi.org/10.1016/j.jad.2012.07.029
80. Liu CH, Jing B, Ma X, Xu PF, Zhang Y, Li F, Wang YP, Tang LR, Wang YJ, Li HY, Wang CY (2014) Voxel-based morphometry study of the insular cortex in female patients with current and remitted depression. Neuroscience 262:190–199. https://doi.org/10.1016/j.neuroscience.2013.12.058
81. Bouckaert F, De Winter FL, Emsell L, Dols A, Rhebergen D, Wampers M, Sunaert S, Stek M, Sienaert P, Vandenbulcke M (2015) Grey matter volume increase following electroconvulsive therapy in patients with late life depression: a longitudinal MRI study. J Psychiatry Neurosci 40(5):140322. https://doi.org/10.1503/jpn.140322
82. Colloby SJ, Firbank MJ, Vasudev A, Parry SW, Thomas AJ, O'Brien JT (2011) Cortical thickness and VBM-DARTEL in late-life depression. J Affect Disord 133(1–2):158–164. https://doi.org/10.1016/j.jad.2011.04.010
83. Koolschijn PC, van Haren NE, Schnack HG, Janssen J, Hulshoff Pol HE, Kahn RS (2010) Cortical thickness and voxel-based morphometry in depressed elderly. Eur Neuropsychopharmacol 20(6):398–404. https://doi.org/10.1016/j.euroneuro.2010.02.010
84. Delaloye C, Moy G, de Bilbao F, Baudois S, Weber K, Hofer F, Ragno Paquier C, Donati A, Canuto A, Giardini U, von Gunten A, Stancu RI, Lazeyras F, Millet P, Scheltens P, Giannakopoulos P, Gold G (2010) Neuroanatomical and neuropsychological features of elderly euthymic depressed patients with early- and late-onset. J Neurol Sci 299 (1–2):19–23. https://doi.org/10.1016/j.jns.2010.08.046
85. Guze BH, Szuba MP (1992) Leukoencephalopathy and major depression: a preliminary report. Psychiatry Res 45(3):169–175
86. Zhang X, Yao S, Zhu X, Wang X, Zhu X, Zhong M (2012) Gray matter volume abnormalities in individuals with cognitive vulnerability to depression: a voxel-based morphometry study. J Affect Disord 136(3):443–452. https://doi.org/10.1016/j.jad.2011.11.005
87. Watanabe K, Kakeda S, Yoshimura R, Abe O, Ide S, Hayashi K, Katsuki A, Umene-Nakano W, Watanabe R, Nakamura J, Korogi Y (2015) Relationship between the catechol-O-methyl transferase Val108/158Met genotype and brain volume in treatment-naive major depressive disorder: voxel-based morphometry analysis. Psychiatry Res 233 (3):481–487. https://doi.org/10.1016/j.pscychresns.2015.07.024
88. Cheng YQ, Xu J, Chai P, Li HJ, Luo CR, Yang T, Li L, Shan BC, Xu XF, Xu L (2010) Brain volume alteration and the correlations with the clinical characteristics in drug-naive first-episode MDD patients: a voxel-based morphometry study. Neurosci Lett 480 (1):30–34. https://doi.org/10.1016/j.neulet.2010.05.075
89. Li C-T, Lin C-P, Chou K-H, Chen IY, Hsieh J-C, Wu C-L, Lin W-C, Su T-P (2010) Structural and cognitive deficits in remitting and non-remitting recurrent depression: a voxel-based morphometric study. NeuroImage 50 (1):347–356. https://doi.org/10.1016/j.neuroimage.2009.11.021
90. Costafreda SG, Chu C, Ashburner J, Fu CH (2009) Prognostic and diagnostic potential of the structural neuroanatomy of depression. PLoS One 4(7):e6353. https://doi.org/10.1371/journal.pone.0006353
91. Ribeiz SR, Duran F, Oliveira MC, Bezerra D, Castro CC, Steffens DC, Busatto Filho G, Bottino CM (2013) Structural brain changes as biomarkers and outcome predictors in patients with late-life depression: a cross-sectional and prospective study. PLoS One 8(11):e80049. https://doi.org/10.1371/journal.pone.0080049
92. Jung J, Kang J, Won E, Nam K, Lee M-S, Tae WS, Ham B-J (2014) Impact of lingual gyrus volume on antidepressant response and neurocognitive functions in major depressive disorder: a voxel-based morphometry study. J Affect Disord 169(0):179–187. https://doi.org/10.1016/j.jad.2014.08.018
93. Yucel K, McKinnon M, Chahal R, Taylor V, Macdonald K, Joffe R, MacQueen G (2009) Increased subgenual prefrontal cortex size in remitted patients with major depressive disorder. Psychiatry Res 173(1):71–76. https://doi.org/10.1016/j.pscychresns.2008.07.013
94. Chen CH, Ridler K, Suckling J, Williams S, Fu CH, Merlo-Pich E, Bullmore E (2007) Brain imaging correlates of depressive symptom severity and predictors of symptom improvement after antidepressant treatment. Biol Psychiatry 62(5):407–414. https://doi.org/10.1016/j.biopsych.2006.09.018
95. Lai CH, Hsu YY (2011) A subtle grey-matter increase in first-episode, drug-naive major

depressive disorder with panic disorder after 6 weeks' duloxetine therapy. Int J Neuropsychopharmacol 14(2):225–235. https://doi.org/10.1017/S1461145710000829

96. Frodl TS, Koutsouleris N, Bottlender R, Born C, Jager M, Scupin I, Reiser M, Moller HJ, Meisenzahl EM (2008b) Depression-related variation in brain morphology over 3 years: effects of stress? Arch Gen Psychiatry 65(10):1156–1165. https://doi.org/10.1001/archpsyc.65.10.1156

97. Pizzagalli DA, Oakes TR, Fox AS, Chung MK, Larson CL, Abercrombie HC, Schaefer SM, Benca RM, Davidson RJ (2004) Functional but not structural subgenual prefrontal cortex abnormalities in melancholia. Mol Psychiatry 9 (4):325, 393-405. https://doi.org/10.1038/sj.mp.4001469

98. Dusi N, Cecchetto F, Brambilla P (2015) Magnetic resonance Spettroscopy Stuides in bipolar disorder patients: focus on potential role of oxidative stress. In: Dieterich-Muszalska A, Grygnon S, Chauhan V (eds) Studies on psychiatric disorders. Springer, New York Heidelberg Dordrecht London, pp 171–196

Check for updates

Chapter 22

Brain Morphometry: Suicide

Savannah N. Gosnell, David L. Molfese, and Ramiro Salas

Abstract

Suicidal behavior has historically been considered a symptom of depressive episodes and a criterion for borderline personality disorder (Association AP, *Diagnostic and Statistical Manual of Mental Disorders*, 2000). Past neuroimaging research has thus treated suicide as an associated symptom of affective, psychotic, and personality disorders or as a side effect of neurologic damage in epilepsy or traumatic brain injury (TBI), rather than as a separate condition. Consequently, individual studies involving suicidal patients have largely focused on a single diagnostic group, such as unipolar depression, rather than studying suicidal behavior independent of diagnosis. As of the fifth edition of the *Diagnostic and Statistical Manual of Mental Disorders* (Association AP, *Diagnostic and Statistical Manual of Mental Disorders*, 2013), suicidal behavior is now considered an independent mental disorder. This reclassification may encourage research aimed at identifying brain morphometry specific to suicidal behavior, even if those characteristics cut across diagnoses. Through isolation of brain morphometry common to suicidal behavior without regard to diagnosis, it may be possible to predict susceptibility to suicidal ideation and future suicide attempt.

Key words Suicide, MRI, Connectivity, Brain morphometry, Brain volume

1 Introduction

Suicide [1, 2] is a pressing public health issue. The World Health Organization estimates the annual incidence of suicide at 16 per 100,000 global deaths [3]. In the United States alone, 36,000 people commit suicide each year [4], making suicide the second leading cause of death among those aged 10–34 [3, 4]. Fully 13.5% of Americans have experienced some form of suicidal ideation during their lifetime, even if they do not attempt suicide [5]. In total, 2.7% of Americans have attempted suicide with intent to die [6]. Suicide attempt is a strong predictor of future suicide, as half of all successful suicides followed an earlier, unsuccessful attempt [7]. Of who previously attempted suicide, 10–20% complete a future suicide attempt [8]. Another strong predictor of suicide attempt is mental disorder. An estimated 90% of suicides are individuals diagnosed with mental disorders [9]. At least some of the one million annual suicides are likely to be individuals with

Gianfranco Spalletta et al. (eds.), *Brain Morphometry*, Neuromethods, vol. 136,
https://doi.org/10.1007/978-1-4939-7647-8_22,

undiagnosed mental disorders, although the size of this population and the suicide rate within that population is difficult to estimate. Identification of a set of brain morphological markers common to individuals who have attempted suicide is thus an important step toward diagnosing and treating this population.

Prior to the invention of noninvasive neuroimaging technologies, morphometric suicide research focused primarily on postmortem examination of tissue. Such research was of limited utility in identifying predictors of suicide attempt. Modern neuroimaging techniques, such as magnetic resonance imaging (MRI) and diffusion tensor imaging (DTI), are capable of identifying changes in gross brain morphology, including differences in the size and density of individual brain regions, the presence of white matter hyperintensities, and alterations in white matter integrity that correlate with suicide attempt. As elaborated in this chapter, such markers may enable more effective diagnosis of suicidal behavior and offer insights into the treatment of this disorder.

Before examining premortem brain morphology in suicide attempters, it is important to operationally define suicide attempt as harming oneself with intent to die. The majority of studies reviewed herein included suicidal individuals with at least one attempt at any point in their lifetimes, without regard to recency of attempt. As of the DSM-5, suicide behavior disorder now requires an attempt within the past 2 years [2]. This change reduces the number of patients classified as suicidal and may reduce the heterogeneity of suicidal populations in research samples. Despite this straightforward definition of suicidal attempt, suicide attempters present across a wide range of ages with comorbid psychiatric and health conditions, each of which likely contributes to overall suicide risk. Further variation in suicidal intent and lethality of attempt suggest suicide attempters are a more heterogeneous population than those who commit suicide [10]. This variation creates challenges for neuroimaging researchers, including difficulty recruiting large, well-matched samples of suicidal patients and the fact that brain morphology varies in relatively minor ways between individuals.

Outside of readily observable brain morphological changes in developmental disorders such as spina bifida or the general enlargement of ventricles in schizophrenia, the size, shape, and connectivity of brain structures in different psychiatric populations are substantially similar. A hypothesis-driven approach to identifying brain structures that are likely to vary in suicidal populations is required to avoid false positives, especially in small sample sizes. Brain structures likely to play a role in suicidal ideation and suicide attempt are those involved in planning and emotional regulation. In general, suicidal patients dwell on past failures and have a perceived inability to anticipate and solve problems [11]. Fronto-striatal-limbic circuitry is thus a promising target for suicide research, as this network is involved in executive functions, planning, and mood or emotional regulation [12–14]. To date, the majority of neuroimaging studies involving suicidal patients have focused on these brain regions. These studies include individual psychiatric populations with and without suicidal attempt along with a limited number of studies focused exclusively on suicidal attempt without regard to psychiatric diagnosis. This chapter reviews morphometric studies of suicidal behavior with a focus on consistencies across diagnoses and structural neuroimaging methodologies. Studies of white and gray matter structure, including white matter hyperintensities and diffusion tensor imaging of white matter connectivity, are considered.

2 White and Gray Matter Structural Imaging

Magnetic resonance imaging (MRI) offers a high-resolution, non-invasive method for visualizing brain structures and obtaining quantitative data such as gray and white matter volume, density, cortical thickness, and surface area.

Volumetric analysis of brain structures may offer insights into psychiatric disorders, including suicidal behavior. To date, more than 20 neuroimaging studies have examined structural differences in psychiatric patients with suicidal ideation/attempt compared to either patients with the same psychiatric diagnosis (but without suicidal ideation/attempt) or to nonpsychiatric healthy controls. Thus far, no consistent pattern of changes in brain structure has emerged as predictive of suicidal ideation/attempt.

The first effort to identify a structural correlate of a neuropsychiatric condition used computed tomography (CT), a low-resolution imaging method, to measure the brains of schizophrenia patients. An inverse correlation between ventricle size and cognitive impairment was observed [15]. Thus it is not surprising that the first structural brain imaging study of suicidality also examined ventricular size [16]. No relationship was found. As neuroimaging methods have improved and image resolution increased, more sophisticated studies of brain structure in suicidal populations have examined white and gray matter volumes in regions of interest (ROIs) believed to be involved in suicidal ideation, such as those underlying cognitive control, response inhibition, sense of self, mood, emotional regulation, and reward processing.

One of the first structural MRI studies of suicidality examined unipolar depression patients with and without suicidal attempt. In a small, all female sample, a reduction in orbitofrontal cortex (OFC) gray matter volume was observed in seven suicidal patients compared to seventeen healthy controls [17]. These results were interpreted as a deficit in decision-making and impulsivity, although patients were not evaluated to determine if any cognitive/

behavioral deficits were present. The study also failed to correct for multiple comparisons across four ROIs (orbitofrontal cortex, cingulate, amygdala, and hippocampus) in each hemisphere (eight total comparisons), and the results did not survive statistical correction. Nonetheless, this study provided a jumping-off point for future research into the relationship between brain structure and suicidality.

Over the past several years, a general reduction in frontal cortex volume has been observed in suicidal psychiatric populations [18], although this reduction is inconsistently reported in the superior/middle [19] and inferior frontal cortex [20, 21]. As a whole, the frontal cortex underlies executive functions such as planning and decision-making [14]. Specifically, the medial and inferior frontal cortices regulate response inhibition [18], which contributes to the regulation of emotionally driven behavior. Impaired planning, decision-making, and inhibition of emotional reactions may contribute to increased risk of suicide [18].

Reductions in brain volume have also been observed in the dorsolateral [22], dorsomedial [18], and ventrolateral prefrontal cortex (PFC) [23] of suicidal patients with affective disorders. The PFC has been implicated in planning and decision-making [24], personality [25], and working memory [26]. Brain volume in suicidal patients with affective and personality disorders is also reduced in the anterior cingulate cortex [21–23, 27], a brain region that subserves emotional awareness [28], pain perception [29], and reward evaluation [30].

Suicidal populations with and without affective disorders exhibit brain volume reductions in the lentiform nucleus (putamen and globus pallidus) [18, 31, 32], caudate [21, 32], and striatum (caudate nucleus, putamen, and nucleus accumbens) [31]. Disruption of reward processing by these structures may lead to discounted delay where patients impulsively prefer immediate, lesser rewards to larger rewards later [33]. While differences in brain volumes alone cannot answer questions related to differences in connectivity between brain regions, it should be noted that the striatum receives inputs from frontal cortex regions, including the anterior cingulate and OFC [31]. Reduced brain volume in frontal cortex regions involved in planning and decision-making as well as striatal regions involved in reward processing and delay of gratification may explain increased impulsivity and emotional dysregulation in suicidal patients, although functional imaging research and neuropsychological assessments will be required to establish a causal link. Lending support for a link between reduced connectivity and increased impulsivity, reduced corpus callosum volume was reported in a sample of suicidal adults with unipolar depression [34]. These patients scored higher on the Barratt Impulsivity Scale, a measure of impulsive behavior [35]. Reduced corpus callosum volume was also observed in a population of suicidal seniors

(71 ± 3.7 yrs), although this may reflect normal aging [36]. It should be noted that fronto-limbic circuitry might be disrupted in suicidal patients with psychosis wherein thalamus volume is decreased [19]. However, thalamic volume is increased in suicidal patients with traumatic brain injury (TBI) [37]; suggesting differences in thalamic volume may not correspond to general suicidal pathology.

A general reduction in temporal lobe volume has also been reported in senior suicidal populations with depression [18] and in suicidal patients with epilepsy [38], a neurological condition that frequently manifests as over-activation of the temporal lobe. Other studies of psychiatric populations have reported reduced brain volume in the inferior and superior temporal lobe of psychotic and bipolar disorder patients with suicidal attempt [19], medial and superior temporal lobe in suicidal borderline patients [39], or in only the superior temporal lobe of suicidal bipolar patients with current depression [22]. Temporal lobe regions are more commonly associated with language and memory functions, rather than with cognitive functions such as planning, emotional regulation, and response inhibition where impairment is more commonly linked to increased risk of suicide.

The observed reductions in temporal lobe volume may reflect an increase in memory deficits among depressed populations, which in turn have increased suicide risk. Indeed volume of hippocampus, a necessary structure for declarative and spatial memory formation, is reduced in suicidal patients with unipolar depression [40]. Reduced amygdala volume may also be present in suicidal patients with unipolar depression, suggesting emotional valence of memories may be affected [21]. It should be noted, however, this study examined the hippocampus/amygdala formation as a whole, not as independent structures; thus, the observed volume reduction may be independent of amygdala. Indeed increases in amygdala volume have been reported in suicidal patients with unipolar depression [17] and schizophrenia [41]. Insula volume is also reduced in suicidal patients with late-onset depression [18], psychotic [19], and personality disorders [39]. The insula plays a role in self-perception and in interpreting how others perceive you [19]. Disruption of one's ability to perceive self in a positive light or properly interpret how the self is viewed by others may contribute to negative self-opinion and lead to suicidal behavior. Alternatively, deficits in social emotional processing (including perception of emotions and faces), functions of the superior temporal cortex, may lead to an increase in social anxiety and suicidal behavior. Corresponding reductions in volume of the fusiform gyrus (face processing), lingual gyrus (visual memory), and parahippocampal gyrus (memory encoding and decoding) have been observed in borderline personality and psychotic patients with a history of high-lethality suicide attempt [19, 39].

Additional reports of reductions in parietal and occipital lobe volume among suicidal patients suggest sensory processes, such as touch and vision, may be altered in this population. A general reduction in parietal lobe volume has been observed in suicidal depressed patients [18, 22], while more specific reductions have been observed in the superior parietal lobe and supramarginal gyrus of psychotic patients with suicidal attempt [19]. In the occipital lobe, suicidal patients with bipolar disorder exhibit decreased volume [22], while cuneus volume was reduced in high-lethality suicidal patients with psychotic and personality disorders [19]. History of self-injurious/parasuicidal behavior in adolescents with borderline personality disorder predicts pituitary gland volume, suggesting the hypothalamic-pituitary-adrenal (HPA) axis may be involved in suicidal behavior in adolescents [42]. Additionally, reductions in midbrain and cerebellum volume have been observed in senior (79.4 ± 5.3 yrs), suicidal populations with late-onset depression, although a lack of similar findings in adult populations suggests these differences may be the result of normal aging [18].

Taken together, volumetric studies have found widespread decreases in white and gray matter. The most consistent findings suggest a role for the fronto-striatal-limbic circuitry in suicidal pathology. Alterations of this circuitry could cause deficits in decision-making and emotional processing, which could result in negative emotional bias and impulsive behavior.

Figure 1 summarizes the main finding of studies investigating gray and white matter structures in suicidal individuals.

3 White and Gray Matter Hyperintensities

While structural MRI images of white and gray matter depend upon longitudinal T1 (spin-lattice) relaxation times for high-resolution structural contrasts, much shorter transverse T2 (spin-spin) relaxation times and fluid attenuation inversion recovery (FLAIR) sequence, which has a long T1 time to remove the signal from fluid, highlight within tissue differences, such as the presence of gray and white matter hyperintensities. Gray matter hyperintensities are attributed to dysfunctional vascular processes, such as subcortical lacunar infarcts that occlude blood supply, resulting in dead tissue. These hyperintensities are interpreted as lesions or neuronal loss in the affected structures [43]. White matter hyperintensities are of two varieties, depending on origin and location: deep and periventricular. Deep white matter hyperintensities result from vascular disruption of blood flow and are similar to gray matter hyperintensities. Periventricular white matter hyperintensities, however, are caused by ependymal loss, cerebral ischemia, or varying levels of myelination and thus may reflect either cell loss or

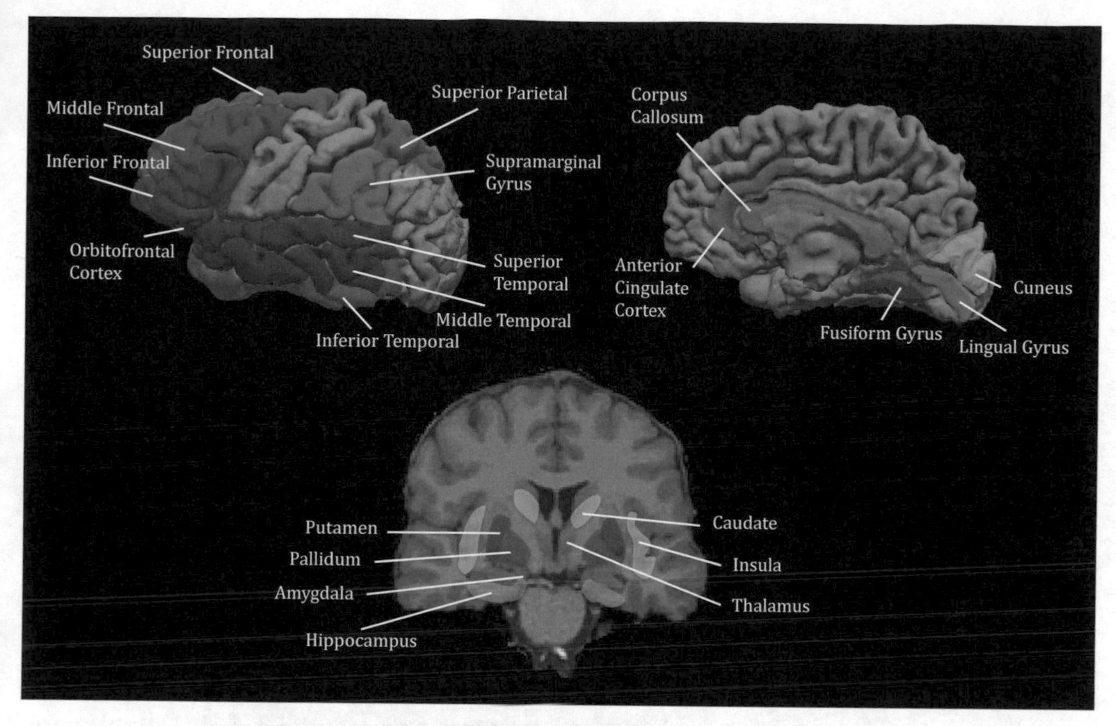

Fig. 1 Studied investigating gray and white matter structures in suicidal individuals reported alterations in every lobe and in several subcortical structures shown above. The most consistent cortical findings showed frontal and temporal reductions, while striatal and limbic structure reductions were the most common subcortical alterations reported

demyelination [44, 45]. Despite general consensus on the etiology of gray and white matter hyperintensities, it is unclear if these morphological changes cause psychiatric disorders, and any effect of hyperintensities likely depends on location.

Five neuroimaging studies of patients diagnosed with unipolar depression and/or bipolar disorder have reported an increase in the total number of hyperintensities compared to non-suicidal patients. Four of these studies identified increased white matter hyperintensities, and the fifth identified an increase in gray matter hyperintensities. In suicidal patients, white matter hyperintensities are between 4.7 [46] and 8 times more common [47]. These hyperintensities are typically located in periventricular white matter hyperintensities areas [47, 48], suggesting brain regions proximal to the lateral ventricles may be more susceptible to vascular incident in suicidal populations. In suicidal adolescents, especially those with unipolar depression, deep white matter hyperintensities are more common, particularly in the right parietal lobe, a region involved in attentional orienting and task switching [49]. Inability to shift between tasks and manage attention could result in feelings of being overwhelmed, leading to depression and suicidal behavior. Patients with unipolar depression also exhibit an increase in subcortical gray matter hyperintensities compared to non-suicidal psychiatric controls [50]. It is unclear if these hyperintensities are the result of prior suicide attempts or an independent pathology specific to suicidal behavior.

A small study of individuals who attempted suicide using carbon monoxide (CO) intoxication ($n = 6$) found an increase in white matter hyperintensities compared to age- and sex-matched healthy controls [51]. This study, conducted without regard to psychiatric diagnosis, identified hyperintensities in the centrum semiovale, a deep white matter structure consisting of cortical projections and association fibers. These findings are consistent with other studies of suicidal populations, although more research is required to determine if the observed white matter hyperintensities were aggravated by CO or if the presence of hyperintensities led to suicide attempt.

While both white and gray matter hyperintensities have been observed in suicidal populations, the underlying mechanism by which this pathology contributes to suicidal behavior remains unclear. In addition to suicidal behavior, an increase in white matter hyperintensities has been observed in patients with late-onset depression, multiple psychiatric admissions, increased risk of relapse, and poorer long-term prognosis [52–55]. Complicating interpretation of these findings is the increase in hyperintensities with age and in cardiovascular disease [56, 57]. Vascular events in otherwise healthy individuals with risk factors for cardiovascular disease can result in both gray and white matter hyperintensities, which may, in turn, lead to depression and suicidal behavior [58, 59]. Alternatively, the observed increase in hyperintensities among suicidal patients may be independent of normal aging and cardiovascular risk factors, thus serving as a reliable marker for suicidal risk [46, 48]. Hyperintensities may even predispose an individual to psychiatric conditions and potentially also toward suicide, given increased risk of psychiatric symptoms in individuals with neonatal white matter injury due to lack of oxygen and hypoxic ischemic insults at birth [60].

Outside of neonatal injury, the impact of hyperintensities on suicidal behavior or other psychiatric disorders likely depends on the recency of ischemic event and also on location [61]. Within suicidal patients, hyperintensities are most commonly observed in brain regions that regulate mood, including the frontal cortex, amygdala, hippocampus, thalamus, basal ganglia, internal capsule, and associated white matter projections [61, 62]. Disruption of these neural circuits is likely to result in mood dysregulation and may increase risk of depression and other psychiatric illnesses [46, 63]. For example, gray matter hyperintensities are increased in the basal ganglia of suicidal patients [50], suggesting disruption of the brain's mood regulation circuitry increases suicide risk. In addition, deep white matter hyperintensities are more common in the right parietal lobe of suicidal adolescents with a primary DSM-III-R or DSM-IV Axis I clinical psychiatric diagnosis [49], suggesting such hyperintensities may be linked to depression [64]. Genetic and environmental stressors may aggravate the impact of hyperintensities by further disrupting neural circuitry responsible for executive planning and mood regulation [47]. Unfortunately, inconsistencies in the location of hyperintensities and the presence of hyperintensities in non-suicidal populations hinder use as a diagnostic predictor of suicidal behavior.

4 White Matter Integrity and Connectivity

As an alternative to measuring gross structural changes in T1- and T2-weighted brain images, the size/extent and myelination of white matter pathways can be assessed using diffusion tensor imaging (DTI). DTI measures diffusion of water molecules. Unlike the isotropic diffusion of water molecules in a liquid-filled sphere, diffusion of water molecules in the brain is constrained by the shape and orientation of brain structures, measurable down to the level of white matter fiber bundles. Diffusion of water along white matter pathways is directional (anisotropic) and is determined by the orientation and thickness of those pathways as well as the extent of myelination. DTI has proven effective in detecting demyelinating disorders such as multiple sclerosis. The technology can also identify disruption of white matter pathways following even mild traumatic brain injury [65]. By examining the disruption of white matter brain pathways, DTI may provide insight into suicidal behavior, including where white matter pathways are disrupted and the potential impact of those disruptions on communication within and between brain regions involved in executive planning, inhibition control, and mood regulation.

The most commonly reported DTI measure is fractional anisotropy (FA), a scalar value of total directional water diffusion. High FA values indicate the presence of thick neural pathways and/or thickly myelinated nerve fibers, whereas low FA values suggest decreased myelin integrity or thinner white matter pathways [66]. FA alone cannot distinguish between large white matter pathways with many fibers and thinner pathways with thick myelination, as both constrain the diffusion of water. Less often reported DTI measures, including radial diffusivity (RD) and axial diffusivity (AD), offer further insights into white matter structure and integrity. RD, a measure of the perpendicular diffusion of water away from white matter pathways, is an indicator of demyelination, while AD, diffusion of water in the direction of white matter pathways, is thought to be a general measure of axonal health with lower values suggesting axonal degeneration [66]. Measures of FA, RD, and AD are susceptible to interference from crossing fibers that cause water diffusion to appear isotropic at points of intersection, making interpretation of DTI findings difficult in areas with a high number of crossing fibers [12].

Mood disorders such as unipolar depression, bipolar disorder, and panic disorder have high rates of suicidal thought and behavior. Disruption of white matter myelination, integrity, or connectivity in brain regions involved in mood regulation may lead to suicidal behavior [67]. Across multiple studies, reduced white matter integrity in the medial PFC and orbitofrontal cortex (OFC) has been reported in suicidal patients with unipolar depression and bipolar disorder [12, 13, 68, 69]. The frontal cortex is of particular interest due to the role this region plays in regulating emotion [68, 70]. It has been hypothesized that disruptions of OFC white matter impair decision-making and emotional processing and may affect impulsivity in depressed and suicidal individuals [13, 68].

A study of suicidal patients with unipolar depression found reduced FA in right hemisphere subgyral frontal cortex white matter, compared to healthy controls [12]. These patients also exhibited increased RD in this region, with no change in AD. An increase in RD with a drop in FA may be interpreted as reduced myelination, rather than general loss of white matter pathways or degeneration of those pathways. Demyelination of the frontal cortex in suicidal patients may impair communication between frontal brain regions involved in executive planning. Decreased FA in the right dorsomedial PFC has also been observed in suicidal patients with unipolar depression relative to non-suicidal unipolar depression patients and healthy controls [69]. The right dorsomedial PFC is believed to play a general role in emotional processing and a specific role in perception of self, including self-worth. Distorted self-perception has been linked with depression [69]. Two additional studies of white matter integrity in suicidal patients with mood disorders found reduced FA in the left OFC [13, 68], suggesting white matter integrity is not consistently altered in the left or right hemispheres of suicidal patients.

Changes in white matter connectivity within subcortical structures involved in emotional regulation have also been reported. In suicidal patients with unipolar depression, decreased FA has been observed in the thalamus and lentiform nucleus, basal ganglia structures involved in mood regulation and emotional processing [12, 13]. Disruption of white matter pathways in the lentiform nucleus may impact regulation of serotonin and dopamine, neurotransmitters that regulate mood and are known to be dysregulated in depression and suicidal behavior [71]. Changes in white matter connectivity within the lentiform nucleus have the potential to disrupt serotonergic and dopaminergic signaling in direct projections to other limbic and striatal structures involved in emotional regulation, such as the caudate and amygdala [12].

In suicidal patients, reduced FA has also been observed in fronto-striatal circuitry, including the left anterior limb of the internal capsule, a structure that connects the frontal cortex to thalamus and lentiform nucleus [12, 13]. Disruption of pathways leading from brain regions involved in executive planning to regions that regulate emotion has the potential to affect suicidal behavior by disinhibiting emotional processes or interfering with rational judgment and planning. Similar reductions in FA have been observed in the centrum semiovale of patients who attempted suicide using CO, although it is unclear if changes in white matter drove the suicide attempt or if CO poisoning directly impacted the centrum semiovale [51].

It is likely that disruption of white matter pathways in frontal and limbic regions affects executive planning, decision-making, mood, and emotional regulation [66]. More specifically, disinhibition of emotional processing or impaired judgment and planning in emotional situations may increase suicide susceptibility [13]. However, reductions in FA among suicide attempters are not consistent across psychiatric diagnoses. Unlike suicidal patients with unipolar depression, suicidal patients with panic disorder or mild traumatic brain injury (TBI) exhibit increased FA in the thalamic radiations, internal capsule, and superior corona radiata, important white matter pathways that connect the cortex to subcortical structures [37, 72]. The discrepancy between increased FA in some suicidal patients and decreased FA in other patients, even within the same frontal striatal white matter pathways (e.g., internal capsule), suggests suicidal behavior does not have a single underlying cause. Further study of frontal striatal white matter pathways is required to explain inconsistencies in FA across suicidal populations. In addition, future DTI studies should include analysis of RD and AD in addition to FA, as these additional measures provide insights into demyelination and disruption of white matter integrity, both of which may be altered in suicidal populations.

Figure 2 summarizes the main findings of studies investigating white matter connectivity in suicidal individuals.

5 Limitations

Structural neuroimaging offers potential insight into the neural circuitry underlying suicidal behavior. However, a general lack of consistency in methods and population characteristics hampers attempts to identify a single set of morphological characteristics predictive of suicide attempt. Morphological studies are correlational and retrospective by nature. The rate at which hyperintensities develop or at which myelination deteriorates and white matter pathways degrade in suicidal populations is unknown. Further complicating assessment of suicide risk, white matter

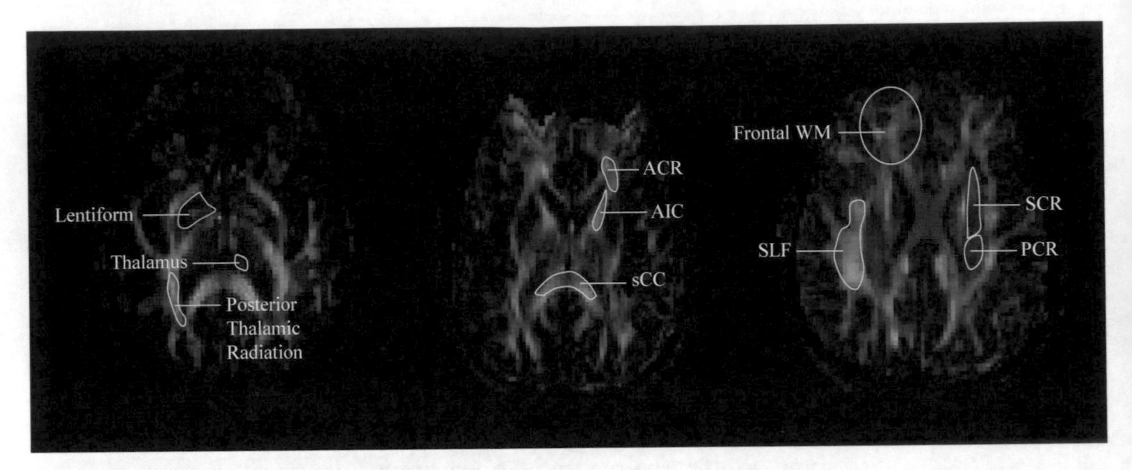

Fig. 2 *AIC* anterior limb of the internal capsule, *ACR* anterior corona radiate, *PCR* posterior corona radiate, *sCC* splenium of the corpus callosum, *SCR* superior corona radiate, *SLF* superior longitudinal fasciculus, *PCR* posterior corona radiate, *WM* white matter. Studies investigating white matter connectivity in suicidal individuals reported alterations in multiple tracts. The most consistent findings showed alterations in connectivity in frontal white matter and thalamic radiations

hyperintensities present from birth may increase risk for psychiatric conditions [60], but may also be related to normal aging and cardiovascular issues independent of suicide risk [58, 59]. Without before- and after-suicide attempt comparisons, it is currently impossible to determine if brain morphological changes observed following suicide predisposed the individual to attempt suicide or if the suicide attempt resulted in the observed brain morphology, particularly in cases of oxygen deprivation, drug overdose, or neurotrauma. By improving sample sizes and through more extensive categorization of suicidal patients, patterns predictive of suicide attempt may emerge.

Untangling the contributions of brain morphology to suicide attempt from morphological changes resulting from suicide attempt is hampered by a general lack of reporting on the seriousness of the attempt and the method of attempt. By definition, the most serious suicide attempts and the most effective methods result in death. Postmortem brain examination using neuroimaging methods is often not possible due to issues such as restricted access to the deceased and degradation of the tissue with time. Among suicide attempters, distinctions between methods of suicide that directly affect the brain, such as oxygen deprivation, drug overdose, or neurotrauma must be made along a continuum to rule out changes in brain morphometry resulting from the attempted suicide versus those present prior to suicide attempt. It should be noted that the method and number of attempts did not correlate with measures of white matter integrity in a small sample of 13 suicidal patients [69]; however, the majority of studies presented in this chapter did not include the method/severity of attempt or number of attempts as variables.

The time between suicide attempt and measurement of any structural or connectivity differences in suicidal patients must also be considered. Hyperintensities or demyelination evaluated immediately after suicide attempt is more likely to have existed prior to the attempt, as cellular death and demyelination in large enough areas to be observable using neuroimaging methods are unlikely to be immediate, unless caused by the method of suicide attempt. Yet of the 32 studies reviewed in this chapter, only three considered the time between most recent suicide attempt and the collection of neuroimaging data [13, 40, 47] and only one study recruited patients from an emergency ward [32]. With DSM-5 reclassification of suicidal behavior as an attempt within the past 2 years, patients who attempted suicide 20 years ago will no longer qualify as suicidal (American Psychiatric Association, 2013). Under the new guideline, sample homogeneity should increase. When paired with clearer definitions of suicidal intent and method, these changes may improve detection of morphological changes present prior to suicidal attempt, and therefore predictive of suicide risk, rather than those resulting from suicide attempt. Researchers with existing data sets should reexamine their patient samples in light of this new 2-year limitation.

The teasing apart of risk factors for suicide attempt is also hampered by small sample sizes and by not dividing patients into different age groups. Adolescents, adults, and elderly populations have different health issues, including cardiovascular and degenerative brain conditions that can result in hyperintensities and demyelination. As such, hyperintensities in a young adult's brain may carry significantly different meaning from those in an elderly person with a history of cardiovascular issues. Likewise, reductions in gray matter volume associated with suicidal behavior are present in cases of childhood trauma [21]; yet relevant personal history information for patients within each study is often missing. It is also important to note that neuroimaging techniques cannot discriminate between vascular and degenerative causes of gray matter reduction [18]. Thus the experimenter must consider likely causes of any observed changes in brain morphology outside of suicide when attempting to identify a set of morphological predictors for suicide. The limited availability of suicidal patients within any single age group is an understandable limitation of research and explains the small, largely heterogeneous samples reviewed herein. To some extent, these variables are controlled for by matching suicidal patients with age- and gender-matched patients with the same psychiatric diagnosis or with matched, healthy controls. However, attempts at matching patients have often resulted in gender neglect. Studies weighted toward or exclusively of one gender should be replicated in complimentary samples using patients of the other gender. Better awareness of gender ratios and larger sample sizes may also facilitate identification of gender-specific brain morphological changes predictive of suicide risk. Such studies may also provide insight into why females with unipolar depression are at greater risk of suicide [46].

6 Conclusions

Numerous studies of suicidal populations with affective, personality, or psychotic disorders; epilepsy; TBI; and aging have utilized MRI as a tool for detecting changes in brain structure, the presence of white matter hyperintensities, and differences in structural connectivity. Across these studies, morphological differences have been found in every lobe of the brain, in several subcortical brain structures, and in major white matter pathways. The reported morphological differences in suicidal patients, compared to either non-suicidal patients with the same diagnosis or to matched healthy controls, vary with suicidal intent and measures of impulsivity. However, intent and impulsivity were not routinely evaluated across studies; most included patients with at least one suicide attempt at any point in their lifetime without regard to time since attempt. While four of the studies reviewed herein divided suicidal patients into high- and low-intent groups [19, 32, 21, 23] and three studies required that the most current (or only) suicide attempt be recent (i.e., within the past 1–6 months; [47, 13, 40]), the majority of studies presented in this chapter made no such restrictions. An underspecified definition of suicide and inconsistencies in observed differences across studies hampers attempts to use morphological brain differences as a diagnostic tool for predicting (and therefore preventing) suicide attempt.

Now that suicidal behavior has been recognized as an independent diagnosis in the DSM-5, future studies should seek to distinguish morphological characteristics unique to suicide from those that may give rise to affective, personality, or psychotic disorders. The most promising area for future suicide research appears to be fronto-striatal circuitry, including white matter pathways connecting brain regions involved in planning, response inhibition, emotional regulation, reward processing, and sense of self [73–77] (Tables 1, 2 and 3).

Table 1
Results from neuroimaging studies exploring gray and white matter structural changes in suicidal individuals

Study	MRI measure	Sample	Definition of suicidal behavior	Results (compared to non-suicidal and/or controls)
Aguilar et al. (2008)	T1	37 (M) schizophrenic: 13 suicidal (37.12 ± 10.02 yr); 24 non-suicidal (42.65 ± 10.19 yr)	At least one attempt	Reduced gray matter density in left superior temporal lobe and left orbitofrontal cortex
Benedetti et al. (2011)	T1	57 bipolar disorder with current depression patients: 19 suicidal (8 lithium, 45.63 ± 11.26 yr.; 5 M, 3 F; 11 no lithium, 43.64 ± 10.36 yr.; 3 M, 7 F), 38 non-suicidal (10 lithium, 46.20 ± 13.27 yr.; 3 M, 8 F; 28 no lithium, 45.93 ± 10.48 yr.; 8 M, 20 F)	At least one attempt	Decreased gray matter volume in dorsolateral prefrontal, orbitofrontal, anterior cingulate, superior temporal, parietal, and occipital cortex; increased gray matter volume in bilateral superior temporal gyrus
Caplan et al. (2010)	T1	51 epilepsy: 11 with suicidal ideation (11.04 ± 2.06 yr.; 6 M, 5 F); 40 without suicidal ideation (9.43 ± 2.07 yr.; 17 M, 23 F)	Presence of suicidal acts and ideations, current and past, abstracted from depression section of schedule for affective disorders and schizophrenia for school-age children—present and lifetime version (K-SADS-PL)	Reduced right orbital frontal gyrus white matter; increased left temporal lobe gray matter
Colle et al. (2015)	T1	63 major depressive disorder: 24 suicidal, including 13 past and 11 acute (44.2 ± 11.9 y; 9 M, 15 F); 39 non-suicidal (47.7 ± 12.6 yr.; 17 M, 22 F)	At least one attempt in lifetime; acute, attempt <1 month; past, attempt >1 month	Reduced hippocampal volume; reduction larger for acute cases
Cyprien et al. (2011)	T1	435 elderly without dementia: 21 suicidal (72.2 ± 4.3 yr.; 5 M, 15 F); 180 non-suicidal, history of depression (71 ± 3.8 yr.; 63 M, 117 F); 234 healthy control (7[illegible] ± 3.8 yr.; 145 M, 89 F)	At least one attempt	Posterior one third of corpus callosum reduced in suicidal group only

(continued)

Table 1 (continued)

Study	MRI measure	Sample	Definition of suicidal behavior	Results (compared to non-suicidal and/or controls)
Dombrovski et al. (2012)	T1	33 elderly with major depression: 13 suicidal (66.0 ± 6.4 yr.; 8 M, 5 F), 20 non-suicidal (67.7 ± 7.0 yr.; 7 M, 13 F); 19 controls (70.5 ± 7.5 yr.; 7 M, 12 F)	At least one attempt and had thoughts of suicide at enrollment	Reduced putamen, associative and ventral striatum, higher delayed discounting; relative to non-suicidal patients
Giakoumatos et al. (2013)	T1	489 schizophrenia, schizoaffective disorder, or psychotic bipolar disorder I: 97 high-lethality attempt (35.6 ± 11.7 yr.; 36 M, 61 F); 51 low-lethality attempt (36.9 ± 12.2 yr.; 14 M, 37 F); 341 non-suicidal (35.9 ± 13.3 yr.; 185 M, 156 F); 262 healthy controls (38.1 ± 12.5 yr.; 129 M, 133 F)	Beck's Lethality Scale: low-lethality (0–1), high medical lethality (2–8)	Decreased gray matter volume in bilateral superior/middle frontal, and inferior/superior temporal regions, left superior parietal and supramarginal regions, and right insula and thalamus. High-lethality patients showed decreased left lingual area and right cuneus compared to low-lethality participants
Goodman et al. (2011)	T1	13 borderline personality disorder and major depressive disorder (15.8 ± 1.1 yr.; 2 M, 11 F); 13 healthy controls (16.2 ± 0.8 yr.; 4 M, 9 F)	At least one attempt	Reduced anterior cingulate cortex (BA24) gray and white matter volume correlated with increase in severity of borderline symptoms and number of suicide attempts
Hwang et al. (2010)	T1	70 (M) late-onset depression (79.4 ± 5.3 yr): 27 history of suicide attempt, 43 no history; 26 healthy controls (79.5 ± 4.3 yr)	At least one attempt	Decreased gray and white matter volume in frontal, parietal, and temporal lobes; insula; lentiform nucleus; midbrain; and cerebellum. Marked reductions in dorsal medial prefrontal cortex
Jovev et al. (2008)	T1	20 adolescent borderline personality disorder patients (17.3 ± 1.1 yr.; 5 M, 15 F)	Semi-structured interview to determine lifetime incidence of self-injurious behavior	Pituitary gland volume predicted by number of parasuicidal behaviors and age

Lijffijt et al. (2014)	T1	93 (F) bipolar disorder I or II; 51 suicidal (36.6 ± 10.7 yr), 42 without suicide attempt (41.1 ± 11.3 yr)	At least one attempt	Deceased PFC gray matter volume in suicidal patients with past psychiatric hospitalization. Increased PFC gray matter volume in suicidal patients without past psychiatric hospitalization. Higher trait aggression predicted attempt history
Matsuo et al. (2010)	T1	47 (F) participants: 10 suicidal bipolar disorder patients (36.2 ± 10.1 yr), 10 non-suicidal bipolar disorder patients (44.2 ± 12.5 yr), 27 healthy controls (36.9 ± 13.8 yr)	At least one attempt	Inverse correlation between volume of anterior genu of the corpus callosum and the Barratt impulsivity scale total, motor, and non-planning scores in suicidal patients
Monkul et al. (2007)	T1	34 (F) participants: 7 suicidal unipolar patients (31.4 ± 13.9 yr); 10 non-suicidal unipolar patients (36.5 ± 7.5); 17 healthy controls (31.3 ± 8.3 yr)	At least one attempt	Reduced orbitofrontal cortex gray matter volume, compared to healthy controls; increased right amygdala volume, compared to non-suicidal patients
Rusch et al. (2008)	T1	105 participants: 10 suicidal schizophrenic patients (30.3 ± 6.5 yr.; 7 M, 3 F), 45 non-suicidal schizophrenic patients (37.3 ± 11.6 yr.; 27 M, 18 F), 50 healthy controls (36.0 ± 9.0 yr.; 31 M, 19 F)	At least one attempt	Increased inferior frontal white matter volume in suicidal patients
Sachs et al. (2014)	T1	246 depressed patients: 23 suicidal (66.74 ± 6.6 yr.; 8 M, 15 F); 223 non-suicidal (69.8 ± 7.5 yr.; 48 M, 175 F)	At least one attempt	Increase in left hemisphere white matter lesions at first MRI scan, compared to scans at 2 and 4 years. Suicide history predicted an increased growth in both left and right white matter lesions over time

(continued)

Table 1 (continued)

Study	MRI measure	Sample	Definition of suicidal behavior	Results (compared to non-suicidal and/or controls)
Soloff et al. (2012)	T1	68 borderline personality disorder patients: 44 suicidal (29.6 ± 8 yr.; 8 M, 36 F) and 24 non-suicidal (25.9 ± 5.7 yr.; 8 M, 16 F), 52 healthy controls (25.9 ± 7.2; 28 M, 24 F)	At least one attempt; lethality defined using Columbia Suicide History Form and Lethality Rating Scale	General decrease in insula gray matter volume relative to non-suicidal patients; high-lethality attempters exhibited decreased orbitofrontal cortex, middle-superior temporal gyrus, insula, fusiform gyrus, lingual gyrus, and parahippocampal gyrus
Spoletini et al. (2011)	T1	50 schizophrenic patients: 14 suicidal (42.9 ± 11.3 yr.; 8 M, 6 F), 36 non-suicidal (39.8 ± 11.4 yr.; 21 M, 15 F), 50 healthy controls (40.0 ± 16.6 yr.; 26 M, 24 F)	At least one attempt	Increased amygdala volume, relative to both non-suicidal patients and healthy controls
Vang et al. (2010)	T1	7 patients recruited from emergency ward following high-lethality suicide attempt (38.1 yr.; 5 M, 2 F); 7 age-/sex-matched healthy controls	Survived a recent, high-lethality suicide attempt	Decreased globus pallidus and right caudate volume. Globus pallidus volumes negatively correlated with solidity (non-impulsive temperament)
Wagner et al. (2011)	T1	60 participants: 15 major depressive disorder, high suicide risk (41.0 ± 12.5 yr.; 4 M, 11 F); 15 major depressive disorder, low suicide risk (34.1 ± 10.5 yr.; 1 M; 14 F); 30 healthy controls (35.1 ± 10.4 yr.; 5 M, 25 F)	At least one attempt or a relative with at least one attempt/completed suicide	Decreased inferior frontal cortex, anterior cingulate cortex, caudate, and amygdala/ hippocampus formation volume relative to healthy controls; decreased anterior cingulate cortex and caudate relative to low suicide risk patients
Wagner et al. (2012)	T1	60 participants: 15 major depressive disorder, high suicide risk (41.0 ± 12.5 yr.; 4 M, 11 F); 15 major depressive disorder, low suicide risk (34.1 ± 10.5 yr.; 1 M; 14 F); 30 healthy controls (35.1 ± 10.4 yr.; 5 M, 25 F)	At least one attempt or a relative with at least one attempt/completed suicide	Decreased ventrolateral PFC, dorsolateral PFC, and anterior cingulate cortex volume in high suicide risk relative to low suicide risk patients and healthy controls

F female, *ROI* region of interest, *M* male, *MRI* magnetic resonance imaging, *PFC* prefrontal cortex

Table 2
Results of neuroimaging studies exploring hyperintensities in suicidal individuals

Study	MRI measure	Sample	Definition of suicidal behavior	Results (compared to non-suicidal and/or controls)
Ahearn et al. (2001)	T2	40 unipolar depression: 20 suicidal (66.0 ± 5.8 yr.; 3 M, 17 F); 20 non-suicidal (66.4 ± 5.7 yr.; 3 M, 17 F)	At least one attempt	Increase in subcortical gray matter hyperintensities, particularly in basal ganglia
Ehrlich et al. (2003)	T2	153 child and adolescent psychiatry inpatients (14.6 ± 3.4 yr.; 114 M, 39 F); with a primary DSM-III-R or -IV Axis I diagnosis; 80 unipolar depression; 43 suicidal	At least one attempt	WM hyperintensities present in suicide attempters. Number of deep white matter hyperintensities in right parietal lobe correlated with suicide risk
Ehrlich et al. (2005)	T2	102 major depressive disorder inpatients (26.7 ± 5.5 yr.; 34 M, 68 F): 62 suicidal; 40 non-suicidal	At least one attempt	Significant increase in periventricular white matter hyperintensities
Lo et al. (2007)	T2	6 suicidal patients (37.3 yr.; 1 M, 5 F); 6 age- and sex-matched healthy controls	Carbon monoxide suicide attempt	Hyperintensities in bilateral periventricular WM and centrum semiovale of suicidal patients. 4 of 6 patients showed bilateral globi pallidi necrosis
Pompili et al. (2007)	T2	65 patients with major depressive disorder or bipolar disorder: 29 suicidal (42.17 ± 13.51 yr.; 5 M, 24 F), 36 non-suicidal (44.61 ± 13.95 yr.; 19 M, 17 F)	At least one attempt	4.7-fold increase in WM hyperintensities
Pompili et al. (2008)	T2	99 bipolar disorder type I, bipolar disorder type II and/or unipolar major depressive disorder inpatients: 44 suicidal (45 57 ± 16.10 yr.; 16 M, 28 F), 55 non-suicidal (47.27 ± 14.54 yr.; 26 M, 29 F)	Attempt within the past 6 months	More periventricular WM hyperintensities. Presence of hyperintensities increased suicide risk eightfold

F female, *M* male, *MRI* magnetic resonance imaging, *WM* white matter

Table 3
Results from neuroimaging studies exploring white matter integrity and connectivity

Study	MRI measure	Sample	Definition of suicidal behavior	Results (compared to non-suicidal and/or controls)
Jia et al. (2010)	DTI	52 major depressive disorder patients: 16 suicidal (34.2 ± 13.7 yr.; 5 M, 11 F); 36 non-suicidal (34.7 ± 12.5 yr.; 20 M, 16 F); 52 healthy controls (37.1 ± 16.0 yr.; 24 M, 28 F)	At least one attempt	Decreased FA and AD in left ALIC; decreased FA, increased RD in right frontal lobe subgyral white matter relative to healthy controls; decreased FA, increased RD in right lentiform nucleus relative to non-suicidal patients
Jia et al. (2013)	DTI	63 major depressive disorder patients: 23 suicidal (36.3 ± 14.5 yr.; 8 M, 15 F), 40 non-suicidal (34.0 ± 14.5 yr.; 21 M, 19 F), 46 healthy controls (33.3 ± 11.4 yr.; 21 M, 25 F)	At least one attempt in past month vs. no attempts	Decreased left ALIC to left orbitofrontal cortex and left thalamus connectivity, relative to non-suicidal patients; decreased FA in left medial frontal cortex, orbitofrontal cortex, thalamus, and fiber number in ALIC, relative to healthy controls
Kim et al. (2015)	DTI	36 panic disorder patients: 12 suicidal (33.4 ± 14.09 yr.; 5 M, 7 F), 24 non-suicidal (34.0 ± 9.38 yr.; 8 M, 16 F)	At least one attempt	Increased FA in internal capsule, splenium, superior and posterior corona radiata, thalamic radiations, sagittal stratum, and superior longitudinal fasciculus. FA for internal capsule and thalamic radiations positively correlated with suicidal ideation
Lo et al. (2007)	DTI	6 suicidal patients (37.3 yr.; 1 M, 5 F); 6 age- and sex-matched healthy controls	Carbon monoxide suicide attempt	Lower FA in centrum semiovale
Lopez-Larson	DTI	59 (M) veterans with mild TBI: 40 non-suicidal (34.60 ± 8.10 yr), 19 suicidal	At least one attempt, interrupted attempt, or aborted attempt	Increased thalamic volume and FA in bilateral thalamic radiations, compared to non-suicidal patients and healthy controls

et al. (2013)		(38.00 ± 7.77 yr), 15 (M) healthy controls (36.47 ± 11.51 yr)		
Mahon et al. (2012)	DTI	44 participants: 14 bipolar disorder with prior attempt (33.3 ± 14.1 yr.; 9 M, 5 F), 15 bipolar disorder without attempt (36.5 ± 12.3 yr.; 9 M, 6 F), 15 healthy controls (33.7 ± 12.6 yr.; 8 M, 7 F)	At least one attempt	Decreased FA in left orbitofrontal cortex WM, compared to non-suicidal patients. Orbitofrontal cortex FA inversely correlated with motor impulsivity in suicidal patients
Olvet et al. (2014)	DTI	52 major depressive disorder patients: 13 suicidal (33.4 ± 13.3 yr.; 6 M, 7 F), 39 non-suicidal (37.1 ± 11.4 yr.; 15 M, 24 F), 46 healthy controls (30.3 ± 9.3 yr.; 25 M, 21 F)	At least one attempt	Decreased FA in right dorsomedial PFC, relative to non-suicidal patients and healthy controls

AD axial diffusivity, *ALIC* anterior limb of internal capsule, *DTI* diffusion tensor imaging, *FA* fractional anisotropy, *F* female, *M* male, *MRI* magnetic resonance imaging, *PFC* prefrontal cortex, *RD* radial diffusivity, *WM* white matter

References

1. Association AP (2000) Diagnostic and statistical manual of mental disorders, Fourth Edition (DSM-IV) edn. AAP, Washington, DC
2. Association AP (2013) Diagnostic and statistical manual of mental disorders. 5th ed. AAP, Washington, DC
3. (WHO) WHO (2008) World Suicide Prevention Day 2008. Around one million people die each year by suicide. Statement 10 September
4. Prevention CfDCa (2011) Web-Based Injury Statistics Query and Reporting System (WISQARS). http://www.cdc.gov/ncipc/dvp/Suicide/suicide_data_sheet.pdf
5. Kessler RC, Borges G, Walters EE (1999) Prevalence of and risk factors for lifetime suicide attempts in the National Comorbidity Survey. Arch Gen Psychiatry 56(7):617–626
6. Nock MK, Kessler RC (2006) Prevalence of and risk factors for suicide attempts versus suicide gestures: analysis of the National Comorbidity Survey. J Abnorm Psychol 115 (3):616–623
7. Isometsa ET, Lonnqvist JK (1998) Suicide attempts preceding completed suicide. Br J Psychiatry 173:531–535
8. Kerkhof AJFM (2000) Attempted suicide: patterns and trends. In: Hawton K, van Herringen K (eds) The international handbook of suicide and attempted suicide. John Wiley & Sons, Ltd., West Sussex, England
9. Reardon S (2013) Suicidal behaviour is a disease, psychiatrists argue. New Scientist, UK, vol May 17 2013, 18:35 17 edn
10. Turecki G (2005) Dissecting the suicide phenotype: the role of impulsive-aggressive behaviours. J Psychiatry Neurosci 30:398–408
11. Desmyter S, Van Heeringen C, Audenaert K (2011) Structural and functional neuroimaging studies of the suicidal brain. Prog Neuro-Psychopharmacol Biol Psychiatry 35 (4):796–808
12. Jia Z, Huang X, Wu Q, Zhang T, Lui S, Zhang J, Amatya N, Kuang W, Chan RC, Kemp GJ, Mechelli A, Gong Q (2010) High-field magnetic resonance imaging of suicidality in patients with major depressive disorder. Am J Psychiatry 167(11):1381–1390
13. Jia Z, Wang Y, Huang X, Kuang W, Wu Q, Lui S, Sweeney JA, Gong Q (2013) Impaired frontothalamic circuitry in suicidal patients with depression revealed by diffusion tensor imaging at 3.0 T. J Psychiatry Neurosci 38 (5):130023
14. Yang Y, Raine A (2009) Prefrontal structural and functional brain imaging findings in antisocial, violent, and psychopathic individuals: a meta-analysis. Psychiatry Res 174 (2):81–88
15. Johnstone E, Frith CD, Crow TJ, Husband J, Kreel L (1976) Cerebral ventricular size and cognitive impairment in chronic schizophrenia. Lancet 308(7992):924–926
16. Schlegel S, Maier W, Philipp M, Aldenhoff JB, Heuser I, Kretzschmar K, Benkert O (1989) Computed tomography in depression: association between ventricular size and psychopathology. Psychiatry Res 29(2):221–230
17. Monkul ES, Hatch JP, Nicoletti MA, Spence S, Brambilla P, Lacerda AL, Sassi RB, Mallinger AG, Keshavan MS, Soares JC (2007) Fronto--limbic brain structures in suicidal and non-suicidal female patients with major depressive disorder. Mol Psychiatry 12(4):360–366
18. Hwang JP, Lee TW, Tsai SJ, Chen TJ, Yang CH, Lirng JF, Tsai CF (2010) Cortical and subcortical abnormalities in late-onset depression with history of suicide attempts investigated with MRI and voxel-based morphometry. J Geriatr Psychiatry Neurol 23 (3):171–184
19. Giakoumatos CI, Tandon N, Shah J, Mathew IT, Brady RO, Clementz BA, Pearlson GD, Thaker GK, Tamminga CA, Sweeney JA, Keshavan MS (2013) Are structural brain abnormalities associated with suicidal behavior in patients with psychotic disorders? J Psychiatr Res 47(10):1389–1395
20. Rusch N, Spoletini I, Wilke M, Martinotti G, Bria P, Trequattrini A, Bonaviri G, Caltagirone C, Spalletta G (2008) Inferior frontal white matter volume and suicidality in schizophrenia. Psychiatry Res 164 (3):206–214. https://doi.org/10.1016/j.pscychresns.2007.12.011
21. Wagner G, Koch K, Schachtzabel C, Schultz CC, Sauer H, Schlosser RG (2011) Structural brain alterations in patients with major depressive disorder and high risk for suicide: evidence for a distinct neurobiological entity? NeuroImage 54(2):1607–1614
22. Benedetti F, Radaelli D, Poletti S, Locatelli C, Falini A, Colombo C, Smeraldi E (2011) Opposite effects of suicidality and lithium on gray matter volumes in bipolar depression. Affect Disord 135(1–3):139–147
23. Wagner G, Schultz CC, Koch K, Schachtzabel C, Sauer H, Schlosser RG (2012) Prefrontal cortical thickness in depressed patients with high-risk for suicidal behavior. J Psychiatr Res 46(11):1449–1455

24. Shimamura A (2000) The role of the prefrontal cortex in dynamic filtering. Psychobiology 28:207–218
25. Damasio A (1994) Descartes' Error. Penguin Putman
26. Baddeley A (1986) Working memory Oxford. Oxford University Press, Oxford
27. Goodman M, Hazlett EA, Avedon JB, Siever DR, Chu KW, New AS (2011) Anterior cingulate volume reduction in adolescents with borderline personality disorder and co-morbid major depression. J Psychiatr Res 45 (6):803–807
28. Lane RD, Reiman EM, Axelrod B, Yun LS, Holmes A, Schwartz GE (1998) Neural correlates of levels of emotional awareness. Evidence of an interaction between emotion and attention in the anterior cingulate cortex. J Cogn Neurosci 10(4):525–535
29. Davis KD, Taylor SJ, Crawley AP, Wood ML, Mikulis DJ (1997) Functional MRI of pain- and attention-related activations in the human cingulate cortex. J Neurophysiol 77:3370–3380
30. Bush G, Vogt BA, Holmes J, Dale AM, Greve D, Jenike MA, Rosen BR (2002) Dorsal anterior cingulate cortex: a role in reward-based decision making. Proc Natl Acad Sci U S A 99(1):523–528
31. Dombrovski AY, Siegle GJ, Szanto K, Clark L, Reynolds CF, Aizenstein H (2012) The temptation of suicide: striatal gray matter, discounting of delayed rewards, and suicide attempts in late-life depression. Psychol Med 42 (6):1203–1215
32. Vang FJ, Ryding E, Traskman-Bendz L, van WD, Lindstrom MB (2010) Size of basal ganglia in suicide attempters, and its association with temperament and serotonin transporter density. Psychiatry Res 183(2):177–179
33. Reynolds B (2006) A review of delay-discounting research with humans: relations to drug use and gambling. Behav Pharmacol 17(8):651–667
34. Matsuo K, Nielsen N, Nicoletti MA, Hatch JP, Monkul ES, Watanabe Y, Zunta-Soares GB, Nery FG, Soares JC (2010) Anterior genu corpus callosum and impulsivity in suicidal patients with bipolar disorder. Neurosci Lett 469(1):75–80
35. Patton JH, Stanford MS (1995) Factor structure of the Barratt impulsiveness scale. J Clin Psychol 51(6):768–774
36. Cyprien F, Courtet P, Malafosse A, Maller J, Meslin C, Bonafé A, Le Bars E, de Champfleur NM, Ritchie K, Artero S (2011) Suicidal behavior is associated with reduced corpus callosum area. Biol Psychiatry 70(4):320–326
37. Lopez-Larson M, King JB, McGlade E, Bueler E, Stoeckel A, Epstein DJ, Yurgelun-Todd D (2013) Enlarged thalamic volumes and increased fractional anisotropy in the thalamic radiations in veterans with suicide behaviors. Front Psych 4:83
38. Caplan R, Siddarth P, Levitt J, Gurbani S, Shields WD, Sankar R (2010) Suicidality and brain volumes in pediatric epilepsy. Epilepsy Behav 18(3):286–290
39. Soloff PH, Pruitt P, Sharma M, Radwan J, White R, Diwadkar VA (2012) Structural brain abnormalities and suicidal behavior in borderline personality disorder. J Psychiatr Res 46(4):516–525
40. Colle R, Chupin M, Cury C, Vandendrie C, Gressier F, Hardy P, Falissard B, Colliot O, Ducreux D, Corruble E (2015) Depressed suicide attempters have smaller hippocampus than depressed patients without suicide attempts. J Psychiatr Res 61:13–18
41. Spoletini I, Piras F, Fagioli S, Rubino IA, Martinotti G, Siracusano A, Caltagirone C, Spalletta G (2011) Suicidal attempts and increased right amygdala volume in schizophrenia. Schizophr Res 125(1):30–40
42. Jovev M, Garner B, Phillips L, Velakoulis D, Wood SJ, Jackson HJ, Pantelis C, McGorry PD, Chanen AM (2008) An MRI study of pituitary volume and parasuicidal behavior in teenagers with first-presentation borderline personality disorder. Psychiatr Res Neuroimaging 162:273–277
43. Boyko OB, Alston SR, Fuller GN, Hulette CM (1994) Utility of postmortem magnetic resonance imaging in neuropathology. Arch Pathol Lab Med 118:219–225
44. Thomas AJ, O'Brien JT, Davis S, Ballard C, Barber R, Kalaria RN, Perry RH (2002) Ischemic basis for deep white matter hyperintensities in major depression: a neuropathological study. Arch Gen Psychiatry 59:785–792
45. Thomas AJ, Perry R, Barber R, Kalaria RN, O'Brien JT (2002) Pathologies and pathological mechanisms for white matter hyperintensities in depression. Ann N Y Acad Sci 977:333–339
46. Pompili M, Ehrlich S, De Pisa E, Mann JJ, Innamorati M, Cittadini A, Montagna B, Iliceto P, Romano A, Amore M, Tatarelli R, Girardi P (2007) White matter hyperintensities and their associations with suicidality in patients with major affective disorders. Eur Arch Psychiatry Clin Neurosci 257:494–499

47. Pompili M, Innamorati M, Mann JJ, Oquendo MA, Lester D, Del Casale A, Serafini G, Rigucci S, Romano A, Tamburello A, Manfredi G, De Pisa E, Ehrlich S, Giupponi G, Amore M, Tatarelli R, Girardi P (2008) Periventricular white matter hyperintensities as predictors of suicide attempts in bipolar disorders and unipolar depression. Prog Neuro-Psychopharmacol Biol Psychiatry 32(6):1501–1507
48. Ehrlich S, Breeze JL, Hesdorffer DC, Noam GG, Hong X, Alban RL, Davis SE, Renshaw PF (2005) White matter hyperintensities and their association with suicidality in depressed young adults. J Affect Disord 86 (2–3):281–287
49. Ehrlich S, Noam GG, Lyoo IK, Kwon BJ, Clark MA, Renshaw PF (2003) Subanalysis of the location of white matter hyperintensities and their association with suicidality in children and youth. Ann N Y Acad Sci 1008 (Dec):265–268
50. Ahearn EP, Jamison KR, Steffens DC, Cassidy F, Provenzale JM, Lehman A, Weisler RH, Carroll BJ, Krishnan KR (2001) MRI correlates of suicide attempt history in unipolar depression. Biol Psychiatry 50(4):266–270
51. Lo CP, Chen SY, Chou MC, Wang CY, Lee KW, Hsueh CJ, Chen CY, Huang KL, Huang GS (2007) Diffusion-tensor MR imaging for evaluation of the efficacy of hyperbaric oxygen therapy in patients with delayed neuropsychiatric syndrome caused by carbon monoxide inhalation. Eur J Neurol 14:777–782
52. Coffey CE, Figiel GS, Djang WT, Weiner RD (1990) Subcortical hyperintensity on magnetic resonance imaging: a comparison of normal and depressed elderly subjects. Am J Psychiatry 147:187–189
53. O'Brien JAD, Chiu E, Schweitzer I, Desmond P, Tress B (1998) Severe deep white matter lesions and outcome in elderly patients with major depressive disorder: follow up study. BMJ 317:982–984
54. Silverstone T, McPherson H, Li Q, Doyle T (2003) Deep white matter hyperintensities in patients with bipolar depression, unipolar depression and age-matched control subjects. Bipolar Disord 5:53–57
55. Yanai I, Fujikawa T, Horiguchi J, Yamawaki S, Touhouda Y (1998) The 3-year course and outcome of patients with major depression and silent cerebral infarction. J Affect Disord 47:25–30
56. Kumar A, Miller D, Ewbank D, Yousem D, Newberg A, Samuels S, Cowell P, Gottlieb G (1997) Quantitative anatomic measures and comorbid medical illness in late-life major depression. Am J Geriatr Psychiatry 5:15–25
57. Awad IA, Spetzler RF, Hodak JA, Awad CA, Carey R (1986) Incidental subcortical lesions identified on magnetic resonance imaging in the elderly. I. Correlation with age and cerebrovascular risk factors. Stroke 17:1084–1089
58. Alexopoulos GS, Meyers BS, Young RC, Campbell S, Silbersweig D, Charlson M (1997) Vascular depression hypothesis. Arch Gen Psychiatry 54:915–922
59. Krishnan KR, Hays JC, Blazer DG (1997) MRI-defined vascular depression. Am J Psychiatry 154:497–501
60. Whitaker AH, Van Rossem R, Feldman JF, Schonfeld IS, Pinto-Martin JA, Tore C, Shaffer D, Paneth N (1997) Psychiatric outcomes in low-birth-weight children at age 6 years: relation to neonatal cranial ultrasound abnormalities. Arch Gen Psychiatry 54:847–856
61. Taylor WD, Payne ME, Krishnan KR, Wagner HR, Provenzale JM, Steffens DC, MacFall JR (2001) Evidence of white matter tract disruption in MRI hyperintensities. Biol Psychiatry 50(3):179–183
62. Soares JC, Mann JJ (1997) The anatomy of mood disorders – review of structural neuroimaging studies. Biol Psychiatry 41(1):86–106
63. Ehrlich S, Noam GG, Lyoo IK, Kwon BJ, Clark MA, Renshaw PF (2004) White matter hyperintensities and their associations with suicidality in psychiatrically hospitalized children and adolescents. J Am Acad Child Adolesc Psychiatry 43(6):770–776
64. Levin RL, Heller W, Mohanty A, Herrington JD, Miller GA (2007) Cognitive deficits in depression and functional specificity of regional brain activity. Cognitive Therapy and Res 31 (2):211–233
65. Toth A (2015) Magnetic resonance imaging application in the area of mild and acute traumatic brain injury. In: Kobeissy FH (ed) Brain Neurotrauma: molecular, neuropsychological, and rehabilitation aspects. CRC Press/Tayor & Francis, Boca Raton (FL)
66. Alexander AL, Lee JE, Lazar M, Field AS (2007) Diffusion tensor imaging of the brain. Neurotherapeutics 4(3):316–329
67. Lippard ETC, Johnston JA, Blumberg HP (2014) Neurobiological risk factors for suicide: insights from brain imaging. Am J Preventive Medicine 47(3):S152–S162
68. Mahon K, Burdick KE, Wu J, Ardekani BA, Szeszko PR (2012) Relationship between suicidality and impulsivity in bipolar I disorder: a

diffusion tensor imaging study. Bipolar Disord 14(1):80–89

69. Olvet DM, Peruzzo D, Thapa-Chhetry B, Sublette ME, Sullivan GM, Oquendo MA, Mann JJ, Parsey RV (2014) A diffusion tensor imaging study of suicide attempters. J Psychiatr Res 51:60–67
70. Kringelbach ML (2005) The orbitofrontal cortex: linking reward to hedonic experience. Nat Rev Neurosci 6:691–702
71. Ryding E, Lindstrom M, Traskman-Bendz L (2008) The role of dopamine and serotonin in suicidal behaviour and aggression. Prog Brain Res 172:307–315
72. Kim B, Oh J, Kim MK, Lee S, Tae WS, Kim CM, Choi TK, Lee SH (2015) White matter alterations are associated with suicide attempt in patients with panic disorder. J Affect Disord 175:139–146
73. American Psychiatric Association (2000) Diagnostic and statistical manual of mental disorders (revised 4th ed.). American Psychiatric Association, Washington, DC
74. World Health Organisation (WHO) Statement 10 September. World Suicide Prevention Day 2008. Around one million people die each year by suicide (2008)
75. Centers for Disease Control and Prevention. Web-Based Injury Statistics Query and Reporting System (WISQARS) (2010)
76. Centers for Disease Control and Prevention. Web-Based Injury Statistics Query and Reporting System (WISQARS) (2011)
77. American Psychiatric Association. Diagnostic and statistical manual of mental disorders (DSM-5®) (2013). American Psychiatric Pub

Chapter 23

Morphological Brain Alterations in Patients with Obsessive–Compulsive Disorder

Premika S.W. Boedhoe and Odile A. van den Heuvel

Abstract

Obsessive–compulsive disorder (OCD) is characterized by repetitive thoughts and behaviors that are experienced as unwanted. Structural magnetic resonance imaging (MRI) studies, using different techniques such as manual tracing, voxel-based morphometry (VBM), cortical thickness analysis, and diffusion tensor imaging (DTI), investigating brain abnormalities in OCD patients have been numerous. These studies have implicated the cortico-striato-thalamo-cortical (CSTC) circuit in the pathophysiology of the disorder. However, results have not always been consistent. Variability in study results may partially be explained by small sample sizes, clinical heterogeneity between patient samples, and methodological differences between studies.

In this chapter, we review the most consistent findings on morphological brain alterations in patients with OCD, and we discuss the relationship within the implicated networks. The reviewed literature shows that the pathophysiology of OCD cannot be explained by alterations in function and structure of the classical CSTC regions exclusively and emphasizes the importance of other fronto-limbic and frontoparietal areas as well as the cerebellum. Moreover, these findings support the notion that the brain alterations found in OCD patients are represented at the network level rather than discrete brain regions. The widespread abnormalities across several different regions and circuits, and their interconnectivity, fit with the complex phenomenology of OCD, which includes different emotional, cognitive, and behavioral domains. A life span perspective on the structural brain abnormalities in OCD is warranted since variation in developmental stage, symptom profile, and disease stage seems to underlie variation in structural abnormalities.

Key words Obsessive–compulsive disorder, Gray matter, White matter, Voxel-based morphometry, Diffusion tensor imaging, Cortical thickness, Brain function

1 Introduction

1.1 Obsessive–Compulsive Disorder

Obsessive–compulsive disorder (OCD) is a relatively common, frequently debilitating neuropsychiatric disorder that affects approximately 1–3% of the population [1–3]. OCD can be characterized by the presence of two sets of symptoms: *obsessions*, which are unwanted and persistent thoughts, urges, or images that produce excessive anxiety or distress, and *compulsions*, which are ritualistic, repetitive, and time-consuming behaviors or mental acts that

Gianfranco Spalletta et al. (eds.), *Brain Morphometry*, Neuromethods, vol. 136,
https://doi.org/10.1007/978-1-4939-7647-8_23,

individuals feel compelled to perform to reduce the anxiety or distress caused by the obsessions [4]. OCD is a clinically heterogeneous disorder with a wide range of symptomatic expression. The age of onset ranges from early childhood into adulthood. In more than 50% of all OCD cases, symptoms emerge during childhood or adolescence [2, 5], and in more than 40% of these early-onset cases, the disorder persists into adulthood [6].

It is becoming increasingly clear that OCD is not a unitary disorder and that it consists of multiple potentially overlapping symptom dimensions [7], which are temporally [8, 9] and transculturally [10] stable. Most studies found that the following four factors best reflect the symptom dimensions in OCD: symmetry obsessions and compulsions; contamination and cleaning; aggressive, sexual, and religious obsessions; and hoarding obsessions and compulsions [11]. Patients may vary widely with respect to symptom profile (e.g., cleaning vs. hoarding), symptom severity, age of symptom onset, and comorbidities (e.g., tic disorders, depression, and anxiety disorders) [12, 13]. The variation in symptom profile indicates the existence of both general and specific etiological factors contributing to the phenomenological heterogeneity. Indeed, several studies using a symptom dimensional approach have suggested that these symptom dimensions may be mediated by partially distinct neural systems [14–17] and may have distinct genetic or etiological origins [10, 18–20].

1.2 Disease Models

The last three decades of investigation into the neurobiological underpinnings of OCD has made substantial progress, mainly since the introduction of in vivo brain imaging. The first neurobiological hypotheses on the etiology of OCD came from descriptions of obsessive–compulsive behavior in patients with subcortical disorders, e.g., Huntington's disease [21], Sydenham's chorea [22], pallidal lesions [23], and frontal lobe lesions [24]. Observations as such, also supported by subsequent animal studies and human neuroimaging research, point to the involvement of parallel, partly segregated, cortico-striato-thalamo-cortical (CSTC) circuits in behavioral control functions involving motor, cognitive, affective, and motivational processes [25–27]. These CSTC circuits describe that innervations from specific frontal regions first project to subregions of the basal ganglia (striatum), travel through direct and indirect circuitries to the thalamus, and finally project back to the frontal regions. Brain imaging studies in OCD supported the CSTC model, which has been the leading disease model for the past two decades. Within these CSTC models of disease, it is hypothesized that an imbalance exists both between the direct and indirect pathways within the specific CSTC circuits and between the parallel CSTC circuits involved in various emotional/motivational, cognitive, and behavioral processes [28–30].

Milad and Rauch (2012) proposed three important CSTC circuits for OCD: the "affective circuit," the "dorsal cognitive circuit," and the "ventral cognitive circuit" [31]. The affective circuit connects the ventromedial prefrontal cortex and the anterior cingulate cortex with the nucleus accumbens and the thalamus and is relevant for affective and reward processing. The dorsal cognitive circuit connects the dorsolateral prefrontal cortex, the caudate nucleus, and the thalamus and is crucial for executive functions such as working memory and planning. The ventral cognitive circuit connects the anterolateral orbitofrontal cortex, the anterior part of the putamen, and the thalamus and is involved in motor preparation and response inhibition [31]. More recently, van den Heuvel et al. (2016) extended the model by including the sensorimotor CSTC related to habitual behavior [32]. The sensorimotor CSTC circuit connects the (pre)motor cortex, putamen, and thalamus and mediates automatic ("overlearned") responding and the transition from goal-directed to habitual behaviors.

Disease models of OCD propose that structural and functional abnormalities in the involved connections play an important role in the pathophysiology of the disease, with specific CSTC circuits mediating specific OCD symptoms and associated neurocognitive dysfunctions [29, 31, 32]. However, based on three decades of brain imaging research in OCD, it is evident that the pathophysiology cannot be explained by alterations in function and structure of the classical CSTC areas exclusively but that fronto-limbic and frontoparietal connections are important as well and that the role of the cerebellum needs more attention in future research [32].

Findings from functional neuroimaging studies of OCD have validated the involvement of fronto-striatal, fronto-limbic, and frontoparietal connections in the pathophysiology of the disease. Structural abnormalities seem to underlie, and probably also result from, these functional abnormalities. In this chapter, we first briefly discuss the history of structural brain imaging techniques in OCD research. Subsequently, we focus on the most consistent findings for specific anatomical regions of interest and discuss their relationship within the associated networks. Lastly, we elaborate on how these findings are implicated in OCD research and might be translated to the clinical field.

1.3 Structural Brain Imaging Techniques Used in OCD Research over the Past Decades

Since the late 1980s, a rapid growth in the number of imaging studies in OCD and improvements in imaging technology and analysis methods have led to considerable advancement in our understanding of the neural substrates of the pathophysiology of OCD. Structural brain abnormalities were first suggested in studies based on qualitative evaluation, whole-brain measurements, and ventricle-to-brain ratios (VBRs) of computed tomography (CT) scans [33, 34]. Subsequent OCD research was predominantly based on volumetric measures of magnetic resonance imaging

(MRI) data, using a region-of-interest (ROI) approach. The ROI-based approach requires manual delineation of cerebral regions in sequential MRI slices, and the areas obtained in each slice are summed up to provide a measure of the volume of the brain structure of interest. In order to minimize observer biases, landmarks and rules for manual tracing must be defined a priori. Given the major emphasis on the orbitofrontal cortex (OFC), striatum, and thalamus in OCD at that time, these studies have primarily focused on anatomical differences in these regions. With the introduction of other imaging modalities, such as positron emission tomography (PET) and functional MRI (fMRI), the list of ROIs extended to limbic structures such as the anterior cingulate cortex (ACC) and more dorsal and lateral prefrontal regions, including frontoparietal regions.

The ROI procedure, however, is laborious, rater-dependent, and limited to a priori hypotheses. Moreover, ROI-based studies are limited in the investigation of neocortical morphology because of the inherent difficulties in defining structurally complex and variable regions of the human cortex. The development of automated techniques in the last decade, such as voxel-based morphometry (VBM), has facilitated systematic morphometry evaluation of the brain as a whole and increased reliability. A key feature of this method is that it examines differences in gray and white matter throughout the brain, without the need to prespecify ROIs.

A disadvantage of VBM is that it is sensitive to a combination of changes in gray matter thickness, intensity, cortical surface area, and cortical folding [35, 36]. It is important to note that cortical surface area and thickness are influenced by different factors during development. Cortical folding measurements are indicative as markers for neurodevelopmental disorders that originate prenatally. In contrast, cortical thickness changes dynamically across the life span as a consequence of development and disease [37]. Thus, analyses of these additional measures are highly relevant given the postulated neurodevelopmental basis for OCD [38]. The most recent advances in neuroimaging techniques using surface-based morphometry allow us to evaluate other structural characteristics like thickness, gyrification, and surface area.

Although research on gray matter abnormalities in OCD has seen considerable progress, the study of white matter (WM) tracts that connect the brain regions implicated in the disorder has received considerably less attention [39]. Growing evidence suggests that OCD symptoms may be at least partly underpinned by reduced WM integrity [39, 40]. Some of the published VBM studies included the information on WM volumes. However, diffusion tensor imaging (DTI) is more suited to study the integrity of WM tracts in OCD. DTI provides a particular unique piece of microstructural information about WM organization and connectivity, which volumetric measurements cannot convey. In the next sections, we discuss the most consistent findings of structural imaging studies in OCD, their relationship within the implicated networks, and their relevance for emotional, cognitive, and behavioral dysfunctions underlying the symptoms.

2 Gray Matter Abnormalities

2.1 Basal Ganglia and Thalamus

Central to the CSTC circuitries are the basal ganglia (BG). Several studies showed increased volume in different components of these deep gray matter (GM) structures in OCD patients compared to healthy controls (Fig. 1) [41–45]. Pujol and colleagues (2004) found that high scores on the "contamination/washing" dimension were associated with the lower volume of the dorsal parts of the bilateral dorsal caudate nucleus, which was replicated by van den Heuvel et al. (2009) [17, 41]. VBM meta-analyses repeatedly, although not consistently, reported increased volumes of the bilateral lenticular nucleus extending to the caudate nucleus [46–49]. The world's largest study to date on subcortical brain volumes by the Enhancing Neuro Imaging Genetics through Meta Analysis (ENIGMA)-OCD consortium, including 622 pediatric subjects and 2967 adult subjects, supports these results. The ENIGMA-OCD consortium reported increased pallidum volume in adult OCD patients compared with controls [50].

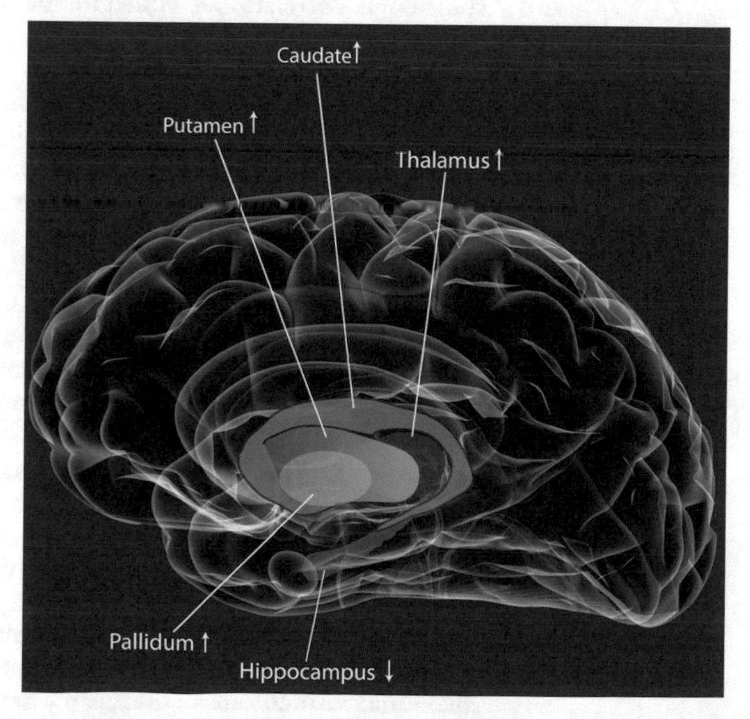

Fig. 1 Subcortical gray matter alteration associated with OCD. The arrows represent either an increase (arrow up) or decrease (arrow down) in gray matter volume

Pujol et al. (2004) showed that the relative enlargement of striatal areas in OCD patients was found to be driven by age and disease duration [41], which was supported by findings of the OCD Brain Imaging Consortium (OBIC) mega-analysis using VBM [51]. These findings suggest that BG alterations progress throughout life. Since these findings were based on cross-sectional data, longitudinal studies are needed to make definite conclusions on the suggested neuroplastic changes during the life span, under the influence of both chronic symptomatology and treatment interventions.

Radua et al. (2010) suggested common as well as distinct neural substrates in OCD and related anxiety disorder patients [48]. The BG seem to be implicated in both OCD and other anxiety disorders. The direction of the findings, however, is diametrically opposite, with OCD patients showing increased BG volume, whereas the other anxiety disorder patients showed decreased BG volume. Since the volume of the BG correlates with the severity of OCD [47] and habitual behavior differentiates OCD from other anxiety disorders, the increased BG volume in OCD may reflect the unique repetitive nature, pathognomonic to OCD.

The investigation of OCD pathophysiology may benefit from studying the illness in childhood, just after the onset of the disorder. In this way, the potentially confounding effects due to plastic changes as a result of chronic symptomatology and long-term treatment effects are minimized, which facilitates to disentangle the neural correlates of vulnerability, chronicity, and treatment effects. So far, pediatric studies were scarce, mostly small sampled, and therefore, the results are not yet conclusive. Szeszko et al. (2008) found increased putamen volume in pediatric OCD patients compared with healthy controls. Additionally, symptom severity correlated significantly with putamen volume in pediatric medication-naïve OCD patients [43] as well as adolescents [45], suggesting that alterations in this region relate to the clinical expression of OCD. In contrast, other pediatric studies reported decreased volume in the putamen [52] and globus pallidus [53], and some studies reported no striatal volume alterations in pediatric OCD [54]. Potential neurodevelopmental factors and limited statistical power due to small sample sizes may contribute these inconsistent results.

Recent advances in neuroimaging analytic techniques allow the definition of surface morphology. It is hypothesized that the shape of specific brain structures may be determined by the physical properties of neural tissue combined with the patterns of neural connectivity [55]. Therefore, shape analysis methods would be more sensitive to detect possible morphological changes in comparison to conventional volumetric measurements. In addition, shape analysis methods might reliably detect subtle volume changes of small regions relevant to OCD, such as the striatal structures

[56]. Vertex-wise shape analyses in adolescents and adults with OCD suggest hypertrophy on the anterior–superior aspect of the caudate [45, 57, 58].

The volume of the thalamus, the key relay station in the CSTC circuits, is shown to be increased in both adult [59, 60] and pediatric OCD patients [61]. Some studies also showed expansion of the surface of the right anterolateral thalamus and loss of typical asymmetry in the pulvinar thalamic nuclei [58, 62]. Additive support came from two meta-analyses by Rotge et al. [49, 63] showing bigger thalamus volume in OCD patients while combining pediatric and adult data. However, more recent meta-analyses did not find altered thalamus volume in OCD patients [46–48]. The OBIC VBM mega-analysis in adult OCD even showed decreased thalamic volume [51]. Recently, the ENIGMA-OCD consortium reported an enlarged thalamus only in pediatric, not in adult, OCD patients [50].

One of the key differences between adult and pediatric OCD seems to be that an increased volume of the caudate, putamen, or pallidum seems to be present in adult OCD (probably related to disease severity, disease chronicity, and/or treatment), whereas an increased volume of the thalamus seems more likely to be implicated in children (probably related to altered neurodevelopment [50]).

2.2 Orbitofrontal Cortex

The OFC is a region that has been consistently implicated in functional neuroimaging studies of OCD and constitutes the basis of the most widely accepted neurobiological model of OCD. Nevertheless, structural neuroimaging findings on the OFC of both adult and pediatric samples have been largely inconsistent. Meta-analyses reported normal [47, 48], greater [49], or smaller [46, 63] OFC volumes in OCD patients (Fig. 2). Sources of inconsistency may be partially explained by the different methods used such as typical ROI-based meta-analyses [63] and voxel-based meta-analytic approaches like anatomical likelihood estimation (ALE [49]) and (effect size) signed differential mapping (ES-SDM [46–48]). Second, the exact location of the changes in this region was heterogeneous across studies. Studies reported abnormalities in different portions (e.g., lateral/medial) of the OFC. The OFC is indeed heterogeneous both in structure and function, and the precise role of its subregions in OCD is still left to be unraveled. Third, the heterogeneity between studies may be attributed, at least in part, to variation in the clinical features of the study populations. For instance, the observed increased regional GM volume of the OFC was more pronounced in patient populations without comorbid depression [64–66]. Furthermore, Rotge et al. (2009) argued, after additional analyses, that disease duration, more specifically the duration of untreated symptoms, and medication may influence the OFC volume in OCD [63]. Recently, OCD studies have been

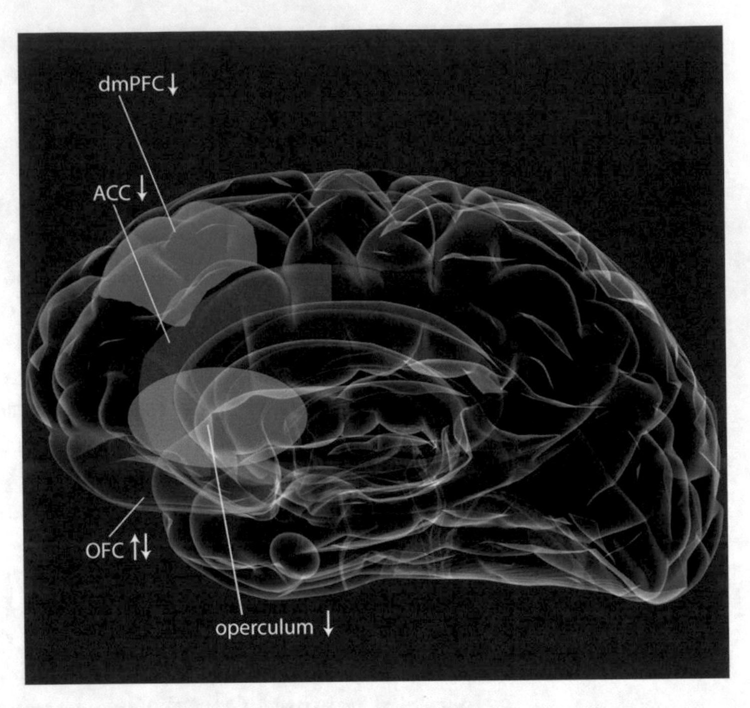

Fig. 2 Cortical gray matter alterations associated with OCD. The arrows represent either an increase (arrow up) or decrease (arrow down) in gray matter volume. In case of inconsistencies in the literature, both an arrow up and down are presented

investigating surface-based measures such as cortical thickness. These studies have found decreased thickness of the orbitofrontal cortex [67–70].

Although the mega-analysis by de Wit et al. (2014) did not report significant group differences in OFC volume either, additional group-by-age interaction analyses showed age-related preservation of orbitofrontal WM volumes in OCD samples compared to controls [51]. Huyser et al. (2013) reported increased orbitofrontal volume after 6 months of CBT in pediatric OCD patients [71]. After a 2-year follow-up of this sample, the OFC volume increase over time sustained, suggesting a differential maturation of the OFC in pediatric OCD compared to controls [72].

These findings, both in adults and in children, highlight the importance of age as a factor in understanding the variability of brain structure alterations in OCD. These age-related neural differences in the orbitofrontal volume in OCD seem to be related to altered neurodevelopment and/or secondary to neuroplastic changes during disease course due to symptom persistence or compensatory processes related to cognitive dysfunction.

2.3 Dorsal Medial Prefrontal Regions and Anterior Cingulate Cortex

Structural abnormalities in the more dorsal parts of the prefrontal cortex (PFC), mainly the ACC, extending into the dorsomedial PFC (dmPFC), appear to be consistent, regardless of the methods utilized. The ACC may be subdivided into a ventral "affective" part and a dorsal "cognitive" part [73]. The affective part of the ACC, through its strong connections with the limbic system, is involved in the assessment of emotional information and the initiation of emotional responses, implicated in the expression of anxiety and distress observed in OCD [74]. The cognitive part of the ACC, in strong connection to the dlPFC, the ventrolateral PFC (vlPFC), and the inferior frontal gyrus (IFG), is implicated in error detection and conflict monitoring [73] and in emotion regulation by cognitive reappraisal [75]. Several meta-analyses including pediatric as well as adult data reported decreased bilateral regional volume in ACC and dmPFC in OCD patients compared with healthy controls (Fig. 2) [46–48].

The involvement of the dorsomedial prefrontal cortex in OCD was confirmed both by the OBIC FreeSurfer mega-analysis [89] and the VBM mega-analysis [51]. The VBM mega-analysis showed that the smaller volume of the dmPFC was most pronounced in patients with additional comorbid anxiety and/or depression [51], which may thus be indicative of a common pathophysiological mechanism across affective disorders related to a shared deficit in emotion regulation [76]. Indeed, Radua et al. (2010) showed that OCD and other anxiety disorders are characterized by overlapping decreased dmPFC/ACC volume [48]. Goodkind et al. (2015) even showed that patients across six diagnostic groups (including schizophrenia, bipolar disorder, depression, addiction, OCD, and anxiety disorders) all show, in comparison with controls, decreased volume of the dmPFC [77]. This suggests that smaller volume of the dmPFC is a more general characteristic of mental disorders, probably related to impaired emotion regulation and other aspects of cognitive control.

2.4 Inferior Frontal Gyrus and Anterior Insula

Other frontal regions, such as the IFG and the anterior insula (i.e., combined the operculum), seem to be affected as well in OCD. Several studies reported decreased GM volume in these regions in adult OCD patients (Fig. 2) [17, 41, 44], which also has been replicated in pediatric OCD patients [54]. Notably, meta-analyses did not find differences between patients and controls [46, 47] or even find increased volume in patients compared with controls [49]. However, the OBIC mega-analysis also found smaller regional GM volume in the IFG extending to the anterior insula in adult OCD patients compared with healthy subjects [51]. The OBIC VBM mega-analysis showed a group-by-age effect for the insular volume, suggesting a relative preservation of insular volume during aging in patients compared with controls [51], who normally show a volume decrease with healthy aging [78]. Additionally, the OBIC FreeSurfer mega-analysis showed decreased cortical thickness of the inferior frontal gyrus consistent with previous cortical thickness work [67–70]. This finding in both the mega-analyses suggests that cortical thinning and GM volume could relate to a similar underlying pathological process responsible for abnormalities in these regions. As reported for the dmPFC, Goodkind et al. (2015) showed that decreased IFG/insula volume is not specific for OCD but also present in many other mental disorders [77].

Besides having a role in cognitive control [79] and attention [80], the IFG/anterior insula is implicated in introceptive awareness [81] and disgust perception [82]. Given that these functions are thought to be abnormal in OCD [30, 83], it may be that the observed aging effect in the anterior insula is related to (compensatory) activation-induced neuroplasticity. Although the direct relationship between the volume and the function of the IFG/anterior insula has not been studied so far, altered activation patterns in this brain area have been published, using various emotional and cognitive fMRI paradigms [79, 84].

2.5 Hippocampus and Amygdala

The ENIGMA-OCD consortium showed a reduction of hippocampal volume in adult OCD patients (Fig. 1) [50]. The effect was notably more pronounced in the medicated patients. The smaller hippocampal volume seemed to be driven, at least partly, by the OCD patients with comorbid depression and late disease onset [50]. These findings are not mutually exclusive, since patients with comorbidities are often the patients who receive medication. Also, Selles et al. (2014) suggested that an increased occurrence of comorbid depression was associated with a late disease onset [85]. Other ENIGMA disease working groups, such as those focusing on major depressive disorder, schizophrenia, and bipolar disorder, also observed smaller hippocampal volume in patients, which suggests that the hippocampal abnormalities are disease non-specific [86–88].

The OBIC FreeSurfer mega-analysis also reported decreased hippocampal volume in OCD patients [89]. Additionally, shape deformity has also been reported for the hippocampus in adult OCD, showing a downward displacement of the head [90]. However, most previous meta-analyses [46–49, 63] and the OBIC VBM mega-analysis [51] did not report abnormalities in the hippocampal complex of OCD patients. A possible reason for the absence of this finding is due to the manner in which FreeSurfer segments subcortical structures. Whereas VBM performs segmentation based on contrast differences between gray and white matter, FreeSurfer segments whole structures based on probabilistic information from a predefined atlas [91]. Therefore, mainly global and regional changes in subcortical structure can be inferred from FreeSurfer, rather than local morphology as with a voxel-based method such as

VBM. It is thus possible that these volume decreases in the hippocampus are not detectable on voxel level, but are detectable when averaging across the whole structure.

De Wit et al. (2014) did show, though, that the (para)limbic part of the medial and lateral temporal cortices, regions that are relatively preserved in healthy aging [92], show greater aging-related volume loss in OCD [51]. The parahippocampal regions are specifically vulnerable to stress-related toxic changes [93]. Greater volume loss in these regions may thus be related to chronic stress and the exaggerated emotional responsiveness seen in OCD [30].

Limbic involvement seems to vary across various subtypes of OCD. Pujol and colleagues (2004) found a relative decrease in right amygdala volume in patients with elevated scores on the "aggressive/checking" dimension [41], and van den Heuvel et al. (2009) reported the "harm/checking" symptom dimension strongly associated with smaller bilateral anterior temporal volumes [17]. The specific association between the checking dimension with amygdala and anterior temporal volume is consistent with some fMRI studies [94]. Volume reduction in similar regions has been described in panic disorder, which, like the OCD subtype with harm/checking symptoms, is characterized by the overestimation of threat [95, 96].

2.6 Cerebellum

Several meta-analyses did not report volumetric differences in the cerebellum [46–49, 63]. This might be explained by the tendency not to list cerebellar findings because of the old but incorrect idea that the cerebellum is not involved in emotion and cognition and thus irrelevant for the understanding of mental disorders. The awareness that the cerebellum is involved in cognitive and emotional processes and therefore indeed relevant for psychiatry recently increased. The recent mega-analysis by de Wit et al. (2014) showed greater cerebellar GM volume bilaterally in patients compared with healthy subjects [51]. The cerebellum is structurally and functionally connected to the parallel CSTC circuits and is thought to integrate cortico-striatal information flow [97]. Aberrant cerebellar activity has been shown in OCD patients during rest and task performance [30].

3 White Matter Abnormalities

Growing evidence exists that OCD symptoms may be partly underpinned by WM abnormalities [39, 40]. While some VBM studies investigated WM volume, DTI is more suited to study the integrity of WM tracts. DTI is a widely used neuroimaging technique to study brain tissue microstructure by quantification of the diffusion characteristics of water molecules [98]. Fractional anisotropy

(FA) is a directionally dependent property of water diffusivity. In healthy WM, the anisotropy is high, reflecting fast diffusivity along the fibers and slow diffusivity perpendicular to them. In WM, an increased FA is possibly related to increased myelination and neuronal remodeling. In diseased WM, abnormally increased FA may lead to functional hyper-connectivity, whereas decreased FA could be a marker of disrupted myelination functionally leading to hypo-connectivity.

Results on WM abnormalities in OCD are largely inconsistent due to several reasons. First, in OCD, WM alterations might be subtle thus difficult to detect. Second, medication effects seem to confound findings [99–102]. Third, since myelination continues into the third decade of life, ongoing brain maturation constitutes confounds in studies with pediatric or adolescent OCD patients as well as young adults [103]. Finally, small samples and inconsistent methodologies have been used [100]. Therefore, conclusions should be made with caution. Below, we discuss the most consistent results regarding WM abnormalities in OCD.

3.1 Corpus Callosum

The corpus callosum connects the left and right cerebral hemispheres and facilitates interhemispheric communication. Its anterior parts connect the right and left prefrontal areas, which play a critical role in the context of internal monitoring, performance control, and inhibitory processes [104]. One of the most consistent DTI findings in the OCD literature describes microstructural abnormalities in the corpus callosum [105, 106].

Radua et al. (2014) performed a multimodal meta-analysis of WM volume and FA in healthy controls and OCD patients. Findings were particularly robust in the anterior midline structures such as the corpus callosum, which showed increased WM volume and decreased FA [100]. Koch et al. (2014) reviewed DTI studies in both adult and pediatric OCD patients. The authors concluded that in adult patients, the corpus callosum is affected by decreased WM integrity [106]. In contrast, pediatric and adolescent findings of the corpus callosum and cingulate bundle point toward increased WM connectivity. Other pediatric studies showed a positive correlation between C-YBOCS scores and FA in the corpus callosum and other white matter tracts such as the cingulum, superior longitudinal fasciculus, inferior longitudinal fasciculus, and inferior frontal-occipital fasciculus [45, 107]. These results suggest that myelination may occur prematurely in patients with early-onset OCD and is most pronounced in the severe cases. This is in line with the neurodevelopmental hypothesis of OCD or at least suggests that early-onset OCD may represent a neurobiologically different neurodevelopmental subtype of the disorder [108]. Koch et al. (2012) investigated the symptom dimension-specific correlates regarding WM integrity [109]. Severity of the “obsession” dimension correlated negatively with WM integrity in the corpus callosum. These results indicate that obsessions, or the patients' inability to control their thoughts, may be related to impaired integrity of the corpus callosum.

3.2 Cingulum

The cingulum interconnects the major components of the fronto-striatal and fronto-limbic circuits such as the cingulate cortex, thalamus, amygdala, and hippocampus [110, 111]. It is mainly involved in emotion processing, nociception, and motor function, but also in higher-level cognitive processes such as attention, conflict or error monitoring, and visuospatial and memory function [112]. Several DTI meta-analyses showed a lower FA in the cingulum [46, 100, 105], possibly related to increased WM volume in this region [100]. In contrast, studies in pediatric and adolescent patients point toward a higher FA in the cingulum [106]. Since this was also the case for the corpus callosum, the findings suggest that increased WM connectivity, potentially due to a premature myelination, may characterize this specific subgroup of young patients [106].

Radua et al. (2014) showed that the decrease in FA of the anterior midline structures such as the cingulum was most prominent in samples that had higher percentages of medicated patients [100]. Although the effects of selective serotonin reuptake inhibitors (SSRIs) on WM parameters are not well understood, these results converge with previous work indicating that compared to drug naïve OCD patients and healthy controls, drug treated OCD patients exhibit significant WM integrity differences [101]. The biological functions of these WM alterations are still unclear. Nevertheless, they are likely to be associated with neural changes within the limbic-cortical networks. For example, the dorsal cingulate cortex and its efferent pathways toward frontal cortical areas are known to be highly important for processes such as performance monitoring and cognitive control [73], and altered connectivity within these circuits exists in OCD [113].

3.3 Superior Longitudinal Fasciculus

The superior longitudinal fasciculus (SLF) is a major association pathway, which links the frontal lobe to the occipital, and part of the parietal and temporal lobes of each cerebral hemisphere. Its connectivity with other cortical regions facilitates higher cognitive functions such as speech processing, attention, working memory, somatosensory monitoring, and visuospatial perception [114–117]. Several meta-analyses described reduced FA in the SLF in OCD patients [46, 100, 105]. In addition, Radua et al. (2014) reported that samples with higher mean YBOCS scores most strongly contributed to this increased FA in the bilateral SLF [100].

Since patients with depression also show decreased FA in the left SLF [118], this finding may be disease non-specific. Disruptions in connectivity of this bundle may result in impaired attention and spatial working memory [119]. The involvement of the frontoparietal connections in the pathophysiology of OCD and other mental disorders seems to be relevant mainly for the understanding of cognitive dysfunctions related to the disease [114–117].

3.4 Inferior Frontal-Occipital Fasciculus and Longitudinal Fasciculus

The inferior frontal-occipital fasciculus (IFOF) is a major cortical association pathway and is known as a direct pathway that connects the occipital, posterior temporal, and the orbitofrontal areas [110, 120]. The inferior longitudinal fasciculus (ILF), on the other hand, is considered to be an indirect pathway essentially connecting similar brain areas, which anteriorly joins the uncinate fasciculus (UF) to relay information to the orbitofrontal brain. Meta-analyses reported that OCD patients have reduced FA in the ILF [46, 100] and higher FA in the left UF [46]. The involvement of both the UF and ILF could be related to the emotional processing deficits seen in OCD patients. Meta-analyses also reported reduced FA in the IFOF of OCD patients [46, 100], an abnormality that is possibly more pronounced in patients with severe ordering symptoms [109].

Microstructural abnormalities in the ILF and IFOF are particularly relevant for the study of OCD, since these bundles represent the main long intra-hemispheric connections to and from the orbitofrontal cortex, which has long been implicated in OCD pathogenesis. Atrophy in the frontal branch of the IFOF is generally associated with the presence of executive problems, apathy, and personality change [105]. Since the parietal lobe is interconnected with the IFOF and SLF, these findings provide additional support to a growing body of evidence that other regions outside the traditional CSTC loops, such as the parietal cortex, are involved in OCD [30, 121].

3.5 Frontal Regions

The OBIC VBM mega-analysis [51] reported decreased WM volumes in frontal regions in the patient group, suggesting abnormalities of WM connections between the prefrontal and subcortical regions. These results are consistent with previous studies on WM volume [17, 122] and converge with a meta-analysis [46] showing altered fronto-striatal WM microstructure in OCD. Ha et al. (2009) showed that patients with predominant contamination/cleaning symptoms exhibited higher FA in the bilateral prefrontal WM [123]. Higher FA in this region, probably reflecting functional fronto-striatal hyper-connectivity, might be related to the impairments in cognitive control and emotion regulation.

4 Discussion

The reviewed literature supports the hypothesis that the structural brain abnormalities found in OCD patients are represented at a system level or network level, rather than discrete brain regions

[30]. The widespread abnormalities across several different regions and circuits might explain the complex phenomenology of OCD, which includes different emotional, cognitive, and behavioral domains. This highlights the importance of considering systems across the entire brain and the interconnectivity between brain regions [94, 124].

Future studies should focus on exploring a multi-model integration of abnormalities in structural and functional connectivity, including its specific relationship to specific emotional, cognitive, and behavioral deficits. Indeed, one of the most recent advances in the field, the Human Connectome Project, aims to combine different imaging modalities, such as resting state fMRI and DTI to acquire information about brain connectivity, task-based fMRI to reveal brain activation patterns during specific emotional and cognitive processes, and structural MRI to capture the shape, volume, and thickness of the cortex and subcortical regions. Results from such multimodal data analyses might help to integrate various findings from previous studies.

Other aspects concerning the pathophysiology such as genetics should be considered as well to facilitate a more holistic view on the neural basis and course of OCD [125]. Menzies et al. (2008) were the first to report that abnormal WM integrity in parietal and frontal regions of OCD patients was also evident in unaffected first-degree relatives of OCD patients [126]. The authors suggested that there are WM endophenotypes representing markers of increased genetic risk for OCD. More recently, neuroimaging studies of twin pairs with and without OCD have begun to disentangle the environmental and genetic contributions to the observed structural and functional brain alterations in OCD [127–129]. den Braber et al. (2010) found that during the performance of a planning task, task-related activation patterns are strongly influenced by genetic risk factors of OCD [127]. While these studies confirm that OCD genetic risk contributes to brain characteristics of the disease, they are unable to identify specific genetic variants that contribute to those changes.

To overcome the limitations of studies on OCD so far, which were predominantly small sampled, separate for pediatric and adult OCD cases, unimodal, and mono-diagnostic, new opportunities now exist within the international platform ENIGMA. ENIGMA is an unprecedented initiative to pool MRI and genetic data of the major mental disorders to perform imaging–genetics meta- and mega-analyses [130]. The overall goal is to unite the imaging and genomics communities to understand mechanisms related to altered brain development across the life span and under the influence of genetic and environmental factors. This will eventually allow us to study disease specificity and neural correlates of the disease across related disorders.

Hibar et al. (2015) recently found significant concordance between OCD risk variants and genetic variants related to increased striatal and thalamus volume [131]. These findings are consistent with the very recent findings from the first ENIGMA-OCD working group meta-analysis, showing that OCD patients have increased volume of the pallidum (adults) and thalamus (children) [50]. These results are largely consistent with current disease models of OCD.

The brain is in continuous development and adaptation, under the influence of genetic, behavioral, and environmental factors. Most imaging results are based on cross-sectional studies in mostly adult patients, which makes it difficult to draw strong conclusions on the neuroplastic changes during the various stages of development and disease. Since the dorsal cognitive CSTC circuit matures only relatively late during adolescence and continues unto early adulthood, individuals seem to be more vulnerable to the development of compulsive behaviors during this developmental stage [132]. Later in life, OCD is associated with age-related increases in the ventral striatal volumes [41, 51] and altered age-related neurodegeneration of the limbic areas [51], probably related to chronic compulsivity, anxiety, and compensatory processes. During the entire life span, the brain circuits remain plastic. The neuroplasticity of the brain is an entrance to therapeutic interventions.

Cognitive behavioral therapy (CBT) is the psychological treatment for OCD and has demonstrated to be highly effective in reducing obsessive–compulsive and related symptoms. The analysis of CBT-related changes at the level of brain areas and circuits may contribute to the elucidation of the neural mechanisms involved in the pathophysiology of OCD. Huyser et al. (2013) reported increased orbitofrontal volume and capsula externa white matter after 6 months of CBT in pediatric OCD patients [71]. After 2 years, patients showed a sustained OFC volume increase, which seems to point to a differential maturation of the OFC in pediatric OCD compared to controls. Findings were most pronounced at younger age, indicating aberrant brain development in pediatric OCD patients in early life. A normalization of reduced GM and WM parietal volumes following combined pharmacological and CBT treatment has also been reported in pediatric OCD patients [133], suggesting some kind of structural plasticity in response to the modulatory effects of CBT.

A number of studies have investigated the effects of pharmacological treatment. Some evidence exists for SSRIs increasing putamen volume and reducing amygdala volume [53, 134]. Gilbert et al. (2000) found a reduction of thalamic volume in pediatric OCD patients after 12-week treatment with the SSRI paroxetine [61], but not following CBT [135]. These findings suggest a differential effect of cognitive and pharmacological treatments on brain structures. Yoo et al. (2007) investigated the potential effects of medication on WM integrity [102]. At baseline, unmedicated adult OCD patients exhibited increased FA in the corpus callosum. After 12 weeks of pharmacotherapy with citalopram, this FA increase had mostly disappeared. This suggests that these alterations in WM microstructure may be amenable to treatment and/or reversible after clinical improvement.

Information on brain structure and function might be useful in the future to predict treatment response. Recent morphological studies have indicated that larger medial prefrontal volume [136] and thinner rostral cingulate cortex [70] were associated with greater CBT-induced symptom reduction in OCD. In contrast, in patients treated with the SSRI fluoxetine, a smaller pretreatment dorsolateral prefrontal volume was associated with greater treatment response, suggesting that brain volume markers might be associated with specific interventions rather than with a non-specific general treatment response. Hashimoto et al. (2014) confirmed the results of Hoexter et al. (2013) [136] and reported smaller volumes of ventral prefrontal and anterior cingulate cortices in OCD non-responders to CBT [137]. Successful CBT seems to rely on a combination of changes in multiple aspects within various interacting CSTC and limbic circuits, involving enhanced effectiveness of the dorsal cognitive control circuit, increased recruitment of the ventromedial prefrontal cortex, and reduction in the pathologically increased activation of both the sensorimotor CSTC circuit and the limbic areas [32].

The brain circuits involved in compulsivity are plastic [32]. Findings of neuroimaging studies largely depend on the developmental stage of the subjects included, the stage of disease and presence of chronicity, and the effects of past and current treatments. In adult studies, it is impossible to disentangle the neural correlates implicated in the cause and consequence of the disease. Therefore, a life span approach is needed to understand how brain changes relate to symptom profile and disease and developmental stage. Insight in the presymptomatic stage of the disease, before the brain has changed by the disease and treatments, will help to understand how risk confers to disease. Only with longitudinal studies following subjects from before the onset of the disease until death, we can truly understand the neurodevelopment aspect of OCD and the neuroplastic changes during the course of the disease.

References

1. Kessler RC, Pethukhova M, Sampson NA, Zaslavsky AM, Wittchen HU (2012) Twelve-month and lifetime prevalence and lifetime morbid risk of anxiety and mood disorders in the United States. Int J Methods Psychiatr Res 21:169–184
2. Ruscio AM, Stein DJ, Chiu WT, Kessler RC (2010) The epidemiology of obsessive-compulsive disorder in the National Comorbidity Survey Replication. Mol Psychiatry 15:53–63

3. Subramaniam M, Abdin E, Vaingankar JA, Chong SA (2012) Obsessive-compulsive disorder: prevalence, correlates, help-seeking and quality of life in a multiracial Asian population. Soc Psychiatry Psychiatr Epidemiol 47:2035–2043
4. American Psychiatric Association (2000) Diagnostic and statistical manual of mental disorders, vol 4. American Psychiatric Association, Washington, DC
5. Nestadt G et al (2000) A family study of obsessive-compulsive disorder. Arch Gen Psychiatry 57:358–363
6. Stewart SE et al (2004) Long-term outcome of pediatric obsessive-compulsive disorder: a meta-analysis and qualitative review of the literature. Acta Psychiatr Scand 110:4–13
7. Mataix-Cols D, Conceicao do Rosario-Campos M, Leckman JF (2005) A multidimensional model of obsessive-compulsive disorder. Am J Psychiatry 162:228–238
8. Mataix-Cols D et al (2002) Symptom stability in adult obsessive-compulsive disorder: data from a naturalistic two-year follow-up study. Am J Psychiatry 159:263–268
9. Rufer M, Grothusen A, Maß R, Peter H, Hand I (2005) Temporal stability of symptom dimensions in adult patients with obsessive-compulsive disorder. J Affect Disord 88:99–102
10. Matsunaga H et al (2008) Symptom structure in Japanese patients with obsessive-compulsive disorder. Am J Psychiatry 165:251–253
11. Bloch MH, Landeros-Weisenberger A, Rosario MC, Pittenger C, Leckman JF (2008) Meta-analysis of the symptom structure of obsessive-compulsive disorder. Am J Psychiatry 165:1532–1542
12. Pallanti S, Grassi G, Sarrecchia ED, Cantisani A, Pellegrini M (2011) Obsessive-compulsive disorder comorbidity: clinical assessment and therapeutic implications. Front Psych 2:1–11
13. Hasler G et al (2005) Obsessive-compulsive disorder symptom dimensions show specific relationships to psychiatric comorbidity. Psychiatry Res 135:121–132
14. An SK et al (2009) To discard or not to discard: the neural basis of hoarding symptoms in obsessive-compulsive disorder. Mol Psychiatry 14:318–331
15. Lawrence NS et al (2007) Neural responses to facial expressions of disgust but not fear are modulated by washing symptoms in OCD. Biol Psychiatry 61:1072–1080
16. Mataix-Cols D et al (2004) Distinct neural correlates of washing, checking, and hoarding symptom dimensions in obsessive-compulsive disorder. Arch Gen Psychiatry 61:564–576
17. van den Heuvel OA et al (2009) The major symptom dimensions of obsessive-compulsive disorder are mediated by partially distinct neural systems. Brain 132:853–868
18. Hasler G et al (2007) Familiality of factor analysis-derived YBOCS dimensions in OCD-affected sibling pairs from the OCD collaborative genetics study. Biol Psychiatry 61:617–625
19. Pinto A et al (2007) Taboo thoughts and doubt/checking: a refinement of the factor structure for obsessive-compulsive disorder symptoms. Psychiatry Res 151:255–258
20. Stewart SE et al (2007) Principal components analysis of obsessive-compulsive disorder symptoms in children and adolescents. Biol Psychiatry 61:285–291
21. Cummings JL, Cunningham K (1992) Obsessive-compulsive disorder in Huntington's disease. Biol Psychiatry 31:263–270
22. Swedo E et al (1989) High prevalence of obsessive-compulsive symptoms in patients with Sydenham's symptoms. Am J Psychiatry 146:246–249
23. Laplane D et al (1989) Obsessive-compulsive and other behavioural changes with bilateral basal ganglia lesions. Brain 112:699–725
24. Eslinger PJ, Damasio AR (1985) Severe disturbance of higher cognition after bilateral frontal lobe ablation: patient EVR. Neurology 35:1731
25. Alexander GE (1986) Parallel organization of functionally segregated circuits linking basal ganglia and cortex. Annu Rev Neurosci 9:357–381
26. Cummings J (1993) Frontal-subcortical circuits and human behavior. Arch Neurol 50:873–880
27. Groenewegen HJ, Uylings HB (2000) The prefrontal cortex and the integration of sensory, limbic and autonomic information. Prog Brain Res 126:3–28
28. Graybiel AM, Rauch SL (2000) Toward a neurobiology of obsessive-compulsive disorder. Neuron 28:343–347
29. Mataix-Cols D, van den Heuvel OA (2006) Common and distinct neural correlates of obsessive-compulsive and related disorders. Psychiatr Clin North Am 29:391–410, viii
30. Menzies L et al (2008) Integrating evidence from neuroimaging and neuropsychological studies of obsessive-compulsive disorder: the

orbitofronto-striatal model revisited. Neurosci Biobehav Rev 32:525–549

31. Milad MR, Rauch SL (2012) Obsessive-compulsive disorder: beyond segregated cortico-striatal pathways. Trends Cogn Sci 16:43–51
32. van den Heuvel OA et al (2016) Brain circuitry of compulsivity. Eur Neuropsychopharmacol 26:810–827
33. Behar D et al (1984) Computerized tomography and neuropsychological test measures in adolescents with obsessive-compulsive disorder. Am J Psychiatry 141:363–369
34. Insel TR, Donnelly EF, Lalakea ML, Alterman IS, Murphy DL (1983) Neurological and neuropsychological studies of patients with obsessive–compulsive disorder. Biol Psychiatry 18:741–751
35. Hutton C, Draganski B, Ashburner J, Weiskopf N (2009) A comparison between voxel-based cortical thickness and voxel-based morphometry in normal aging. NeuroImage 48:371–380
36. Voets NL et al (2008) Evidence for abnormalities of cortical development in adolescent-onset schizophrenia. NeuroImage 43:665–675
37. Frye RE et al (2010) Surface area accounts for the relation of gray matter volume to reading-related skills and history of dyslexia. Cereb Cortex 20:2625–2635
38. Rosenberg DR, Keshavan MS (1998) Toward a neurodevelopmental model of obsessive-compulsive disorder. Biol Psychiatry 43:623–640
39. Fontenelle LF et al (2009) Is there evidence of brain white-matter abnormalities in obsessive-compulsive disorder?: a narrative review. Top Magn Reson Imaging 20:291–298
40. Douzenis A et al (2009) Obsessive-compulsive disorder associated with parietal white matter multiple sclerosis plaques. World J Biol Psychiatry 10:956–960
41. Pujol J et al (2004) Mapping structural brain alterations in obsessive-compulsive disorder. Arch Gen Psychiatry 61:720–730
42. Gilbert AR et al (2008) Brain structure and symptom dimension relationships in obsessive-compulsive disorder: a voxel-based morphometry study. J Affect Disord 109:117–126
43. Szeszko PR et al (2008) Gray matter structural alterations in psychotropic drug-naive pediatric obsessive-compulsive disorder: an optimized voxel-based morphometry study. Am J Psychiatry 165:1299–1307
44. Yoo SY et al (2008) Voxel-based morphometry study of gray matter abnormalities in obsessive-compulsive disorder. J Korean Med Sci 23:24–30
45. Zarei M et al (2011) Changes in gray matter volume and white matter microstructure in adolescents with obsessive-compulsive disorder. Biol Psychiatry 70:1083–1090
46. Peng Z et al (2012) Brain structural abnormalities in obsessive-compulsive disorder: converging evidence from white matter and grey matter. Asian J Psychiatr 5:290–296
47. Radua J, Mataix-Cols D (2009) Voxel-wise meta-analysis of grey matter changes in obsessive-compulsive disorder. Br J Psychiatry 195:393–402
48. Radua J, van den Heuvel OA, Surguladze S, Mataix-Cols D (2010) Meta-analytical comparison of voxel-based morphometry studies in obsessive-compulsive disorder vs other anxiety disorders. Arch Gen Psychiatry 67:701–711
49. Rotge J-Y et al (2010) Gray matter alterations in obsessive-compulsive disorder: an anatomic likelihood estimation meta-analysis. Neuropsychopharmacology 35:686–691
50. Boedhoe PSW et al (2017) Distinct subcortical volume alterations in pediatric and adult OCD: a worldwide meta- and mega-analysis. Am J Psychiatry 174:60–70
51. de Wit SJ et al (2014) Multicenter voxel-based morphometry mega-analysis of structural brain scans in obsessive-compulsive disorder. Am J Psychiatry 171:340–349
52. Rosenberg DR et al (1997) Frontostriatal measurement in treatment-naive children with obsessive-compulsive disorder. Arch Gen Psychiatry 54:824–830
53. Szeszko PR et al (2004) Brain structural abnormalities in psychotropic drug-naive pediatric patients with obsessive-compulsive disorder. Am J Psychiatry 161:1049–1056
54. Carmona S et al (2007) Pediatric OCD structural brain deficits in conflict monitoring circuits: a voxel-based morphometry study. Neurosci Lett 421:218–223
55. van Essen DC (1997) A tension-based theory of morphogenesis and compact wiring in the central nervous system. Nature 385:313–318
56. Levitt JJ et al (2004) Shape of caudate nucleus and its cognitive correlates in neuroleptic-naive schizotypal personality disorder. Biol Psychiatry 55:177–184
57. Choi JS et al (2007) Shape deformity of the corpus striatum in obsessive-compulsive disorder. Psychiatry Res 155:257–264

58. Shaw P et al (2015) Subcortical and cortical morphological anomalies as an endophenotype in obsessive-compulsive disorder. Mol Psychiatry 20:224–231
59. Atmaca M et al (2006) Volumetric MRI assessment of brain regions in patients with refractory obsessive-compulsive disorder. Prog Neuro-Psychopharmacol Biol Psychiatry 30:1051–1057
60. Atmaca M, Yildirim H, Ozdemir H, Tezcan E, Kursad Poyraz A (2007) Volumetric MRI study of key brain regions implicated in obsessive-compulsive disorder. Prog Neuro-Psychopharmacol Biol Psychiatry 31:46–52
61. Gilbert R et al (2000) Decrease in thalamic volumes of pediatric patients with obsessive-compulsive disorder who are taking paroxetine. Arch Gen Psychiatry 57:449–456
62. Kang D-H et al (2008) Thalamus surface shape deformity in obsessive-compulsive disorder and schizophrenia. Neuroreport 19:609–613
63. Rotge J-Y et al (2009) Meta-analysis of brain volume changes in obsessive-compulsive disorder. Biol Psychiatry 65:75–83
64. Kim J-J et al (2001) Grey matter abnormalities in obsessive—compulsive disorder: statistical parametric mapping of segmented magnetic resonance images. Br J Psychiatry 179:330–334
65. Christian CJ et al (2008) Gray matter structural alterations in obsessive-compulsive disorder: relationship to neuropsychological functions. Psychiatry Res 164:123–131
66. Valente AA et al (2005) Regional gray matter abnormalities in obsessive-compulsive disorder: a voxel-based morphometry study. Biol Psychiatry 58:479–487
67. Shin Y-W et al (2007) Cortical thinning in obsessive compulsive disorder. Hum Brain Mapp 28:1128–1135
68. Kuhn S et al (2013) Reduced thickness of anterior cingulate cortex in obsessive-compulsive disorder. Cortex 49:2178–2185
69. Venkatasubramanian G et al (2012) Comprehensive evaluation of cortical structure abnormalities in drug-naïve, adult patients with obsessive-compulsive disorder: a surface-based morphometry study. J Psychiatr Res 46:1161–1168
70. Fullana MA et al (2014) Brain regions related to fear extinction in obsessive-compulsive disorder and its relation to exposure therapy outcome: a morphometric study. Psychol Med 44:845–856
71. Huyser C et al (2013) Increased orbital frontal gray matter volume after cognitive behavioural therapy in paediatric obsessive compulsive disorder. World J Biol Psychiatry 14:319–331
72. Huyser C et al (2014) A longitudinal VBM study in paediatric obsessive-compulsive disorder at 2-year follow-up after cognitive behavioural therapy. World J Biol Psychiatry 15:443–452
73. Bush G, Luu P, Posner M (2000) Cognitive and emotional influences in anterior cingulate cortex. Trends Cogn Sci 4:215–222
74. Aouizerate B et al (2004) Pathophysiology of obsessive-compulsive disorder: a necessary link between phenomenology, neuropsychology, imagery and physiology. Prog Neurobiol 72:195–221
75. Ochsner KN, Gross JJ (2005) The cognitive control of emotion. Trends Cogn Sci 9:242–249
76. van Tol M et al (2010) Regional brain volume in depression and anxiety disorders. Arch Gen Psychiatry 67:1002–1011
77. Goodkind M et al (2015) Identification of a common neurobiological substrate for mental illness. JAMA Psychiat 72:305–315
78. Ziegler G et al (2012) Brain structural trajectories over the adult lifespan. Hum Brain Mapp 33:2377–2389
79. Tops M, Boksem MA (2011) Potential role of the inferior frontal gyrus and anterior insula in cognitive control, brain rhythms, and event-related potentials. Front Psychol 2:1–14
80. Corbetta M, Shulman GL (2002) Control of goal-directed and stimulus-driven attention in the brain. Nat Rev Neurosci 3:201–215
81. Critchley HD, Wiens S, Rotshtein P, Ohman A, Dolan RJ (2004) Neural systems supporting interoceptive awareness. Nat Neurosci 7:189–195
82. Calder AJ, Lawrence AD, Young AW (2001) Neuropsychology of fear and loathing. Nat Rev Neurosci 2:352–363
83. Paulus MP, Stein MB (2010) Interoception in anxiety and depression. Brain Struct Funct 214:451–463
84. de Wit SJ et al (2012) Presupplementary motor area hyperactivity during response inhibition: a candidate endophenotype of obsessive-compulsive disorder. Am J Psychiatry 169:1100–1108
85. Selles RR, Storch EA, Lewin AB (2014) Variations in symptom prevalence and clinical correlates in younger versus older youth with obsessive–compulsive disorder. Child Psychiatry Hum Dev 45:666–674
86. Schmaal L et al (2016) Subcortical brain alterations in major depressive disorder:

findings from the ENIGMA major depressive disorder working group. Mol Psychiatry 21:806–812

87. van Erp TGM et al (2016) Subcortical brain volume abnormalities in 2028 individuals with schizophrenia and 2540 healthy controls via the ENIGMA consortium. Mol Psychiatry 21:547–553
88. Hibar DP et al (2016) Subcortical volumetric abnormalities in bipolar disorder. Mol Psychiatry 21:1710–1716
89. Fouche J-P, du Plessis S, Hattingh C, et al (2017) Cortical thickness in obsessive-compulsive disorder: multisite mega-analysis of 780 brain scans from six centres. Br J Psychiatry 210:67–74
90. Hong SB et al (2007) Hippocampal shape deformity analysis in obsessive-compulsive disorder. Eur Arch Psychiatry Clin Neurosci 257:185–190
91. Fischl B et al (2002) Whole brain segmentation: automated labeling of neuroanatomical structures in the human brain. Neuron 33:341–355
92. Grieve SM, Clark CR, Williams LM, Peduto AJ, Gordon E (2005) Preservation of limbic and paralimbic structures in aging. Hum Brain Mapp 25:391–401
93. Kassem MS et al (2013) Stress-induced grey matter loss determined by MRI is primarily due to loss of dendrites and their synapses. Mol Neurobiol 47:645–661
94. Harrison BJ et al (2009) Altered corticostriatal functional connectivity in obsessive-compulsive disorder. Arch Gen Psychiatry 66:1189–1200
95. Massana G et al (2003) Parahippocampal gray matter density in panic disorder: a voxel-based morphometric study. Am J Psychiatry 160:566–568
96. Massana G et al (2003) Amygdalar atrophy in panic disorder patients detected by volumetric magnetic resonance imaging. NeuroImage 19:80–90
97. Middleton FA, Strick PL (2000) Basal ganglia output and cognition: evidence from anatomical, behavioral, and clinical studies. Brain Cogn 42:183–200
98. Le Bihan D et al (2001) Diffusion tensor imaging: concepts and applications. J Magn Reson Imaging 13:534–546
99. Fan Q et al (2012) Abnormalities of white matter microstructure in unmedicated obsessive-compulsive disorder and changes after medication. PLoS One 7(4):e35889
100. Radua J et al (2014) Multimodal voxel-based meta-analysis of white matter abnormalities in obsessive–compulsive disorder. Neuropsychopharmacology 39:1547–1557
101. Benedetti F et al (2013) Widespread changes of white matter microstructure in obsessive-compulsive disorder: effect of drug status. Eur Neuropsychopharmacol 23:581–593
102. Yoo SY et al (2007) White matter abnormalities in drug-naïve patients with obsessive-compulsive disorder: a diffusion tensor study before and after citalopram treatment. Acta Psychiatr Scand 116:211–219
103. Peters BD et al (2012) White matter development in adolescence: diffusion tensor imaging and meta-analytic results. Schizophr Bull 38:1308–1317
104. Schmahmann JD et al (2007) Association fibre pathways of the brain: parallel observations from diffusion spectrum imaging and autoradiography. Brain 130:630–653
105. Piras F, Piras F, Caltagirone C, Spalletta G (2013) Brain circuitries of obsessive compulsive disorder: a systematic review and meta-analysis of diffusion tensor imaging studies. Neurosci Biobehav Rev 37:2856–2877
106. Koch K, Reeß TJ, Rus OG, Zimmer C, Zaudig M (2014) Diffusion tensor imaging (DTI) studies in patients with obsessive-compulsive disorder (OCD): a review. J Psychiatr Res 54:26–35
107. Gruner P et al (2012) White matter abnormalities in pediatric obsessive-compulsive disorder. Neuropsychopharmacology 37:2730–2739
108. Fontenelle LF, Mendlowicz MV, Marques C, Versiani M (2003) Early- and late-onset obsessive-compulsive disorder in adult patients: an exploratory clinical and therapeutic study. J Psychiatr Res 37:127–133
109. Koch K et al (2012) White matter structure and symptom dimensions in obsessive-compulsive disorder. J Psychiatr Res 46:264–270
110. Catani M, Howard RJ, Pajevic S, Jones DK (2002) Virtual in vivo interactive dissection of white matter fasciculi in the human brain. NeuroImage 17:77–94
111. Wakana S, Jiang H, Nagae-Poetscher LM, van Zijl PCM, Mori S (2004) Fiber tract-based atlas of human white matter anatomy. Radiology 230:77–87
112. Devinsky O, Morrell MJ, Vogt BA (1995) Contributions of anterior cingulate cortex to behaviour. Brain 118:279–306
113. Schlösser RGM et al (2010) Fronto-cingulate effective connectivity in obsessive compulsive disorder: a study with fMRI and dynamic

causal modeling. Hum Brain Mapp 31:1834–1850

114. Bernal B, Altman N (2010) The connectivity of the superior longitudinal fasciculus: a tractography DTI study. Magn Reson Imaging 28:217–225
115. Hoeft F et al (2007) More is not always better: increased fractional anisotropy of superior longitudinal fasciculus associated with poor visuospatial abilities in Williams syndrome. J Neurosci 27:11960–11965
116. Karlsgodt KH et al (2008) Diffusion tensor imaging of the superior longitudinal fasciculus and working memory in recent-onset schizophrenia. Biol Psychiatry 63:512–518
117. Madhavan KM, McQueeny T, Howe SR, Shear P, Szaflarski J (2014) Superior longitudinal fasciculus and language functioning in healthy aging. Brain Res 1562:11–22
118. Murphy ML, Frodl T (2011) Meta-analysis of diffusion tensor imaging studies shows altered fractional anisotropy occurring in distinct brain areas in association with depression. Biol Mood Anxiety Disord 1(1):3
119. Makris N et al (2005) Segmentation of subcomponents within the superior longitudinal fascicle in humans: a quantitative, in vivo, DT-MRI study. Cereb Cortex 15:854–869
120. Matsumoto R et al (2010) Reduced gray matter volume of dorsal cingulate cortex in patients with obsessive-compulsive disorder: a voxel-based morphometric study. Psychiatry Clin Neurosci 64:541–547
121. Piras F et al (2015) Widespread structural brain changes in OCD: a systematic review of voxel-based morphometry studies. Cortex 62:89–108
122. Togao O et al (2010) Regional gray and white matter volume abnormalities in obsessive-compulsive disorder: a voxel-based morphometry study. Psychiatry Res 184:29–37
123. Ha TH et al (2009) White matter alterations in male patients with obsessive-compulsive disorder. Neuroreport 20:735–739
124. Subira M et al (2015) Structural covariance of neostriatal and limbic regions in patients with obsessive-compulsive disorder. J Psychiatry Neurosci 41:115–123
125. Pauls DL, Abramovitch A, Rauch SL, Geller DA (2014) Obsessive-compulsive disorder: an integrative genetic and neurobiological perspective. Nat Rev Neurosci 15:410–424
126. Menzies L et al (2008) White matter abnormalities in patients with obsessive-compulsive disorder and their first-degree relatives. Am J Psychiatry 165:1308–1315
127. den Braber A et al (2010) Brain activation during cognitive planning in twins discordant or concordant for obsessive-compulsive symptoms. Brain 133:3123–3140
128. den Braber A et al (2011) White matter differences in monozygotic twins discordant or concordant for obsessive-compulsive symptoms: a combined diffusion tensor imaging/voxel-based morphometry study. Biol Psychiatry 70:969–977
129. den Braber A et al (2012) Brain activation during response interference in twins discordant or concordant for obsessive compulsive symptoms. Twin Res Hum Genet 15:372–383
130. Thompson PM et al (2014) The ENIGMA consortium: large-scale collaborative analyses of neuroimaging and genetic data. Brain Imaging Behav 8:153–182
131. Hibar DP et al (2015) Common genetic variants influence human subcortical brain structures. Nature 520:224–229
132. Arnsten AFT, Rubia K (2012) Neurobiological circuits regulating attention, cognitive control, motivation, and emotion: disruptions in neurodevelopmental psychiatric disorders. J Am Acad Child Adolesc Psychiatry 51:356–367
133. Lázaro L et al (2009) Brain changes in children and adolescents with obsessive-compulsive disorder before and after treatment: a voxel-based morphometric MRI study. Psychiatry Res 172:140–146
134. Hoexter MQ et al (2012) Gray matter volumes in obsessive-compulsive disorder before and after fluoxetine or cognitive-behavior therapy: a randomized clinical trial. Neuropsychopharmacology 37:734–745
135. Rosenberg DR, Benazon NR, Gilbert A, Sullivan A, Moore GJ (2000) Thalamic volume in pediatric obsessive-compulsive disorder patients before and after cognitive behavioral therapy. Biol Psychiatry 48:294–300
136. Hoexter MQ et al (2013) Differential prefrontal gray matter correlates of treatment response to fluoxetine or cognitive-behavioral therapy in obsessive-compulsive disorder. Eur Neuropsychopharmacol 23:569–580
137. Hashimoto N et al (2014) Brain structural abnormalities in behavior therapy-resistant obsessive-compulsive disorder revealed by voxel-based morphometry. Neuropsychiatr Dis Treat 10:1987–1996

Chapter 24

Personality Is Reflected in Brain Morphometry

Laura Petrosini, Debora Cutuli, Eleonora Picerni, and Daniela Laricchiuta

Abstract

To fully characterize the relationship between structure and function as it relates to personality measures, techniques are needed that can distinguish among the different structural compartments of the gray and white matter contributing to the measures of size (cortical thickness and volume). By using structural neuroimaging techniques, such as region of interest (ROI)- and voxel-based morphometry (VBM), several studies addressed the associations between personality factors and morphometric measures that allowed characterizing the subtle brain structural differences in relation to different temperamental traits. To address brain-trait relationships, global measures, as total intracranial volume, total brain volume, total gray or white matter volumes, as well as regional measures, as gray or white matter volumes of specific brain areas have been investigated in relation to the specific dimensions of personality. Understanding anatomic variations as they relate to personality traits may help putting the functional findings in context and pave the way for studying micro-structural influences on personality.

Key words ROI- and voxel-based morphometry, Diffusion Tensor Imaging, Temperamental traits, Big Five model, Extraversion, Neuroticism, Cloninger's psychobiological model, Novelty seeking, Harm avoidance, Emotion regulation

1 Introduction

Personality is a stable, organized collection of psychological traits and mechanisms that influences interactions with and modifications to the psychological, social, and physical environment surrounding the individual. Personality traits represent tendencies to manifest habitual patterns of cognition, emotion, motivation, and behavior in response to a variety of eliciting stimuli. According to this perspective and in accordance with the American Psychiatric Association [1], personality traits are considered "enduring patterns of perceiving, relating to, and thinking about the environment and oneself that are exhibited in a wide range of social and personal contexts." It is retained that the development of temperamental traits is based on the joint influence of genetics, biology, and environment and arises from the continuous functioning of specific brain systems which themselves are the result of genetic factors

Gianfranco Spalletta et al. (eds.), *Brain Morphometry*, Neuromethods, vol. 136,
https://doi.org/10.1007/978-1-4939-7647-8_24,

[2]. Studies of the biological components of the personality have been conducted with respect to the heritability of traits, the role of neurotransmitters, and the identification of neural structures that mediate trait-typical behaviors. Recent advances in brain imaging have now opened the gates for novel and interdisciplinary approaches to neuroscience of the personality. One approach is to correlate temperamental traits and neurobiological data derived from experimental methods, such as brain imaging.

Various models and taxonomies have been proposed to describe the diversity of personality characteristics with a finite number of independent dimensions. From a biological perspective, the personality traits can be traced back to specific brain structures and neural mechanisms. The most influential biology-based personality theories include Eysenck's personality model [3], Gray's reinforcement sensitivity theory [4], McCrae and Costa's Big Five model [5], and Cloninger's psychobiological model [6]. Despite the somewhat divergent views on the personality structure, highly significant associations have been reported among factors from different models, suggesting that different models may measure the same construct [7]. Main personality traits are briefly described In Table 1.

Table 1
Biology-based personality theories include McCrae and Costa's Big Five model, Eysenck's personality model, Gray's Reinforcement Sensitivity Theory, and Cloninger's psychobiological model

Model by	Traits	Features: tendency to be
McCrae and Costa	Openness to experience	Imaginative, artistic, emphatic, creative, curious, and liberal
	Conscientiousness	Efficient, organized, meticulous, scrupulous, ambitious, mindful of details, self-disciplined, and careful
	Agreeableness	Trusting, honest, generous, forgiving, humble, merciful, altruist, kind, friendly, and cooperative
McCrae and Costa; Eysenck	Extraversion	Warm, sociable, dominant, assertive, active, fun-loving, cheerful, energized, and optimistic
	Neuroticism	Mentally instable, anxious, moody, irritable, gloomy, self-conscious, impulsive, fragile, and sad
Eysenck	Psychoticism	Aggressive, hostile, lacking in empathy, cruel, and troublesome
Cloninger	Harm avoidance	Excessively worried, pessimistic, shy, fearful, doubtful, and easily fatigued
	Novelty seeking	Explorative, active in response to novelty, impulsive, extravagant in approach to reward, quick-tempered, fickle, excitable, and avoidant of frustration
	Reward dependence	Responsive to reward, social approval, social support, and sentiment
	Persistence	Perseverant in spite of fatigue or frustration
Gray	Behavioral inhibition system	Anxious and sensitive to punishment, boring things, or negative events
	Behavioral activation system	Hopeful, happy, impulsive and sensitive to reward

2 Brain Regions and Personality Traits

The mapping of psychological processes onto specific brain regions might be retained a form of "new phrenology" [8], as if it were possible that complex psychological processes are represented by a single brain region, for example, "fear is processed by the amygdala." Conversely, one of the most striking aspects of all studies analyzing the brain morphometry of personality traits is the high number of regions exhibiting significant (positive or negative) correlations with a specific temperamental trait, emphasizing how distributed the neural representation of personality traits is likely to be. For instance, criticizing the isomorphic views of Eysenck and Gray, Zuckerman [9] states that each trait can be linked to multiple cerebral systems and each system can contribute to more than one personality trait.

The opinion that multiple brain structures contribute to personality traits stems not only from neuropsychological and physiological data but also from studies showing that individual differences in temperamental traits covary with morphometric data (e.g., cortical thickness and gray and white matter volumes) in distributed brain regions [10]. In fact, structural brain scans provide an important way to analyze brain-personality relationship. If a trait is hypothesized to be associated with the functions of given brain regions, the finding that this trait is associated with structural variations in those regions supports the hypothesis.

Recently, several studies addressed the associations between personality factors and morphometric measures by using structural neuroimaging techniques, such as region of interest (ROI)- and voxel-based morphometry (VBM) that allowed characterizing subtle brain structural differences and mapping different temperamental traits onto in vivo human brain. To address brain-trait relationships, researches have investigated global measures, as total intracranial volume, total brain volume, total gray or white matter volumes, as well as regional measures, as gray or white matter volumes of specific brain areas in relation to the specific dimensions of personality.

3 Some Preliminary Methodological Issues

Although the results of the volumetric personality analyses may be model-dependent [11], before analyzing the complex associations between brain regions and personality traits, some intrinsic questions as *allometry of measures* and *functional differences* have to be taken into account. In fact, evolutionary brain allometric studies indicate that the white matter (WM) disproportionately increases as a function of gray matter (GM) to the power of 0.99–1.37

[12]. The size of several brain regions disproportionately increases as a function of total brain size with different rates of power, with the neocortex that expands at the highest power in comparison to deep gray matter structures [13]. Thus, correcting each regional brain volume by the same scaling factor appears to be an oversimplified approach.

Furthermore, global or regional brain measures appear to be related to the functional differences. It is known that total brain volume is larger in autistic children and smaller in children with attention-deficit/hyperactivity disorder (ADHD) in comparison to typically developing children [14–17]. Also differences in regional volumes may reflect differences in functional capacity, as demonstrated in the presence of specific pathologies [18]. For example, in ADHD GM reduction in the anterior cingulate cortex has been reported to correlate with scores of selective inattention [19, 20]. Thus, questions on whether regional volumes of the individual brain structures scale uniformly or nonuniformly to global brain measures and on whether global or regional brain measures may be directly related to functional differences are the first questions to be addressed in morphometric studies.

4 Personality Traits and Brain Volume, Age, and Sex

The intricate relationship among regional brain volume, age, and sex may represent another confounding factor regarding the complex issue of brain morphometry and personality traits. In fact, many studies indicated that the brain tends to shrink with increasing age, although it shrinks at different rates in different brain regions [21–26], in different age cohorts [22], and for different sex groups [27, 28]. Moreover, sex dimorphism in regional brain volumes has been described [29, 30]. As a further difficulty, personality traits themselves are differently associated with sex and age factors. For example, studies on temperamental traits of novelty seeking and harm avoidance, as defined in the Temperament and Character Inventory (TCI) by Cloninger [6], report that female individuals had harm avoidance scores higher than male subjects and younger individuals had novelty seeking scores higher than older subjects [31–33]. Also the personality traits of the Big Five [5] exhibit a clear sex dimorphism, being the levels of neuroticism and the prevalence of anxiety and mood disorders higher in females than males [24, 34, 35]. With respect to age, neuroticism consistently decreases and agreeableness increases with increasing age [36–39]. Thus, the intricate relationships among personality traits, global brain measures, regional brain volumes, age, and sex may bias the results of morphometric studies on personality, making them a great challenge.

Furthermore, in analyzing the correlations between brain morphometry and Big Five personality traits, it has been emphasized that different combinations of confounding factors, such as years of education, socioeconomic status, and Beck Depression Inventory scores besides the abovementioned age, sex, and global brain measures, can induce marked differences in the results of structural studies on personality [40]. Since it would be not enough to perform only a regression model with one combination of factors, it has been suggested the incorporation of "structure equation modeling (e.g. the automated specification search tool) as an additional method to narrow the possibility of well-fitted models" [40].

5 "Larger Is More Powerful"?

In investigating the brain regions more likely associated from a structural point of view with a personality scale, a preliminary question is how the brain structure—specifically, the relative volume—relates to function. A greater-than-average volume may signify greater-than-average power to carry out specific functions, on the assumption that larger populations of neurons can produce larger outputs and can therefore be more influential than smaller populations of neurons. Nevertheless, a smaller-than-average volume may indicate increased efficiency or that the structure is reorganized to perform a particular function. Human and experimental evidence favors the "larger-is-more-powerful" position: training on particular tasks or experiencing complex environment does increase the volume of the functionally related brain structures [41, 42]. Moreover, studies on the relationship between cortical thickness and intelligence indicated that intelligence in children and adults correlates positively with brain volumes [43]. Thus, it may be assumed that volume tends to positively covary with function.

Furthermore, variance in the normal range of expression of personality traits appears to be linked to structural variance in specific brain structures [10, 44, 45]. In particular, VBM studies indicate that individual differences in personality traits covary with brain morphometric measures, as cortical thickness and GM volume of specific brain regions.

6 Morphometric Correlates of Big Five Dimensions

Each of the dimensions of the Big Five model (extraversion, neuroticism, conscientiousness, agreeableness, and openness to experience) defines a personality dimension that is anchored by an opposite pole (e.g., the opposite pole of extraversion is introversion; the opposite of neuroticism is emotional stability). Extraversion and neuroticism are of particular interest since their structural bases appear to contribute to the predisposition toward mood and anxiety disorders [46]. Although structural imaging methods (VBM, Diffusion Tensor Imaging—DTI—and cortical thickness analysis) have yielded some inconsistent and even sometimes contradictory results, the behavioral phenotypes associated to extraversion and neuroticism traits are reflected in structural differences mainly within the cortical networks, suggesting specific modifications in information processing and transmission [47].

6.1 Extraversion and Brain Volumes

When the Big Five personality traits are associated with the morphometry of specific brain regions, extraversion covaries positively with volume and thickness of the orbitofrontal cortex, a region involved in processing the reward value of stimuli and approach-related behaviors and, in the precuneus, involved in reflective self-awareness [10, 47–49]. Furthermore, extraversion positively correlates with volume in the anterior cingulate gyrus, a critical area for decision-making in social situations and for guiding social behavior; with the middle temporal gyrus (MTG), a region involved in social cognition [50]; and with the medial prefrontal cortex [50–52], a region involved in monitoring actions on complex social goals and in self-reflective processing, which enables the emotional understanding of others' intentional acts [53–55]. Higher extraversion scores have been also associated with lower cortical volume in the superior temporal gyrus [47]. Moreover, extravert individuals show greater frontal activation at rest [56], and positive correlations are described between extraversion levels and activations of the amygdala and prefrontal cortex in response to positive stimuli [49, 57]. In accordance highly significant positive correlations emerge between extraversion and perfusion measures in the basal ganglia, thalamus, and cerebellum [58], suggesting that neural systems associated with personality traits are not confined to higher-level executive brain regions, but rather they are represented at all levels, cortical and subcortical, of neural processing.

The two core aspects of extraversion (positive affect and tendency to seek out social situations) may be related to differences in cortical thickness of extravert vs. introvert individuals [47]. Specific prefrontal cortical regions in extravert individuals are smaller in the right vs. left hemisphere, while a thinner cortical GM ribbon in the fusiform gyrus is described in extravert in comparison to introvert individuals [59]. The relative regional thinning in the prefrontal cortex could represent the neural substrate for the diminished social inhibition and increased approach behavior found in extraversion, whereas thickening in this area may relate to social wariness and avoidance described in introversion (Fig. 1).

Consistently with these indications, using 3-Tesla proton magnetic resonance spectroscopy, increased glutamate levels in the dorsolateral prefrontal cortex were found in introverts when compared with extraverts [60]. Furthermore, transcranial direct current

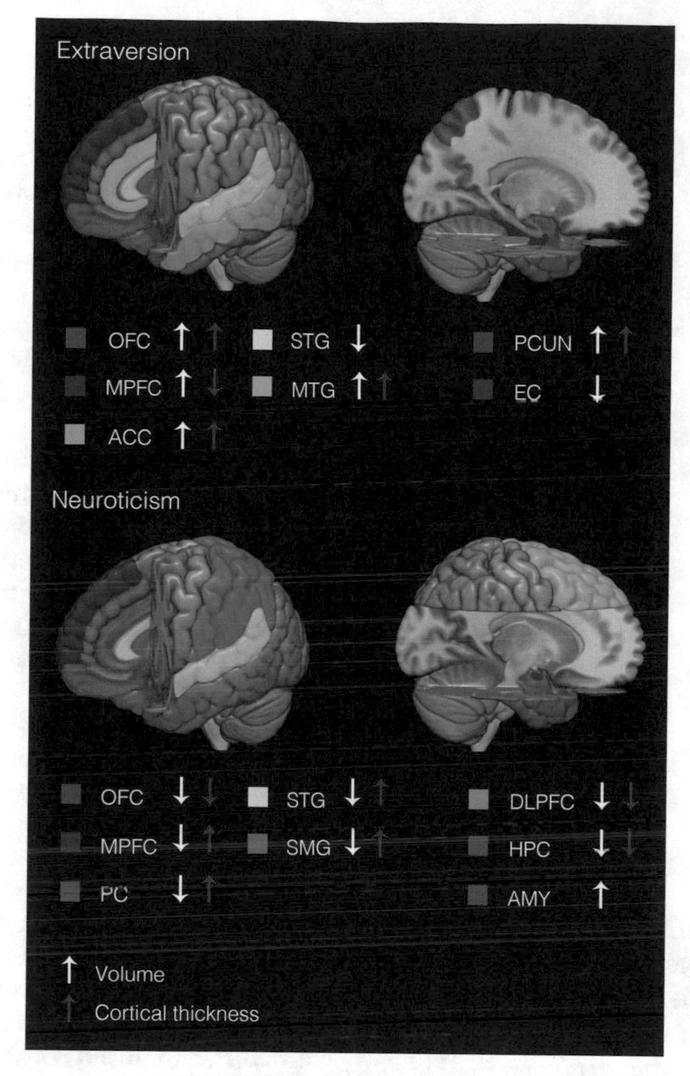

Fig. 1 Brain structural variations in extraversion and neuroticism. Areas significantly associated with extraversion or neuroticism were highlighted in different colors. *OFC* orbitofrontal cortex, *MPFC* medial prefrontal cortex, *ACC* anterior cingulate cortex, *STG* superior temporal gyrus, *MTG* middle temporal gyrus, *PCUN* precuneus, *EC* entorhinal cortex, *PC* parietal cortex, *SMG* supramarginal gyrus, *DLPFC* dorsolateral prefrontal cortex, *HPC* hippocampus, *AMY* amygdala. White arrow = variation in volumes; red arrow = variation in cortical thickness

stimulation (tDCS) over the left dorsolateral prefrontal cortex results in increased ratings during negative emotional picture processing, and this effect is stronger in individuals with low extraversion scores [61]. Thus, the left dorsolateral prefrontal cortex seems to contribute to extraversion since extravert individuals have a greater ability to regulate positive life events important for enhancing social functioning.

6.2 Neuroticism and Brain Volumes

Neuroticism that reflects the propensity toward experiencing negative emotions and thoughts is linked to structural variations in distributed brain circuits, as reflected by various measures (GM and WM volume, thickness, fractional anisotropy, mean diffusivity, and interregional anatomical connectivity strength) (Fig. 1). In individuals with high levels of neuroticism, morphometric studies revealed reduced volume and thickness in the orbitofrontal cortex, in the dorsolateral prefrontal cortex, and in the posterior hippocampus, regions associated negative affect processing and the reappraisal of the emotional experience [10, 44, 45, 50, 59]. Recently, it has been reported that individuals who score high on neuroticism are characterized by thicker cortex, smaller area, and folding in prefrontal-temporal regions [47]. Interestingly, individuals scoring high on neuroticism exhibit these morphological differences in a left lateralized circuit, with GM volumes larger in the amygdala and smaller in the orbitofrontal cortex [46]. These observations are in agreement with the findings described in the orbitofrontal cortex in anxiety trait [62] and depression [63, 64]. Furthermore, neuroticism may be linked to sustained medial prefrontal cortex response to emotional facial expressions [65].

Also DTI studies indicate reduced WM integrity in the uncinate fasciculus connecting the orbitofrontal cortex and amygdala/hippocampal regions, and in the cingulum bundle [66, 67], and in the more direct pathway from the amygdala to prefrontal cortex [68] in individuals with high neuroticism scores.

6.3 The Influence of Age on the Relationship Big Five Personality Traits-Brain Volumes

The structural features described in young adults are in contrast with the changes found in adolescents and elderly adults. Namely, in adolescent females extraversion correlates negatively with GM volume of the medial frontal gyrus, and neuroticism correlates positively with GM volume and cortical thickness of the *subgenual* anterior cingulate cortex. In males the correlations between GM volumes and personality traits show an opposite effect, suggesting a neuro-maturational divergence during adolescence [69]. In aging adults, extraversion is associated with increased thickness in both the right and left prefrontal cortex, and neuroticism is associated with decreased thickness of the right prefrontal cortex [70]. Interestingly, inferior and posterior regions of the bilateral prefrontal cortices demonstrate significant thinning with increasing age apart from the thickness of the medial orbitofrontal cortex that increases with age [23, 71]. Taken together, these findings indicate that the neural correlates of extraversion at prefrontal level are altered with aging, perhaps in relation to the atrophy of these regions that are important for regulation of social contacts. However, age-related anatomical changes, life experiences, or both may alter the direction and laterality of the structure-function relationships.

7 Morphometric Correlates of TCI Temperamental Traits

The four personality traits (novelty seeking, harm avoidance, reward dependence, and persistence) described in Temperament and Character Inventory (TCI) [31] are relatively stable over time and potentially associated with specific genetic and biological substrates. Even in Cloninger's model, two temperamental traits (novelty seeking and harm avoidance) received particular attention (Fig. 2).

It has been demonstrated that novelty seeking correlates positively with volume of the frontal and posterior cingulate cortex; harm avoidance negatively correlates with volume of orbitofrontal, occipital, and parietal structures as well as GM volume in the right hippocampus [45]. As further structural correlates, the strength of fiber tracts from the hippocampus and amygdala to striatum predicts individual differences in novelty seeking [72], the microstructural integrity of WM in cortico-limbic circuit is negatively associated with harm avoidance scores [32], and striatal activity is correlated with novelty-based choices [73]. Thus, novelty seeking and harm avoidance traits involve cortical and subcortical structures. In fact, these temperamental traits are associated with motivational and emotional processing, attentional focus, inhibitory control, and reward sensitivity [74, 75], all functions mediated not only by the cortex but also by basal ganglia and limbic system. Through the cerebral cortex and limbic structures, the information about emotionally significant stimuli is conveyed to basal ganglia that mediate the autonomic and somatic components of arousal and action [76] and serve as gating system allowing to select among a variety of available behaviors [77, 78]. This gating system applies to a large range of decisions made on the basis of motivation to approach what is found rewarding and/or to avoid what is found negative. More specifically, increased bilateral caudate and pallidum volumes are positively correlated with higher novelty seeking scores, and increased mean diffusivity in the bilateral putamen is associated with higher harm avoidance scores [79]. These morphometric data indicate that individuals with a micro-structure of putamen characterized by higher mean diffusivity values will be more vulnerable in experiencing negative emotional states and tendencies to withdrawal and inhibition. In contrast, individuals with larger volumes of caudate and pallidum will be more vulnerable in experiencing positive emotional states and tendencies to approach. Thus, both macro- and micro-structural variations of basal ganglia contribute to explain the biological variance associated with personality phenotypes. Notably, subjects characterized by relatively low striatal dopaminergic receptor density are reported to score lowest on novelty seeking and highest on harm avoidance [80]. Interestingly, harm avoidance scores correlate negatively with

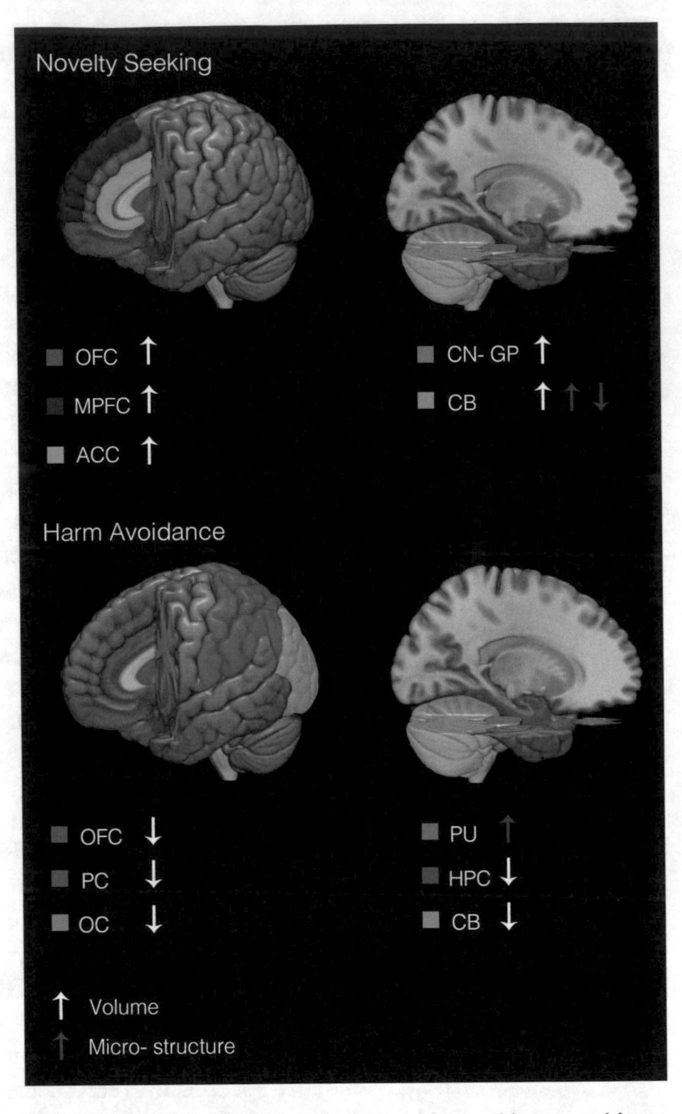

Fig. 2 Brain structural variations in novelty seeking and harm avoidance. Areas significantly associated with novelty seeking or harm avoidance were highlighted in different colors. *OFC* orbitofrontal cortex, *MPFC* medial prefrontal cortex, *ACC* anterior cingulate cortex, *CN* caudate nucleus, *GP* globus pallidum, *CB* cerebellum, *PC* parietal cortex, *OC* occipital cortex, *PU* putamen, *HPC* hippocampus. White arrow = variation in volumes; blue arrow = variation in micro-structure

fractional anisotropy and positively with mean diffusivity measures in WM cortico-limbic tracts [32]. Overall these data emphasize the relationship between anxiety-related personality trait and DTI-derived indices of WM and GM integrity.

In addition to the macro- and micro-structural characteristics in the above-described brain regions, even at cerebellar level, it is possible to evidence morphometric features related to personality traits. Starting from neuroimaging studies showing structural and functional abnormalities in the cerebellum of patients with personality disorders or depression [81, 82] and perfusion values at cerebellar level that correlated positively with novelty seeking and negatively with harm avoidance [58], neuroimaging studies have evidenced a cerebellar role in personality characteristics. By using a ROI-based approach, in healthy subjects of both sexes, WM and GM cerebellar volumes have been found to be associated positively with novelty seeking and negatively with harm avoidance [33]. These data appear fully consistent with the different engagement that subjects with different styles of personality require to their cerebellar circuitries [83]. Whether as hypothesized "larger is more powerful," the higher requests the high-scored novelty seeking subjects make to their cerebellum could enlarge it, and vice versa the lower requests the high-scored harm avoidance subjects make to their cerebellum could reduce it. As already described, among healthy individuals, the major factors contributing to personality traits are sex and age. In particular, harm avoidance and other anxiety-related traits are linked to sex-related differences. In males a negative association between harm avoidance scores and cerebellar volumes has been found, indicating that to reach higher harm avoidance scores males had to have lower cerebellar volumes. As harm avoidance scores are related to sex differences, novelty seeking scores are related to age-related differences. Interestingly, cerebellar cortical volumes are negatively correlated with age, whereas cerebellar WM volumes do not exhibit age-related differences [33].

The associations between specific cerebellar regions and TCI temperamental traits have been further studied, using VBM and DTI analyses [84]. We found positive associations between novelty seeking scores and volumes of the limbic posterior vermis (lobules VIIb, VIII, and Crus 2), a cerebellar area particularly involved in the regulation of emotion and affect and in higher-level cognitive processes [85], due to its well-known projections to the Brodmann's area 46 in the dorsolateral prefrontal cortex [86]. The relationship between novelty seeking scores and cerebellar structures was found not only at macro- but also at micro-structural level, as indicated by mean diffusivity and fractional anisotropy data. Novelty seeking scores were associated positively with fractional anisotropy in lobules IV, V, VI, and IX and negatively with mean diffusivity in lobules IV, V, VI, VIII, IX, and Crus 1 and 2. In healthy subjects, a triad including increased volumes, decreased mean diffusivity, and increased fractional anisotropy indicates high integrity and efficiency and advanced organization of the structure. Thus, the increased cerebellar gray matter volumes, the increased white matter fractional anisotropy, and the decreased gray matter mean diffusivity associated with high novelty seeking scores emphasize that the macro- and micro-structure of posterior vermis strongly supports the behaviors of novelty seeking.

To sum up, morphometric data of cortical and subcortical structures and their looped architectural connections (as the cortico-striatal-cerebellar-thalamic-cortical circuit) appear to sustain the ability to form intentions and to bring them to fruition, and thus they result strongly associated to construction of normal or abnormal personality.

8 Theoretical Questions

8.1 How Do Variations in Brain Size Correlate with Variations in Brain Activity?

All these findings indicate that variations in global and regional brain measures may take on a behavioral meaning for the expression of the personality dimensions. It should be remembered, however, that a decrease or an increase in size (volume or thickness) in a given structure does not necessarily mean a decrease or an increase in activity of that region. For example, the lateral prefrontal cortex of individuals with high extraversion scores, as stated previously, exhibits decreased thickness; in parallel, the same structure shows a greater activation to positive stimuli in comparison to negative stimuli in functional magnetic resonance imaging (fMRI) studies [51]. It has been suggested that the cortical thinning could be associated with lower resting baseline, and this would allow for a greater differential response to extraversion-relevant (positive) stimuli. Alternatively, it has to be also considered that if the thinning was due to fewer inhibitory interneurons or dendritic pruning related to enhanced processing efficiency, a thinner cortex could result in increased activity. To characterize the relationship between structure and function in relation to personality dimensions, it is necessary to distinguish among the different structural compartments of the GM (neurons, interneurons, glia, and neuropil) that contribute to the thickness and volume of the brain regions and to analyze the changes that could differently influence the brain activity in relation to personality.

8.2 A Cellular View of Changes of Brain Volumes in Relation to Temperamental Traits

It is intriguing to advance that macroscopic changes of brain volumes associated to temperamental traits we have just described could stem from differences in number, density, and morphology of various neuronal and glial populations as well as in the degree of connectivity of circuits related to personality traits. Identifying the micro-structural features of specific personality-related brain regions could shed light on discrepancy in cognitive-emotional processing characterizing the different temperamental traits.

It is generally retained that the role played by a specific brain region in the neural representation of a personality trait (or of a psychological function) may depend on the neuronal activity of many other regions at that same time, a concept that has been named "neural context," "functional connectivity" or "effective connectivity," or more recently "embodiment" [87–89]. For example, neuroimaging studies indicate that the connectivity among different brain regions can vary as a function of psychological factors, as attention or learning [90, 91], and suggest that obtaining best performances depends on the highest degree of functional interactions among brain areas [92].

Recent evidence showed the direct involvement of glial cells (astrocytes, oligodendrocytes, and microglia) in information processing [93–96]. Namely, in response to cognitive, sensory, or emotional stimuli, the astrocytes modulate excitatory and inhibitory transmission between neurons even at distant sites through intercellular long-range propagating calcium waves, and thus they mediate the Hebbian learning at synaptic level and the formation and consolidation of long-term memories by reinforcing or weakening specific neuronal patterns [93]. Oligodendrocytes, which form the myelin sheath surrounding the axons, configure their electrical and structural properties (diameter, spacing and clustering of ion channels at the nodes) [95]. It can be suggested thus that functions of neural networks underlying the regulation of affect and emotion depend on the multifaceted balance of neuronal and glial functions.

According to their own personality trait, individuals tend to be more or less emotionally stable. The two emotional conditions are associated to opposite morphological changes of dendrites and axonal branches, alterations of neuropil, and thus variations of GM volume in specific brain areas. Consistently with these cellular mechanisms, it has been advanced that in depressed individuals the sustained activity in reverberatory circuits in the amygdala and augmented compensatory regulation demanded in the orbitofrontal and anterior cingulate cortex can cause an excessive glutamate release and overstimulation of the NMDA receptors. The excitotoxicity can lead to neuronal and glial damage and death, associated with reduced GM volume as occurring in the amygdala and orbitofrontal cortex in the presence of depression [81, 97]. Although the knowledge of cellular processes underpinning personality traits is at its first steps, the proposed mechanisms could contribute to the variegated brain morphometry of personality traits.

As a final note however, it is necessary to mention that different patterns of cellular alterations can lead to similar behavioral phenotypes and that similar patterns of cellular alterations can lead to different phenotypes. This heterogeneity is reflected in the heterogeneity of the structural neuroimaging findings in healthy individuals.

9 Conclusions

Independently from the cellular mechanisms leading to increase or decrease of brain volumes in relation to personality traits, it is stimulating to wonder whether personality determines the size of brain regions or, conversely, the differently sized brain regions determine personality. Limited data are available to clarify which comes first. In fact, it may be that specific brain regions play a specific role in determining aspects associated with temperamental traits. Conversely, it is possible that the anatomical effects are secondary to the individual behavioral patterns that lead to the particular personality trait. If so, the effects of being extravert or neurotic, seeker or avoiding, might lead to specific variations in regional brain morphometry via different life experiences (more or less social contact, more or less negative affect experienced, more or less impulsiveness, more or less anxiety-related behaviors, etc.). It has been ascertained that the development and the organization of neural circuitries is influenced both by genetic predisposition and environmental events, factors that govern the neuroplastic responses to life experiences and determine how neurons are connected and communicate each other. It is important to remember that when genes and environmental influences have lasting effects on shaping our personality, they must do so by changing the brain [98]. In a virtuous loop, the changes of the brain in turn maintain the personality features (Figs. 1 and 2).

References

1. American Psychiatric Association (2000) Diagnostic and statistical manual of mental disorders, 4th edn. AAA Publisher, Washington
2. DeYoung CG, Gray JR (2009) Personality neuroscience: explaining individual differences in affect, behavior, and cognition. In: Corr PJ, Matthews G (eds) The Cambridge handbook of personality psychology. Cambridge University Press, New York, pp 323–346
3. Eysenck HJ, Eysenck MW (1985) Personality and individual differences: a natural science approach. Plenum, New York
4. Grey JA (1982) The neuropsychology of anxiety: an enquiry into the functions of the septo-hippocampal system, 1st edn. Oxford University Press, Oxford
5. McCrae RR, Costa PT Jr (1987) Validation of the five-factor model of personality across instruments and observers. J Pers Soc Psychol 52(1):81–90
6. Cloninger CR (1987) A systematic method for clinical description and classification of personality variants: a proposal. Arch Gen Psychiatry 44:573–588
7. Zuckerman M, Cloninger CR (1996) Relationships between Cloninger's, Zuckerman's, and Eysenck's dimensions of personality. Pers Individ Dif 21:283–285
8. Uttal WR (2001) A credo for a revitalized behaviorism: characteristics and emerging principles. Behav Process 54(1–3):5–10
9. Zuckerman M (2008) Zuckerman-Kuhlman personality questionnaire: an operational definition of the alternative five factorial model of personality. In: Boyle GJ, Matthews G, Saklofske DH (eds) Personality theory and assessment, vol 2. Sage, Los Angeles, pp 219–238
10. DeYoung CG, Hirsh JB, Shane MS et al (2010) Testing predictions from personality neuroscience. Brain structure and the big five. Psychol Sci 21:820–828
11. O' Brien LM, Ziegler DA, Deutsch CK et al (2006) Adjustment for whole brain and cranial size in volumetric brain studies: a review of

common adjustment factors and statistical methods. Harv Rev Psychiatry 14(3):141–151
12. Changizi MA (2001) Principles underlying mammalian neocortical scaling. Biol Cybern 84(3):207–215
13. Finlay BL, Darlington RB, Nicastro N (2001) Developmental structure in brain evolution. Behav Brain Sci 24:263–308
14. Hazlett HC, Poe MD, Gerig G et al (2011) Early brain overgrowth in autism associated with an increase in cortical surface area before age 2 years. Arch Gen Psychiatry 68 (5):467–476
15. Courchesne E, Pierce K, Schumann CM et al (2007) Mapping early brain development in autism. Neuron 56(2):399–413
16. Bralten J, Greven CU, Franke B et al (2015) Voxel-based morphometry analysis reveals frontal brain differences in participants with ADHD and their unaffected siblings. J Psychiatry Neurosci 41(2):140377
17. Maier S, Perlov E, Graf E et al (2016) Discrete global but no focal Gray Matter volume reductions in Unmedicated adult patients with attention-deficit/hyperactivity disorder. Biol Psychiatry 80(12):905–915. https://doi.org/10.1016/j.biopsych.2015.05.012
18. Rogers JC, De Brito SA (2016) Cortical and subcortical Gray Matter volume in youths with conduct problems: a meta-analysis. JAMA Psychiat 73:64–72
19. Silk TJ, Vilgis V, Adamson C et al (2016) Abnormal asymmetry in frontostriatal white matter in children with attention deficit hyperactivity disorder. Brain Imaging Behav 10 (4):1080–1089. https://doi.org/10.1007/s11682-015-9470-9
20. Bonath B, Tegelbeckers J, Wilke M et al (2016) Regional Gray Matter volume differences between adolescents with ADHD and typically developing controls: further evidence for anterior cingulate involvement. J Atten Disord. pii: 1087054715619682
21. Good CD, Johnsrude IS, Ashburner J et al (2001) A voxel-based morphometric study of ageing in 465 normal adult human. NeuroImage 14:21–36
22. Raz N, Gunning-Dixon F, Head D et al (2004) Aging, sexual dimorphism, and hemispheric asymmetry of the cerebral cortex: replicability of regional differences in volume. Neurobiol Aging 25(3):377–396
23. Salat DH, Buckner RL, Snyder AZ et al (2004) Thinning of the cerebral cortex in aging. Cereb Cortex 14(7):721–730
24. Schmitt DP, Realo A, Voracek M et al (2008) Why can't a man be more like a woman? Sex differences in big five personality traits across 55 cultures. J Pers Soc Psychol 94(1):168–182
25. Sowell ER, Peterson BS, Thompson PM (2003) Mapping cortical change across the human life span. Nat Neurosci 6(3):309–315
26. Tisserand DJ, Pruessner JC, Sanz Arigita EJ et al (2002) Regional frontal cortical volumes decrease differentially in aging: an MRI study to compare volumetric approaches and voxel-based morphometry. NeuroImage 17:657–669
27. Cowell PE, Sluming VA, Wilkinson LD et al (2007) Effects of sex and age on regional prefrontal brain volume in two human cohorts. Eur J Neurosci 25(1):307–318
28. Murphy DG, Decarli C, McIntosh AR et al (1996) Sex differences in human brain morphometry and metabolism: an in vivo quantitative magnetic resonance imaging and positron emission tomography study on the effect of aging. Arch Gen Psychiatry 53(7):585–594
29. Goldstein JM, Seidman LJ, Horton NJ (2001) Normal sexual dimorphism of the adult human brain assessed by in vivo magnetic resonance imaging. Cereb Cortex 11(6):490–497
30. Luders E, Narr KL, Thompson PM et al (2006) Gender effects on cortical thickness and the influence of scaling. Hum Brain Mapp 27(4):314–324
31. Cloninger CR, Svrakic DM, Przybeck TR (1993) A psychobiological model of temperament and character. Arch Gen Psychiatry 50:975–990
32. Westlye LT, Bjørnebekk A, Grydeland H et al (2011) Linking an anxiety-related personality trait to brain white matter microstructure: diffusion tensor imaging and harm avoidance. Arch Gen Psychiatry 68:369–377
33. Laricchiuta D, Petrosini L, Piras F et al (2014) Linking novelty seeking and harm avoidance personality traits to cerebellar volumes. Hum Brain Mapp 35:285–296
34. Chapman BP, Duberstein PR, Sörensen S et al (2007) Gender differences in five factor model personality traits in an elderly cohort: extension of robust and surprising findings to an older generation. Pers Individ Dif 43(6):1594–1603
35. Costa PT Jr, Terracciano A, McCrae RR (2001) Gender differences in personality traits across cultures: robust and surprising findings. J Pers Soc Psychol 81(2):322–331
36. Donnellan MB, Lucas RE (2008) Age differences in the Big Five across the life span: evidence from two national samples. Psychol Aging 23(3):558–566
37. Lüdtke O, Trautwein U, Husemann N (2009) Goal and personality trait development in a

transitional period: assessing change and stability in personality development. Personal Soc Psychol Bull 35(4):428–441

38. Rantanen J, Metsäpelto RL, Feldt T (2007) Long-term stability in the big five personality traits in adulthood. Scand J Psychol 48 (6):511–518
39. Roberts BW, Walton KE, Viechtbauer W (2006) Patterns of mean-level change in personality traits across the life course: a meta-analysis of longitudinal studies. Psychol Bull 132(1):1–25
40. Hu X, Erb M, Ackermann H et al (2011) Voxel based morphometry studies of personality: issue of statistical model specification-effect of nuisance covariates. NeuroImage 54:1994–2005
41. Boyke J, Driemeyer J, Gaser C et al (2008) Training-induced brain structure changes in the elderly. J Neurosci 28:7031–7035
42. Di Paola M, Caltagirone C, Petrosini L (2012) Prolonged rock climbing activity induces structural changes in cerebellum and parietal lobe. Hum Brain Mapp 4:2707–2714
43. McDaniel L (2005) In search of higher education. Biomed Instrum Technol 39(6):451–453
44. Gardini S, Cloninger CR, Venneri A (2009) Individual differences in personality traits reflect structural variance in specific brain regions. Brain Res Bull 79:265–270
45. Yamasue H, Abe O, Suga M et al (2008) Gender-common and specific neuroanatomical basis of human anxiety-related personality traits. Cereb Cortex 18:46–52
46. Mincic AM (2015) Neuroanatomical correlates of negative emotionality-related traits: a systematic review and meta-analysis. Neuropsychologia 77:97–118
47. Riccelli R, Toschi N, Nigro S et al (2017) Surface-based morphometry reveals the neuroanatomical basis of the five-factor model of personality. Soc Cogn Affect Neurosci 12 (4):671–684. pii: nsw175
48. Rauch SL, Milad MR, Orr SP et al (2005) Orbito-frontal thickness, retention of fear extinction, and extraversion. Neuroreport 16:1909–1912
49. Cremers H, van Tol MJ, Roelofs K et al (2011) Extraversion is linked to volume of the orbitofrontal cortex and amygdala. PLoS One 6(12): e28421
50. Kapogiannis D, Sutin A, Davatzikos C et al (2013) The five factors of personality and regional cortical variability in the Baltimore longitudinal study of aging. Hum Brain Mapp 34(11):2829–2840
51. Canli T, Zhao Z, Desmond JE et al (2001) An fMRI study of personality influences on brain reactivity to emotional stimuli. Behav Neurosci 115(1):33–42
52. Coutinho JF, Sampaio A, Ferreira M et al (2013) Brain correlates of pro-social personality traits: a voxel-based morphometry study. Brain Imaging Behav 7(3):293–299
53. Amodio DM, Frith CD (2006) Meeting of minds: the medial frontal cortex and social cognition. Nat Rev Neurosci 7:268–277
54. Mitchell JP, Banaji MR, Macrae CN (2005) General and specific contributions of the medial prefrontal cortex to knowledge about mental states. NeuroImage 28(4):757–762
55. Benoit RG, Gilbert SJ, Volle E et al (2010) When I think about me and simulate you: medial rostral prefrontal cortex and self-referential processes. NeuroImage 50 (3):1340–1349
56. Johnson DL, Wiebe JS, Gold SM et al (1999) Cerebral blood flow and personality: a positron emission tomography study. Am J Psychiatry 156(2):252–257
57. Canli T (2004) Functional brain mapping of extraversion and neuroticism: learning from individual differences in emotion processing. J Pers 72(6):1105–1132
58. O'Gorman RL, Kumari V, Williams SC et al (2006) Personality factors correlate with regional cerebral perfusion. NeuroImage 31 (2):489–495
59. Wright CI, Williams D, Feczko E et al (2006) Neuroanatomical correlates of extraversion and neuroticism. Cereb Cortex 16(12):1809–1819
60. Grimm S, Schubert F, Jaedke M et al (2012) Prefrontal cortex glutamate and extraversion. Soc Cogn Affect Neurosci 7:811–818
61. Peña-Gómez C, Vidal-Piñeiro D, Clemente IC et al (2011) Down-regulation of negative emotional processing by transcranial direct current stimulation: effects of personality characteristics. PLoS One 6:e22812
62. Roppongi T, Nakamura M, Asami T et al (2010) Posterior orbitofrontal sulcogyral pattern associated with orbitofrontal cortex volume reduction and anxiety trait in panic disorder. Psychiatry Clin Neurosci 64 (3):318–326
63. van Tol MJ, van der Wee NJ, van den Heuvel OA (2010) Structural MRI correlates for vulnerability and resilience to major depressive disorder. J Psychiatry Neurosci 36(1):15–22
64. Lacerda AL, Keshavan MS, Hardan AY et al (2004) Anatomic evaluation of the orbitofrontal cortex in major depressive disorder. Biol Psychiatry 55(4):353–358

65. Haas BW, Constable RT, Canli T (2008) Stop the sadness: neuroticism is associated with sustained medial prefrontal cortex response to emotional facial expressions. NeuroImage 42 (1):385–392
66. Baur V, Hänggi J, Jäncke L (2012) Volumetric associations between uncinate fasciculus, amygdala, and trait anxiety. BMC Neurosci 13:4
67. Bjørnebekk AM, Fjell KB, Walhovd H (2013) Neuronal correlates of the five factor model (FFM) of human personality: multimodal imaging in a large healthy sample. NeuroImage 65:194–208
68. Kim MJ, Whalen PJ (2009) The structural integrity of an amygdala-prefrontal pathway predicts trait anxiety. J Neurosci 29:11614–11618
69. Blankstein U, Chen JY, Mincic AM (2009) The complex minds of teenagers: neuroanatomy of personality differs between sexes. Neuropsychologia 47(2):599–603
70. Wright CI, Feczko E, Dickerson B et al (2007) Neuroanatomical correlates of personality in the elderly. NeuroImage 35:263–272
71. Grieve SM, Clark CR, Williams LM (2005) Preservation of limbic and paralimbic structures in aging. Hum Brain Mapp 25:391–401
72. Cohen MX, Schoene-Bake JC, Elger CE et al (2009) Connectivity-based segregation of the human striatum predicts personality characteristics. Nat Neurosci 12:32–34
73. Wittmann BC, Daw ND, Seymour B et al (2008) Striatal activity underlies novelty-based choice in humans. Neuron 58:967–973
74. Goldsmith HH, Lemery KS (2000) Linking temperamental fearfulness and anxiety symptoms: a behavior-genetic perspective. Biol Psychiatry 48(12):1199–1209
75. Goldsmith HH, Davidson RJ (2004) Disambiguating the components of emotion regulation. Child Dev 75(2):361–365
76. Cain CK, LeDoux JE (2007) Escape from fear: a detailed behavioral analysis of two atypical responses reinforced by CS termination. J Exp Psychol Anim Behav Process 33(4):451–463
77. McNab F, Klingberg T (2008) Prefrontal cortex and basal ganglia control access to working memory. Nat Neurosci 11(1):103–107
78. Koziol LF, Budding DE, Chidekel D (2010) Adaptation, expertise, and giftedness: towards an understanding of cortical subcortical, and cerebellar network contributions. Cerebellum 9:499–529
79. Laricchiuta D, Petrosini L, Piras F et al (2014) Linking novelty seeking and harm avoidance personality traits to basal ganglia: volumetry and mean diffusivity. Brain Struct Funct 219:793–803
80. Montag C, Markett S, Basten U et al (2010) Epistasis of the DRD2/ANKK1 Taq Ia and the BDNF Val66Met polymorphism impacts novelty seeking and harm avoidance. Neuropsychopharmacology 35(9):1860–1867
81. Fitzgerald PB, Laird AR, Maller J et al (2008) A meta-analytic study of changes in brain activation in depression. Hum Brain Mapp 29:683–695
82. Liu Z, Xu C, Xu Y et al (2010) Decreased regional homogeneity in insula and cerebellum: a resting-state fMRI study in patients with major depression and subjects at high risk for major depression. Psychiatry Res 182:211–215
83. Schutter DJ, Koolschijn PC, Peper JS et al (2012) The cerebellum link to neuroticism: a volumetric MRI association study in healthy volunteers. PLoS One 7(5):e37252
84. Picerni E, Petrosini L, Piras F et al (2013) New evidence for the cerebellar involvement in personality traits. Front Behav Neurosci 7:133
85. Stoodley CJ, Schmahmann JD (2010) Evidence for topographic organization in the cerebellum of motor control versus cognitive and affective processing. Cortex 46:831–844
86. Kelly RM, Strick PL (2003) Cerebellar loops with motor cortex and prefrontal cortex of a nonhuman primate. J Neurosci 23:8432–8444
87. McIntosh AM, Bastin ME, Luciano M (2013) Neuroticism, depressive symptoms and white-matter integrity in the Lothian birth cohort 1936. Psychol Med 43:1197–1206
88. Friston KJ, Buechel C, Fink GR et al (1997) Psychophysiological and modulatory interactions in neuroimaging. NeuroImage 6 (3):218–229
89. Friston KJ, Harrison L, Penny W (2003) Dynamic causal modelling. NeuroImage 19 (4):1273–1302
90. Büchel C, Coull JT, Friston KJ (1999) The predictive value of changes in effective connectivity for human learning. Science 283 (5407):1538–1541
91. Büchel C, Friston K (2000) Assessing interactions among neuronal systems using functional neuroimaging. Neural Netw 13(8–9):871–882
92. Canli T, Amin Z, Haas B et al (2004) A double dissociation between mood states and personality traits in the anterior cingulate. Behav Neurosci 118(5):897–904
93. Pereira A, Furlan FA (2010) Astrocytes and human cognition: modeling information, integration and modulation of neuronal activity. Prog Neurobiol 92:405–420

94. Barres BA (2008) The mystery and magic of glia: a perspective on their roles in health and disease. Neuron 60:430–440
95. Edgar N, Sibille E (2012) A putative functional role for oligodendrocytes in mood regulation. Transl Psychiatry 2:e109
96. Molofsky AV, Krencik R, Ullian EM (2012) Astrocytes and disease: a neurodevelopmental perspective. Genes Dev 26:891–907
97. Sacher J, Neumann J, Fünfstück T et al (2012) Mapping the depressed brain: a meta-analysis of structural and functional alterations in major depressive disorder. J Affect Disord 140 (2):142–148
98. De Young CG (2010) Personality neuroscience and the biology of traits. Soc Personal Psychol Compass 10:1165–1180

Chapter 25

Brain Morphometric Techniques Applied to the Study of Traumatic Brain Injury

Elisabeth A. Wilde, Brian A. Taylor, and Ricardo E. Jorge

Abstract

Traumatic brain injury (TBI) occurs when an external mechanical force causes brain dysfunction, and it is a major public health concern.

Definitive biomarkers of TBI have not been identified yet, but recent advances in neuroimaging shed new light on TBI pathophysiology. The chapter will first review the different approaches adopted in TBI assessment. It will then focus on the morphometric and volumetric alterations in TBI across various injury stages (severe and acute phases); severities, beginning with mTBI and progressing across the range of severity to severe TBI; and mechanisms.

All the present evidence will be finally discussed with emphasis on the limitations of the currently published research literature, as well as on future directions.

Key words Traumatic brain injury, TBI, Brain morphometry, Brain atrophy, Concussion, Ventricle enlargement

1 Introduction

Traumatic brain injury (TBI) occurs when an external mechanical force causes brain dysfunction [1]. TBI is a global public health concern, as international incidence and prevalence rates of TBI, particularly injuries classified as mild (mTBI) such as concussion, have increased since 2000 [2]. The symptoms of TBI are highly variable from one person to another, and understanding of the effects of TBI, and perhaps especially mTBI, is largely complicated by the fact that a definitive biomarker of TBI has not been identified [3–5]. However, significant advances in basic and translational neuroscience and in neuroimaging brought novel insights into the pathophysiology of TBI. In particular, advanced magnetic resonance imaging (MRI) data acquisition and analysis techniques have increased our understanding of the morphological and structural alterations that can occur following TBI. The current chapter addresses morphometric and volumetric alterations in TBI across

Gianfranco Spalletta et al. (eds.), *Brain Morphometry*, Neuromethods, vol. 136,
https://doi.org/10.1007/978-1-4939-7647-8_25, © Springer Science+Business Media, LLC 2018

various injury severities and mechanisms. We will review the literature in adult brain injury, beginning with mTBI and progressing across the range of severity to severe TBI. Finally we will review studies that have analyzed subjects with diverse TBI severity. To conclude, an overall discussion and summary of the field will be provided, with emphasis on the limitations of the currently published research literature as well as on future directions.

2 TBI Assessment

Treatment and medical management of TBI require accurate diagnosis as well as a reliable classification of injury severity. Until the advent of modern neuroimaging advances, classification of TBI as being mild, moderate, or severe was based on neurological or behavioral manifestations subsequent to the event. Although advanced clinical neuroimaging methodologies have allowed for more direct assessment approaches of injury severity, identification and classification of TBI remains relatively complicated, particularly when considering mild injuries.

2.1 Behavioral Approaches

TBI has historically been classified on the basis of the presence and duration of loss of consciousness (LOC), alteration of consciousness (AOC), and/or post-traumatic amnesia (PTA). Whereas LOC describes a period of unconsciousness (blackout) as a result of the traumatic event, AOC involves marked confusion or ambiguity regarding the event itself or the period of time subsequent to the event. PTA has been defined as the interval during which loss of memory for the event itself as well as the events immediately following the injury is experienced. The Glasgow Coma Scale (GCS) is the most widely used measure for the clinical classification of TBI [6]. The GCS defines mild injuries as being characterized by scores between 13 and 15, moderate TBI as characterized by scores between 9 and 12, and severe TBI as characterized by GCS scores between 3 and 8 [7]. Given the limitations of the GCS [8–10], PTA has also been investigated as a criterion for TBI severity. Clinical classification using PTA characterizes mild injuries as having experienced PTA up to 24 h, moderate between 1 and 7 days, and severe TBI when PTA is greater than 7 days [11]. Despite greater accuracy than GCS scores in predicting later functional status [12], identification of occurrence and duration of PTA is often difficult to establish in clinical contexts.

2.2 Neuroimaging Approaches

Different neuroimaging techniques are used to assess severity of TBI. Computed tomography (CT) continues to be frequently utilized in emergency departments to evaluate the need for immediate neurosurgical intervention in the presence of intracranial hematoma, contusion, or subarachnoid hemorrhage [11]. In addition, it can reveal indicators of diffuse axonal injury as well as evidence of edema and mass effect. However, TBI can result in wide ranging alterations of the brain that can be subtle, and clinical CT results can underestimate extent of damage [13]. Modern neuroimaging techniques such as magnetic resonance imaging (MRI) allows for increased signal contrast among tissue types and better spatial resolution.

It is noteworthy that while modern classification systems have been developed for use in civilians [14] and in military personnel [see review by [15]], global acceptance of a specific TBI nomenclature is still in progress.

3 Volumetric and Morphometric Analyses in Adult Traumatic Brain Injury

Volumetric and morphometric analyses allow for the quantification of brain tissue volume, thickness, area, curvature, shape, and other characteristics in addition to cerebrospinal fluid and total intracranial volumes, generally utilizing 3D volumetric high-resolution T1-weighted imaging or combinations of complementary MR sequences. These techniques have become increasingly automated, and several forms of both commercial software packages and publically available software exist at no cost for the quantitation and rendering of morphometric data derived from MRI. In addition to group analyses, such software may be useful for rendering data in an individual patient or subject.

Despite differences in magnetic field strength, acquisition parameters, and analytic procedures across studies, a growing body of research suggests that mild TBI may be associated with both global and regional volumetric alterations. We begin with discussion of community populations with mild TBI and also present data from special populations, including athletes with sports-related concussion and service members and veterans with mild TBI. Please *see* Table 1 for a summary of these studies.

3.1 Studies in Mild Traumatic Brain Injury in Community Samples

3.1.1 Findings in the Acute Stage

Acute-phase alterations in cortical and subcortical morphology are observed in mTBI patients as early as 7 days post-injury compared to controls [16]. Although the whole-brain analysis failed to show differences among patients and controls in cortical morphology, Dall'Acqua et al. [16] observed significantly reduced cortical surface area in the lateral prefrontal cortex, postcentral gyrus, inferior temporal gyrus, and the insula in a subgroup of mTBI patients with severe post-concussive symptoms (PCS) compared to controls. In this sample of patients with mTBI, prefrontal cortical surface area correlated with indices of white matter (WM) disruption and measures of cognitive control.

Subcortical morphological alterations are also observed in acute mTBI. Holli et al. [17] found greater asymmetry of left and

Table 1
Studies in mild traumatic brain injury in community samples

Author	TBI participant characteristics and comparison		Cohort	Time post-injury	Imaging acquisition details	Analysis method	Summary of findings
	Patients	Controls					
Hofman et al. (2001)	21 uncomplicated mTBI (GCS 14–15, $M = 14.48 \pm 0.6$) Sex: 12 male Age: 22.8 ± 7.65 years		Consecutive emergency department patients	Acute/subacute and chronic (initial scan 2–5 days post-injury; follow-up at 6 months post-injury)	1.5 T Philips ACS; axial dual T2-weighted fast spin echo TR/TE/NEX = 3000/23–120/2; echo train length = 12; matrix = 225 × 3 × 186; slice thickness = 5 mm; slices = with 24; 10% interslice gap	Volumetric analysis	57% of patients had abnormal MRI findings that related to longitudinal brain atrophy; lesions unrelated to neurocognitive performance, which was in average range
Cohen et al. (2007)	20 mTBI (GCS 13–15) Sex: 11 male Age: median 35 ± 10 years	19 controls matched on age and sex Sex: 11 male Age: median 39 ± 11 years	Trauma centers and databases	Acute to chronic (7 patients ≤9 days; 13 patients $M = 4.6$ years [range1.2–31.5 months] post-injury)	1.5 T; sagittal MPRAGE; TR/TE/TI = 14.7/7.0/300 ms; 128 slices; slice thickness = 1.5 mm; matrix = 256 × 256; FOV = 210 × 210 mm^2	Lesion detection and volumetric analysis (MIDAS)	30% of patients exhibited lesions; increased global and GM atrophy observed in patients with and without lesions compared to controls
Holli et al. (2010)	42 mTBI (GCS 13–15) Sex: 17 male Age: 38.8 ± 13.6 years	10 controls Sex: 4 male Age: 39.8 ± 12.9 years	Hospital emergency department recruitment	Acute (≤3 weeks post-injury)	1.5 T Siemens MAGNETOM Avanto; 3D MPRAGE; TR/TE/TI = 1910/3.1/1100; slice thickness/gap = 1.0/0; matrix = 256 × 256; FOV = 250; flip angle = 15°	Texture analysis (MaZda)	Significantly greater asymmetry observed in left and right mesencephalon, WM of the corona radiata, and portions of CC in patients compared to controls
Ross et al. (2012)	16 mTBI patients (GCS $M = 14.4$, median = 15.0 [11–15]) Sex: 6 male Age: 48.1 ± 10.9 years Silver et al. (2011) criteria used	20 controls Sex: 10 male Age: 68.3 ± 3.6 years	Neuropsychiatry clinic outpatients	Chronic (initial scan $M = 2 \pm 1.4$ [0.4–5.7] years post-injury; follow-up $M = 1.0 \pm 0.4$ [0.4–2.6] years later)	3 T GE/Siemens/Philips protocol detailed on the NeuroQuant website (http://www.cortechs.netproducts/neuroquant.php)	NeuroQuant automated brain segmentation and volumetric analysis	Significant longitudinal reduction in total brain, frontal, cerebral WM, and cerebellum volumes was observed in patients compared to controls and related to functional outcomes (e.g., vocational status) at follow-up
Zhou et al. (2013)	28 mTBI patients with PTSD symptoms (GCS 13–15) Sex: 22 male Age: 34 ± 11.5 years Subset of mTBI patients ($n = 19$) and controls ($n = 12$) underwent follow-up MRI	22 controls matched on age, sex, education, and handedness Sex: not reported Age: 35.1 ± 11.3 years	Level 1 trauma center emergency department	Subacute and chronic ($M = 23$ [3–53] days post-injury; 1 year follow-up)	3 T Siemens Tim Trio; 3D MPRAGE; TR/TE/TI = 2300/2.98/900 ms; flip angle = 9°; voxel = 1 mm^3	Automated segmentation, voxel-based morphometry and volumetric analysis (FreeSurfer)	Average longitudinal global atrophy rate of 7.6 cm^3 observed in patients compared to controls; supratentorial alterations strongly correlated with global atrophy rate of whole-brain and segmented volume and of the ventricles and segmented ventricular volumes Longitudinal WM volume reductions observed in bilateral rostral anterior cingulum and left caudal anterior cingulum and cingulate gyrus isthmus in mTBI patients compared

							to controls; longitudinal GM volume reductions observed in the right precuneus and right inferior and medial orbital olfactory frontal regions in mTBI; results accounted for differences in acute MRI days post-injury No group differences in volumes observed at initial scan; at time 2, volumetric differences observed in rostral anterior cingulum bilaterally, in the left caudal anterior cingulum, in the left cingulate gyrus isthmus WM, and in the right precuneal GM in mTBI patients compared to controls Volumetric alterations correlated with PCS severity and anxiety ratings, as well as with neuropsychological performance in the patients
Lopez-Larson et al. (2013)	40 veterans with mTBI without suicidal behavior Sex: 40 male Age: 34.60 ± 8.10 years 19 veterans with mTBI and suicidal behavior Sex: 19 male Age: 38.00 ± 7.77 years (diagnosed with evidence of AOC or LOC at the TOI)	15 controls Sex: 15 male Age: 36.47 ± 11.51 years	Veterans/military VA and community convenience sample	Not reported	3 T Siemens Trio; 3D sagittal MPRAGE GRAPPA; 12-channel head coil; TE/TR/TI = 3.38 ms/2.0 s/1.1 s; flip angle = 8°; matrix = 256 × 256; FOV = 256 mm^2; slices = 160; slice thickness = 1.0 mm	Automated segmentation and volumetric analysis (FreeSurfer)	mTBI groups similar on lifetime number of TBI, major depression, or PTSD, although veterans with mTBI and suicidal behavior significantly more depressed than veterans with mTBI without suicidal behavior Comorbid mTBI and suicidal behavior related to significantly disproportionately larger left thalamus volume compared to controls independently and when collapsed with mTBI without suicidal behaviors; and significantly disproportionately larger right thalamus volume compared to controls and mTBI without suicidal behaviors when examined independently and when collapsed across groups, which did not differ
Toth et al. (2013)	14 mTBI patients (GCS = 15, LOC < 1 min; PTA < 30 min; normal post-traumatic CT) Sex: 9 male Age: 34.9 ± 18.4 years	12 controls matched on age and sex Sex: 9 male Age: 35.8 ± 18.5 years		Acute and subacute (initial scan ≤72 h post-injury, *M* = 48 h [12–72 h]; follow-up scan, *M* = 35 days post-injury [28–43 days])	3 T Siemens MAGNETOM Tim Trio; 12-channel head coil; 3D MPRAGE; TR/TI/TE = 1900/900/3.41 ms; flip angle = 9°; FOV = 210 × 240 mm^2; slice thickness/gap = 0.94/0 mm; matrix = 244 × 256; axial	Automated segmentation and volumetric analysis (FreeSurfer; longitudinal volumetric changes using FSL SIENA)	Significant longitudinal reduction in cortical GM volume at 1 month than 72 h post-injury (mean loss of 1.02%); and significant longitudinal enlargement in ventricular (more pronounced in lateral ventricle) and extracerebral CSF volumes at 1 month than 72 h post-injury in

(continued)

Table 1
(continued)

Author	TBI participant characteristics and comparison: Patients	Controls	Cohort	Time post-injury	Imaging acquisition details	Analysis method	Summary of findings
					slices = 160; receiver bandwidth = 180 HZ/pixel		patients compared to controls (gain $M = 3.4\%$) Outward longitudinal edge displacement observed in several voxels lining third and lateral ventricles in patients compared to controls
Maller et al. (2014)	14 mTBI patients with major depression (GCS 13–14) Sex: 6 male Age: 48.0 ± 9.92 years 12 mTBI patients without major depression (GCS 13–14) Sex: 10 male Age: 33.08 ± 12.69 years Comparisons: 26 patients with major depression without histories of head injury Sex: 17 male Age: 44.08 ± 12.99 years	23 controls Sex: 9 male Age: 38.35 ± 13.00 years	Clinical referral	Chronic (6 weeks–10 years)	1.5 T GE Signa; AC-PC aligned sagittal SPGR; TR/TE = 100/450 ms; matrix = 224 × 224; NEX = 1; slice thickness/gap = 1.4/0 mm, in-plane resolution = 0.94 mm^2	Automated segmentation and volumetric analysis (FreeSurfer)	Significantly reduced volume in the left inferior parietal cortex and right insula and hippocampal regions in mTBI patients (with and without depression) compared to controls; significantly reduced inferior temporal and parietal regions and right lingual region in the mTBI patients who developed depression compared to those without depression; significantly reduced the right inferior temporal region was observed in the mTBI patients who developed depression compared to comparisons; significantly reduced left lingual and right hemispheric regions extending along the lateral and medial occipital regions to the inferior and superior parietal lobe and the cuneus were observed in the comparisons compared to controls
Tate et al. (2014)	12 active duty military with blast-related mTBI Sex: 12 male Age: 26.53 ± 6.16 years	11 active duty military with no histories of TBI matched on sex Sex: 11 male Age: 27.73 ± 2.41 years	Active duty military personnel (OEF/OIF); DVBIC	Subacute ($M = 105 \pm 51$ days post-injury)	3 T Siemens Tim Trio; MPRAGE; TR/TE = 2200/2.8 msec; slice thickness = 0.8; in-plane resolution = 0.8 × 0.8; FOV = 256 × 256 × 192; flip angle = 13°	Automated segmentation and volumetric analysis (FreeSurfer)	Significant cortical thinning in blast-related TBI group in the left superior temporal/transverse and superior frontal gyri compared to controls after FDR correction; no significant associations observed among time since injury and cortical thickness after FDR correction; thickness of the left superior temporal and transverse and superior frontal gyri related with audition and speech/language

							abilities, related to audiology problems like tinnitus, ruptured tympanic membrane, conductive and/or sensorineural hearing loss and reduction
Depue et al. (2014)	21 previously deployed veterans with comorbid mTBI and PTSD (GCS 13–15; LOC < 5 min in 52.4%) Sex: 20 male Age: 29 (23–43) years OSU TBI-ID criteria used	16 previously deployed veterans without PTSD or mTBI Sex: 14 male Age: 28 (24–45)	Convenience sample of OEF/OIF/OND veterans	Chronic (but not reported)	1.5 T Philips Achieva; 16-channel head coil; 3D sagittal TFE; TR/TE = 7100/3.2 ms; flip angle = 8°; FOV = 240 × 240 mm; voxel size = 1 × 1 × 1.03 mm^3; slices = 160	Voxel- and surface-based morphometry (FSL)	Disproportionately reduced bilateral amygdala volume observed in veterans with mTBI/PTSD compared to comparisons in whole-brain VBM analysis; similar results obtained through SBM analysis in bilateral anterior amygdala in mTBI/PTSD veterans; disproportionate left amygdala volume reductions associated with increased commission errors during go/no-go task and increased impulsivity ratings, and disproportionate right amygdala reductions associated with symptoms of PTSD in veterans with mTBI/PTSD (not observed in comparisons)
Corbo et al. (2014)	43 veterans with histories of mTBI and early life trauma (i.e., prior to being 18 years old/pre-deployment; 4 veterans had pre-deployment TBI that was moderate [n = 3] and severe [n = 1]) Sex: 35 male Age: 35.64 ± 1.32 years	65 veterans with histories of mTBI without early life trauma Sex: 59 male Age: 33.35 ± 1.23 years	Veterans/military personnel (OEF/OIF); DVBIC	Chronic (but not reported)	3 T Siemens Tim Trio; 12-channel head coil; 3D MPRAGE; TR/TE = 2530/3.32 ms; flip angle = 7°; FOV = 256 × 256; slice thickness = 1 mm	Automated segmentation and volumetric analysis (FreeSurfer)	Veterans with histories of TBI and early life trauma significantly greater number of lifetime TBI compared to comparisons PTSD symptom severity negatively correlated with cortical thickness in controls and positively correlated with cortical thickness in the veterans with TBI and early life trauma in the left paracentral and posterior cingulate gyrus; this pattern also observed with volume of the right posterior cingulate cortex Greater cortical thickness observed in bilateral posterior cingulate in comparisons compared to veterans with TBI and early life trauma Lifetime TBI trended toward being negatively associated with right postcentral gyrus cortical thickness; trend-level interactions also observed between lifetime TBI and thickness of the right superior parietal cortex and left isthmus

(continued)

Table 1
(continued)

Author	TBI participant characteristics and comparison		Cohort	Time post-injury	Imaging acquisition details	Analysis method	Summary of findings
	Patients	Controls					
							cingulate, which were negatively associated with comparisons not unrelated in veterans with TBI and early life trauma; later age of first trauma associated with greater cortical thickness in the right pars triangularis of the inferior frontal gyrus Groups did not significantly differ on volumes of the amygdala or hippocampus; left amygdala volume correlated with PTSD severity in the veterans with TBI and early life trauma that was not observed in comparisons; right hippocampus volume correlated with lifetime TBI history
Little et al. (2014)	43 mTBI patients (GCS $\geq$ 13; 29 patients with only 1 mTBI; 14 patients with histories of >1 mTBI) Sex: 23 male Age: 34 (20–58) years Diagnosis based on ACRM criteria (1993)	37 controls Sex: 16 male Age: 33 (19–60) years	Convenience community sample	Chronic (6 months to 28 years post-injury, $M = 5.3$ years)	Not reported	Voxel-based morphometry (SPM5)	Whole-brain volumetric analysis indicated group effect on total intracranial GM, WM, and CSF volumes above total intracranial volume Observed reduced WM density in mTBI patients in the right internal capsule and right ventrolateral prefrontal compared to controls; observed altered GM density in the somatosensory, parietal, temporal, parahippocampal, and cerebellar regions compared to controls; observed reduced density in the temporal lobes, parahippocampal gyri, ventrolateral prefrontal regions, external capsule, and cerebellum in patients with multiple mTBI compared to patients with one mTBI In mTBI patients, regardless of single or multiple injuries, executive function related with decreased density in the

							posterior cingulate and increased density in the internal capsule, whereas attention related with density in the cingulate, parietal, and occipital WM and temporal GM and memory relate with reduced density in the parahippocampal gyri, anterior temporal lobe, and internal capsule
Dean et al. (2015)	16 mTBI patients (GCS 13–15) Sex: 7 male Age: 27.0 ± 1.6 years ICD-10 criteria used	9 controls Sex: 4 male Age: 21.8 ± 1.5 years		Chronic (> 1 year post-injury)	3 T Siemens Trio; HD 3D MPRAGE; TR/TE/TI = 1830/ 4.43/1100 ms; flip angle = 11°; FOV = 256 mm^2; slices = 176; voxel = 1 mm^3; matrix = 256 × 256	Automated segmentation and voxel-based morphometry (FreeSurfer; DARTEL in SPM8)	No significant differences in cortical thickness observed between patients and controls; cortical thickness unrelated with PCS symptoms Disproportionate reduction of GM within bilateral prefrontal areas observed in patients compared to controls; greater reduction of GM in bilateral medial temporal lobe, inferior parietal lobe, and right precuneus associated with greater PCS symptom severity in patients
Killgore et al. (2016)	26 mTBI (GCS not reported) Sex: 11 male Age: 23.38 ± 5.23 years	12 controls matched on age and education Sex: 4 male Age: 25.0 ± 6.55 years		Subacute and chronic (2 weeks to1 year post-injury)	3 T Siemens Tim Trio; 32-channel head coil; 3D sagittal MPRAGE; TR = 2.1 s; TE = 2.3 ms; flip angle = 12°; slices = 176; matrix = 256 × 256; slice thickness = 1 mm; voxel size = 1 mm^3	Voxel-based morphometry (SMP8)	Longer post-injury assessment time related to larger GM volume in the ventromedial prefrontal cortex and right fusiform gyrus, which were significantly increased in TBI patients at the latest follow-up compared to controls; ventromedial prefrontal cortex volume positively correlated with nonverbal fluency and negatively correlated with anxiety scores; fusiform gyrus volume related with design fluency and psychomotor vigilance performance
Dall'Acqua et al. (2016)	51 mTBI patients (GCS 13–15) Sex: 32 male Age: 34.5 ± 12.4 years All patients had normal radiological findings European Federation of Neurological Societies guidelines criteria used	53 controls matched on age, sex, and education Sex: 33 male Age 34.2 ± 12.1 years	Emergency departments	Acute (< 7 days post-injury $M = 4.9 \pm 1.47$)	3 T Philips Ingenia; 15-channel SENSE head coil; voxel = 1 mm^3; matrix = 240 × 240; slices = 160; reconstructed resolution = of 0.94 × 0.94 × 1.0 mm^3 (reconstruction matrix 256 × 256 pixels, 160 sagittal slices); FOV = 240 × 240 mm^2, TR/TE = 8.14/3.70 ms; flip	Automated segmentation and surface-based morphometry (FreeSurfer) Graph theory analysis of DTI	Cortical surface area in bilateral, lateral, medial, and orbitofrontal, prefrontal, and anterior cingulate cortices, bilateral postcentral gyrus and central sulcus, and right inferior temporal gyrus negatively correlated with severity of injury symptoms in mTBI patients; cortical volume in bilateral postcentral gyri, left lateral prefrontal cortex, right insula and inferior temporal gyrus negatively correlated with severity of injury

(continued)

Table 1
(continued)

Author	TBI participant characteristics and comparison		Cohort	Time post-injury	Imaging acquisition details	Analysis method	Summary of findings
	Patients	Controls					
					angle $\alpha = 8°$; SENSE factor R = 1.8; time = 7:29 min		symptoms; cortical thickness not associated with severity of PCS symptoms in patients No significant whole-brain analysis differences observed in cortical surface area, volume, and thickness between patients and controls; ROI analysis indicated left parietal and left temporal reductions in patients compared to controls; a subgroup of mTBI patients with severe PCS symptoms showed significantly reduced cortical surface area in the lateral prefrontal cortical, central sulcus/postcentral gyrus, inferior temporal gyrus, and the insula compared to controls Cortical surface area (i.e., GM) alterations in the five identified clusters correlated with greater DTI-derived WM subnetwork disconnections in mTBI patients; group differences observed in WM and GM were exacerbated by greater PCS symptoms Reduced cortical surface area in frontal regions related to impairments in go/no-go performance in mTBI patients
Da costa et al. (2016)	25 mTBI patients (GCS 13–15) Sex: 18 male Age: 42.7 ± 16.3 years A subset ($n = 19$ completed follow-up scanning American Academy of Neurology criteria used	18 controls matched on demographic characteristics Sex: 11 male Age: 38.7 ± 12.6 years	Recently diagnosed patients	Subacute and chronic (initial scan 63.5 ± 42 days post-injury; follow-up 180 ± 38 post-injury)	GRE-EPI 3D axial SPGR; TR/TE = 2000/30 ms; 255 volumes; flip angle = 90°; matrix = 64 × 64; voxel = 3.6 × 3.6 × 3 mm^3; time = 8:38 min	Automated segmentation and volumetric analysis (FSL)	No significant volumetric differences observed in subacute mTBI patients compared to controls; longitudinal reductions in GM volumes observed in mTBI patients compared to controls; GM volume correlated with concussion symptom ratings in mTBI patients but not controls; WM volume negatively correlated with WM cerebrovascular response in mTBI patients

Zagorchev et al. (2016)	44 mTBI patients Sex: 26 male Age: 31.0 ± 13.9 years Diagnosis based on ACRM criteria (1993)	29 community controls Sex: 18 male Age: 29.4 ± 11.7 years	Hospital level 1 trauma center	Subacute and chronic (2 months [$M = 60$ days] post-injury and 1 year follow-up [$M = 384.3 \pm 38.9$ days])	3 T Philips Achieva; 8-channel SENSE head coil; sagittal MPRAGE; 140 contiguous 1.2 mm slices; TR/TE/TI = 6.8/3.3/852.9 ms; NEX = 1; flip angle = 8°; BW/Pixel = 241; FOV = 256 mm; matrix = 256; resolution = 1 × 1 mm; with TFE prepulse delay; 8:55 min	Shape-constrained deformable modeling	Significantly smaller volumes of the caudate, putamen, and right thalamus observed in MTBI patients 2 months post-injury compared to controls that persisted at the 1 year follow-up; significantly reduced amygdala volumes were also observed in mTBI patients at the 1 year follow-up compared to controls; significant longitudinal reductions were also noted in the amygdala, hippocampus, brainstem, and cerebellum in the mTBI patients compared to the controls
Mac Donald et al. (2013)	4 military personnel with primary blast-related TBI Sex: 3 male Age: median = 30 (23–36) years	18 military personnel matched on age and sex Sex: 18 male Age: median = 31 (19–49) years	Military personnel convenience sample	Chronic (2–4 years post-injury)	1.5 T Siemens Avanto; voxel = 1 mm^3	Volumetric analysis (manual tracing and automated)	Middle cerebellar peduncle significantly reduced in military personnel with blast-mTBI compared to controls in both manual ROI traced and manual segmentation analyses; other group differences did not survive FDR correction
Lao et al. (2015)	10 male athletes of contact sports (1 with known concussion) Sex: 10 male Age: not reported	None	Collegiate sports athletes	(19 scans total: 10 pre- and 9 postseason; 2 with no postseason MRI and 1 with no preseason MRI)	3 T GE HDxT; not reported	Manual segmentation	Morphometric results did not detect significant whole structure differences between preseason and postseason in the corpus callosum in the contact sport players; although fused morphometric and diffusion tensor imaging revealed significant group differences in the ventral genu, isthmus, and splenium of the corpus callosum
Jarrett et al. (2016)	45 collegiate athletes with ($n = 11$) and without diagnosed concussion Sex: 25 male Age: 21.2 ± 3.1 years	15 controls from the same university Sex: 12 male Age: 22.9 ± 2.3 years	Collegiate ice hockey athletes	Baseline, acute, and subacute (preseason; 72 h, 2 weeks, and 22 months post-injury)	3 T Philips Achieva; 8-channel SENSE head coil; sagittal 3D; TR/TE = 8.1/3.7 ms; flip angle = 6°; matrix = 256 × 256 × 160; FOV = 256 × 256 × 160 mm^3; voxel = 1 mm^3; SENSE factor of 2 along the left-right direction	Automated brain segmentation and volumetric analysis (FSL SIENA)	Longitudinal % BVC observed in the concussed athletes were not significantly reduced 72 h post-injury, but were significantly reduced in concussed athletes 2 weeks and months post-injury compared to controls; full season % BVC in non-concussed athletes was significantly reduced compared to controls but did not differ from concussed athletes; results were similar for male and female athletes; longitudinal % BVC in the concussed athletes were attributable to time since injury and not age, sex, cognitive performance, or WM hyperintensity load

right mesencephalon and WM of the corona radiata and corpus callosum in acute-phase patients with mTBI compared to controls. The basal ganglia and thalamus volumes have also been reported to be reduced in the acute- and subacute-phase mTBI [18]. These differences were persistent at 1 year post-injury, at which time hippocampal and amygdala atrophy was also observed [18].

It is important to note that morphological alterations have not been identified during the early phases of injury in all studies of mTBI [19] and that the former might not be detectable until later in the clinical course. For instance, altered gray matter (GM) volume in the ventromedial prefrontal cortex and right fusiform gyrus was more pronounced among patients assessed at longer post-injury intervals [20]. Thus, the most robust findings in mTBI have included cross-sectional studies performed in the chronic phase [21–23] or prospective longitudinal studies beginning at the acute period [18, 19, 24–26].

3.1.2 Findings in the Chronic Stage

Global changes in GM, WM, and central spinal fluid (CSF) occur in chronic mTBI [23]. Increased global and GM atrophy is observed in subacute and chronic mTBI patients compared to controls regardless of the presence of lesions [22]. In mTBI, specific regional atrophy has been particularly evinced in the anterior cingulate gyrus and the right precuneus gray matter [25]. GM differences are also shown in frontal, parietal, temporal, hippocampal, parahippocampal, and cerebellar regions in chronic mTBI patients relative to controls [21, 23, 27]. Injury characteristics predict chronic-phase changes. GM density reductions are exacerbated in patients with histories of multiple as opposed to a single mTBI [23]. Greater GM atrophy in bilateral medial temporal lobe, inferior parietal lobe, and right precuneus is associated with more severe PCS symptoms [21]. In chronic-phase patients, altered WM/GM density in the posterior cingulate, internal capsule, and temporal and parietal regions relates to executive dysfunction and inattention, while memory impairment is associated with reduced cortical density in the parahippocampal gyri, anterior temporal lobe, and internal capsule [23]. In addition, Maller et al. [27] reported significantly reduced left inferior parietal and right insular cortices compared to controls and further implicated reduced inferior temporal and parietal regions in the development of major depressive disorder. On the other hand, Dean et al. [21] did not observe altered cortical thickness in chronic mTBI patients.

3.1.3 Findings of Longitudinal Studies

Atrophy following mTBI is largely supported by longitudinal studies. The presence of identifiable lesions 2–5 days post-injury has been related to longitudinal brain atrophy [28]. Significant longitudinal reductions in cortical GM volume and significant enlargement of CSF and ventricular volumes were reported in patients as early as 1 month post-injury [26]. Longitudinal GM atrophy is observed in mTBI patients after 3 months post-injury compared to controls and is correlated with severity of PCS [19]. At 1 year post-injury, GM, WM, and global brain atrophy is reported in mTBI patients compared to controls [25]. Longitudinal WM and GM volumetric reductions in mTBI are particularly observed in anterior brain regions, including the cingulum and cingulate gyrus and right frontal regions [25]. Ongoing frontal, cerebral WM, and cerebellum atrophy has been supported along the chronic phase of mTBI and is related to poor functional outcomes (e.g., vocational status) [24].

3.1.4 Sports-Related Concussion

Sports-related concussion is often considered a special population for concussion or mild TBI and has been gaining increasing attention. Morphometric analyses in sports-related concussion have produced mixed results. Lao et al. [29] did not detect significant structural differences between preseason and postseason in the corpus callosum of contact sport players. However, results from multimodal morphometric and diffusion tensor imaging techniques revealed significant postseason group differences in the ventral genu, isthmus, and splenium of the corpus callosum. More important, only one athlete was known to have had a concussion during the season. Although Jarrett et al. [30] did not observe brain volume changes in concussed collegiate ice hockey athletes 72 h post-injury, significant reductions were observed in the concussed athletes 2 weeks and 2 months post-injury compared to controls. Longitudinal percent brain volume changes in the concussed athletes were attributable to time since injury and not age, sex, cognitive performance, or white matter hyperintensity volume [30]. Furthermore, full season percent brain volume change in non-concussed athletes was significantly reduced compared to controls but not statistically different from concussed athletes.

3.1.5 Mild TBI in Military and Veteran Populations

There is a higher frequency of mTBI among veterans and military personnel of the recent Iraq and Afghanistan conflicts in comparison to previous conflicts [31]. Since 2000, an estimated 300,000–500,000 deployed active duty service members suffered closed TBI as the result of exposure to blasts. Most (82.4%) of these injuries involved blast-related concussion or mTBI (blast-mTBI), alone or in combination with blunt trauma (for further review, see [15]). Traditional classification systems of TBI fail to accurately characterize the majority of brain injuries among military personnel [32]. However, some morphological and volumetric alterations are reported in cortical and subcortical brain regions, particularly at later post-injury assessments, suggesting the presence of neuropathological changes resulting from blast exposure.

Thinning of cortical gray matter is observed in veterans and active duty service members, although it is not clear whether causation can be attributed to the effects of mTBI specifically. Tate et al. [33] demonstrated significant thinning in the left but not the right Heschl's gyrus in a small cohort of blast-injured service members who had substantial hearing/auditory loss. Veterans with comorbid PTSD and mTBI showed greater and more diffuse thinning that was associated with elevated measures of stress [34]. Overall findings suggest that mTBI may exacerbate the vulnerability of the brain to stressful events, although it is clear that additional studies are warranted.

Mac Donald et al. [35] reported that military personnel with blast-mTBI showed altered volume of the medulla compared to military personnel with blast exposure who had no histories of subsequent TBI symptoms, although specifics of this difference were not provided. Depue et al. [36] observed reduced bilateral amygdala volume in veterans with mTBI and PTSD compared to comparison veterans without PTSD or histories of TBI. Disproportionate reductions in left amygdala volume were associated with increased commission errors during go/no-go task and increased impulsivity ratings.

3.2 Studies in Moderate to Severe TBI

Volumetric analysis has also been informative in populations of subjects with more severe TBI as it may allow not only measurement of volumetric change between groups of subjects but may be useful in documenting both deleterious changes over time as well as potential recovery or response to treatment. General findings in studies of patients with moderate to severe TBI have been summarized in Table 2.

3.2.1 Global and Regional Brain Atrophy

Volumetric changes are frequently observed in moderate to severe TBI patients [37, 38]. Global brain volume reductions are related to alterations in cerebral metabolism and blood flow in the frontal, temporal, and parietal lobes, which were implicated as having the greatest atrophy [37]. Widespread GM and WM reduction [38, 39] has been reported in most, but not all case control studies of patients with moderate to severe TBI [40]. GM volumetric reductions may be especially observed in the frontal and temporal lobes, although WM changes show greater specificity for frontal cortical regions [41]. Xu et al. [37] also indicated that primary contusions can result in increased localized, regional atrophy in patients with moderate to severe TBI.

3.2.2 Ventricular Expansion

Ventricular enlargement is a rather consistent finding in moderate to severe injuries [39] associated with brainstem and global GM atrophy [38]. Although enlarged CSF has been reported in chronic patients with moderate to severe TBI [41] as well as in a longitudinal study of this type of cases [42], there are studies that did not find differences in CSF volume in their sample of patients compared to controls [43]. Figure 1 shows classic expansion of the ventricular system in a patient with TBI (right) as compared to an uninjured healthy comparison individual.

Table 2
Studies with mild TBI patients

Author	TBI participant characteristics and comparison: Patients	Controls	Cohort	Time post-injury	Imaging acquisition details	Analysis method	Summary of findings
Ariza et al. (2006)	20 moderate to severe TBI (GCS ≤ 12; 4 moderate TBI [GCS 9–12]; 16 severe TBI [GCS ≤8]) Sex: 16 male Age: 25 ± 7 years	20 controls matched on age and education Sex: 16 male Age: 25 ± 7 years	Hospital neurotrauma unit patients	Chronic (≥6 months post-injury; $M = 246 \pm 31$ days)	1.5 T GE Signa; SPGR, IR-PREPPED; coronal collection; TR/TI/TE = 12.1/300/5.2 ms; FOV = 240 × 240 mm^2; matrix = 256 × 256; flip angle = 20°; slice thickness/gap = 1/0 mm	Manual volumetric analysis	Significantly reduced bilateral hippocampus volumes observed in patients compared to controls; greatest atrophy observed in the hippocampal head than other segments; left and right hippocampal volume, respectively, correlated with episodic verbal memory and nonverbal memory in TBI patients but were unrelated with injury-related clinical outcomes (e.g., GCS, PTA)
Kim et al. (2008)	29 moderate to severe TBI patients (GCS ≤ 12) Sex: 21 male Age: 36.9 ± 11.4 years	20 controls Sex: 17 male Age: 34.9 ± 9.8 years	Rehabilitation clinics	Chronic (≥3 months post-injury)	3 T Siemens Trio; 3D MPRAGE; TR/TE/TI = 1620/3/950 ms; flip angle = 15°; slice thickness/gap = 1/0 mm; slices = 160; FOV = 192 × 256 mm^2; matrix = 192 × 256; 1 NEX = 1; scan time = 6 min; voxel size = 1 mm^3	Tensor-based morphometry	Reductions in widespread areas of both GM and WM observed in patients compared to controls; regions with greatest atrophy included the thalamus, midbrain, cingulate cortex, corpus CC, cerebellum, caudate, and frontal and temporal cortices; ventricular expansion observed in TBI patients; results maintained after excluding TBI patients ($n = 12$) with lesion load >1.5 cm^3
Ng et al. (2008)	14 moderate to severe TBI patients (GCS ≤ 12 or PTA ≥ 1 h) Sex: 10 male Age: 35.1 years	None Some comparisons to a normative sample for imaging volumes conducted	Hospital neuro-rehabilitation program	Subacute/chronic (4.5 ± 0.5 months post-injury; 29.3 ± 4.1 months post-injury; time between scans $M = 24.8 \pm 4.4$ months)	1.5 T Signa Echospeed; 8-channel head coil; sagittal TR/TE = 300/13 ms; slice thickness/gap = 5/2.5 mm;	Automated and manual segmentation and volumetric analysis (INSECT, DISPLAY)	Significant longitudinal enlargement of CSF observed in patients compared to normative sample; significant longitudinal volumetric reduction observed in

(continued)

Table 2
(continued)

Author	TBI participant characteristics and comparison		Cohort	Time post-injury	Imaging acquisition details	Analysis method	Summary of findings
	Patients	Controls					
					matrix = 256 × 128; axial GRE; TR/TE = 450/20; flip angle = 20°; slice thickness/gap = 3/0 mm; matrix = 256 × 192 3D IR prepped radio-frequency spoiled-GRE images: TI/TR/TE = 300/12/5; flip angle = 20°; slice thickness = 1 mm no gap; matrix = 256 × 256; FOV = 25 cm		bilateral hippocampi in TBI patients, which differed from the normative sample Quantitative MRI data superior to that of visual radiological inspection for detecting longitudinal brain changes
Xu et al. (2010)	32 moderate to severe TBI patients (GCS ≤ 8 or of 9–15 with intracranial bleeds per CT) Sex: 28 male Age: 33.4 ± 15.0 years	12 controls matched on age Sex: 12 male Age: 31.8 ± 9.7 years	Research center	Acute and chronic (initial scan ≤1 week post-injury, median = 8 days; follow-up 6 months post-injury, median = 195 days)	1.5 T Siemens Sonata; MPRAGE; TR/TE = 1900/3.5 ms; FOV = 256 × 256; slice thickness = 1 mm	Semiautomated segmentation and volumetric analysis (ImageJ and BrainSuite)	Significant reduction observed in all lobular volumes in acute phase and at 6 months post-injury in TBI patients compared to controls; greatest volumetric reduction observed in temporal lobes followed by the parietal and frontal lobes compared to the occipital lobe; lobular atrophy significantly greater in lobes with a primary contusion compared to lobes without the primary hemorrhagic contusion and with greater initial total hemorrhagic contusion volume; global

							atrophy associated with increased ICP as well as with craniotomy for mass lesion removal Significant longitudinal reduction in total brain volume observed in TBI patients at 6 months post-injury compared to both acute-phase volumes and controls, which did not differ; significant longitudinal reduction in GM volumes observed in TBI patients at 6 months post-injury compared to controls; significant longitudinal elevations of CSF volume observed in TBI patients at 6 months post-injury compared to acute-phase volumes Regional brain atrophy correlated with metabolic parameters (from PET) and with CBF in the frontal, temporal, and parietal lobes; occipital lobe atrophy was not associated with metabolic parameters
Hudak et al. (2011)	25 moderate to severe TBI patients (initial GCS median = 8 [3–14]) Sex: 22 male Age: 23 (19–37) years	None	Convenience sample	Acute and chronic (initial scan median = 3 [1–8] days post-injury; follow-up scan median = 8 [7–9] months)	3 T GE Signa Excite ($n = 15$); 3D fast spoiled gradient-recalled acquisition in the steady state (GRASS); matrix = 256 × 192; FOV = 240; slices = 130; slice thickness/gap = 1.3/0; TE = 2.4 ms; flip angle = 25°; NEX = 2; time = 6 min 3 T Siemens ($n = 10$);	Automated segmentation and volumetric analysis (FreeSurfer)	Significant correlations among % volume change of the left rostral anterior cingulate and left and right lateral orbitofrontal cortices with depression symptoms observed in TBI patients; % volume change unrelated with neuropsychological outcomes in TBI patients

(continued)

Table 2
(continued)

Author	TBI participant characteristics and comparison		Cohort	Time post-injury	Imaging acquisition details	Analysis method	Summary of findings
	Patients	Controls					
					3D MPRAGE; FOV = 240 mm; slice thickness/ gap = 1/0 mm; TR/TE/ TI = 2250/4.0/ 900 ms; time = 5 min 36 s		
Lutkenhoff et al. (2013)	25 moderate to severe TBI (GCS 3–14, median = 7, mode = 3) Sex: 21 male Age: 35.6 ± 15.25 years Inclusion criteria of admission GCS ≤ 8, or admission GCS 9–14 with CT evidence of intracranial bleeding	None	Convenience sample	Acute and chronic (acute scans 1–32 days post-injury [72% scanned within 5 days]; chronic scans 138-420 days post-injury [72% before 202 days]; median inter-scan interval = 183 days)	3 T Siemens Tim Trio; 3D MPRAGE; TR/TE = 2250/ 2.99 ms; flip angle = 9°; FOV = 256 × 240 × 160 mm^3; matrix = not reported; slices = 160; resolution = 1 mm^3; no interslice gap	Tensor-based morphometry (LONI, FSL SIENA)	Significant longitudinal volumetric reduction in the anterior and lateral dorsal along with significant longitudinal enlargement of the ventral posterolateral and medial geniculate nuclei of the thalamus observed; greater atrophy of the anterior nucleus atrophy significantly positively associated with worse 6 month GOS-E scores; greater medial geniculate nucleus enlargement positively associated with worse 6 month GOS-E scores; % BVC indicative of atrophy in TBI patients and positively associated with 6 month GOS-E scores; significant longitudinal atrophy localized in anterodorsal limbic and mediodorsal association nuclei of the thalamus that predicted 6 months post-injury behavioral outcomes Decision tree achieved 84%

							accuracy in differentiating patients who were and were not in a vegetative state at 6 months using physiological measures (i. e., indexed atrophy of the anterior thalamic nuclei) followed by behavioral measures (i.e., post-resuscitation GCS score)
Brezova et al. (2014)	62 moderate to severe TBI patients (GCS ≤ 13, $M = 9.1 \pm 3.4$; 36 moderate TBI [GCS $M = 11.5 \pm 1.6$]; 26 severe TBI [GCS $M = 5.7 \pm 1.8$]) Sex: 45 male Age: 32.5 ± 15.2 years	None	Hospital admittance	Acute/subacute and chronic (initial scan between 1 and 26 days post-injury; 2 follow-up scans at 3 and 12 months post-injury)	1.5 T Siemens Symphony Sonata; 3D sagittal MPRAGE; TR/TE/TI = 7.1/3.5/1000 ms; flip angle = 7°; FOV = 256 × 256; matrix = 256 × 192 × 128; reconstructed to 256 × 256 × 128, giving a reconstructed voxel resolution of 1.00 × 1.00 × 1.33 mm^3	Automated segmentation and volumetric analysis (Neuro-Quant)	Significant, longitudinal cortical GM volume reductions observed at 3 months post-injury were maintained at 12 months post-injury; longitudinal volumetric reductions in the hippocampus and lenticular nucleus were observed but did not reach significance at 12 months post-injury; significant longitudinal reductions in lobar WM volume were observed at 12 months post-injury compared to 3 months post-injury; steady longitudinal reductions in brainstem volume were observed across time points, with significant reductions observed between the acute injury phase and at 3 months post-injury and between 3 months post-injury and 12 months post-injury; significant longitudinal ventricular enlargement was observed in the acute phase with further enlargement observed at 12 months post-injury and related to both the volumetric alterations in the brainstem and in total

(continued)

Table 2
(continued)

Author	TBI participant characteristics and comparison		Cohort	Time post-injury	Imaging acquisition details	Analysis method	Summary of findings
	Patients	Controls					
							cortical GM but not WM; longitudinal reductions in total brain volume also observed between the acute phase and at 3 months post-injury and between the acute phase and at 12 months post-injury; DAI grade was unrelated with lobar WM volume
Kim et al. (2014)	25 moderate to severe TBI patients with confirmed or probable DAI (initial GCS < 12; PTA > 1 h; or identified abnormality on radiological evaluation) Sex: 13 male Age: 29.1 ± 11.50 (16–60) years	18 controls matched on age and education Sex: 9 male Age: 31.9 ± 8.4 years	Outpatient rehabilitation program	Chronic ($M = 36.9 \pm 68$ months post-injury)	3 T Siemens Trio; 3D MPRAGE; TR/TI/TE = 1620/950/3 ms; flip angle = 15°; slice thickness/gap = 1/0 mm; slices = 160; FOV = 192×256 mm^2; matrix = 192×256; NEX = 1; time = 6 min; voxel = 1 mm^3	Automated segmentation and volumetric analysis (FreeSurfer) Graph theory network analysis	Connection-wise analysis indicated reduced connectivity in TBI patients compared to controls, especially in those originating from subcortical nodes including the thalamus, caudate, and hippocampus; increased shortest path length measure in TBI patients indicated reduced network efficiency; modularity and transitivity similar between groups In TBI patients, shortest path length negatively related with executive function and verbal learning as well as with family- and self-reported executive function

Hillary et al. (2014)	21 moderate to severe TBI patients (GCS 3–12, $M = 7.1$; unavailable for 7 patients) Sex: 18 male Age: 27.9 ± 9.7 years Patients had resolved PTA at assessments	15 controls Sex: 9 male Age: 28.8 ± 11.9 years		Chronic (completed two scans: At 3 [$M =$ 113.6 ± 32.3 days] and 6 months post-injury [time between scans $M = 106.2 \pm 24.6$ days])	3 T Philips Achieva with 6-channel head coil or 3 T Siemens MAGNETOM Trio with 8-channnel head coil; 3D MPRAGE; TR/TE = 468.45/ 16.1 ms; flip angle = 18°; FOV = 250 × 200 mm^2; matrix = 256 × 180; voxel = 1.2 mm^3	Voxel-based morphometry (SBM8) Graph theory network analysis	No differences observed on global WM, GM, or CSF volumes in TBI patients compared to controls Graph theory results indicated hyperconnectivity during early chronic (first scan) phases of injury but did not persist 3 months later; network strength, number of links, mean correlation coefficient among nodes, clustering coefficients, and small-worldness in the TBI patients significantly differed from controls at 3 months but not 6 months post-injury; most highly connected nodes at time 1 were observed within the right executive control, anterior and posterior salience, language, ventral and dorsal default mode, sensorimotor, and auditory networks; at follow-up only the language, sensorimotor, executive control, anterior salience, and ventral and dorsal default mode network regions remained highly connected; in controls, only the right executive control network was highly connected at both scans Low to moderate associations among connectivity within the right executive control and anterior insula networks with executive function task performance were observed in the TBI patients

(continued)

Table 2
(continued)

Author	TBI participant characteristics and comparison: Patients	Controls	Cohort	Time post-injury	Imaging acquisition details	Analysis method	Summary of findings
Leunissen et al. (2014)	25 moderate to severe TBI patients (LOC $M = 17.3 \pm 11.5$ days; PTA $M = 22.6 \pm 12.3$ days; GCS not reported) Sex: 18 male Age: 24.6 (16–34) years Used the Mayo classification system (Malec et al. 2007)	21 controls Sex: 11 male Age: 26 (21–35) years		Chronic (1–10 years post-injury, $M = 4.8 \pm 2.6$)	3 T Siemens MAGNETOM Trio; 8-channel head coil; sagittal MPRAGE; TR/TE = 2300/2.98 ms; voxel = $1.1 \times 1 \times 1$ mm^3; FOV = 240×256; slices = 160	Automated segmentation (FSL FIRST)	Reduced volume of the left caudate, bilateral putamen, and bilateral thalamus observed in TBI patients compared to controls with localized caudate and putamen atrophy in the limbic, executive, and rostral-motor areas and widespread thalamus atrophy Global caudate volume positively related to local WM microstructure; putamen and thalamus volume significantly positively related to local and global WM microstructure Global putamen volume was negatively associated with switching performance; right ventro-anterior caudate volume was negatively associated with slower reaction and the lateral and medial wall of the caudate body and head and lateral and medial wall of the right caudate tail were negatively associated with switching performance; reduced volume of the lateral-posterior wall of the left putamen and the medial wall of the right putamen related to switching performance; the ventro-

							and lateral-anterior as well as the medial posterior regions of the right thalamic nucleus related to switching performance
Konstantinou et al. (2016)	17 moderate to severe male TBI patients (initial GCS < 12) Sex: 17 male Age: 31.9 ± 10 years	15 controls matched on age and education Sex: 15 male Age: 33.8 ± 10.3 years	Physician referral	Chronic (2–22.8 years post-injury, $M = 8.36 \pm 6.34$)	3 T Philips Achieva; 8-channel head coil; 3D MPRAGE; TR/TE = 25/1.85 ms; flip angle 30°; voxel size = 1 mm^3	Voxel-based morphometry (DARTEL)	Significantly reduced GM volumes observed in the TBI patients compared to controls in the orbitofrontal cortex (extended across the superior and middle orbital gyrus), bilateral putamen, thalamus, insula and temporal lobe, and left inferior frontal gyrus; observed significantly reduced WM volumes in the TBI patients in the bilateral frontal cortex, thalamus, and cerebellum; significantly increased CSF volume observed in the TBI patients GM and WM alterations of the fronto-subcortical network correlated, but were not related with CSF volume across groups and in the controls; relations not significant in the TBI patients; GM and WM volume positively correlated and CSF negatively correlated with cognitive performance of verbal memory, visual memory, executive function, and attention/organization and increased detection of TBI patients

(continued)

Table 2 (continued)

Author	TBI participant characteristics and comparison: Patients	Controls	Cohort	Time post-injury	Imaging acquisition details	Analysis method	Summary of findings
Warner et al. (2010b)	25 patients with diffuse axonal injury (i.e., TAI) (GCS 3–15); Sex: 18 male Age: 26.8 ± 11.3 years	22 controls Sex: 14 male Age: 32.4 ± 13.5 years	Hospital recruitment	Acute (initial scan median = 1 (0.5–9) day post-injury; follow-up median 7.9 (6–14) months post-injury)	3 T GE Signa Excite; MPRAGE using fast spoiled gradient-recalled; matrix = 256 × 92; FOV = 240 mm; slices = 130; slice thickness/ gap = 1.3/0 mm; TE = 2.4 ms; flip angle = 25°; NEX = 2; time = 6 min	Automated segmentation and volumetric analysis (FreeSurfer)	Major global brain atrophy observed (mean whole-brain parenchymal volume loss of 4.5%) in TBI patients compared to controls; significantly decreased volume observed in the amygdala, hippocampus, thalamus, CC, putamen, precuneus, postcentral gyrus, paracentral lobule, and parietal and frontal cortices in patients compared to controls; neither the caudate nor inferior temporal cortex showed atrophy in TBI patients; whole-brain parenchymal volume and inferior parietal cortex, pars orbitalis, pericalcarine cortex, and supramarginal gyrus atrophy predicted long- term disability in TBI patients
Warner et al. (2010a)	24 patients with diffuse axonal injury (GCS $M = 6.4 \pm 4.2$) Sex: 16 male Age: 27.2 ± 11.4 years	None	Hospital Department of Neurological Surgery referrals	Acute (initial scan $M = 2.2 \pm 2.4$ days post-injury; follow-up $M = 7.7 \pm 1.9$ months post-injury)	3 T GE Signa Excite; 3D fast spoiled gradient-recalled acquisition in the steady state (GRASS); FOV = 240 mm; slice thickness/ gap = 1.3/0 mm; slices = 130; TE = 2.4 ms; flip angle = 258°;	Automated segmentation and volumetric analysis (FreeSurfer)	Amygdala, hippocampus, and cortical region volumes correlated with learning and memory performance in TBI patients; thalamus, superior frontal, superior parietal, and precuneus cortices correlated with executive function; bilateral thalamus volumes correlated with processing speed; subcortical and

					NEX = 2; matrix = 256 × 92; time = 6 min		cortical brain volumes associated with learning and memory and processing speed, but not executive function
Palacios et al. (2013)	26 severe diffuse TBI (GCS ≤ 8, *M* = 5.19 ± 1.70) Sex: 16 male Age: 27.40 ± 5.15 years (only ≤40 years included) All patients showed microbleeds as sign of diffuse pathology on T2* and FLAIR sequences, but had no large lesions	22 controls matched on age, sex, handedness, and education Sex: 12 male Age: 30.26 ± 12.48 years	Recruited from head injury rehabilitation programs	Chronic (≥2 years post-injury, *M* = 4.20 ± 1.14)	3 T Siemens MAGNETOM TrioTim syngo; 3D MPRAGE sagittal scan; TR/TI/TE = 2300/900/3 ms; flip angle = 12°; FOV = 244 × 244 mm^2; slice thickness/gap = 1/0 mm; voxel = 1 mm^3	Automated segmentation and cortical thickness analysis (FreeSurfer, FSL FIRST, SIENAX)	TBI patients showed bilateral cortical atrophy compared to controls in the rostral and middle frontal, superior and middle temporal (including anterior temporal lobe regions), superior and inferior parietal, precentral and postcentral, precuneus, parahippocampal, lingual, pericalcarine, isthmus-cingulate, and anterior and posterior cingulate cortical regions; in TBI patients, cortical thickness correlated with global FA in several regions; in TBI patients, left superior frontal and inferior and superior parietal cortical thickness positively correlated with memory scores; in controls, cortical thickness in the rostral middle and superior frontal cortices negatively correlated with memory scores Significant differences in ratio of global hippocampal volumes and in specific significant reductions in the left hippocampal head and tail per shape analysis were observed in TBI patients compared to controls

(continued)

Table 2 (continued)

Author	TBI participant characteristics and comparison		Cohort	Time post-injury	Imaging acquisition details	Analysis method	Summary of findings
	Patients	Controls					
Kim et al. (2013)	4 diffuse TBI patients (GCS not reported) Sex: not reported Age: 38.3 years	None		Chronic (initial scan $M = 6.6$ months post-injury; follow-up scan $M = 15.8$ months later)	Not reported	Voxel-based morphometry (multivariate pattern analysis)	Longitudinal atrophy observed in TBI patients in cortical areas including posterior temporal lobes, posterior cingulate, and superior parietal lobe (atrophy rate $M = 4.2 \pm 1.8\%$); WM and deep GM atrophy observed in the thalamus, primary motor tract, and mid- and posterior bodies of the corpus callosum (atrophy rate $M = 7.3 \pm 3.9\%$)
Uruma et al. (2015)	29 patients with diffuse axonal injury per (Gennarelli et al. 1998) with independent ADL at home but without social activities due to cognitive deficit (GCS 3–8, $M = 6.5 \pm 1.8$) Sex: 23 male Age: 34.2 ± 5.8 years No patients had intracranial hyperintensities on T2-weighted MRI; 20 patients had traumatic microbleeds and 21 patients had ventricular enlargement	None		Chronic (18 to 163 months post-injury, $M = 60.4 \pm 42.8$)	1.5 T Seimens MAGNETOM Avanto; sagittal GRE; TR/TE = 544/2.20 ms; NEX = 1; flip angle = 15°; FOV = 23 × 23 cm; slice thickness = 1 mm	Voxel-based morphometry (SPM8, DARTEL)	Regional WM volume reduction observed in corpus callosum, fornix, internal capsule, corona radiata, sagittal stratum, external capsule, cingulum, and superior and inferior fronto-occipital fasciculi Regional WM volume reduction in the splenium of the corpus callosum correlated with performance IQ and processing speed scores; high area under the curve analysis values indicated that extent of reduction in the splenium distinguished among patients with DAI with marked performance IQ and processing speed impairments

Ubukata et al. (2016)	10 male patients with diffuse axonal injury per Gennarelli with independent ADL at home but lacked social activities as a result of cognitive deficit (PTA 3–150 days, $M = 62.5 \pm 40.4$) Sex: 10 male Age: 30.8 ± 10.5 years	12 control matched on age and sex Sex: 12 male Age: 29.8 ± 6.3 years (controls more education)		Chronic (12–262 months post-injury, $M = 106.9 \pm 79.4$	3 T Siemens MAGNETOM Trio; 3D MPRAGE; TR/TI/TE = 2000/990/4.38 ms; flip angle = 15°; FOV = 225 × 240 mm^2; matrix = 240 × 256; voxel size = 0.9375 × 0.9375 × 1 mm^3; slices = 208; no interslice gap	Voxel-based morphometry (SPM8, MATLAB VBM8.1, DARTEL)	Compared to controls, patients showed GM reductions in widespread brain regions including the left parahippocampal gyrus, bilateral amygdala, left insula, right putamen, and thalamus; WM reductions in the splenium of the corpus callosum that correlated with fractional anisotropy as well as with processing speed scores

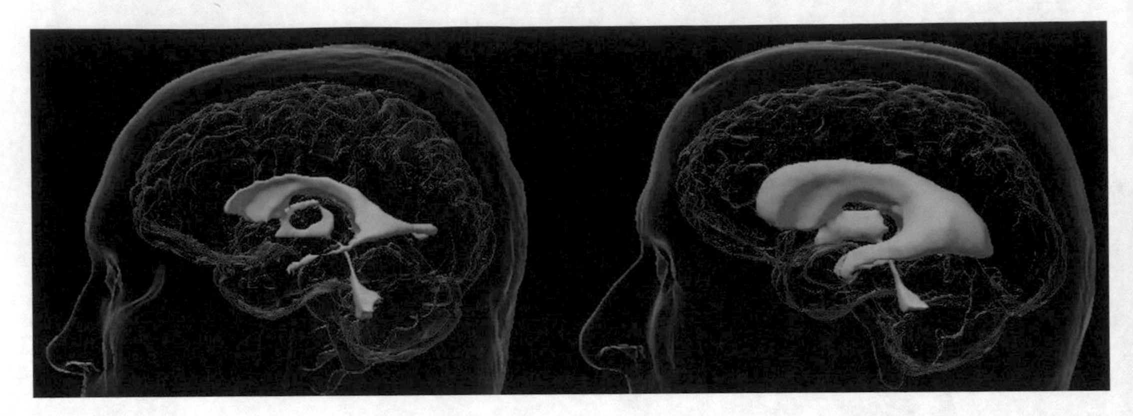

Fig. 1 Ventricular expansion in a patient with TBI (right) as compared to an uninjured healthy comparison individual

3.2.3 Cortical Changes

TBI is associated with cortical GM reductions [38] that are especially observed in the frontal lobe, specifically in the orbitofrontal cortex, [41, 44], and in the temporal lobe [39]. In one study that examined depression in moderate to severe TBI, the degree of atrophy in the left rostral anterior cingulate cortex and the right lateral orbitofrontal cortex was associated with depression symptom severity but not with neuropsychological performance [44]. Lobar and callosal [39] WM atrophies have been noted in moderate to severe TBI, with lobar reductions not related with the extent of observable diffusion axonal injury (DAI) [38].

In moderate to severe TBI patients, the hippocampus is particularly affected [42]. One study reported that the greatest amount of atrophy occurs in the head of the hippocampus [45] and was not related to clinical markers of injury severity (e.g., PTA or GCS). However, it was associated with memory performance.

3.2.4 Subcortical Changes

Moderate to severe injuries have been associated with volumetric reductions of the thalamus and specific thalamic nuclei [46]. The presence of thalamic atrophy was related to poorer long-term functional outcomes and poorer neuropsychological performance [47]. The striatum appears to be also affected in moderate to severe injuries. Regional caudate and putamen atrophy [41] provide anatomic evidence of the disruption of corticostriatal networks involved in motor, executive, and motivational functions [46, 48].

The brainstem and the cerebellum are also volumetrically reduced in moderate to severe TBI [38, 41, 48].

3.3 Studies in Severe TBI

3.3.1 Global and Regional Brain Atrophy

Global brain atrophy is observed in subacute and chronic patients with severe TBI and correlates with both PTA and functional independence in the subacute stage and with PTA, coma duration, and functional outcomes at 1 year post-injury [49]. Substantial progressive subcortical GM and WM atrophy is also observed in severe TBI patients, particularly in the frontal lobes [49]. Ubukata and colleagues [50] observed widespread GM reductions in their sample of severe TBI patients. Though most studies report significant associations between atrophic changes, behavioral deficits, and neuropsychological performance, it is noteworthy that Shah et al. [51] did not observe significant correlations between brain volumes and cognition.

3.3.2 Cortical Changes

Several studies have noted significantly reduced cortical volume in frontal, temporal, and parietal regions among chronic patients with severe, diffuse TBI [52]. In patients with severe TBI with coexistent mass lesions, Wang et al. [53] observed decreased cortical thickness that related to general outcome scores. The authors also observed a significant association between the type (i.e., hemorrhagic or nonhemorrhagic) and volume of lesion with acute and long-term outcome scores. Another recent article reported significant reductions in the right caudal anterior cingulate cortex that correlated with executive dysfunction in patients with severe TBI [54].

Structural disruption of the hippocampus is a consistent finding in the literature. The hippocampus is not only reduced in severe TBI, but shape analysis also indicated alterations in the head and tail of the hippocampus in patients compared to controls [52, 55–57]. In severe TBI patients, reductions in both hippocampus [52, 55] and fornix volumes correlate with coma duration and neuropsychological function [56, 57]. Di Paola and colleagues [57] observed a trend for greater bilateral hippocampal reduction in severe TBI patients with episodic memory deficits compared to severe TBI patients with executive dysfunction.

3.3.3 Subcortical Changes

Regional subcortical WM reductions are evinced in a broad number of regions including corpus callosum, cingulum, and motor pathways in patients with severe TBI [58]. Reduced corpus callosum volume is associated with coma duration and neuropsychological function [56], whereas specific reductions in the splenium of the corpus callosum relate to indices of WM microstructural disruption and neuropsychological function in severe, chronic TBI [50, 58].

Volumetric reductions are also observed in limbic system structures, basal ganglia, and thalamus of patients with severe and chronic TBI [50]. Within the limbic system, amygdala volumes also show general reduction in severe TBI patients, which were related to neurophysiological indicators of arousal in one study [55]. The left nucleus accumbens was also found to be disproportionately reduced in one sample of patients with chronic severe TBI [51]. Finally, both the tectum and the cerebellum showed progressive volumetric reductions among patients with severe TBI [49] (see Table 3).

Table 3
Studies in severe TBI

Author	TBI participant characteristics and comparison: Patients	TBI participant characteristics and comparison: Controls	Cohort	Time post-injury	Imaging acquisition details	Analysis method	Summary of findings
Tomaiuolo et al. (2004)	19 severe TBI patients (GCS < 8) Sex: 12 male Age: 33.58 ± 14.27 years	19 controls matched on sex Sex: 12 male Age: 35.47 ± 14.65 years	Hospital rehabilitation participants	Chronic (3–133 months post-injury, $M = 10.16 \pm 28.9$)	1.5 T Siemens Vision MAGNETOM; TR/TE = 11.4/4.4 ms; flip angle = 15°; slice thickness = 1 mm	Volumetric analysis	Significant volumetric reductions observed in the hippocampus, fornix, and CC in patients compared to controls; CC volume correlated with coma duration and immediate verbal memory; fornix volume correlated with verbal and nonverbal memory; right hippocampus volume correlated with nonverbal memory
Sidaros et al. (2009)	24 severe TBI (GCS < 8) Sex: 18 male Age: 33.2 ± 13.5 years	14 controls Sex: 9 male Age: 31.2 ± 8.1 years	Hospital brain injury unit	Subacute and chronic (initial scan ≤12 weeks post-injury; follow-up 12 months post-injury; days between scans $M = 343 \pm 47$)	1.5 T Siemens MAGNETOM; 3D sagittal; MPRAGE, TR/TE/TI = 13.5/7/100 ms, flip angle = 15°; voxel = 1 mm^3	Tensor-based morphometry and automated segmentation and volumetric analysis (FSL SIENA and SIENAX)	Significantly disproportionate TBV reductions observed in subacute TBI patients compared to controls that correlated with PTA duration and subacute functional independence, but not coma duration; longitudinal % BVC significantly greater in TBI patients compared to controls and correlated with both PTA and coma duration as well as functional outcomes at 12 months post-injury Significant longitudinal

							volume reductions observed in a cluster that encompassed the brainstem and cerebellar peduncles, bilateral thalamus, internal and external capsules, putamen, inferior and superior longitudinal fasciculi, genu, splenium and body of the CC, and corona radiata in TBI patients compared to controls; significant longitudinal reductions also observed within the cerebellum and frontal lobes; significant longitudinal expansion observed in the ventricles and within the subarachnoid space including at the fundus of the intraparietal sulcus; maximal longitudinal volumetric loss observed in the tectum in TBI patients compared to controls
Shah et al. (2012)	14 severe TBI patients (GCS 3–7, $M = 5.29 \pm 1.54$) Sex: 12 male Age: 22.0 ± 4.62 years	15 controls generally matched on age, race, and ethnicity Sex: 13 male Age: 22.2 ± 5.93 years		Chronic (5.6–6.9 months post-injury; $M = 197.43 \pm 11.75$ days)	3 T Philips Intera; 3D sagittal; TR/TE = 8.6/3.9 ms; slice thickness/gap = 1/0 mm	Automated segmentation and volumetric analysis (FreeSurfer)	Disproportionate reduction observed in the left nucleus accumbens in the TBI patients compared to controls; no significant differences in bilateral frontal subcortical or WM volume or in bilateral ventral

(continued)

Table 3
(continued)

Author	TBI participant characteristics and comparison						
	Patients	Controls	Cohort	Time post-injury	Imaging acquisition details	Analysis method	Summary of findings
							diencephalon volume between TBI patents and controls observed; no significant associations of brain volumes with cognitive performance observed
Wang et al. (2013)	5 patients with severe TBI with large lesions (GCS not reported) Sex: not reported Age: not reported	None	Not reported	Acute and chronic (participants scanned 5 days post-injury and again at 6 months post-injury)	Not reported	Spatiotemporal 4D modeling (ARCTIC)	Whole-brain analysis indicated longitudinal trend of decreased cortical thickness, which correlated with change in clinical GOS scores; regional analysis findings that longitudinal WM volume increase and GM volume decrease related to difference in acute and chronic GOS scores Jacobian determinant of deformation field, volume of nonhemorrhagic lesion and volume of hemorrhagic lesion correlated with difference in acute and chronic GOS scores; volume of nonhemorrhagic lesion correlated with acute GCS scores; volume of

							hemorrhagic lesion correlated with acute GOS scores
Merkley et al. (2013)	12 severe TBI (GCS < 9 or LOC > 6 h and PTA > 7 days) Sex: 8 male Age: 23.0 years	18 controls Sex: 10 male Age: 21.0 years		Post-acute and chronic (2–18 months post-injury, $M = 8.9 \pm 4.5$)	3 T Siemens Allegra; 3D MPRAGE; TR/TE = 2000/4.13 ms; flip angle = 8°; FOV = 24 × 24 cm; slice thickness/gap = 1/0 mm; matrix = 512 × 512 × 160; slices = 160	Automated segmentation and volumetric analysis (FreeSurfer)	Cortical volume reductions in TBI compared to controls in bilateral frontal, temporal, and inferior parietal regions; patients also showed specific regional reduction of the right caudal anterior cingulate, which related to executive dysfunction
Di Paola et al. (2015)	15 severe TBI patients with selective cognitive deficits of episodic memory impairment ($n = 8$) or executive dysfunction ($n = 7$) (GCS < 8; dys-episodic severe TBI coma duration $M = 32.75 \pm 23.7$ [8–78] days; dysexecutive severe TBI coma duration $M = 22.71 \pm 10.16$ [10–42] days) Sex: 7 male Age: 31.66 ± 12.67 years Dys-episodic patients had no focal brain lesions; dysexecutive patients showed focal lesions: 3 patients with right and 2 patients with left frontal lobe lesions	16 controls Sex: 16 male Age: 37.12 ± 6.94 years	Post-coma unit	Chronic (≥60 days post-injury; dys-episodic TBI patients $M = 183.25 \pm 102.91$ [60–403] days post-injury; dysexecutive TBI patients $M = 444.43 \pm 339.64$ [65–887] days post-injury)	3 T Siemens Allegra; birdcage head coil; 3D sagittal plane using a modified driven equilibrium Fourier transform sequence; TR/TE/TI = 7.92/2.4/910 ms; flip angle = 15°; voxel = 1 mm^3	Automated segmentation and volume analysis (FSL)	Reduced bilateral hippocampus volume in the dys-episodic and reduced right hippocampus volume observed in the dysexecutive TBI patients, which were statistically similar, compared to controls; follow-up analysis indicated trend of largest hippocampal volumes in controls, then the dysexecutive TBI patients, with the smallest volumes in the dys-episodic TBI patients Left hippocampus volume associated with episodic memory performance

(continued)

Table 3
(continued)

Author	TBI participant characteristics and comparison		Cohort	Time post-injury	Imaging acquisition details	Analysis method	Summary of findings
	Patients	Controls					
Fisher et al. (2015)	19 severe TBI patients (PTA $M = 66.89 \pm 50.45$ [5–189] days) Sex: 15 male Age: 44.89 ± 13.76 years	19 controls matched on age, education, and sex Sex: 15 male Age: 43.95 ± 15.15 years	Outpatient records of hospital brain injury units	Chronic ($\geq$1 year post-injury; $M = 12.37 \pm 7.99$ years)	3 T Philips Achieva; 8-channel head coil: 3D high-resolution TFE using inversion prepulse; shot interval = 1800 ms; TR/TE = 6.4/2.8 ms; flip angle = 8°; voxel = 1 mm^3; slices = 200; FOV = 256	Automated brain segmentation and volumetric analysis (FSL SIENAX and FIRST)	Significant atrophy in the right insula, bilateral amygdala, and bilateral hippocampus in patients compared to controls; larger left insula and bilateral amygdala volumes correlated positively with EEG alpha power and suppression, and with skin conductance levels indicating involvement during arousal

3.4 Studies in Heterogeneous Samples of Patients with Different TBI Severity

With a few exceptions, the literature examining a spectrum of patients with mild to severe TBI comprise either longitudinal or cross-sectional studies during the chronic stage of TBI (see Table 4). Studies during the acute injury phase have generally not observed early morphological alterations, although most have reported progressive atrophy globally and in specific brain regions (e.g., [59, 60]).

3.4.1 Global and Regional Atrophy

Global cerebral atrophy is common in TBI and is particularly evinced in more severe injuries of a diffuse nature (e.g., [59–67]). Post-traumatic cerebral atrophy is predicted by extent of acute DAI [61] as well as with other factors related to injury severity [60, 62, 64, 68, 69] and relates to functional outcomes (e.g., [61, 70–72]). Current and past history of alcohol misuse exacerbates general brain atrophy and relates to post-injury intellectual and cognitive abilities in TBI [73].

Total and regional GM and WM atrophy is also demonstrated in study groups collapsed across injury severities (e.g., [67, 74–76]). Atrophic changes may be worse in patients with focal mass and correlate with lesion volume [61, 71]. The frontal and temporal lobes appear to be particularly susceptible to effects of TBI [59, 67, 70, 77]. The extent of GM and WM atrophy predicts functional and cognitive outcomes in TBI [70, 77–79]. Jorge et al. analyzed the effect of mood disorders on volumetric measures and reported significantly reduced left frontal GM volumes related to major depressive disorder [80]. Furthermore, they observed a synergistic effect of mood disorders and TBI severity to reduce hippocampal volume [81] .

3.4.2 Ventricle Enlargement

Increased lateral ventricles, temporal horns of the lateral ventricles, and/or ventricle-to-brain ratio are reported in TBI [62, 66, 69, 82–84]. Ventricular enlargement largely relates to extent of post-injury cerebral atrophy in TBI [85] and is related to injury severity [86] and is worsened in patients with lesions [62]. Exacerbated enlargements also occur with current and past histories of moderate to heavy alcohol use [73]. Enlarged lateral ventricular volume was a robust predictor of cognitive outcomes in one study [87], although other results have not supported such an association [66, 86].

3.4.3 Cortical Changes

Widespread reductions in cortical GM volume [72, 75, 88], thickness [78, 89], and density [63] are observed in TBI across severities. Reduced cortical GM volumes related to shorter time since injury and are also correlated with cognitive performance [75] and rehabilitation outcomes across injury severity in chronic patients with TBI [88]. Frontal, temporal, and parietal cortical gray matter regions appear to be particularly affected [63, 65, 76, 79, 88–90]. The cingulate cortex has also been frequently implicated

Table 4
Studies in TBI across severities

Author	TBI participant characteristics and comparison		Cohort	Time post-injury	Imaging acquisition details	Analysis method	Summary of findings
	Patients	Controls					
Anderson et al. (1996)	35 mild to severe TBI with lesions (GCS 3–15, $M = 8.14 \pm 3.80$) Sex: 26 male Age: 29.43 ± 11.70 28 mild to severe TBI without lesions (GCS 3–15, M = with ($n = 35$) and without ($n = 28$) frontal lesions (GCS 3–15, $M = 8.14 \pm 3.80$) Sex: 19 male Age: 31.33 ± 13.16	33 controls Sex: 24 male Age: 29.24 ± 9.67	Hospital neuroradiological database	Subacute (≥6 weeks post-injury)	Magnet not specified; slice thickness/gap = 5/2 mm; procedures fully outlined elsewhere (Blatter et al. 1995)	Volumetric analysis (ANALYZE)	Significantly greater VBR observed in TBI patients with lesions compared to both TBI patients without lesions and controls; significantly reduced thalamus volumes observed in TBI patients with lesions compared to TBI patients without lesions (even when matched on GCS), which had significantly reduced volumes compared to controls Greater injury severity associated with greater VBR and reduced thalamus volumes; larger VBR associated with greater lesion volume and lower thalamic volumes and GCS in TBI patients; GCS, VBR, lesion volume, and thalamus volume correlated, and most robust relation observed among GCS and VBR
Johnson et al. (1996)	97 mild to severe TBI (GCS 3–15, $M = 8.3 \pm 3.8$) Sex: 66 male Age: 29.52 ± 16.64 years	166 controls Sex: 79 male Age: 39 ± 13 years	Hospital level 1 trauma center	Chronic (≥ 90 days post-injury, $M = 770 \pm 754$ days)	1.5 T GE Signa; sagittal images; TR/TE/NEX = 500/11/2; matrix = 256 × 192; FOV = 24 cm; slice thickness/gap = 5/1 mm	Volumetric analysis	Significantly greater atrophy observed in the genu and isthmus of the CC than other (i.e., four) callosal regions in TBI patients compared to controls; significant

							relation observed among the splenium and cognitive performance in women but not male patients with TBI
Bigler et al. (1997)	94 mild to severe TBI (GCS 3–15) Sex: 59 male Age: 27 ± 9 years	96 controls Sex: 37 male Age: 31 ± 8 years		Subacute and chronic (44 scans ≤100 days post-injury; 55 scans >100 days post injury)	1.5 T; quadrature head coil; sagittal images; TR/TE/NEX = 500/11/2 Coronal intermediate and T_2-weighted (3800/21,105/2) fast spin-echo images; section thickness = 3 mm; matrix = 512 × 256; FOV = 22 cm; flow compensation, with an inferior saturation pulse, and variable bandwidth used Axial intermediate and T_2-weighted (3000/31,90/1) standard spin-echo images, section thickness = 5 mm; intersection gap = 2 mm; FOV = 22 cm; matrix = 256 × 192	Volumetric analysis (ANALYZE)	Significantly reduced hippocampal volume and enlarged temporal horn observed in patients compared to controls that correlated with cognitive outcome At >70 days post-injury, temporal horn volume correlated with IQ; hippocampal volume correlated with verbal memory performance in patients compared to controls
Tate and Bigler (2000)	86 mild to severe TBI (GCS 3–15, $M = 8.4 \pm 3.93$) Sex: 58 male Age: 30.0 ± 11.73 years	46 controls Sex: 31 male Age: 37.21 ± 13.08 years		Subacute and chronic (≥ 2 months post-injury)	1.5 T; quadrature head coil; sagittal images; TR/TE/NEX = 500/11/2 Coronal intermediate and T_2-weighted (3800/21,105/2) fast spin-echo images; section thickness = 3 mm; matrix = 512 × 256; FOV = 22 cm; flow compensation, with an inferior saturation pulse, and variable bandwidth used Axial intermediate and T_2-weighted (3000/31,90/1) standard spin-echo images; section thickness = 5 mm; intersection gap = 2 mm; FOV = 22 cm; matrix = 256 × 192	Volumetric analysis	Significantly reduced volumes of the fornix and hippocampus observed in TBI patients compared to controls; fornix and hippocampus volumes correlated with injury severity; hippocampus volume correlated with memory performance

(continued)

Table 4 (continued)

Author	TBI participant characteristics and comparison		Cohort	Time post-injury	Imaging acquisition details	Analysis method	Summary of findings
	Patients	Controls					
MacKenzie et al. (2002)	14 mild to moderate TBI (GCS 9–15: 11 mTBI [GCS 13–15]; 3 moderate TBI [GCS 9–12]) Sex: 13 male Age: 36.1 ± 12.1 years	10 controls Sex: not reported Age: 34.9 ± 7.1 years (all had 1 MRI scan; 4 underwent 2 MRI scans over 3 months apart)	Tertiary care record review	Subacute and chronic (14 scans >3 months post-injury; 7 patients completed 2 scans >3 months apart)	1.5 T GE Signa; quadrature transceiver-receiver head coil; axial images; TR/TE = 2700/16; echo train length = 80; echo train length = 8; FOV = 22 cm; matrix = 256 × 192; slice thickness/gap = 3 or 5/0 mm; images interpolated to a 256 × 256 matrix	Volumetric analysis	No significant differences in global brain, CSF, or % VBP observed between TBI patients at the subacute time point and controls; significantly greater longitudinal decline in % VBP and change in % VBP observed in TBI patients compared to controls that were exacerbated by LOC
Yount et al. (2002)	27 mild to severe TBI (GCS 5-14, $M = 8.33 \pm 3.17$; 8 mTBI [GCS 13–15]; 14 moderate to severe TBI [GCS 7–12]; 5 severe TBI [GCS 3–6]) Sex: 18 male Age: 26 ± 8.0 years	12 controls matched on mean age and total ICV and sex ratio Sex: 8 male Age: not reported	Hospital level 1 trauma center	Chronic ($M = 22.8$ [3.07–81.37] months post-injury)	1.5 T GE Signa; sagittal images; RT/TE/NEX = 500/11/2; matrix = 256 × 192 pixel; FOV = 2 cm; slice thickness/gap = 5/1 mm Axial intermediate and T2-weighted (3000/31, 90/1) spin-echo images; matrix = 256 × 192; FOV = 22 cm; slice thickness = 5 mm; interspace gap = 2 mm	Volumetric analysis (ANALYZE)	Significant atrophy observed in the posterior CG in TBI patients compared to controls that correlated with injury severity; reduced cross-sectional surface areas of CC and thalamus observed in TBI patients; increased lateral ventricular volume and VBR as well as reduced TBV observed in TBI compared to controls Comparison of injury severity in TBI patients indicated significant atrophy of posterior CG, thalamus, and CC, and increased lateral ventricle volumes in TBI patients with moderate to severe (combined) injuries

							compared to controls; and significant atrophy of posterior CG, anterior CG, and thalamus, and increased lateral ventricle volumes in moderate to severe TBI patients compared to mTBI patients, which did not differ from controls Neither neuropsychological performance nor mood ratings related to CG morphometry in TBI patients
Kesler et al. (2003)	25 mild to severe TBI patients with high and low (<90, $M = 76 \pm 11.4$; at least 90, $M = 101 \pm 7.9$) IQ scores (GCS 3–14, $M = 7.5 \pm 3.4$) Sex: 13 male Age: 26.16 ± 8.8 years	None	Research database	Chronic (not reported)	1.5 T GE Signa; quadrature head coil; sagittal images; TR/TE/NEX = 500/11/2	Volumetric analysis	Significantly reduced TICV observed in TBI patients with high IQ compared to TBI patients with low IQ (even after accounting for GCS); no group differences observed for VBR; significantly greater change in IQ scores observed in TBI patients with low IQ; TICV significantly correlated with FSIQ and IQ change; VBR only correlated with GCS; TICV and education predicted IQ group
Bergeson (2004)	75 mild to severe TBI (GCS 3–14, $M = 7.2 \pm 4.3$; 9 mTBI [GCS 13–15]; 9 moderate TBI [GCS 9–12]; 53 severe TBI [GCS 3–8]) Sex: 50 male Age: 32.9 ± 11.8 years	75 controls matched on age and sex Sex: 50 male Age: 31.4 ± 12.2 years		Subacute and chronic (≥90 days post injury, $M = 33.4 \pm 22.8$ months)	1.5 T GE Signa or 1.5 T Picker scanner; parameters not reported	Volumetric analysis (ANALYZE)	Significantly increased atrophy observed in frontal and temporal lobes in TBI patients compared to controls; atrophy correlated with poorer executive function and memory performance

(continued)

Table 4
(continued)

Author	TBI participant characteristics and comparison		Cohort	Time post-injury	Imaging acquisition details	Analysis method	Summary of findings
	Patients	Controls					
Wilde et al. (2004)	77 mild to severe TBI (GCS 3–15); 25 with positive blood alcohol level and 52 with negative blood alcohol level Sex: 50 male Age: 32.9 ± years	None	Hospital level 1 trauma center	Subacute and chronic (≥90 days post-injury)	1.5 T GE Signa; T_1-weighted sagittal images: TR/TE/NEX = 500/11/2; T2-weighted spin-echo axial intermediate images = 3000/31, 90/1, thickness/gap = 5/1.5 or 5/2.0 mm, FOV = 22 cm, matrix = 256 × 192 (reported in Blatter et al. 1997)	Volumetric analysis (ANALYZE)	Increased general brain atrophy, CSF and total ventricular volume and VBR observed in patients with a positive BAL and/or a history of moderate to heavy alcohol use compared to TBI patients with negative BAL; positive BAL associated with lower IQ scores and poorer cognitive performance, which was moderated by history of moderate to heavy alcohol use
Gale (2005)	9 mild to severe TBI (GCS 5–15, $M = 9.1 \pm 3.4$) Sex: 8 male Age: 29.1 ± 7.8 years	9 controls Sex: 8 male Age: 28.8 ± 8.4 years	Inpatient hospital and medical centers	Chronic ($M = 10.6 \pm 2.4$ months post-injury)	1.5 T GE Signa; high-resolution SPGR gradient echo pulse coronal images; TR/TE/NEX = 24/6; 1 flip angle = 40°; slice thickness/gap = 1.5/0 mm; FOV = 240 × 240 mm; resolution = 0.9375 mm^2	Voxel-based morpho-metry	Significantly reduced GM density observed in frontal and temporal cortices, CG, subcortical GM, and the cerebellum in TBI patients compared to controls; significantly reduced GM density correlated with poorer attention performance and with lower GCS
Himanen et al. (2005)	61 mild to severe TBI* (PTA < 1 h to >7 days; 17 mTBI [PTA < 1 h]; 12 moderate TBI [PTA 1-24 h]; 11 severe [PTA 1-7 days; 21 very severe	None	Hospital neurology referral	Chronic ($M = 30$ years post-injury)	1.5 T Siemens MAGNETOM; 3D MPRAGE; TR/TE = 10/4; flip angle = 10°; matrix = 192 × 256; slice thickness/gap = 1.5/0 mm For volume measurements, oblique	Manual volumetric analysis	Reduced hippocampal and increased lateral ventricular volumes significantly associated with impaired memory and executive function,

	[PTA > 7 days]) Sex: 41 male Age: 29.4 ± 10.8 years Defined TBI severity by post-traumatic amnesia (PTA) duration				coronal slices (thickness = 3 mm) oriented perpendicularly to its long axis of hippocampus (hippocampal volumes) and axial slices (thickness = 3 mm)		and with memory complaints; lateral ventricular volume shown as best predictor of cognitive outcomes in TBI patients; modest relation among injury severity and cognitive performance observed
Isoniemi et al. (2006)	58 mild to severe TBI with ($n = 19$) and without the APOE4 genotype Sex: 39 male Age: 60.2 ± 10.2 years PTA used to diagnose TBI		Hospital neuropsychological evaluation for TBI	Chronic ($M = 31.3 \pm 3.9$ years post-injury)	1.5 T Siemens MAGNETOM; parameters not reported	Volumetric analysis	No significant differences in examined regional brain volumes including hippocampal and lateral ventricular volumes observed between TBI patients with and without the *APOE4* genotype; groups did not significantly differ on the presence or amount of focal visually detected MRI abnormalities, whether defined as contusions or signs of TIA
Wilde et al. (2006)	60 mild to severe TBI (GCS 3–15, $M = 8.71 \pm 3.22$) Sex: 38 male Age: 28.6 ± 8.2 years	None	Consecutive trauma I center inpatient rehabilitation unit admissions	Subacute and chronic (≥90 days post-injury, $M = 3.04 \pm 1.82$ years post-injury)	1.5 T GE Signa Excite; FOV = 22 cm; matrix = 256 × 192; slice thickness/gap = 5.0/1.5 or 2 mm	Volumetric analysis	Longer PTA duration predicted increased brain atrophy (VBR); each additional day of PTA association with 6% odds increase in later development of atrophy
Jorge et al. (2007)	37 mild to severe TBI patients with ($n = 19$) and without ($n = 18$) mood disorders (GCS 3–15; 13 mTBI [GCS 13–15]; 24 moderate to severe TBI [GCS 3–12]) Sex: 19 male Age: 36.3 ± 14.3 years	None	Post-hospital admittance	Acute, subacute, and chronic (evaluated at baseline and then at 3, 6, and 12 months post-injury)	1.5 T GE Signa; TR/TE/NEX = 24/5/2; flip angle = 40°; FOV = 26; matrix = 256 × 192; coronal slice thickness = 1 mm	Automated segmen-tation and volumetric analysis (BRAIN2)	Significantly reduced left, right, and total hippocampal volumes observed in moderate and severe TBI patients compared to mTBI patients; at 3 months post-injury, mood disorder developed in 51.3% of TBI patients, regardless of injury severity; significantly

(continued)

Table 4 (continued)

Author	TBI participant characteristics and comparison						
	Patients	Controls	Cohort	Time post-injury	Imaging acquisition details	Analysis method	Summary of findings
							reduced left frontal GM volumes observed in TBI patients with mood disorders compared to TBI patients without mood disorders; significantly reduced left and right hippocampal volumes observed in TBI patients with mood disorders compared to TBI patients without mood disorders and were exacerbated by increased injury severity such that moderate to severe TBI patients with mood disorders had significantly reduced hippocampal volumes compared to moderate to severe TBI patients without mood disorders Hippocampal volumes unrelated with memory or executive function and did not predict functional or cognitive outcomes 1 year post-injury Logistic regression results indicated that reduced hippocampal volumes and the presence of mood disorders and alcohol use predicted decreased likelihood of returning to productive activity 1 year post-injury

Trivedi et al. (2007)	37 mild to severe TBI (GCS < 15, M GCS-6 score = 121.7 ± 170.5 h; 11 mTBI [GCS 13–15]; 10 moderate TBI [GCS 9–12]; 16 severe TBI [GCS ≤ 8]) Sex: 27 male Age: 29.3 ± 10.9 years	30 controls Sex: 13 male Age: 24.5 ± 6.4 years	Department of neurosurgery, trauma, and/or rehabilitation	Subacute and chronic (initial scan $M = 79$ days post-injury; follow-up $M = 409$ days post-injury; M time between scans = 329.5 ± 54.5 days)	3 T GE 3 Signa; axial plane SPGR; structural images; TR/TE/TI = 9/1.8/600 ms; flip angle = 20°; matrix = 256 × 192 × 124 (axial 256 × 192 in-plane, interpolated to 256 × 256); FOV = 240 mm; slice thickness = 1.2 mm (124 slices); ±16 kHz receiver bandwidth; acquisition time ~ 7.5 min	Volumetric analysis (SIENA)	Significant longitudinal reduction in % BVC observed in TBI patients compared to controls that correlated with post-injury coma duration
Bendlin et al. (2008)	35 mild to severe TBI (GCS ≤ 13) Sex: 26 male Age: 30.45 ± 11.37 years	36 controls Sex: 18 male Age: 28.47 ± 9.78 years	Department of neurosurgery, trauma and/or rehabilitation referrals	Subacute (8–12 weeks post-injury, $M = 56$ days)	3 T GE Signa; 3D inversion recovery prepared fast gradient echo pulse sequence; TR/TE/TI = 9/1.8/600 ms; flip angle = 20°; matrix = 256 × 192 × 124; interpolated to 256 × 256 × 124; FOV = 240 mm; slice thickness = 1.2 mm (124 slices); receiver bandwidth = ± 16 kHz; acquisition time ~7.5 min	Voxel-based morpho-metry (SPM5)	Significantly reduced frontal WM (anterior to left lateral ventricle) observed in TBI patients compared to controls; significantly reduced GM volume observed in insula, caudate, anterior cingulum, pulvinar nucleus, thalamus, parahippocampal, the cerebellum (particularly in the vermis), and in the medial frontal GM in TBI patients compared to controls; acute-phase and longitudinal atrophy observed in TBI patients in WM (corona radiata, corpus callosum [especially the splenium], internal capsule, external capsule, the superior and inferior longitudinal fascicules, cingulum, inferior fronto-occipital fasciculus, corticospinal tract, superior, middle, and inferior cerebellar peduncles, and small

(continued)

Table 4
(continued)

Author	TBI participant characteristics and comparison: Patients	Controls	Cohort	Time post-injury	Imaging acquisition details	Analysis method	Summary of findings
							regions of cerebellar WM; longitudinal decline in WM volume observed in large regions of frontal, temporal, parietal, and occipital regions, and in small cerebellar areas; significant longitudinal declines in GM volume observed in a large area of medial frontoparietal GM, a cluster in superior frontal GM, thalamus, basal ganglia, nucleus accumbens, thalamus, and small portions of the cerebellum compared to controls
Marcoux et al. (2008)	15 mild to severe TBI (GCS $\leq$ 8 or GCS 9–14 with demonstrated intracranial bleed on CT) Sex: 12 male Age: 33.1 $\pm$ 10 years	30 controls Sex: 13 male Age: 24.5 $\pm$ 6.4 years		Acute and chronic (initial scan 4–19 days post-injury; follow-up scan 6 months post-injury)	1.5 T GE Signa lx; coronal 3D SPGR; TR/TE = 24/3 ms; matrix = 256 $\times$ 256; slice thickness/gap = 1.875/0 mm	Volumetric analysis	Frontal lobe and global brain atrophy observed in TBI patients 6 months post-injury compared to controls; elevated lactate/ pyruvate ratio correlated with extent of frontal lobe atrophy
Fujiwara et al. (2008)	58 mild to severe TBI (12 mTBI [GCS 13–15]; 27 moderate TBI [GCS 9–12]; 19 severe TBI [GCS 3–8]); 18 with focal cortical contusions and 40 with diffuse injury only Sex: 37 male Age: 31.24 $\pm$ 10.45 years	25 controls Sex: 9 male Age: 27.72 $\pm$ 7.93 years (did not undergo MRI; underwent behavioral testing)		Chronic (M = 1 year post-injury)	1.5 T GE Signa; 3D SPGR; TR/TE = 35/5 ms; NEX = 1; flip angle = 35°; FOV = 22 $\times$ 16.5 cm; resolution = 0.859 $\times$ 0.859; slice thickness = 1.2-1.4 mm	Semiauto-mated segmen-tation (SABRE)	Patterns of regional volume loss associated with functional performance; GM loss in the ventral frontal cortex correlated with functional outcomes; the Smell Identification Test was most sensitive to ventral frontal cortex

							damage in TBI patients, regardless of the presence of focal lesions, and related to temporal lobe and posterior cingulate/retrosplenial volumes; executive function associated with superior medial frontal volume
Levine et al. (2008)	69 mild to severe TBI: 13 mTBI (GCS $M = 14.6 \pm 0.7$) Sex: 8 male Age: 33.7 ± 13.1 years 30 moderate TBI (GCS $M = 11.1 \pm 2.0$) Sex: 16 male Age: 32.0 ± 11.1 years 26 severe TBI (GCS $M = 6.7 \pm 2.6$) Sex: 19 male Age: 28.9 ± 8.0 years	12 controls matched on age and sex Sex: 7 male Age: 27.2 ± 3.3 years	Level 1 trauma center patients as part of the Toronto TBI study	Chronic (≥ 1 year post-injury; $M = 1.37 \pm 0.86$)	1.5 T GE Signa; 3D sagittal; TR/TE =35/5 ms; flip angle = 35°; NEX = 1.0; FOV = 220 × 220 mm; matrix = 256 × 256 × 124: slices = 124; slice thickness = 1.3 mm	Semiautomated segmen-tation (SABRE)	Stepwise dose-response relationship between parenchymal volume loss and TBI severity observed; brain volume alterations differentiated moderate to severe TBI patients from mTBI patients as well as between TBI patients and controls; volume loss in widespread WM regions and sulcal/subdural CSF correlated with TBI severity; robust alterations observed in frontal, temporal, and cingulate regions; greater volume loss observed in frontal and temporal regions in TBI patients with focal lesions compared to TBI patients without focal lesions and in TBI patients with diffuse injury compared to controls
Schonberger et al. (2009)	98 mild to severe TBI (GCS 13–15; $M = 9.0 \pm 4.2$) Sex: 74 male Age: 34.5 ± 14 years Injury severity determined on PTA duration	None	Recruited from hospital rehabilitation programs	Chronic ($M = 2.3 \pm 1.5$ [0.3–5.7] years post-injury)	1.5 T Philips Intera; high-resolution axial 3D fast-field echo sequence; TR/TE = 92 ms; flip angle = 2°; slice thickness/gap = 1.5/0 mm; matrix = 256 × 256; FOV = 250 mm^2	Lesions volume' automated segmen-tation and volumetric analysis (SPM5)	Older age and longer PTA associated with larger lesion volumes in both GM and WM in nearly brain regions; older age associated with smaller GM volumes in most cortical regions; longer

(continued)

Table 4 (continued)

Author	TBI participant characteristics and comparison: Patients	Controls	Cohort	Time post-injury	Imaging acquisition details	Analysis method	Summary of findings
							PTA associated with smaller WM volumes in most brain regions; older age worsened deleterious effects of TBI on neural tissue
Vannorsdall et al. (2010)	14 mild to severe TBI (GCS 3–15) Sex: 17 male Age: 38.8 ± 13.6 years	28 controls matched on age, sex, race, education, estimated premorbid IQ, and total ICV Sex: 4 male Age: 39.8 ± 12.9 years	Hospital and rehabilitation centers	Chronic (≥ 18 months post-injury; *M* = 122 [18–366] months)	1.5 T GE Signa; 3D coronal SPGR; TR/TE =35/5 s; flip angle = 45°; matrix =256x256; slice thickness/gap = 1.5/0 mm; slices = 124	Volumetric analysis and voxel-based morpho-metry (SPM2; MATLAB)	Total and disproportionate WM volume reduction (atrophy) observed in TBI patients compared to controls; significant reduction of WM density in the frontal lobes, limbic region, corpus callosum, cingulate, thalamus, parahippocampal region, and cerebellum in TBI patients compared to controls; significantly reduced GM density observed in temporal poles, and central regions of the frontal, parietal, and occipital lobes, cingulate cortex, thalamus, and cerebellum; no regions of increased WM or GM density observed in TBI patients compared to controls

Strangman et al. (2010)	50 mild to severe TBI with documented memory impairment (GCS <9–15, LOC $M = 12.0 \pm 23.5$ days; 12 mTBI [LOC ≤ 30 min or GCS 13–15]; 12 moderate TBI [LOC 30 min to 24 h or GCS 9–12]; 24 severe TBI [LOC > 24 h or GCS < 9]; 2 n/a) Sex: 36 male Age: 47.2 ± 11.4 years	None	Convenience community sample	Chronic ($M = 11.5 \pm 9.3$ years post-injury)	1.5 T Siemens Avanto; 8-channel Tim head coil; two MPRAGE sequences; TR/TE/TI = 1.91 s/4.13 ms/1.1 s; flip angle = 15°; slices = 120; matrix = 128 × 128; voxel = 1 mm^3	Automated segmen-tation and volumetric analysis (FreeSurfer)	Volume reduction observed in the hippocampus, lateral prefrontal cortex, thalamus, and several subregions of the cingulate cortex and predicted rehabilitation outcome in TBI patients
Schonberger et al. (2011)	54 mild to severe TBI patients with ($n = 13$) and without depressive disorder (GCS 3–15, $M = 9.3 \pm 4.2$) Sex: 41 male Age: 35.0 ± 15.2 years	None	Hospital rehabilitation program	Chronic ($M = 2.2$ [0.3–5.7] years post-injury)	1.5 T Philips Intera; axial 3D fast-field echo; TR/TE = 9/2 ms; flip angle = 20°; slice thickness/gap = 1.5/0 mm; matrix = 256 × 256; FOV = 250 mm	Manual segmen-tation and volumetric analysis (SPM5)	A substantial number of TBI patients evinced GM or WM lesions in the frontal (80%), temporal (65%), and parietal (50%) lobes as well as in the sublobar (67%) and limbic (65%) regions that did not predict post-injury depression; significantly increased left-right volume asymmetry of the frontal lobes was observed in the TBI patients (right > left) with post-injury depression compared to TBI patients without depression; significantly decreased left-right volume asymmetry of the parietal lobes was observed in the TBI patients with post-injury depression (right > left) compared to TBI patients without depression; both volume ratios of left and right frontal and parietal lobes significantly predicted depression in TBI patients

(continued)

Table 4 (continued)

Author	TBI participant characteristics and comparison		Cohort	Time post-injury	Imaging acquisition details	Analysis method	Summary of findings
	Patients	Controls					
Tate et al. (2011)	65 mild to severe TBI patients (GCS 3–15, $M = 8.85 \pm 3.22$) Sex: 44 male Age: 27.66 ± 7.58 years	None	Inpatient rehabilitation unit of level 1 trauma center	Chronic ($M = 34$ [3.2–91.63] months post-injury)	1.5 T GE; TR/TE = 3000/90; FOV = 24 cm; slice thickness/gap = 5/1 mm	Manual segmen-tation and volumetric analysis (ANALYZE)	Significant correlations among corrected TBV, ventricle-to-cranial ratio, VBR, and brain to total ICV ratio with neuropsychological performance were observed in TBI patients; TBV highly correlated with ICV; corrected TBV, ventricle-to-cranial ratio, VBR, parenchymal volume loss, and brain to total ICV ratio highly correlated and more strongly associated with neuropsychological performance than both TBV and ICV; corrected brain volumes provided global indicators of cerebral atrophy in relation to neuropsychological performance in TBI patients
Yurgelun-Todd et al. (2011)	15 veterans with one or more TBI (14 with ≥1 mTBI; 1 moderate to severe TBI [GCS not reported]; 14 normal on clinical MRI) Sex: 15 male Age: 34.9 ± 9.71 years (only 14 had MRI) Injury severity classification according to the OSU-TBI criteria	17 controls Sex: 17 male Age 34.0 ± 10.59 years	Veterans	Chronic ($M = 95.4110.5$ months)	3 T Siemens MAGNETOM Trio; 3D sagittal MPRAGE GRAPPA; 12-channel head coil; TE/TR/TI = 3.38 ms/2.0 s/1.1 s; flip angle = 8°; matrix = 256 mm^2; FOV = 256; slices = 160; slice thickness/gap = 1/0 mm	Automated segmen-tation and volumetric analysis (FreeSurfer)	Trend-level reductions observed in right, left, and total frontal lobe volume in veterans with TBI compared to controls

Farbota et al. (2012)	17 mild to severe TBI patients (GCS at admittance 3–15; $M = 7.2$; unavailable for 7 patients) Sex: 14 male Age: 34.5 ± 12.0 years Patients had resolved PTA at assessments	13 controls Sex: 5 male Age: 26.8 ± 8.9	Level 1 trauma center	Subacute and chronic (<3 months post-injury at initial scan)	3 T GE; quadrature birdcage head coil; 3D inversion-prepped spoiled GRE; TR/TI/TE = 9/600/1.8 ms; flip angle = 20°; matrix = 256 × 256 × 124; FOV = 240 × 240 mm; slice thickness = 1.2 mm; slices = 124; receiver bandwidth = ± 16 Hz; time = 7.5 min; reconstructed voxel size = 0.94 × 0.94 × 1.2 mm^3	Tensor-based morpho-metry and warping analysis (SPM8)	Longitudinal effects significantly differed for TBI patients and controls in clusters in large surface regions of the frontal, temporal, occipital, and posterior/inferior parietal cortices as well as in a medial interhemispheric region that included anterior and superior corona radiata, superior longitudinal fasciculus, posterior inferior longitudinal fasciculus, genu and splenium of the corpus callosum, forceps minor, right internal capsule, superior external capsule, and sagittal stratum; greater volume loss during the first interval was observed in the TBI patients in the left frontal, temporal, posterior parietal, occipital, and interhemispheric cortices, parahippocampal gyrus, lateral cerebellum, brainstem, thalamus, external capsule, and splenium of the corpus callosum; during the second interval volume loss was observed in the TBI patients in the left parahippocampal gyrus, brainstem, thalamus, inferior longitudinal fasciculus, superior longitudinal fasciculus, external capsule, forceps minor, superior and

(continued)

Table 4
(continued)

Author	TBI participant characteristics and comparison		Cohort	Time post-injury	Imaging acquisition details	Analysis method	Summary of findings
	Patients	Controls					
							anterior corona radiata and splenium, body and genu of the corpus callosum; total brain volume loss not observed across time points in TBI patients compared to controls Neuropsychological performance correlated with the brainstem and cerebellum volume during time 1 and with the anterior cingulate cortex at follow-up
Levine et al. (2013)	63 mild to severe TBI patients with diffuse or focal ($n = 20$) damage (13 mTBI [GCS $M = 14.6 \pm 0.7$]; 29 moderate TBI [GCS $M = 11.0 \pm 2.1$]; 21 severe TBI [GCS $M = 6.3 \pm 2.5$]) Sex: 37 male Age: 31.28 ± 10.66 years	27 controls Sex: 11 male Age: 27.7 ± 7.6 years	Level 1 trauma center patients as part of the Toronto TBI study; see Levine et al. [68]	Chronic (1 year post-injury, $M = 1.10 \pm 0.33$)	1.5 T GE Signa; 3D sagittal; TR/TE =35/5 ms; flip angle = 35°; NEX = 1.0; FOV = 220 mm^2; matrix = 256 × 256; slices = 124; slice thickness = 1.3 mm	Semiautomated segmen-tation (SABRE)	Processing speed scores correlated with longitudinal BVC in TBI patients across severities and in TBI patients with diffuse injuries only; tests of processing speed, working memory, and verbal learning and memory robustly associated with distributed volumetric reductions among temporal, ventromedial prefrontal, right parietal, and cingulate regions that worsened in TBI patients with focal lesions, but was also observed in the diffuse TBI patients; GM generally more affected than WM; clinical definitions poorer predictors of neuropsychological

							outcomes than volumetric brain data
Takayanagi et al. (2013)	10 patients with TBI diagnosed with personality change secondary to TBI, apathetic subtype per DSM-IV criteria (2 mTBI; 2 moderate TBI; 6 severe TBI; based on GCS and LOC) Sex: 8 male Age: 43.7 ± 11.1 years *Clinical comparisons*: 15 patients with schizophrenia according to DSM-IV criteria Sex: 12 male Age: 33.5 ± 10.6 years	47 controls Sex: 38 male Age: 41.4 ± 12.2 years	Recruited from hospital clinics	Chronic (≥ 18 months post-injury, $M = 72 \pm 47$ [18–157])	1.5 T GE Signa; 3D coronal SPGR; TR/TE = 35/5 s; flip angle = 45°; FOV = 210 × 240 mm^2; slice thickness/gap = 1.5/0 mm; matrix = 256 × 256; and 124 coronal slices; receiver bandwidth = 180 HZ/pixel	Automated segmen-tation and volumetric and cortical thickness analysis (FreeSurfer)	Significant group effect on volumes of bilateral hippocampi, right thalamus, and the brainstem that were reduced in TBI patients compared to controls and the patients with schizophrenia, which did not differ Significant group effects on cortical thickness in several regions; TBI patients and controls did not differ; TBI patients had thicker cortices in bilateral pars orbitalis and lateral orbitofrontal regions compared to schizophrenic patients In the TBI patients, higher apathy rating scores related with worse cognitive performance on selected tasks
Spitz et al. (2013)	69 mild to severe TBI (PTA 0- > 7 days; 8 mTBI [PTA ≤ 24 h]; 16 moderate TBI [PTA 1–7 days]; 45 severe TBI [PTA > 7 days]) Sex: 56 male Age: 34.74 ± 12.41 years (range 17–69 years) GCS not reported; subsample underwent cognitive testing	25 controls Sex: 11 male Age: 30.26 ± 12.48 years	Recruited from head injury rehabilitation programs	Chronic ($M = 19.64 \pm 15.56$ months post-injury [2.76–64.59])	3 T Siemens MAGNETOM Verio; 3D sagittal scan; TR/TI/TE = 1930/1100/248 ms; flip angle = 12°; FOV = 256 × 256 mm^2; slice thickness/gap = 1/0 mm; matrix = 256 × 256 × 160; 160 sagittal slices	Automated segmen-tation and volumetric analysis (FreeSurfer)	Reduced cortical volume associated with older age and shorter time post-injury in TBI patients; older age associated with widespread cortical volume reductions, with largest clusters observed in temporal and parietal regions; shorter time since injury associated with widespread cortical volume reductions, with strong associations observed in frontal, occipital, parietal, and

(continued)

Table 4
(continued)

Author	TBI participant characteristics and comparison		Cohort	Time post-injury	Imaging acquisition details	Analysis method	Summary of findings
	Patients	Controls					
							temporal lobe regions; relations were weak in controls and only observed in left hemisphere, with largest cluster observed in the insula; group × age interaction observed in right hemisphere superior parietal and superior temporal regions whereby increasing age resulted in significantly lower volume in the TBI patients compared to controls
							Reduced cortical volume in left and right hemispheres observed in TBI patients, regardless of injury severity (per PTA), with largest clusters observed in the left cuneus and right inferior parietal and frontal regions; compared to controls, mTBI patients showed significant regional variation in left hemisphere GM, including the pericalcarine region of the parietal lobe; the moderate TBI patients did not differ from controls; compared to controls, severe TBI

							patients showed significant bilateral reductions of regional cortical GM, including in the left cuneus and superior frontal regions and right superior and rostral middle frontal regions Alterations in cortical GM volumes correlated with cognitive performance in the TBI patients across injury severity
Green et al. (2014)	56 mild to severe patients (GCS $M = 6.19 \pm 3.42$; 69.6% severe TBI [GCS ≤ 8]) Sex: 41 male Age: 40.16 ± 15.63 years	12 controls Sex: 6 male Age: 36.3 ± 12.5 years	Hospital rehabilitation centers	Chronic (initial scan $M = 5.2 \pm 1.15$ months post-injury; follow-up $M = 20 \pm 4.7$ months post-injury; months between scans $M = 25.4 \pm 10.0$)	1.5 T GE Signa Echospeed; 8-channel head coil; $TR/TE = 300/13$ ms; slice thickness/gap = 5/2.5 mm; matrix = 256 × 128; axial GRE: $TR/TE = 450/20$, flip angle = 20°, slice thickness = 3 mm no gap, matrix 256 × 192	Semiautomated and manual segmen-tation and ROI volumetric analysis	Significant (> 2 z-scores below controls) longitudinal atrophy was observed in at least one brain region in 96% of TBI patients, with 75% of TBI patients showing declines in at least 3–4 of the examined brain regions; observed reduction in ventricle-to-brain ratio in mTBI patients was significantly associated with atrophy in the corpus callosum and right hippocampus, which were not associated; significantly less atrophy in the genu of the corpus callosum was observed in the mTBI patients compared to the body or splenium of the corpus callosum

(continued)

Table 4
(continued)

	TBI participant characteristics and comparison						
Author	Patients	Controls	Cohort	Time post-injury	Imaging acquisition details	Analysis method	Summary of findings
Guild and Levine (2015)	60 mild to severe TBI patients (GCS 3–15; 13 mTBI [GCS 13–15]; 26 moderate TBI [GCS 9–12]; 21 severe TBI [GCS < 8]) Sex: 34 male Age: 30.8 ± 10.5 years	8 controls matched on age and education Sex: 4 male Age: 29.3 ± 8.2 years	Research study	Chronic (1 year post-injury, $M = 12.7 \pm 4.6$ months)	1.5 T GE Signa; 3D sagittal; TR/TE =35/5 ms; flip angle = 35°; NEX = 1.0; FOV = 220 × 220 mm^2; matrix = 256 × 256; slices = 124; slice thickness = 1.3 mm	Semiautomated segmen-tation and volumetric analysis (SABRE)	GCS and outcome measures differentiated between patients and controls, which was largely driven by the severe TBI patients compared to the other two TBI patient groups, regardless of the presence or not of lesions Volume loss greatest in the bilateral middle medial frontal regions and the cingulate gyrus (right greater than left); the right inferior parietal, bilateral aspects of the basal ganglia/thalamus, bilateral medial temporal and right lateral frontal regions also showed significant volume loss that related to post-TBI outcome; volumetric alterations were associated with physical dependency outcomes
Bigler et al. (1996)	TBI patients (GCS 3–15): 29 subacute TBI patients (GCS $M = 8.1 \pm 3.5$; age 28.9 ± 12.0 years) 54 chronic TBI patients (GCS $M = 8.6 \pm 4.0$; age 28.6 ± 11.4 years) Sex: 57 male 16 TBI patients in both groups	72 controls Sex: 26 male Age: 37.9 ± 11.5 years	Hospital trauma unit and inpatient rehabilitation	Subacute and chronic (initial scans ≤90 days post-injury, $M = 43.9 \pm 26.9$; follow-up >90 days post-injury, $M = 857.3 \pm 828.8$)	1.5 T sagittal; quadrature head coil; SPGR 3D TR/TE/NEX = 500/11/2); slice thickness = 1.5 mm; T2-weighted and coronal intermediate (3800/3, 105/1) fast spin-echo; interleaved slices; thickness = 3 mm; matrix 512 × 256; FOV = 22 cm; axial intermediate and T2-weighted	Volumetric analysis (ANALYZE)	Significant enlargement of the temporal horn bilaterally, VBR, and third ventricle observed in subacute TBI patients compared to controls; significant longitudinal enlargement of the ventricle-to-brain ratio as well as reduction of the total brain volume

(3000/30,90/2) standard spin-echo images slice thickness = 5 mm; interspace gap = 1.5 or 2 mm; FOV = 22 cm; matric = 256 × 192 (not part of standard clinical protocol)	and left hippocampus observed in chronic TBI patients compared to both acute TBI patients and controls; significant longitudinal enlargement of the third ventricle and bilateral temporal horns observed in chronic TBI patients compared to controls Hippocampal volumes (left > right) correlated with memory performance in chronic TBI patients but not in subacute TBI patients

as being altered in TBI [79, 88], with reductions being associated with diminished cognitive outcomes [63].

3.4.4 Subcortical Changes

Several studies have also supported reduction in subcortical white matter subsequent to TBI [67, 90]. Reduced volume and surface area are observed in the corpus callosum [62, 65, 66, 84] . However, regional specificity of callosal disruption is suggested. The genu appears to be particularly affected by TBI in general [84, 91], although the literature has produced mixed results regarding regional susceptibility. For instance, Kim et al. [76] observed the greatest longitudinal atrophy in the middle and posterior bodies of the corpus callosum. Callosal alterations correlate with neuropsychological outcomes in chronic TBI patients. Sex differences have also been reported such that splenium volumes were associated with cognitive performance only in the female patients with TBI.

Subcortical GM regions may be especially susceptible to the mechanical effects of TBI [62]. This is particularly true for deep subcortical structures [70]. One study indicated putamen reductions in patients with TBI relative to controls [65]. Reduced volumes of the thalamus [62, 65, 66, 76, 88] and hippocampus [65, 83, 87–89, 92, 93] are largely supported by the published research literature and are thought to be worse in more severe injuries as well as in patients with relative to without lesions [62]. While reductions relate to injury severity and correlate with poor neuropsychological performance [65, 72, 83, 92, 93], there is no support for the additive effect of early aging in TBI patients with the Alzheimer's dementia genotype [94]. Within the limbic system, additional subcortical gray matter structures including the amygdala [65, 72] and fornix [62, 93], along with distinct regions of the cingulate gyrus [63, 68], are volumetrically reduced in patients with TBI and perhaps especially in more severe as opposed to milder injuries [66]. The brainstem is also reduced in TBI in general [89], and alterations in this region correlate with neuropsychological performance [90].

Cerebellar reductions occur in the subacute and chronic injury phases in patients with TBI and predict neuropsychological outcomes [61, 62, 90]. Cerebellar gray and white matter atrophy is supported in subacute patients compared to controls in several regions, but especially in the vermis [74]. Reduced densities of cerebellar white and gray matter are also observed in chronic-phase TBI [63, 67].

4 Comment

Traditionally, neuroimaging for TBI has relied upon CT and MRI for determining injury severity, examining the distribution and nature of lesions, making the decision for immediate surgical intervention, and monitoring lesion change over time. Recently, however, advanced imaging techniques are being evaluated and developed for use in the clinical management and treatment planning beyond their current role as research applications.

Importantly, structural MRI techniques including volumetric studies offer a potential "biomarker" for the diagnosis of TBI which otherwise relies on subjective assessments and self-reports of symptoms. In addition, these techniques have been used to evaluate the interplay between TBI and commonly associated comorbidities such as post-traumatic stress disorder (PTSD), depression, and substance misuse that may result on cumulative effects in areas affected by each of these disorders. Finally structural imaging could be used to examine long-term changes associated with TBI, particularly neurodegenerative conditions such as chronic traumatic encephalopathy.

Brain volumetric studies among TBI patients generally show atrophy of both gray and white matter structures with increasing severity. The vulnerability of the frontal and temporal lobes is especially well-demonstrated with this modality [66, 95], including a trajectory of atrophy that may continue for some time after injury [74, 90].

We have emphasized that structural changes associated with the milder forms of TBI are far more difficult to observe and measure using conventional imaging methods. Furthermore, structural imaging methods should not be examined in isolation and should be combined with other sequences, sensitive to complementary tissue damage, including diffusion tensor, metabolic, and functional imaging to improve sensitivity and specificity. This is important as no one imaging technique provides complete information about the brain or brain injuries. Newer, more automated methods for examining volumetric and lesion data are constantly under development, and utilization of these newer methods may prove to be important ways of further characterizing TBI abnormalities.

Overall, advanced MRI morphometric techniques hold great promise in refining the diagnosis and prognosis of both acute and chronic TBI, particularly in cases where more conventional MR sequences are unrevealing. Furthermore, these techniques may significantly contribute to a clearer conceptualization of the natural course of recovery from TBI, may increase our understanding of the potential contribution of trauma to later neurodegenerative disease, and may enable evaluation of pharmacologic and rehabilitative interventions.

We should emphasize that, as with other advanced imaging across the field of TBI and other disease processes, current barriers for application in clinical practice include the lack of standardization in acquisition and analysis tools and a lack of well-considered normative data. Furthermore, consideration is warranted regarding how to best apply appropriate control groups that are truly comparable in terms of background characteristics and exposure to stressors.

References

1. Centers for Disease Control and Prevention (2015) Report to congress on traumatic brain injury in the United States: understanding the public health problem among current and former military personnel. Concussion and Traumatic Brain Injury, CDC Injury Center. N. C. f. I. P. a. C. D. o
2. Cassidy JD, Carroll LJ, Peloso PM et al (2004) Incidence, risk factors and prevention of mild traumatic brain injury: results of the WHO collaborating Centre task force on mild traumatic brain injury. J Rehabil Med 36:28–60. https://doi.org/10.1080/16501960410023732
3. Bigler ED, Bazarian JJ (2010) Diffusion tensor imaging: a biomarker for mild traumatic brain injury? Neurology 74:626–627. https://doi.org/10.1212/WNL.0b013e3181d3e43a
4. Tate DF, Shenton ME, Bigler ED (2012) Introduction to the brain imaging and behavior special issue on neuroimaging findings in mild traumatic brain injury. Brain Imaging Behav 6:103–107. https://doi.org/10.1007/s11682-012-9185-0
5. Tate DF, Wilde EA, Bouix S, McCauley SR (2015) Introduction to the brain imaging and behavior special issue: mild traumatic brain injury among active duty service members and veterans. Brain Imaging Behav 9:355–357. https://doi.org/10.1007/s11682-015-9445-x
6. Teasdale G, Maas A, Lecky F et al (2014) The Glasgow coma scale at 40 years: standing the test of time. Lancet Neurol 13:844–854. https://doi.org/10.1016/S1474-4422(14)70120-6
7. Teasdale G, Jennett B (1974) Assessment of coma and impaired consciousness. A practical scale. Lancet 304:81–84. https://doi.org/10.1016/S0140-6736(74)91639-0
8. Reith FCM, Brennan PM, Maas AIR, Teasdale GM (2016) Lack of standardization in the use of the Glasgow coma scale: results of international surveys. J Neurotrauma 33:89–94. https://doi.org/10.1089/neu.2014.3843
9. Edwards SL (2001) Using the Glasgow coma scale: analysis and limitations. Br J Nurs 10:92–101. 10.12968/bjon.2001.10.2.5391
10. Gabbe BJ, Cameron PA, Finch CF (2003) The status of the Glasgow coma scale. Emerg Med (Fremantle) 15:353–360. https://doi.org/10.1111/j.1742-6723.2006.00867.x
11. Saatman KE, Duhaime AC, Bullock R et al (2008) Classification of traumatic brain injury for targeted therapies. J Neurotrauma 25:719–738. https://doi.org/10.1089/neu.2008.0586
12. Nakase-Richardson R, Sepehri A, Sherer M et al (2009) Classification schema of posttraumatic amnesia duration-based injury severity relative to 1-year outcome: analysis of individuals with moderate and severe traumatic brain injury. Arch Phys Med Rehabil 90:17–19. https://doi.org/10.1016/j.apmr.2008.06.030
13. Azouvi P (2000) Neuroimaging correlates of cognitive and functional outcome after traumatic brain injury. Curr Opin Neurol 13:665–669. https://doi.org/10.1097/00019052-200012000-00009
14. Ruff RM, Iverson GL, Barth JT et al (2009) Recommendations for diagnosing a mild traumatic brain injury: a national academy of neuropsychology education paper. Arch Clin Neuropsychol 24:3–10. https://doi.org/10.1093/arclin/acp006
15. Helmick KM, Spells CA, Malik SZ et al (2015) Traumatic brain injury in the US military: epidemiology and key clinical and research programs. Brain Imaging Behav 9:358–366. https://doi.org/10.1007/s11682-015-9399-z
16. Dall'Acqua P, Johannes S, Mica L et al (2016) Connectomic and surface-based morphometric correlates of acute mild traumatic brain injury. Front Hum Neurosci 10:127. https://doi.org/10.3389/fnhum.2016.00127
17. Holli KK, Harrison L, Dastidar P et al (2010) Texture analysis of MR images of patients with mild traumatic brain injury. BMC Med Imaging 10:8. https://doi.org/10.1186/1471-2342-10-8
18. Zagorchev L, Meyer C, Stehle T et al (2015) Differences in regional brain volumes two months and one year after mild traumatic brain injury. J Neurotrauma 34:1–24. https://doi.org/10.1089/neu.2014.3831
19. da Costa L, van Niftrik CB, Crane D et al (2016) Temporal profile of cerebrovascular reactivity impairment, gray matter volumes, and persistent symptoms after mild traumatic head injury. Front Neurol 7:70. https://doi.org/10.3389/fneur.2016.00070
20. Killgore WDS, Singh P, Kipman M et al (2016) Gray matter volume and executive functioning correlate with time since injury following mild traumatic brain injury. Neurosci Lett 612:238–244. https://doi.org/10.1016/j.neulet.2015.12.033

21. Dean PJA, Sato JR, Vieira G et al (2015) Long-term structural changes after mTBI and their relation to post-concussion symptoms. Brain Inj 29:1211–1218. https://doi.org/10.3109/02699052.2015.1035334
22. Cohen BA, Inglese M, Rusinek H et al (2007) Proton MR spectroscopy and MRI-volumetry in mild traumatic brain injury. AJNR Am J Neuroradiol 28:907–913. doi: 28/5/907 [pii]
23. Little DM, Geary EK, Moynihan M et al (2014) Imaging chronic traumatic brain injury as a risk factor for neurodegeneration. Alzheimers Dement 10:S188–S195. https://doi.org/10.1016/j.jalz.2014.04.002
24. Ross DE, Ochs AL, Seabaugh JM et al (2012) Progressive brain atrophy in patients with chronic neuropsychiatric symptoms after mild traumatic brain injury: a preliminary study. Brain Inj 26:1500–1509. https://doi.org/10.3109/02699052.2012.694570
25. Zhou Y, Kierans A, Kenul D et al (2013) Mild traumatic brain injury: longitudinal regional brain volume changes. Radiology 267:880–890. https://doi.org/10.1148/radiol.13122542
26. Toth A, Kovacs N, Perlaki G et al (2013) Multi-modal magnetic resonance imaging in the acute and sub-acute phase of mild traumatic brain injury: can we see the difference? J Neurotrauma 30:2–10. https://doi.org/10.1089/neu.2012.2486
27. Maller JJ, Thomson RHS, Pannek K et al (2014) Volumetrics relate to the development of depression after traumatic brain injury. Behav Brain Res 271:147–153. https://doi.org/10.1016/j.bbr.2014.05.047
28. Hofman PA, Stapert SZ, van Kroonenburgh MJ et al (2001) MR imaging, single-photon emission CT, and neurocognitive performance after mild traumatic brain injury. AJNR AmJ Neuroradiol 22:441–449
29. Lao Y, Law M, Shi J et al (2015) A T1 and DTI fused 3D corpus callosum analysis in pre- vs. -post-season contact sports players. Proc SPIE Int Soc Opt Eng 9287:1–6. https://doi.org/10.1117/12.2072600
30. Jarrett M, Tam R, Hernandez-Torres E et al (2016) A prospective pilot investigation of brain volume, white matter hyperintensities, and hemorrhagic lesions after mild traumatic brain injury. Front Neurol 7:11. https://doi.org/10.3389/fneur.2016.00011
31. Owens BD, Kragh JF, Wenke JC et al (2008) Combat wounds in operation Iraqi freedom and operation enduring freedom. J Trauma 64:295–299. https://doi.org/10.1097/TA.0b013e318163b875
32. Hoge CW, McHurk D, Thomas JL et al (2008) Mild traumatic brain injury in U.S. soldiers returning from Iraq. N Engl J Med 358:453–463. https://doi.org/10.1056/NEJMoa072972
33. Tate DF, York GE, Reid MW et al (2014) Preliminary findings of cortical thickness abnormalities in blast injured service members and their relationship to clinical findings. Brain Imaging Behav 8:102–109. https://doi.org/10.1007/s11682-013-9257-9
34. Corbo V, Salat DH, Amick MM et al (2014) Reduced cortical thickness in veterans exposed to early life trauma. Psychiatry Res Neuroimaging 223:53–60. https://doi.org/10.1016/j.pscychresns.2014.04.013
35. Mac Donald C, Johnson A, Cooper D et al (2013) Cerebellar white matter abnormalities following primary blast injury in US military personnel. PLoS One 8(2):e55823. https://doi.org/10.1371/journal.pone.0055823
36. Depue BE, Olson-Madden JH, Smolker HR et al (2014) Reduced amygdala volume is associated with deficits in inhibitory control: a voxel- and surface-based morphometric analysis of comorbid PTSD/mild TBI. Biomed Res Int 2014:691505. https://doi.org/10.1155/2014/691505
37. Xu Y, McArthur DL, Alger JR et al (2010) Early nonischemic oxidative metabolic dysfunction leads to chronic brain atrophy in traumatic brain injury. J Cereb Blood Flow Metab 30:883–894. https://doi.org/10.1038/jcbfm.2009.263
38. Brezova V, Goran Moen K, Skandsen T et al (2014) Prospective longitudinal MRI study of brain volumes and diffusion changes during the first year after moderate to severe traumatic brain injury. NeuroImage Clin 5:128–140. https://doi.org/10.1016/j.nicl.2014.03.012
39. Kim J, Avants B, Patel S et al (2008) Structural consequences of diffuse traumatic brain injury: a large deformation tensor-based morphometry study. NeuroImage 39:1014–1026. https://doi.org/10.1016/j.neuroimage.2007.10.005
40. Hillary FG, Rajtmajer SM, Roman CA et al (2014) The rich get richer: brain injury elicits hyperconnectivity in core subnetworks. PLoS One 9(11):e113545. https://doi.org/10.1371/journal.pone.0104021
41. Konstantinou N, Pettemeridou E, Seimenis I et al (2016) Assessing the relationship between neurocognitive performance and brain volume in chronic moderate-severe traumatic brain injury. Front Neurol 7:29. https://doi.org/10.3389/fneur.2016.00029

42. Ng K, Mikulis DJ, Glazer J et al (2008) Magnetic resonance imaging evidence of progression of subacute brain atrophy in moderate to severe traumatic brain injury. Arch Phys Med Rehabil 89(12 Suppl):S35–S44. https://doi.org/10.1016/j.apmr.2008.07.006
43. Braskie MN, Klunder AD, Hayashi KM et al (2010) Plaque and tangle imaging and cognition in normal aging and Alzheimer's disease. Neurobiol Aging 31:1669–1678
44. Hudak A, Warner M, Marquez de la Plata C et al (2011) Brain morphometry changes and depressive symptoms after traumatic brain injury. Psychiatry Res - Neuroimaging 191:160–165. https://doi.org/10.1016/j.pscychresns.2010.10.003
45. Ariza M, Serra-Grabulosa JM, Junqué C et al (2006) Hippocampal head atrophy after traumatic brain injury. Neuropsychologia 44:1956–1961. https://doi.org/10.1016/j.neuropsychologia.2005.11.007
46. Leunissen I, Coxon JP, Caeyenberghs K et al (2014) Subcortical volume analysis in traumatic brain injury: the importance of the fronto-striato-thalamic circuit in task switching. Cortex 51:67–81. https://doi.org/10.1016/j.cortex.2013.10.009
47. Lutkenhoff ES, McArthur DL, Hua X et al (2013) Thalamic atrophy in antero-medial and dorsal nuclei correlates with six-month outcome after severe brain injury. NeuroImage Clin 3:396–404. https://doi.org/10.1016/j.nicl.2013.09.010
48. Kim J, Parker D, Whyte J et al (2014) Disrupted structural Connectome is associated with both psychometric and real-world neuropsychological impairment in diffuse traumatic brain injury. J Int Neuropsychol Soc 20:887–896. https://doi.org/10.1017/S1355617714000812
49. Sidaros A, Skimminge A, Liptrot MG et al (2009) Long-term global and regional brain volume changes following severe traumatic brain injury: a longitudinal study with clinical correlates. NeuroImage 44:1–8. https://doi.org/10.1016/j.neuroimage.2008.08.030
50. Ubukata S, Ueda K, Sugihara G et al (2016) Corpus callosum pathology as a potential surrogate marker of cognitive impairment in diffuse axonal injury. J Neuropsychiatry Clin Neurosci 28:97–103. https://doi.org/10.1176/appi.neuropsych.15070159
51. Shah S, Yallampalli R, Merkley TL et al (2012) Diffusion tensor imaging and volumetric analysis of the ventral striatum in adults with traumatic brain injury. Brain Inj 26:201–210. https://doi.org/10.3109/02699052.2012.654591
52. Palacios EM, Sala-Llonch R, Junque C et al (2013) Long-term declarative memory deficits in diffuse TBI: correlations with cortical thickness, white matter integrity and hippocampal volume. Cortex 49:646–657. https://doi.org/10.1016/j.cortex.2012.02.011
53. Wang B, Prastawa M, Irimia A, et al (2013) Analyzing imaging biomarkers for traumatic brain injury using 4d modeling of longitudinal MRI. In: 2013 I.E. 10th Int. Symp. Biomed. Imaging. pp 1392–1395
54. Merkley TL, Larson MJ, Bigler ED et al (2013) Structural and functional changes of the cingulate gyrus following traumatic brain injury: relation to attention and executive skills. J Int Neuropsychol Soc 19:899–910. https://doi.org/10.1017/S135561771300074X
55. Fisher AC, Rushby JA, McDonald S et al (2015) Neurophysiological correlates of dysregulated emotional arousal in severe traumatic brain injury. Clin Neurophysiol 126:314–324. https://doi.org/10.1016/j.clinph.2014.05.033
56. Tomaiuolo F, Carlesimo GA, Di Paola M et al (2004) Gross morphology and morphometric sequelae in the hippocampus, fornix, and corpus callosum of patients with severe non-missile traumatic brain injury without macroscopically detectable lesions: a T1 weighted MRI study. J Neurol Neurosurg Psychiatry 75:1314–1322. https://doi.org/10.1136/jnnp.2003.017046
57. Di Paola M, Phillips O, Costa A et al (2014) Selective cognitive dysfunction is related to a specific pattern of cerebral damage in persons with severe traumatic brain injury. J Head Trauma Rehabil 30:1. https://doi.org/10.1097/HTR.0000000000000063
58. Uruma G, Hashimoto K, Abo M (2015) Evaluation of regional white matter volume reduction after diffuse axonal injury using voxel-based Morphometry. Magn Reson Med Sci 14:183–192. https://doi.org/10.2463/mrms.2014-0104
59. Marcoux J, McArthur DA, Miller C et al (2008) Persistent metabolic crisis as measured by elevated cerebral microdialysis lactate-pyruvate ratio predicts chronic frontal lobe brain atrophy after traumatic brain injury. Crit Care Med 36:2871–2877. https://doi.org/10.1097/CCM.0b013e318186a4a0
60. MacKenzie JD, Siddiqi F, Babb JS et al (2002) Brain atrophy in mild or moderate traumatic brain injury: a longitudinal quantitative analysis. AJNR Am J Neuroradiol 23:1509–1515. https://doi.org/10.1007/BF01402368
61. Ding K, Marquez de la Plata C, Wang JY et al (2008) Cerebral atrophy after traumatic white

matter injury: correlation with acute neuroimaging and outcome. J Neurotrauma 25:1433–1440. https://doi.org/10.1089/neu.2008.0683

62. Gale SD, Johnson SC, Bigler ED, Blatter DD (1995) Nonspecific white matter degeneration following traumatic brain injury. J Int Neuropsychol Soc 1:17–28. https://doi.org/10.1017/S1355617700000060
63. Gale SD, Baxter L, Roundy N, Johnson SC (2005) Traumatic brain injury and grey matter concentration: a preliminary voxel based morphometry study. J Neurol Neurosurg Psychiatry 76:984–988. https://doi.org/10.1136/jnnp.2004.036210
64. Trivedi MA, Ward MA, Hess TM et al (2007) Longitudinal changes in global brain volume between 79 and 409 days after traumatic brain injury: relationship with duration of coma. J Neurotrauma 24:766–771. https://doi.org/10.1089/neu.2006.0205
65. Warner MA, Youn TS, Davis T et al (2010) Regionally selective atrophy after traumatic axonal injury. Arch Neurol 67:1336–1344. https://doi.org/10.1001/archneurol.2010.149
66. Yount R, Raschke KA, Biru M et al (2002) Traumatic brain injury and atrophy of the cingulate gyrus. J Neuropsychiatry Clin Neurosci 14:416–423
67. Vannorsdall TD, Cascella NG, Rao V et al (2010) A morphometric analysis of Neuroanatomic abnormalities in traumatic brain injury. J Neuropsychiatry Clin Neurosci 22:173–181. https://doi.org/10.1176/jnp.2010.22.2.173
68. Levine B, Kovacevic N, Nica EI et al (2008) The Toronto traumatic brain injury study: injury severity and quantified MRI. Neurology 70:771–778. https://doi.org/10.1212/01.wnl.0000304108.32283.aa
69. Wilde EA, Bigler ED, Pedroza C, Ryser DK (2006) Post-traumatic amnesia predicts long-term cerebral atrophy in traumatic brain injury. Brain Inj 20:695–699. https://doi.org/10.1080/02699050600744079
70. Guild EB, Levine B (2015) Functional correlates of midline brain volume loss in chronic traumatic brain injury. J Int Neuropsychol Soc 21(8):650–655. https://doi.org/10.1017/S1355617715000600
71. Levine B, Kovacevic N, Nica EI et al (2013) Quantified MRI and cognition in TBI with diffuse and focal damage. NeuroImage Clin 2:534–541. https://doi.org/10.1016/j.nicl.2013.03.015
72. Warner MA, de la Plata CM, Spence J et al (2010) Assessing spatial relationships between axonal integrity, regional brain volumes, and neuropsychological outcomes after traumatic axonal injury. J Neurotrauma 27:2121–2130. https://doi.org/10.1089/neu.2010.1429
73. Wilde EA, Bigler ED, Gandhi PV et al (2004) Alcohol abuse and traumatic brain injury: quantitative magnetic resonance imaging and neuropsychological outcome. J Neurotrauma 21:137–147. https://doi.org/10.1089/089771504322778604
74. Bendlin BB, Ries ML, Lazar M et al (2008) Longitudinal changes in patients with traumatic brain injury assessed with diffusion-tensor and volumetric imaging. NeuroImage 42:503–514. https://doi.org/10.1016/j.neuroimage.2008.04.254
75. Spitz G, Bigler ED, Abildskov T et al (2013) Regional cortical volume and cognitive functioning following traumatic brain injury. Brain Cogn 83:34–44. https://doi.org/10.1016/j.bandc.2013.06.007
76. Kim J, Avants B, Whyte J, Gee JC (2013) Methodological considerations in longitudinal morphometry of traumatic brain injury. Front Hum Neurosci 7:52. https://doi.org/10.3389/fnhum.2013.00052
77. Bergeson AG, Lundin R, Parkinson RB et al (2004) Clinical rating of cortical atrophy and cognitive correlates following traumatic brain injury. Clin Neuropsychol 18:509–520. https://doi.org/10.1080/1385404049052414
78. Schönberger M, Ponsford J, Reutens D et al (2009) The relationship between age, injury severity, and MRI findings after traumatic brain injury. J Neurotrauma 26:2157–2167. https://doi.org/10.1089/neu.2009.0939
79. Fujiwara E, Schwartz ML, Gao F et al (2008) Ventral frontal cortex functions and quantified MRI in traumatic brain injury. Neuropsychologia 46:461–474. https://doi.org/10.1016/j.neuropsychologia.2007.08.027
80. Jorge RE, Robinson RG, Moser D et al (2004) Major depression following traumatic brain injury. Arch Gen Psychiatry 61:42–50. https://doi.org/10.1001/archpsyc.61.1.42
81. Jorge RE, Acion L, Starkstein SE, Magnotta V (2007) Hippocampal volume and mood disorders after traumatic brain injury. Biol Psychiatry 62:332–338. https://doi.org/10.1016/j.biopsych.2006.07.024
82. Anderson CV, Bigler ED, Blatter DD (1995) Frontal lobe lesions, diffuse damage, and neuropsychological functioning in traumatic brain-injured patients. J Clin Exp Neuropsychol 17:900–908. https://doi.org/10.1080/01688639508402438

83. Bigler ED, Johnson SC, Anderson CV et al (1996) Traumatic brain injury and memory: the role of hippocampal atrophy. Neuropsychology 10:333–342. https://doi.org/10.1037/0894-4105.10.3.333
84. Green RE, Colella B, Maller JJ et al (2014) Scale and pattern of atrophy in the chronic stages of moderate-severe TBI. Front Hum Neurosci 8:67. https://doi.org/10.3389/fnhum.2014.00067
85. Tate DF, Khedraki R, Neeley ES et al (2011) Cerebral volume loss, cognitive deficit, and neuropsychological performance: comparative measures of brain atrophy: II. Traumatic brain injury. J Int Neuropsychol Soc 17:308–316. https://doi.org/10.1017/S1355617710001670
86. Kesler SR, Adams HF, Blasey CM, Bigler ED (2003) Premorbid intellectual functioning, education, and brain size in traumatic brain injury: an investigation of the cognitive reserve hypothesis. Appl Neuropsychol 10:153–162. https://doi.org/10.1207/S15324826AN1003_04
87. Himanen L, Portin R, Isoniemi H et al (2005) Cognitive functions in relation to MRI findings 30 years after traumatic brain injury. Brain Inj 19:93–100. https://doi.org/10.1080/02699050410001720031
88. Strangman GE, O'Neil-Pirozzi TM, Supelana C et al (2010) Regional brain morphometry predicts memory rehabilitation outcome after traumatic brain injury. Front Hum Neurosci 4:182. https://doi.org/10.3389/fnhum.2010.00182
89. Takayanagi Y, Gerner G, Takayanagi M et al (2013) Hippocampal volume reduction correlates with apathy in traumatic brain injury, but not schizophrenia. J Neuropsychiatry Clin Neurosci 25:292–301. https://doi.org/10.1176/appi.neuropsych.12040093
90. Farbota KDM, Sodhi A, Bendlin BB et al (2012) Longitudinal volumetric changes following traumatic brain injury: a tensor-based Morphometry study. J Int Neuropsychol Soc 18:1–13. https://doi.org/10.1017/S1355617712000835
91. Johnson SC, Pinkston JB, Bigler ED, Blatter DD (1996) Corpus Callosum morphology in normal controls and traumatic brain injury: sex differences, mechanisms of injury, and neuropsychological correlates. Neuropsychology 10:408–415. https://doi.org/10.1037/0894-4105.10.3.408
92. Bigler ED, Blatter DD, Anderson CV et al (1997) Hippocampal volume in normal aging and traumatic brain injury. AJNR Am J Neuroradiol 18:11–23
93. Tate DF, Bigler ED (2000) Fornix and hippocampal atrophy in traumatic brain injury. Learn Mem 7:442–446. https://doi.org/10.1101/lm.33000
94. Isoniemi H, Kurki T, Tenovuo O et al (2006) Hippocampal volume, brain atrophy, and APOE genotype after traumatic brain injury. Neurology 67:756–760. https://doi.org/10.1212/01.wnl.0000234140.64954.12
95. Wilde EA, Hunter JV, Newsome MR et al (2005) Frontal and temporal morphometric findings on MRI in children after moderate to severe traumatic brain injury. J Neurotrauma 22:333–344. https://doi.org/10.1089/neu.2005.22.333

Index

Gianfranco Spalletta et al. (eds.), *Brain Morphometry*, Neuromethods, vol. 136,
https://doi.org/10.1007/978-1-4939-7647-8,

U

V

W

Printed in the United States
By Bookmasters